FLOW THROUGH POROUS MEDIA

Based on the Sixth
State-of-the-Art Symposium,
sponsored by INDUSTRIAL AND
ENGINEERING CHEMISTRY and
the Division of Industrial
and Engineering Chemistry of
the American Chemical Society,
Washington, D.C., June 9-11, 1969.
Richard J. Nunge, Chairman.
Reprinted from INDUSTRIAL AND
ENGINEERING CHEMISTRY,
September 1969-October 1970

AMERICAN CHEMICAL SOCIETY PUBLICATIONS WASHINGTON, D.C.

ISBN 842-0111-0
Library of Congress Catalog Card Number 78-146798
PRINTED IN THE UNITED STATES OF AMERICA

CONTENTS

PREFACE

All the papers in this volume were originally presented at the Sixth State of the Art Summer Symposium sponsored by the ACS Division of Industrial and Engineering Chemistry in June 1969. All have also been published in Industrial and Engineering Chemistry at intervals between October 1969 and October 1970. They are brought together in this volume for greater convenience as a reference source and teaching aid.

Although the phenomenological approach to description of flow through porous media provides a basis that is satisfactory, for some purposes the real challenge in this area is to find fundamental, underlying principles that will make empiricism unnecessary. The authors of these papers lay the groundwork for the fundamental approach. We hope that workers in such diverse fields as petroleum production, polymer processing and water desalination will find in this volume a springboard from which they can launch more accurate attacks on their particular flow-through-porous-media problems.

In addition to the obviously invaluable work of the symposium chairman and authors, many of the ACS Publications staff have contributed to the readability and attractiveness of this volume. They include Gloria Dinote, who handled manuscript editing, and Leroy Corcoran and Judy Bitting of the ACS Production staff.

D. H. Michael Bowen
ACS PUBLICATIONS

INTRODUCTION

The unifying theme of the review papers comprising the Symposium on Flow Through Porous Media is the movement of material in and through both natural and man-made porous beds of solids. Because of the occurrence of porous media in a wide variety of important practical applications, such flows are the concern of workers in several diverse areas. By assembling a group of review papers relating mainly to the basic phenomena involved rather than papers on isolated applications, the symposium attempts to be general in scope so that the needs of as broad an audience as possible might be served. Thus the overall objective of the symposium is to provide a critical survey of the techniques, both experimental and analytical, which are useful in treating many of the problems which involve flow through porous media.

For the most part, the symposium papers are concerned with factors affecting the "flow phenomenon" aspect of the transport processes occurring in porous media, although some applications in separation processes are described. A partial list of the topics discussed includes: methods of determining pore structure of porous media, anisotropy, diffusion and flow of gases, non-Newtonian fluid flow through porous media, multiphase flow in liquid-liquid and solid-liquid systems, physics and thermodynamics of capillary action, flow of suspensions, dispersion and miscible displacement, theoretical framework for coalescence, and reverse osmosis.

Perhaps if one point is made clear in the symposium, it is that little can as yet be predicted from first principles about problems related to flow in porous media. The phenomenological approach, as exemplified by Darcy's law, has long proved useful in handling such problems and will undoubtedly continue to do so in the future. However, inroads are being made toward more complete descriptions so that working equations may be extrapolated to new situations with confidence. But, in the final analysis, theory must be closely integrated with experiment for verification and evaluation of undetermined parameters. In this sense, problems involving flow through porous media are akin to problems involving turbulent flow. While each area has been studied over a long time period, there still exists the possibility of fundamental breakthroughs in the theory. This, then, summarizes the critical and challenging properties of the subject area, namely to find better working correlations to deal with the everyday practical applications and simultaneously to search for a deeper fundamental understanding.

The symposium reflects on both the theoretical and applied aspects of the state of development. Because this is primarily a review symposium, one should find much that is well-known to the experienced worker gathered together and hence it should prove useful for instruction in basic concepts. At the same time, current developments are described so that the symposium papers should serve as a guide to reading the recent literature and to evaluating the future literature. It is believed that this book will prove to be a valuable reference source to practicing engineers, scientists, educators, and students.

1

Determination of Structure of Porous Media

Determination of the Structure of Porous Media

FRANCIS A. L. DULLIEN V. K. BATRA

Progress made in the field of the structure of porous media is reviewed and discussed

Porous materials occur in great variety, both in nature and in industry. Living organisms are all porous; their life functions would not be possible without pores in them. Pores make breathing possible as well as the circulation of natural fluids in both plant and animal life. Porous structures also serve as heat insulators in nature—*e.g.*, the fur of animals and the feathers of birds.

In the inanimate world porous structures are just as widespread and important. Soil is porous and so are most natural rocks, to a varying degree. Graphite, mica, sandstone, and limestone are just a few examples. Ground water, petroleum, and natural gas are among the important substances that are contained in the pore spaces provided by various geological formations.

Among industrial products, porous materials are again numerous and of great practical value. Materials of construction such as ceramics, concrete, and timber are porous. Our clothing is made of porous textiles; paper is made from fibers. In addition, porous materials play an important role in many process industries as adsorbents, such as silica gel, active charcoal, zeolites-molecular sieves, a large variety of contact catalysts, and filters. Electrodes in batteries and electrolysis plants are often porous. Many commercial products are granular or porous, and the technology of drying such materials has to consider their porous nature. The pore structure of ores is important in process metallurgy.

The variety of different pore shapes and sizes is just as great as the types and origins of various porous materials. There are pores ranging in size from caverns to molecular interstices. The shapes of pores can be spherical, bubble-like—*e.g.*, the pores in concrete—or flat and slit-shaped, as in mica or various clays; however, most pores have very irregular shapes for which there are no dictionary definitions. Often, pores were formed by the packing of some granular or fibrous material; often, they were the product of chemical action. Their shapes are different every time, depending on their origin. Pores can also be classified by considering the degree to which they are interconnected with each other. A pore inside a sample is "interconnected" if it is accessible from the outside through both ends, and it is "isolated" if it is not accessible at all. "Dead-end" pores are connected to the outside through one end only. The physical properties of materials are strongly influenced by the number, size, and often, shape of pores in them, and also by the degree to which the pores are interconnected. The effect pore structure has on the properties of materials is one of the reasons for our interest in pore structure analysis. Apart from the thermally insulating properties of porous substances, our main interest centers around the passage of materials, gaseous or liquid, into, out of, and through the porous spaces. Knowledge of the internal surface area of porous adsorbents is also of great practical importance because of the role the internal surface area plays in determining the effectiveness of the material as an adsorbent or a contact catalyst.

The variety of porous materials and their practical significance, along with the shapes, sizes, and nature of the pores is so great, that no comprehensive treatment of pore structure analysis has ever been attempted. The closest work to a comprehensive treatise is represented by the Proceedings of the Tenth Symposium of the Colston Research Society held in the University of Bristol in 1958 and entitled "The Structure and Properties of Porous Materials, Butterworths Scientific Publications, London, 1958." It consists of a collection of papers presented at the Symposium, which cover a wide variety of porous materials and different approaches to pore structure analysis. Even this big contribution to the subject matter of pore structure analysis fails to be completely representative, however, and it does not contain a critical discussion of the state of the art in general. In the intervening ten years a great deal of additional original work has been done in the field of pore structure analysis. In this paper an attempt has been made to review the progress made in the entire field in the past decade and discuss it critically. While the material covered in this review may appear quite voluminous, there can be no claim made to completeness. Any omissions from the review are due purely to oversight by the authors.

The Objectives of Pore Structure Analysis

The structure of most porous media is far too irregular and complicated to allow a rigorously correct geometrical description. Even if it were possible somehow to enter all pores with a tiny probe and determine the coordinates of every point on the bounding surfaces of the pores and store this information in the memory of some giant computer, this kind of three-dimensional map of the pore structure would only be of limited usefulness, without appropriate equations capable of giving the mass transport and the capillary behavior for such highly irregular channels. Such equations, however, have not been developed, much the same way as an accurate mapping of the pore structure has not been possible up to the present time. We are quite far removed from the realization of these ideal goals, and this will probably remain true for a long time to come, notwithstanding the possibilities to be opened up by the advent of bigger and faster computers.

More realistic goals of pore structure analysis will be, as they have been in the past, to develop improved mathematical models of the pore structure and to determine experimentally the parameters that are necessary to compare different samples in terms of the models. In this sense the ideal goal of pore structure analysis is to develop mathematical models that will account for all the important properties, such as effective diffusivity, permeability, relative permeability, capillary hysteresis, dispersion coefficient, etc., of real porous media, quantitatively. It can be expected that different types of porous media will require quite different models to achieve this goal.

For the vast majority of porous media, our present knowledge of the size, shape, and degree of interconnectedness of the pores is inadequate to permit the development of an ideally satisfactory model.

The model of pore structure that is used almost universally at the present time consists of straight, nonintersecting cylindrical tubes, the diameters of which are distributed according to some distribution function. Whereas this model can be fitted to account for certain properties of porous samples, it cannot even qualitatively account for two-phase flow behavior and capillary hysteresis.

To develop more accurate models of real porous media, a great deal more information is required on the shape, true size, and degree, as well as the type of interconnectedness of the pores in such materials. The best way this kind of information can be obtained, and suitable mathematical models developed, is by a combination of properly designed experiments and a careful mathematical analysis of the problems.

It is apparent from reviewing the literature of pore structure analysis that, notwithstanding the tremendous amount of ingenuity expended in this field, the goal of developing ideally satisfactory mathematical models for porous media is not within our reach yet. We have to be content with models of more limited scope, which are adequate to account only for some of the properties of porous media. The significance of such models should not be underestimated, however, since they are very useful in their intended field of applications, and this is often all that is required of them.

The various contributions to pore structure analysis can be divided roughly into two categories: (1) Development of mathematical models to be used in correlating certain properties of certain types of porous media, and (2) Basic studies aimed at an improved understanding of pore structure. In some contributions, both objectives have been pursued simultaneously; however, in the vast majority of cases the contributions fall quite naturally into one or the other of the above categories.

The methods of pore structure investigation can be conveniently classified as follows: vapor sorption, diffusion, fluid flow, capillarity, micrography, radiography, and miscellaneous methods. There are very few instances where the pore structure of one and the same substance has been investigated by a variety of different methods. In the majority of the cases only one method has been used by individual research groups, and often little attention has been paid to results obtained by others, using different methods. The fact that various researchers are interested in different applications of porous media is at least partly responsible for this state of affairs. Another important factor is that often some porous materials are more readily investigated by a certain method, while in other cases different methods are more convenient to apply. The emphasis that is being placed on ease and speed of operation is also indicated by the recent trend to make the most popular methods as completely automatic and computerized as possible.

Considering the fact that pore structure researchers are, in the majority of the cases, method oriented, it seemed logical to treat the subject, also in the present review, by discussing the various different methods one by one. At the same time, however, every effort has been made to refer to the results obtained by the other methods whenever information was available.

Investigation of Pore Structure Using Sorption Isotherms

The study of internal surface area and pore structure of porous solids in the pore size range of about 10–1000 Å has great technological importance, mainly because of applications in the fields of heterogeneous catalysis and industrial adsorption of vapors. There is a large amount of literature on these fields, and they have occupied the attention of many eminent researchers for several decades.

The work done in pore structure research in the past decade, using sorption isotherms, resulted in a considerable evolution of our understanding of adsorption and desorption processes, sorption hysteresis, the relationship between pore structure and the foregoing phenomena, and our appreciation of the problems yet unsolved in this field. Simultaneously, important progress has been made in the methods of computing pore size distributions based on the pore model of straight cylindrical nonintersecting capillary tubes. The most significant recent development in technical facilities is represented by completely automatic and computerized equipment and procedures.

Much of the recent development finds its origin in a suggestion made by Wheeler (*270*), which pointed out the necessity of taking into account the thickness of the adsorbed multilayer when applying the Kelvin equation of capillary condensation. It is necessary to do this because in very small pores the thickness of the adsorbed multilayer becomes a significant fraction of the pore radius.

AUTHORS *F. A. L. Dullien is a Professor and V. K. Batra is a graduate student in the Department of Chemical Engineering, University of Waterloo, Waterloo, Ontario, Canada.*

Methods using simple pore models. A large number of authors (*15, 22, 23, 61, 78, 128, 236, 242, 257*) have devised various computational methods to calculate pore size distributions from desorption isotherms, by using either the cylindrical or the parallel-plate pore model.

One of the first to develop a method based on Wheeler's suggestion was Shull (*236*), who assumed that the pore size distribution may be represented by simple analytical forms of Maxwellian or Gaussian type. These were substituted into Wheeler's equation

$$V_s - V = \pi \int_r^\infty (r - t)^2 L(r) dr \tag{1}$$

where V_s is volume of gas adsorbed at saturation pressure, V is the volume of gas adsorbed at intermediate pressure p, $L(r)dr$ is the total length of pores whose radii fall between r and $r + dr$, r is the corrected Kelvin radius, and t is the multilayer thickness that is built up at pressure p. This equation merely states that the volume of gas $V_s - V$, not yet adsorbed at pressure p, is equal to the total volume of pores that have not been filled. Here

$$r = t - \frac{2\sigma v}{R_g T \ln (p/p_0)} \tag{2}$$

where σ is the surface tension, v is the molar volume of the adsorbent, R_g is the universal gas constant, T is the absolute temperature, and p_0 is the saturation pressure.

With the simple functional forms used Equation 1 could be integrated analytically, resulting in two- or three-parameter expressions for $V_s - V$. Families of standard isotherms were prepared, and the values of the parameters were found by comparing the experimental isotherms with the standard ones. The values of the parameters, of course, determined the pore size distribution function directly. To evaluate Equation 2, t had to be calculated as a function of p/p_0. This was obtained by Shull from nine published isotherms (nitrogen gas) on crystalline materials of large crystal size, by assuming the thickness of one monolayer to be 4.3 Å for nitrogen.

In subsequent years, the *t*-curve, as it is now called, became a very important tool in pore structure analysis, particularly in the hands of deBoer and coworkers. A detailed discussion of their work will be presented later in this section.

Voigt and Tomlinson (*257*) extended the approach of Shull also by applying it to the "ink bottle"-type pore model. This is a significant improvement, because hysteresis can be taken into account. Voigt and Tomlinson used their method to determine the pore size distribution of porous Vycor glass where the ink bottle model represented the isotherms better than the cylindrical one.

Numerous attempts have been made (*15, 22, 23, 61, 78, 128, 242*) to develop more general solutions to the pore size integral. Shortly after Shull's paper was published, Barrett *et al.* (*15*) worked out a computational method that has achieved a great deal of success.

The total volume of adsorbate lost in a desorption step can be expressed as $\Delta V = \Delta V_c + \Delta V_m$, where ΔV_c is capillary desorption, and ΔV_m is multilayer desorption. If it is assumed for the moment that ΔV_m may be calculated, the ΔV_c for the desorption step is obtained. This volume ΔV_c is the capillary liquid lost between the beginning and the end of a step between two relative pressures p_1/p_0 and p_2/p_0.

The actual volume of the pores emptied in the desorption step is $\Delta V_p = R_n \Delta V_c$, where $R_n = r_p/(\bar{r} + \Delta t)^2$, r_p is the actual pore radius, $\bar{r}$ is the mean capillary condensate radius for the two relative pressures p_1/p_0 and p_2/p_0, and Δt is the decrease in multilayer thickness for the desorption step.

The major problem involved in this calculation is to accurately and conveniently find ΔV_m. The value of ΔV_m, for a particular step, would appear to be $\Delta t \Sigma S_p$, where ΣS_p is the sum of the surface areas over all steps up to and including the preceding one. However, this would assume the area ΣS_p to be planar, whereas it is actually made up of the curving walls of pores. For each step S_p is obtained, assuming cylindrical pores, by $S_p = 2\Delta V_p/r_p$.

To allow for the curvature, Barrett *et al.* used a factor C, equal to $(r_p - t)/r_p$, and assigned a choice of constant values to it, to be chosen according to the approximate range of sizes of pores expected. This procedure gives reasonably accurate results for pores down to radii of 35 Å. However, for pores of smaller diameter the value of C changes with pore size too rapidly to be correctly assigned a constant value.

The final equation developed by Barrett *et al.* is

$$\Delta V_p = R_n(\Delta V - C\Delta t \Sigma S_p) \qquad (3)$$

Cranston and Inkley (61) made the B.J.H. (Barrett, Joyner, and Halenda) method more precise by using a variable C both for the step being calculated and for each of the pore sizes. They multiplied S_p for each pore size range by its individual value of C instead of summing S_p. While more precise, their method is also far more laborious than the B.J.H. method.

The model used in both the B.J.H. and the Cranston and Inkley methods consists of straight cylindrical capillaries of uniform cross section that are closed at one end. This is a highly idealized model. In the majority of the cases the choice of the adsorption, rather than the desorption isotherm, for the calculations would result in very different pore size distribution curves. Because of the crudeness of the model used, any increased precision of the calculations can be expected to result in very little real improvement in the pore size distribution obtained. Only by using more realistic pore models can one hope to calculate more accurate pore size distributions.

Steggerda (*242*) and Innes (*128*) gave methods for parallel-plate pore model (slit-shaped pores). Owing to the simpler geometry, the computation is less tedious than in the cylindrical case.

The computational method of Dollimore and Heal (*78*) is more refined, mathematically, than either the B.J.H. or the Cranston and Inkley method. They used the equation of Halsey (*116*)

$$t = 4.3\left(\frac{5}{\ln p/p_0}\right)^{1/3} \qquad (4)$$

to calculate t as a function of p/p_0, resulting in variable t from one step to the next. By using this equation there was no need to use a correction factor to allow for the varying curvature of the capillary. The method is fast; however, it depends on the validity of the Halsey equation and, much like the B.J.H. and the Cranston and Inkley method, it is based on the cylindrical pore model.

Bernardini and Collepardi (*22*) modified the method of Dollimore and Heal (*78*), and eliminated the relative pressure by combining the Halsey equation (Equation 4) and the Kelvin-Wheeler equation (Equation 2), which resulted in a direct relationship between the thickness of the adsorbed layer t and the pore radius r_p.

With the exception of Voigt and Tomlinson (*257*), the above authors did not make a serious attempt to include into their considerations the problem presented by pore shapes deviating from the simple open cylindrical or slit shapes. They usually compared the total surface areas obtained by their respective techniques with the value given by the BET method as a check.

Shull (*236*) used the desorption isotherm for silica gel samples and found reasonable agreement with BET surface areas. Barrett *et al.* (*15*) also used the desorption isotherms and obtained good agreement with BET surface areas in most cases. Cranston and Inkley (*61*), however, performed a more comprehensive study of this question by comparing the surface areas calculated both from the adsorption and the desorption branch of the isotherms with the BET areas. These comparisons showed, on the average, better agreement in the case of adsorption isotherms. The surface areas calculated by Bernardini *et al.* (*23*) were strikingly different from the corresponding BET areas. Voigt and Tomlinson (*257*) used the adsorption branch for porous Vycor glass where the pores were mostly of the "ink-bottle" type. They obtained poor agreement with the BET surface area. They also demonstrated a considerable variation in the calculated surface area, depending on the substance adsorbed.

It is clear from the experimental evidence that it depends on the type of porous material whether the adsorption or the desorption branch is preferable to use for pore size distribution calculations. Also, agreement with BET surface areas does not necessarily prove that the pore size distribution is accurate. Indeed, in general, use of neither the adsorption nor the desorption branch may be expected to result in correct pore size distributions, because it has been demonstrated that irreversible phenomena may occur along both the adsorption and the desorption isotherm. The most important factor in determining these irreversibilities is the shape of the pores.

Adsorption hysteresis and pore structure. The problem of a relationship between the type of hysteresis loop and the corresponding capillary shape was approached in complete generality by deBoer (*67*), who distinguished five types of hysteresis loops. Fifteen shape groups of capillaries were considered by him, and an analysis was made of the type of hysteresis loop that may result from capillary condensation in them. The reader is referred to the original paper for details of this enlightening treatise; the nature of the problem can be appreciated by considering capillaries of varying width, open at both ends, as an example. (See Figure 1.) One can expect capillary condensation to take place first

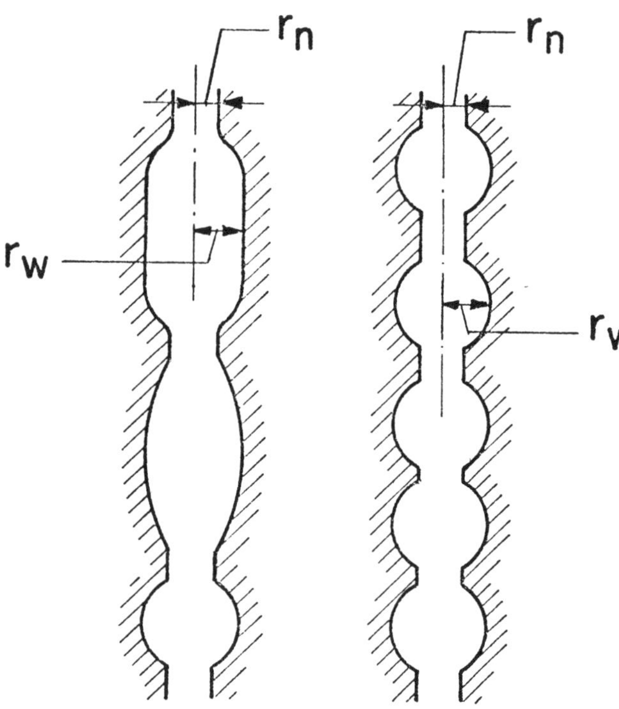

Figure 1. Tubular capillaries with slightly widened parts

(From deBoer, "Structure and Properties of Porous Materials," Butterworth, London, 1958, by permission)

in the narrow parts at a relative pressure, corresponding to the mean (or effective) radius r_m of an inscribed cylinder into the narrow part, as given by the Kelvin equation

$$p/p_0 = \exp\left(-\frac{2\sigma v}{R_g T r_m}\right) \tag{5}$$

where, for a cylinder of radius r_n,

$$\frac{1}{r_m} = 1/2\left(\frac{1}{r_n} + \frac{1}{\infty}\right) = \frac{1}{2\,r_n} \tag{6}$$

As soon as the narrow parts are filled with liquid, however, spherical menisci are formed for which the mean radius r_m is

$$\frac{1}{r_m} = 1/2\left(\frac{1}{r_w} + \frac{1}{r_w}\right) = \frac{1}{r_w} \tag{7}$$

Providing that $r_w \leq 2\,r_n$, the equilibrium relative pressure corresponding to $r_m = r_w$ is lower than the prevailing pressure, and as a result the wider parts of the capillary are filled in a nonequilibrium process at a constant relative pressure corresponding to $r_m = 2\,r_n$.

On the other hand, the entire capillary is emptied in the desorption process, at the relative pressure corresponding to the mean radius $r_m = r_n$. The desorption from the wider parts is also a nonequilibrium process, because the relative pressure corresponding to $r_m = r_n$ is lower than the value corresponding to $r_m = r_w$. The result is a hysteresis loop and a great deal of uncertainty as far as the actual pore size is concerned. The analysis of other pore shapes causing capillary hysteresis is similar, although the appearance of the hysteresis loop varies from one pore shape to another.

DeBoer and coworkers used these ideas to analyze hysteresis loops of a variety of substances in a long series of papers starting in 1963. These papers represent the most consistent effort so far to investigate pore structure in depth by analyzing sorption isotherms. In the first three papers (*69*, *164*, *165*) aluminum oxide systems were studied. Slit-shaped pores were assumed and the method developed by Steggerda (*242*) was used to calculate pore size distributions.

A somewhat modified *t*-curve was introduced (*164*) by letting

$$t = (X/S) \times 10^4\ \text{Å} \tag{8}$$

where X is the adsorbed volume in ml of liquid adsorbate, and S is the specific surface area in m^2/g of adsorbent. DeBoer and coworkers (*164*) treat the adsorbate consistently as a close-packed liquid and let $S = S_{BET}$. This results in the equation

$$t = 15.47(V_a/S_{BET}) = 3.54(V_a/V_m)\ \text{Å} \tag{9}$$

where V_a is the adsorbed volume of the adsorbate in ml gas STP/g of adsorbent, and V_m is the volume of gas in ml STP/g of adsorbent that should be able to cover the whole surface with a unimolecular layer. The value

3.54 Å for the statistical thickness of a monolayer differs from the value 4.3 Å used by Shull and Barrett *et al.*

The desorption isotherms were used (*165*) to calculate pore size distribution curves. The cumulative surface areas S_{cum} were compared with S_{BET}. In some cases the ratio S_{cum}/S_{BET} was greater than one by as much as 30 to 40%. This is due, in part, to the existence of pores with widened parts, whereas the volume of these parts is attributed to pores that are narrower. For many other samples S_{cum} was smaller, sometimes much smaller than S_{BET}. These samples apparently contained a number of narrow pores, the size distribution of which could not be calculated on the basis of the Kelvin equation.

In a subsequent paper (*68*), deBoer *et al.* showed that for a gibbsite sample containing "ink bottle"-type pores the adsorption branch of the isotherm gave agreement with S_{BET}, whereas use of the desorption branch gave a cumulative surface area which was almost 50 times the BET value. The reason for this behavior is that, if the desorption branch is used, the volume of the "bottles" is mistakenly connected with the size of the "neck," which leads to a far too large surface area. In the same paper it was demonstrated that in the case of slit-shaped pores of another gibbsite sample, use of the desorption branch resulted in excellent agreement with S_{BET}, whereas the adsorption branch gave a value about 50% too low. The reason is that in the adsorption process no meniscus can be formed between parallel plates forming slits until the adsorbed layers on the pore walls make contact with each other. This results in measured pore sizes that are too large and which, in turn, give too low surface areas.

It is evident from these results that only an appreciation of the relationship between the type of hysteresis loop and the shapes of capillaries may permit the calculation of correct pore size distributions. For a detailed discussion of this subject the reader is referred to deBoer's original paper (*67*). It is also important to realize (*68*) that basically there are two causes of "permanent," or "reversible" hysteresis. [DeBoer and coworkers use the term reversible for the capillary hysteresis, which is not due to hysteresis of the contact angle (*280*). Since hysteresis of any kind involves thermodynamically irreversible processes, this terminology may be misunderstood. The term permanent hysteresis used by Everett (*97*) may be preferable]. One is the presence of pores with wide bodies, closed on all sides except for a narrow neck opening—the so-called ink-bottle theory, first suggested by Kraemer (*148*) and popularized by McBain (*179*). The most important feature of this type of hysteresis is the delayed, nonequilibrium desorption from wide pores, which takes place at a relative pressure corresponding to the size of the narrow opening. The other mechanism is characteristic of pores open on all sides. In such pores no meniscus can form in the adsorption process until the adsorbed film grows thick enough to block the pore somewhere (delayed meniscus formation theory). This theory is from Foster (*101*). In the ink bottle-type pores the adsorption process is closer to equilibrium, whereas in pores open to all sides it is the desorption process that is closer to equilibrium. Everett (*97*) pointed out that the "delayed meniscus formation" mechanism is always present in nonuniform capillaries.

DeBoer and coworkers developed the *t*-curve into a powerful tool of pore structure analysis. By using Equation 9, t was calculated as a function of the relative pressure from experimental results on a variety of different poreless samples to avoid the influence of capillary condensation (*70*). Inasmuch as one can be certain that capillary condensation was absent, the *t*-curve appears to be "universal"—*i.e.*, no appreciable influence of the chemical nature of the surface could be observed.

The universal t *vs.* p/p_0 relationship has been used by deBoer and coworkers (*71, 166*) to investigate pore structure of unknown samples as follows. The measured adsorbate volume V_a is converted into a function of t, and the V_a *vs.* t plot is analyzed. As long as the multilayer is formed unhindered a straight line is obtained that goes through the origin, and its slope is a measure of surface area

$$S_t = 15.47\ V_a/t \qquad (10)$$

One might expect S_t to be equal to S_{BET}; however, this is only approximately true because the *t*-curve is an average of various samples. At higher relative pressures, deviation from a straight line may occur that is probably due to capillary condensation.

At a certain pressure, capillary condensation may occur in pores of certain shapes and dimensions, the material takes up more adsorbate than corresponds to the volume of the multilayer, the adsorption branch lies above the *t*-curve, and the slope of the V_a *vs.* t plot increases.

In pores that are open on all sides—*e.g.*, slit-shaped pores—where meniscus cannot form until the pores are completely filled by the adsorbed layers on both parallel walls, surface area becomes gradually unavailable for adsorption. At the point in the adsorption branch where this process becomes noticeable, the V_a *vs.* t plot will start having a smaller slope corresponding to the surface area still accessible. Application of these ideas to various samples gave a picture of the pore structure, which was consistent with the observed capillary hysteresis behavior (*71, 166*).

The case of "cylindrical" pores open at both ends is quite different from the case of "slit-shaped" pores open on all sides. In the latter case, the two flat surfaces facing each other are separated by a gap and there is no direct connection between the multilayers adsorbed on them; hence, these layers can build up on both sides until they eventually bridge the gap between them and touch at a point. For the cylindrical pores open at both ends, however, the adsorbed multilayers form a closed entity of cylindrical shape and, therefore, capillary condensation generally occurs at some point before the capillary is completely filled with multilayers. In fact, the filling of the capillary should occur when the surface of the multilayer starts behaving like a liquid surface.

The problem of explaining, quantitatively, the phenomenon of capillary condensation in cylindrical capillaries open at both ends has attracted considerable attention. In 1938, Cohan (*57*) suggested that capillary condensation takes place in the cylindrical film at the walls of the pores, and consequently the pores are filled by capillary condensate. A thermodynamic analysis of this problem was presented by Broekhoff and deBoer (*35*).

The main idea in the treatment of Broekhoff and deBoer is that the thermodynamic potential of the adsorbed multilayer depends, in addition to the radius of curvature of the capillary, on the thickness t of the layer. Therefore, for a cylinder with a given diameter, the thermodynamic potential of the adsorbed multilayer will vary continuously with t. Their equilibrium condition can be written as

$$R_gT \ln p_0 - \frac{\sigma v}{r - t_e} - F(t)_{t=t_e} = R_gT \ln p \tag{11}$$

where p_0 is the equilibrium vapor pressure over the flat surface of the bulk liquid at temperature T, and p is the equilibrium vapor pressure over the cylindrical layer of a mean radius $r - t_e$ and of a thickness t_e. The function $F(t) \geq 0$ represents the effect of the surface forces of the solid walls on the thermodynamic potential of the adsorbed multilayer. Its value goes to zero for very large t, and it increases with decreasing values of t. The function $F(t)$ is given by the t-curve determined on "flat" surfaces, which was expressed in analytical form by curve-fitting the data points (*36*). The physical meaning of Equation 11 is evidently that p_0, the vapor pressure of the bulk liquid with a flat surface, is decreased to the value p because the surface of the adsorbate is curved, and because of the proximity of the solid walls of the adsorbent.

By using the analytical expressions for $F(t)$, Equation 11 has been solved, with nitrogen as adsorbate, for t_e as a function of p/p_0 and r, and the results tabulated. For a given p/p_0, t_e increased with decreasing r, a fact which is already evident from the physical content of Equation 11.

The left side of Equation 11 may be regarded as μ_a, the chemical potential of the adsorbed layer, whereas the right side is μ_g, the chemical potential of the vapor. Therefore, when dN moles of the adsorbate are transferred from the vapor to the adsorbed layer, at constant temperature, and pressure p of the vapor phase, the change in the Gibbs free energy is

$$dG_{T,P} = (\mu_a - \mu_g)\, dN \tag{12}$$

While the radius of curvature is changing, the pressure in the condensed phase cannot simultaneously remain constant; however, the effect of this pressure change on G can be neglected. Equation 11 is the special case of Equation 12 when the system is in equilibrium, or $dG_{T,P} = 0$. For a fixed value of r there are an infinity of equilibrium situations, each characterized by a pair of values $(p/p_0, t_e)$. The second derivative of G is

$$\frac{d^2G}{dN^2} \geq 0 \tag{13}$$

for stable equilibrium. Under the conditions of an adsorption experiment, it is justified to assume that the pressure p in the vapor phase remains constant, while dN moles of vapor are transferred from the vapor to the adsorbed phase. Therefore, the condition of stable equilibrium from Equations 13 and 11 becomes

$$-\,dF(t)/dt - v/(r - t)^2 \geq 0 \tag{14}$$

where the substitution

$$dN = -\frac{dA}{v}(r - t) = \frac{2\pi L}{v}(r - t)\, dt \tag{15}$$

has been made. dA is the change in surface area accompanying the adsorption of dN moles. With the aid of the analytical expression for $F(t)$ Equation 14 can be evaluated for various values of t_e, for a fixed value of r. The particular value of t_e, for which the left side of Equation 14 exactly equals zero, is the "critical thickness" t_{cr}, corresponding to incipient metastable equilibrium. This value of the equilibrium thickness, and the corresponding value of p_a/p_0, define the point where, for a given r, capillary condensation takes place and the whole pore spontaneously fills with capillary condensate.

Equation 14 has been solved, giving a relationship between t_{cr} and r. At the same time Equation 11 provides a relation between t_{cr}, r, and p_a/p_0. By eliminating t_{cr} between these two equations, r has been calculated as a function of p_a/p_0. For each p_a/p_0, a larger value of r was obtained than when Cohan's equation was used. The physical meaning of this result is that the surface forces help to bring about capillary condensation.

The above treatment lends itself immediately to pore size distribution calculation from the adsorption branch of the isotherm by a technique similar to that used by Barrett *et al.* (*15*). An important difference is that t now depends on r, as well as on p/p_0. An exact computational method, as well as a simplified technique, has been developed by Broekhoff and deBoer (*36*).

The above outlined thermodynamic analysis has been extended (*37*) to "ink bottle"-type pores with essentially spherical shapes.

The desorption condition from cylindrical capillaries was also obtained by Broekhoff and deBoer (*35, 40*), using two different approaches. The first, and somewhat indirect approach, consisted of integrating Equation 12 from equilibrium thickness t_e of the multilayer to $t = r$, the pore radius, and letting $\Delta G_{T,P}$ equal zero, resulting in a relationship between t_e, r, and p_0. Simultaneous solution of this equation with the equilibrium condition, Equation 11, gave the desired relation between r and p_D, the desorption pressure. This approach is indirect, however, because it approaches desorption from the opposite direction—*i.e.*, adsorption. Indeed, when integrating Equation 12 at $p = p_D$, from t_e to $t_e \leq t \leq r$, ΔG had a hump between $t = t_e$ and $t = r$, which represents the thermodynamic reason for the absence of capillary condensation (complete filling) at the pressure p_D. This will only occur at a higher pressure p_a as already discussed above. The argument, that from the desorption direction the hump in ΔG does not exist, is intuitive.

In the more direct approach (*40*), the authors analyzed the change in shape of the meniscus during evaporation from a cylindrical capillary that was originally completely filled with condensate. The geometrical condition of desorption is that the meniscus is tangential to the adsorbed multilayer. By using the thermodynamic analysis discussed above, an expression was obtained for the slope of the meniscus at a distance t from the pore wall. Introducing the geometrical condition of desorption into this equation, the same relationship between r, t_e, and p_D was obtained as by the above analysis of the adsorption process. A similar equation was derived earlier by Deryagin (*73, 75*), by making use of the concept of disjoining pressure. His equations, however, did not find practical application in the calculation of pore distributions from sorption isotherms.

An interesting consequence of the analysis of Broekhoff and deBoer is that the shape of the meniscus in a pore may not be expected to be hemispherical, but rather conical. Also, the emptying of the pores was an intrinsically continuous process, rather than the classical discontinuous evaporation step.

Using the analytical expressions (*36*) obtained for $F(t)$, the pore radius r was calculated (*38*) as a function of p_D/p_0. For any given p_D/p_0, a greater value of r was obtained than by the form of the Kelvin equation given by Equation 2. The physical meaning of this result is, again, that the surface forces emanating from the walls of the adsorbent have the effect of holding the capillary condensate in the pores.

With the theoretical apparatus developed by Broekhoff and deBoer, it was an easy matter to calculate, for open-ended cylindrical pores, the theoretical width of the hysteresis loop, $p_a/p_0 - p_D/p_0$, as a function of p_a/p_0. The calculated values were compared with the experimental ones for a variety of adsorbents (*38*). In many cases good agreement was obtained, which was taken as an indication of the type of pore structure of the adsorbents.

Pore size distributions were calculated (*38*) from the desorption branch by a completely analogous technique to the one given in (*36*), and compared with the curves obtained from the adsorption branch of the isotherms. The good agreement obtained for samples where the shape of the hysteresis loop indicated a pore structure consisting approximately of open-ended cylinders was a major achievement by these authors.

The treatment of desorption from cylindrical capillaries was extended (*39*) to slit-shaped pores.

In another paper, deBoer *et al.* (*72*) considered adsorption on small spherical particles where the effect of curvature makes the equilibrium value of t smaller than that measured on a flat surface at the same p/p_0. Geometrical as well as thermodynamical analysis of the physical situation resulted in a quantitative explanation of the observed t-values.

Considering the accomplishments of deBoer and coworkers, it seems proper to quote some of their views on the state-of-the-art of pore structure analysis by sorption isotherms (*38*):

"The calculations of pore distributions are based on a number of simplifications and idealizations, one of which is the assumption of highly idealized pore models. Ideal pore shapes will hardly ever be encountered in porous materials found in practice."

"The significance of certain discrepancies between cumulative surface areas and, for example, B.E.T. surface areas should not be overstressed, as a complete concordance is not to be expected in view of the inherent approximate character of every method of pore size distribution calculations. Moreover, it is not pretended that the results of cumulative calculations lead to certainty with regard to the shape of the pores present."

"It is very difficult, except in exceptional cases, to evaluate the pore shape with certainty from sorption data alone."

"For tubular pores the results from the adsorption branch and from the desorption branch should be reasonably consistent, whereas for 'ink bottle' like pores such a consistency should be absent. In this last case the results obtained from the adsorption branch should be preferred."

The problem of capillary hysteresis, as related to vapor sorption, received considerable attention from a number of authors, in addition to the fundamental work of deBoer and coworkers, which has already been discussed above. For a comprehensive discussion of adsorption hysteresis the interested reader is referred to the treatise of Everett (*97*) on this subject.

The domain theory of capillary condensation hysteresis was discussed by Everett in detail. A convenient definition of a pore domain is a region of pore space accessible from neighboring regions through pore constrictions. A given element of volume in an isolated pore domain has the property that during an adsorption process a liquid-vapor interface will sweep through it at a certain relative pressure X_{12}, and during the desorption the interface will pass back through it, leaving it empty, at X_{21}. Hysteresis is due to the fact that $X_{12} \geq X_{21}$.

In the independent domain theory it is assumed that each pore domain behaves as though it were interacting with the vapor as an isolated pore. Often pore domains are not independent, but they interact with each other. Pores of different sizes are interconnected and, therefore, one pore may block the emptying or filling of other pores. Consequently, condensation-evaporation effects are controlled not only by the pore size distribution, but to a major extent by the topographical sequence of these spaces.

The independent domain theory has been worked out for the interpretation of experimentally observed hysteresis loops and scanning curves. An interesting example for the application of this theory is shown in Figure 2. The same point Q can be reached from either branch of the primary hysteresis loop. After reversing both paths at Q, as shown in the Figure, both the ascending curve from Q arising from (i) and the descending curve arising from (ii) have the same slope, as required by theory (*25*).

In discussing the thermodynamical basis for capillary condensation, Everett made the following comment on

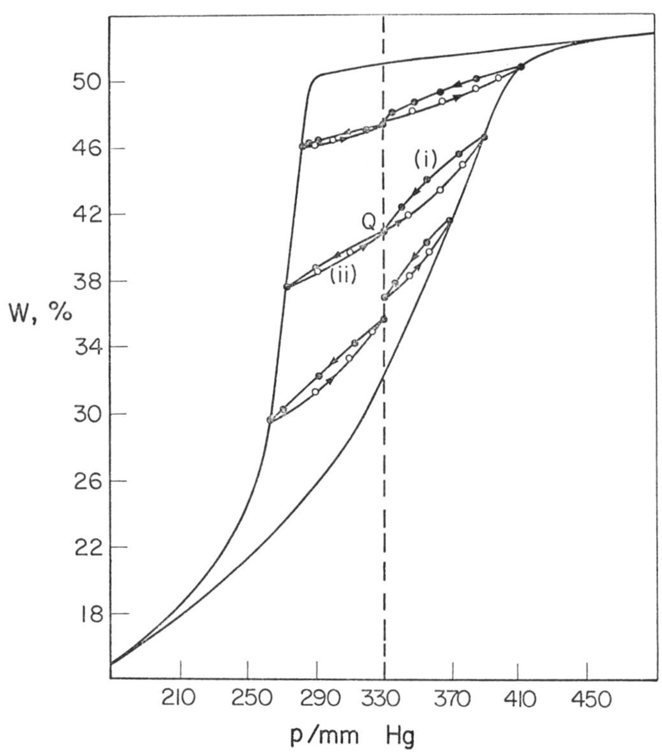

Figure 2. Slopes of adsorption and desorption curves immediately after reversal at a given point

(From Everett, "The Solid/Gas Interface," Vol. 2, Marcel Dekker, New York, 1967, by permission)

the applicability of the Kelvin equation to the case in which the radii approach a few times the molecular diameter: "Although there will, no doubt, be quantitative deviations from this equation (and these deviations may be quite large), the fundamental molecular kinetic factors that lead to this equation for macroscopic systems will still manifest themselves in microscopic systems." It seems difficult to argue with the soundness of this statement.

Everett (*97*) has also given a lucid discussion of the thermodynamics of adsorption hysteresis, showing the entropy production in the various nonequilibrium steps that may occur in either the adsorption or the desorption process.

Among other works on adsorption hysteresis, we mention the paper by Ekedahl and Sillen (*91*) who developed a mathematical model of sorption hysteresis using load functions. They discuss the relationship between their approach and that of Everett (*97*).

The approach of Ekedahl and Sillen, as well as the domain theory, are purely mathematical in character, and their success in predicting the observed sorption behavior depends entirely on whether their premises correspond to reality. While these may be oversimplified for many real sorption isotherms, the powerful mathematical theory is of great value, nevertheless, in explaining complex processes that would be impossible to grasp intuitively.

Quinn and McIntosh (*207*) studied the hysteresis loop in adsorption isotherms on porous Vycor glass. Diagrams of domain complexions for pores of the ink bottle-type were given. Evidence was presented showing the interdependence of domains or voids in Vycor glass. The compressibility of the porous adsorbent was evaluated and shown to differ with different adsorbates, suggesting distributions of adsorbed matter specific to the adsorbate.

Kington and Smith (*136*) studied the differential heat of adsorption by argon on some materials. They obtained the heat of adsorption by calorimetric method and also by the application of the Clapeyron equation. The Clapeyron heat was greater than the calorimetric heat in the desorption branch, indicating that there is justification for assuming that the desorption process is accompanied by irreversible change. They interpreted this conclusion as support for any mechanistic theory of capillary condensation based on a neck controlled desorption process, at least in porous Vycor glass.

Micropores or "ultramicropores." The very small pores, having radii of about 16 Å or less, present special difficulties in pore structure analysis. The most conspicuous phenomenon in the region of small pore sizes is the consistent absence of capillary hysteresis. Dubinin and coworkers have maintained for many years (*e.g.*, *19*) that capillary condensation cannot occur in what they call "micropores."

It is quite true that (macroscopic) thermodynamics cannot be expected to apply to systems consisting of only a few hundred or even fewer molecules. Similar approaches, as the ones used in the study of nucleation phenomena, could also be appropriate in the study of adsorption at low values of p/p_0. A complicating factor is the presence of the attractive forces emanating from the surface of the adsorbent. Broekhoff and deBoer have pointed out (*38*) that in the case of capillaries open at both ends, for small values of r the width of the Gibbs free energy barrier at the desorption pressure p_D is very narrow in terms of t. It follows from this that if, owing to statistical fluctuations, the thickness t momentarily builds up to a value that is on the "filling" side of the barrier, spontaneous filling of the pore will occur. This would explain the absence of hysteresis in very small pores notwithstanding the occurrence of capillary condensation.

Brunauer *et al.* (*42*) have developed a method for micropore analysis (MP-method), using the deBoer t-curve method. They plotted the adsorbed volume (as liquid) *vs.* t from t = 35 Å upward. In microporous adsorbents, they found that the slope of the curve soon started to decrease, which they attributed to the progressive elimination of the surface of the smallest pores owing to complete filling. The slopes were accepted as a measure of the available surface area of the sample. From the surface area of the pores filled, as well as the corresponding value of t, the size as well as the volume of that class of pores was calculated in an elementary fashion. The cumulative pore surfaces agreed with the BET surfaces, and the cumulative pore volumes agreed with the total pore volumes accessible to nitrogen.

According to Dubinin *et al.* (*81*), however, it is meaningless to talk about layerwise adsorption on micropore surface, and it cannot even be possible to determine the surface areas of such pores. The first part of this claim is very likely true in the case of some so-called "porous crystals"—*e.g.*, zeolites and bentonites—where the pores are of approximately the same size as the molecules of the adsorbate (*e.g.*, *12*), but it is not obvious why the same claim should also apply to pores whose diameters are a few times that of the adsorbate molecules, as Dubinin appears to maintain. As pointed out by Barrer, porous crystal sorbents differ from microporous carbons or silica gels in that the latter normally exhibit a spectrum of pore sizes, whereas in crystals the pore systems are often quite rigid and all of one type, having diameters of molecular dimensions. Accordingly, porous crystal sorbents can behave toward appropriate mixtures of molecules as perfect molecular sieves.

Kiselev and Lopatkin (*141*) calculated geometrical constants, including surface areas, of so-called "hollows" in *X*-zeolites. The pore structure in these consists of hollows connected by "windows," much like in packs of spheres.

The nature of adsorption in micropores probably also depends on their shape. Also, a certain surface roughness on the molecular scale constitutes a class of micropores in which a process analogous to completely reversible capillary condensation may occur—much the same as in bigger capillaries of conical shape with one end closed.

Carman and Raal (*50*) demonstrated that adsorption on an uncompressed nonporous powder exhibits reversible multilayer adsorption, whereas the same powder in the form of a compressed pellet shows hysteresis of the capillary condensation type. Experiments that are essentially the converse of those of Carman and Raal have been conducted by Bailey (*8*) who has shown that hysteresis loops exhibited by porous carbons and coconut shell charcoals are eliminated, if the carbon is finely ground in an agate ball mill to pass 300 mesh. The adsorption isotherm became reversible on the ground sample, and it fell slightly above that shown by the original unground sample.

Because of the somewhat contradictory evidence, it is probably not warranted at this stage to try to reach definite conclusions as to the true nature of adsorption processes that take place in micropores.

Miscellaneous. Several other interesting developments took place that are all pertinent to the problems of pore structure analysis. The papers by Kuhn (*156*) stressed an important point by focussing attention on surface sites with different adsorption potentials. There can be little doubt that this picture corresponds to physical reality, and it may have a fundamental role in determining rate processes that take place on the surface of an adsorbent. From the point of view of equilibrium studies, however, this is probably not of decisive importance, and Kuhn's results can be obtained without resorting to the concept of heterogeneous adsorbent surface. He has used the Polanyi adsorption potential as a function of one characteristic "radius." The physical meaning of this is not quite satisfactory since, for the case of capillaries with curved walls, one would expect the "adsorption potential" to depend on a number of factors. For the most general case these should include the statistical thickness t of the adsorbed layer (or, alternatively the radius of the adsorbed cluster, or, drop on the site), the mean radius of curvature of the capillary (a dip in the surface should possess a higher adsorption potential than a flat surface and this, in turn, should be more adsorptive than a hump), and the mean radius of curvature of the interface between the adsorbed layer and the vapor phase (the symmetry relations of the environment of a molecule at the interface influence the thermodynamic potential of the adsorbed layer). In fact, deBoer and coworkers took all but the second of these factors into account, which probably becomes increasingly significant in very small pores. The fact that Kuhn obtained a very simple equation for isotherms, which can represent all main isotherm types principally through the variation of a single constant, is interesting and may be useful.

Another interesting development is from Brunauer *et al.* (*42*). They developed and applied a computational method of pore structure analysis, without a pore shape model, by monitoring both the volume desorbed (or adsorbed) in an elementary step, and the surface area that is liberated (or occupied) simultaneously. By dividing the volume increment with the area increment, they obtain the hydraulic radius of the pore elements involved in the process. To correct for the thickness, t, of the adsorbed multilayer, they must assume some kind of a pore shape. The calculation of the surface area increment is based on a simple thermodynamic relationship, which has been used by Kiselev (*140*).

In another paper, Brunauer and coworkers explain that the choice of cylindrical shape for pore model is a fortunate one in the majority of cases because, after all, most pores are "tubular," in the widest sense of the word (*43*). This is even more true if only elemental segments of pores are considered. These authors have also demonstrated that the parallel plate and the spherical pore models do not yield very different results either.

Karnaukhov (*134*) stressed the need for combined investigation of the pore structure of catalysts. In particular, electron microscopic studies were recommended, along with sorption data, to establish the type of pore structure. He provided a table comparing pore radii obtained by the Kelvin equation with the actual pore radii for a number of different pore types. The general conclusion from this comparison was that the desorption branch will yield the correct radii in the majority of the cases. However, whenever the pore considered is accessible only through a throat, it is the radius of the throat that will be obtained. Use of the adsorption branch will result in values that are greater than the actual radii. He presented experimental evidence to illustrate the difference between the distributions of some silica gels calculated from the adsorption and desorption branches, respectively. Karnaukhov also pointed out that the greater the depression of the adsorbate over the concave meniscus with a given pore size, the more accurately the structural properties of the catalyst are reflected. Therefore, the nature of the adsorbate is important. He claimed that carbon tetrachloride, *n*-hexane, and benzene are more advantageous than nitrogen or argon for sorption studies.

Zhdanov (*279*) prepared artificial mixtures of porous glasses with known different pore sizes and measured the adsorption-desorption isotherms. The pore size distribution was then back-calculated. The one calculated from the adsorption branch gave no indication that the pore size distribution was bidisperse, whereas the one obtained from the desorption branch did.

Dubinin (*79*) mentioned in one of his earlier papers that electron photomicrographs of coarse-pored silica gel were used to obtain pore size distributions, which were compared with pore size distributions derived from capillary condensation theory with satisfactory results. Electron photomicrographs or optical photomicrographs have rarely been used to evaluate pore size distribution.

Spencer and Fereday (*240*) pointed out that the pore size distributions calculated from sorption isotherms depend very strongly on the values for the liquid parameters used in the Kelvin equation. They claimed that this fact makes the absolute value of the pore sizes obtained quite uncertain.

Many more methods have been worked out for pore size distribution and surface area calculations (*4*, *27*, *44*, *51*, *63*, *64*, *92*, *146*, *162*, *208*, *210*, *214*, *223*, *256*). Most of these methods use shortcuts or claim better correlation of experimental data. The adsorption methods of measuring the specific surface and structure of catalyst pores have been reviewed by Karnaukhov (*133*) up to 1962.

A large number of papers have applied the established methods of pore structure investigation from sorption isotherms to a variety of specific porous materials. An incomplete sampling of these is contained in (*6*, *80*, *102*, *129*, *137*, *138*, *139*, *142*, *144*, *155*, *157*, *182*, *191*, *237*, *253*, *260*, *273*, *274*, *277*, *281*).

One of the more recent features in the area of the technology of pore structure investigations was the appearance of automatic adsorption apparatuses, and computerized calculations of surface areas and pore size distributions from the adsorption isotherms (*10*, *11*, *54*, *81*, *99*, *111*, *176*, *225*, *243*).

A few dynamic methods have also been developed for surface area and pore size distribution measurement. They are all based on making a mixture of a nonadsorbable and an adsorbable gas flow either through or past the porous sample. The main advantages claimed are experimental convenience and speed of obtaining results (*45*, *59*, *98*, *115*, *189*). The reverse problem of predicting the rate of adsorption of a gas by a porous body has also been studied (*203*, *230*, *245*).

Relationship between Diffusion and Flow Through Porous Media and the Pore Structure

The movement of fluids in porous media and the quantitative laws governing this transport have become a field of study covering large areas of science and technology, the most important of which may be soil science, petroleum production, and contact catalysis.

Permeabilities and effective diffusivities have been correlated in terms of simplified models of pore structure; however, the success of such correlations often cannot be taken as evidence of the correctness of the pore model. Tortuosity has often been used as an adjustable parameter to bridge the gap between calculated and observed rates of transport.

Our understanding of the mechanism of diffusion and flow gases through capillaries and porous media has advanced considerably in the past decade. Probably one of the most important advances took place in developing a theoretically well-founded relationship for the simultaneous diffusion and flow of gases through straight cylindrical tubes that would apply from high pressure to practically zero pressure.

The problem of the "viscous slip" in flow of fluids through capillaries was recognized long ago. The concept of "diffusion slip" is relatively new. Phenomenologically speaking, viscous slip means the effect that volumetric flow rate *vs.* mean pressure plots do not extrapolate into the origin of the coordinate system, but rather have a positive intercept of flow rate at zero mean pressure. To bring Poiseuille's law in line with the experimental facts, a constant term (slip flow term) equal to the magnitude of this intercept had to be added to it. Again, phenomenologically speaking, the diffusion slip amounts to the following: when gases counterdiffuse in the absence of any hydrodynamic pressure gradient, there is a net mass flow of the gases relative to the walls of the capillary, except if the two gases have exactly the same molecular weights.

Studies of flow and diffusion in capillaries. Maidanik (*170*) developed a semiempirical theory for the flow of fluids through capillaries, powder beds, and porous media. The theory bridges the gap between hydrodynamic region and diffusion region. In addition to the usual factors, the flow depends on the exchange of momentum between the molecules and the surface of the sample. The molecules colliding with the surface can be divided on a statistical basis into two distinct fractions. A fraction of the molecules undergoes diffuse scattering on striking the surface, and the rest undergo specular reflection. It is, however, assumed that various collision processes (*i.e.*, molecule-molecule collisions and molecule-surface collisions) are independent of the momentum and independent of each other. In certain cases sorption of molecules may occur. Wicke and Hugo (*271*) gave an expression for the diffusion slip and used the equation to interpret the results they obtained in experiments on counter-diffusion of two gases through fritted glass. The ratio of the molar fluxes of the two diffusing gases was equal to the inverse ratio of the square root of the molecular weights of the gases. Similar results had been reported by Hoogschagen (*123*), and Evans *et al.* (*95*). The results of these studies are consistent with the existence of diffusion slip (or diffusion drift). Correct theoretical equations for the counter-diffusion of gases through porous media were presented down to zero pressures by Evans *et al.* (*96*), Scott and Dullien (*235*), and Rothfield (*222*). When the pressures become sufficiently low, the mean free path of the gas molecules becomes much greater than the dimensions of the capillaries. Under such conditions the diffusing gas molecules do not interact with each other at all, and therefore diffuse completely independently of each other (Knudsen diffusion). Essentially this means that the frequency of collisions of molecules with one another is small in comparison with their collisions with the surface of the porous material. It is well known that for Knudsen diffusion the ratio of the molecular fluxes of the diffusing species is equal to the square root of the inverse ratio of the molecular weights of the gases. The much more recent experimental fact that the same ratio of the molecular fluxes applies at higher pressures (where the diffusing molecules interact with each other) has also greatly facilitated the derivation of the general diffusion equations. In two other publications (*82*, *234*), Scott and Dullien derived diffusion equations for the case of cylindrical tubes, using elementary momentum theory. It was assumed that in the intermediate pressure range, where the intermolecular collisions and the wall collisions are of comparable frequencies, the total pressure drop of a component gas in a mixture can be considered as being due to the sum of two momentum transfer processes, the rate of momentum transfer to the wall, and the rate of momentum transfer owing to unlike molecular collisions.

Dullien and Scott (*82*) derived an equation that related the flux ratio N_A/N_B to the molecular weight ratio M_A/M_B, as a function of the dimensions of the capillary tube, mean molecular speed, diffusion coefficient, and the composition of the mixture. This equation shows that N_A/N_B may vary from $(M_B/M_A)^{1/2}$ to M_B/M_A, depending on the conditions. The result $N_A/N_B = M_B/M_A$ would correspond to a very short capillary having a very large diameter (or diffusion in free space).

Scott and Dullien (*234*) also derived a theoretical equation for the flow of gases in capillaries. This equation reproduces the experimental results of Knudsen and others over the complete range of laminar, slip, and Knudsen flow, including the minimum in the specific flow rate *vs.* mean pressure curve observed by Knudsen at very low pressures. Wakao *et al.* (*262*), using some of the results obtained earlier by Scott and Dullien, derived equations for the effect of total pressure gradient on flow and diffusion of a binary gas system in a fine capillary. Dyer and Sunderland (*88*), and Gunn and King (*113*) derived bulk and diffusional flow equations to predict mass transport in freeze dried meat.

All the diffusion and flow equations were derived for a case of pore structure entirely different from the pore structure of most real porous media. The equations, however, can be used as interpolation formulas to describe the diffusion and flow through porous samples without resorting to any extra adjustable parameters.

The "dusty gas" model. Mason and coworkers (*96*) based their derivations on a special model of a porous medium consisting of a random distribution of spheres fixed in space. These spheres were treated as giant molecules of another gaseous component having very high molecular weight. This technique greatly simplified the mathematical manipulations. Actually this model, popularly known as "dusty gas" model of porous medium, had been recommended earlier by Deryagin and Bakanov (*76*) who have, in addition, shown theoretically that a minimum occurs in the curve of permeability *vs.* average pressure for extremely porous bodies. This result has been verified experimentally for materials like steel wool (*246*).

Models consisting of straight cylindrical micropores and macropores. In the Wakao and Smith model the porous medium is supposed to consist of macropores in series, micropores in series, and combination of micropores and macropores in series. Wakao and Smith (*261*) used the diffusion equations derived in (*96, 222, 235*) and applied them to their model. According to their theory, the total rate of diffusion is the sum of the three constituent contributions. Pore size distribution data obtained by nitrogen adsorption and mercury porosimetry results were used in conjunction with the theory. Experimental data could be fitted very well in terms of this model for high alumina pellets.

Robertson and Smith (*215*) calculated both diffusion and flow measurements on some alumina catalyst samples. They used a pore model consisting of a set of macropores, and a set of micropores of straight cylindrical shapes. No micropores were in parallel with macropores. The path lengths were evaluated from the flow measurements and, assuming that the diffusion path length is the same as the flow path lengths, the diffusion rates were predicted. The measured rates were 40 to 80% of the calculated values.

Smith and coworkers used the equations derived by them in (*262*) to predict diffusion and flow rates measured in porous catalysts (*195*). They also studied the decrease in the effectiveness factor, owing to the existence of a pressure drop from the center of the catalyst pellet toward the exterior surface, as a result of a change in moles in the reaction. They found significant decrease only for low density type catalyst containing large macropores (*194*).

Henry *et al.* (*118*) checked the predictions of Wakao and Smith over a thousandfold pressure range with satisfactory results. In another paper Cunningham and Geankoplis (*62*) tested the Wakao and Smith model by varying the macropore radius over a wide range. They derived new equations for pore structure containing many sizes of macropores, and for one containing two micropores; they obtained good fits to experimental results.

Models consisting of a bundle of cylindrical tubes or capillaries. Hewitt (*120*), in an extensive review article, reported on a test of pore structure model consisting of a number of capillaries having one fixed diameter that pass through the medium in a tortuous path, such that the true path length along the capillary is X_e, in a distance X through the medium. The ratio X_e/X is called the tortuosity. For this model

$$\varphi = 1.14\ K_0^2/B_0 \tag{16}$$

where all the three parameters can be determined experimentally. (φ is a constant less than unity and is equal to the ratio of the gas diffusion coefficient in the medium to the free gas diffusion coefficient; B_0 is the permeability, and K_0 is a constant representing the viscous slip effect.) For a number of British nuclear graphites it was found that the value of φ calculated from the above equation was, on the average, five times as large as the measured value, which implies that the model is seriously wrong.

According to Hewitt, refinement over the above model is obtained if one takes into account the distribution of pore sizes. The expression obtained this way for B_0, K_0, and φ can be used together with mercury porosimeter data to serve as a useful frame of reference to compare experimental results.

Schlosser (*229*) developed a model of capillaries with different diameters. This equation allows for diffusion and molecular flow on the one hand, and viscous and slip flow on the other. The method was recommended to obtain pore size distribution.

Nicholson and Petropoulos (*190*) used a model of porous medium consisting of a bundle of cylindrical tubes, each of which is constructed from a series of sections with radii chosen from a distribution function. Structure factors introduced by Barrer and Gabor (*13*) were calculated.

Models consisting of convergent-divergent sections in series. Houpeurt (*124*) pointed out that pore structure consists of convergent-divergent sections in series (see Figure 3). Hence, much of the pressure drop in the pores is due to the loss of kinetic energy that is dissipated because of sudden variations in cross-sections of the pores. This is also consistent with the results calculated by Petersen (*198*), who demonstrated that in the case of diffusion, convergent-divergent pore structure (Figure 4) will give rise to an effective tortuosity. Batra *et al.* (*17*) studied the laminar flow of liquids through periodically convergent-divergent tubes and channels. The value of the friction factor increases with decreasing wavelength-to-diameter ratios (<0.5) by as much as 120% of the uniform tube value (wave amplitude to diameter ratios were in the range 0.13–0.25). The value of critical Reynolds number of transition to turbulence decreases with decreasing wavelength-to-diameter ratio.

Foster and Butt (*103*) devised a computational model for the structure of porous materials. Two major arrays of pores were assumed: a centrally convergent and a centrally divergent one. The two are interconnected at specified intervals. The shape of the arrays is calculated from the pore size distribution. By assuming a proper amount of mixing, the model could be fitted to experimental results.

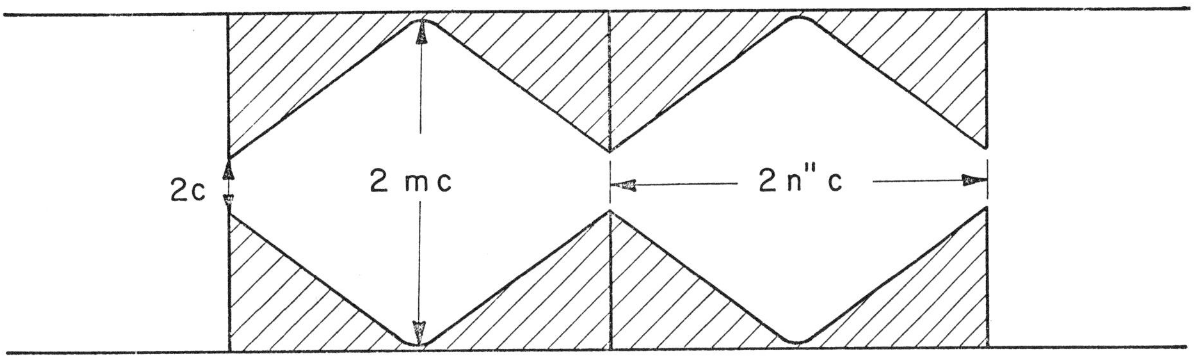

Figure 3. Pore model with periodic "orifice plates." Vein section varies from thresholds c to enlargements mc

(From Houpeurt, Rev. Inst. Fr. Petrole Ann. Combust. Liquides, ***14****, 1468 (1959), by permission)*

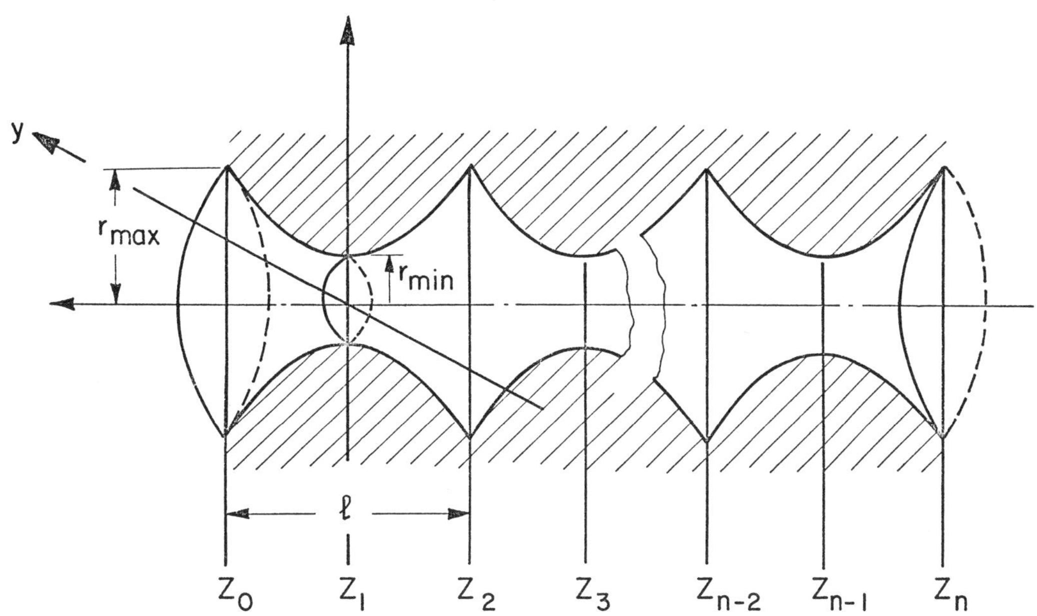

Figure 4. Model of pore with periodic pore constrictions

(From Petersen, A.I.Ch.E. J., ***4****, 343 (1958), by permission)*

Miscellaneous studies. In a number of papers particular emphasis was placed on the relationship between pore structure and mass transport through the porous medium. In a series of papers (*251*, *252*) Turner proposed a frequency response method to analyze the physical structure of a packed bed or a porous solid. Wyllie and Gregory (*276*) tested the well-known Kozeny-Carman equation in the case of unconsolidated aggregates composed of particles, and found that the shape factor appeared to be between 2 and 3, as claimed by Carman. (It is known, however, that there are serious problems with the Kozeny-Carman equation in heterogeneous packs containing many small particles that contribute an "excessive" amount to the specific surface area of the aggregate.)

Marshall (*174*) used probability considerations to develop a series for the permeability in terms of the mean pore radii in each of the *n* equal fractions of the total pore space. Calculated and measured permeabilities were found to agree over a wide range of values.

Millington and Quirk (*184*), using a different statistical model, obtained a generalized relationship between the porosity and the cross-sectional area controlling flow in a porous material. Pore size distribution data are used to predict both saturated and unsaturated permeabilities.

Barrer and Gabor (*13*) described a four parameter representation of mass transport properties of porous materials. Unlike most other authors, they allowed for the occurrence of surface diffusion or surface flow, in addition to volume diffusion. Grubner (*112*) claims good agreement between the predictions of steady-state gas flow technique and the structural information obtained by other methods. Greenburg and Weger (*109*) obtained good correlation between mean cross-section diameters obtained from mercury porosimetry measurements and experimentally observed permeabilities. Johnson and Stewart (*131*) predicted effective diffusion coefficients with fair accuracy from pore size distribution data.

Pope (*202*) measured the gas-phase molecular flow of hydrogen or helium through porous media of known structure containing adsorbed sulphur dioxide molecules. The observed permeabilities depended markedly on the adsorbate concentration. This the author explained in terms of blocking of pores by the adsorbate. It was suggested that similar experiments would be useful for pore structure investigation of samples with unknown pore structure.

A rather different method of pore size distribution determination was used by Ergun and Owen (*94*) who subjected the particles to progressive grinding and determined the aerodynamic densities of the batches by gas flow method. The change in particle density could be related to the pore size distribution.

An interesting phenomenon, the natural consequence of diffusional slip or drift, is the so-called pressure effect, or Kirkendal effect, in gas diffusion in capillaries or porous media. This effect is observable when the capillary connects two closed vessels in which the different gases are initially at the same pressure. Owing to the drift, a pressure difference will develop between the two containers, which in turn will set up a compensating hydrodynamic or bulk flow. Some of the work that has been done in this field is described in (*244*, *259*, *263*).

A great deal of material on diffusion in porous media up to about 1960 can be found in a review paper by Barrer (*14*). This author devoted considerable attention to the phenomenon of surface diffusion. In Barrer's opinion the physically most satisfactory method of analyzing the surface fluxes is in terms of a random walk concept.

It is not easy to understand, however, why such molecules as spend part of the time adsorbed on the solid surface contribute more to the mass transport than those molecules that stay practically all the time in the gas phase. The mobility of molecules in condensed phases is normally much less than that of molecules in a gas phase. Hence, one would think that the adsorbed molecules can be expected to have decreased mobility, as compared with those that are not adsorbed. If this view is taken, the only way to explain the experimentally observable excess mass transport in the case of adsorbing vapors is to postulate a driving force very much in excess of that in the equilibrium vapor phase. This hydrodynamic explanation of surface flow was advocated among others by Flood and Huber (*100*, *125*).

In a somewhat different approach (*87*), experimentally observed "surface diffusion" could be accounted for by using the concept of surface "roughness" on a molecular scale. Joint application of expressions having the same form as the Laplace and the Kelvin equations give, for very small surface capillaries, relatively very great pressure drops in the condensed phase, as compared with the pressure drop in the equilibrium vapor phase. This pressure drop provides the necessary driving force for surface "diffusion." At very low values of the relative pressure, below the monolayer coverage, "patches" of adsorbed molecules on the surface can be assumed to have statistical existence. Owing to the presence of a pressure differential between the upstream and downstream end of a "patch," these molecules travel in the same direction as the vapor, and thus give rise to enhanced mass transport through the sample.

Various aspects of diffusion and flow of gases through porous media are discussed in (*2*, *5*, *21*, *107*, *110*, *119*, *149*, *169*, *218*, *268*).

Klinkenberg (*143*) calculated the pore size distribution of a sandstone sample from the results of displacement experiments with miscible liquids, using the customary pore model consisting of straight cylindrical capillaries. Mercury porosimetry gave an entirely different pore size distribution on the same sandstone. The cylindrical capillary model does not appear to be equally suited to account for the results of different displacement processes.

DeBethune and Rowell (*66*) pointed out that for powder packs four independent (mean) pore radii can be determined experimentally: the "Adzumi radius" (obtained from the ratio of Poiseuille flow to Knudsen flow), the "hydraulic radius" (obtained from the ratio of porosity to specific surface), the "Knudsen radius" (obtained from the ratio of Knudsen flow to porosity), and the "sigman radius" (obtained from the specific surface). The degree of agreement of these various radii is a measure of the homogeneity of the pack. The "double-broad Gaussian" distribution (a four parameter distribution) is thought to be the most promising one for the characterization of powder packs.

Flow of polymer solutions through capillaries and porous media has received increasing attention. Bauer *et al.* (*18*) found, in the case of capillary flow of aluminum dilaurate toluene gels, a family of flow curves of nominal wall rate of shear *vs.* wall shearing stress, depending on the dimensions of the capillary used. Lifson (*163*) found that the intrinsic viscosity of polymer solutions varied with the width of pores when this width is not much larger than the macromolecule in solution.

Skibitzke *et al.* (*238*) described a radio tracer technique for the study of flow of liquids in saturated porous materials. Brigham *et al.* (*34*) and Blackwell (*24*) found, in miscible displacement experiments, that the rate of mixing increased with decreasing particle-size and increasing degree of heterogeneity of the medium. They suggest that beds consisting of smaller beads are more heterogeneous, and that this is the probable cause of the increased mixing in such beds. Johansen and Dunning

(*130*) studied homogeneous fluid flow through consolidated porous media and found that both the fluid saturation and the permeability were affected by the wettability of the solid by the fluid. Atkins and Saki (*7*) investigated the flow of liquid helium through microporous glass, and found that the onset temperature for super-fluid flow was markedly less than the λ point of the bulk liquid.

Wyllie and Gregory (*275*) determined the formation factor (the ratio of the resistivity of a solution containing dielectric solids to the resistivity of a similar solution not containing solids) experimentally for a variety of particle shapes and cementations.

Volarovich and Churaev (*258*) determined the pore size distributions of peat, clay, and sand by a miscible displacement method, using a radioactive tracer indicator. The cylindrical capillary model was assumed.

Blokh and coworkers also used the miscible displacement method with KCl solutions of different concentrations to determine the pore size distribution of some ceramic products (*26*).

Capillarity and Pore Structure

Capillarity plays an extraordinarily important role in science and technology. Soil science and groundwater engineering depend altogether on the pore structure and capillary characteristics of soil. Similarly, in petroleum reservoir engineering calculations of capillary pressures are basic required data. A relationship between pore structure of reservoir rocks and the recoverability of petroleum from the rocks would be of great value to the petroleum industry. Drying is yet another important technology where capillarity and pore structure are of fundamental importance. Capillary behavior of various coatings play a very important role from the point of view of their protective effectiveness. The significance and control of their pore structure has not been adequately studied yet. The importance of pore structure and capillary behavior of concrete and fire-bricks has been realized for a long time. A great deal of research could be done in this respect, also, in the paper and textile industry, with beneficial results. The pore structure and wetting characteristics of porous electrodes plays a decisive role in the effectiveness of batteries. Also, the detailed relationships have not been developed in this field either. Porous plastics are used for numerous purposes as intermediates as well as final products. Everywhere detailed knowledge of pore structure and the relationship between this and the desired properties is required for better design.

The basis of capillarity is that in equilibrium the surface free energy is minimum. A particular form of this condition is the well-known Laplace equation

$$P_c = \sigma \left(\frac{1}{r_1} + \frac{1}{r_2} \right) \tag{17}$$

where P_c is the capillary pressure, σ is the surface tension, and r_1 and r_2 are the two principal radii of curvature of the surface at any point. The statement that the harmonic average of the two principal radii of curvature is the same at every point in the equilibrium surface is a mathematical consequence of the fact that the surface area is minimum. All methods of pore size determination that make use of measurement of capillary pressure are based on the claim that the harmonic average of the principal radii of curvature of an equilibrium liquid surface can somehow be related to the pore radius. This relationship is very difficult to establish for just about any pore shape other than the simplest geometrical shapes. For the case of straight cylindrical capillaries Laplace's equation will give

$$P_c = \frac{2\sigma \cos \theta}{r} \tag{18}$$

where r is now the radius of the capillary tube and θ is the contact angle of wetting. This equation has been used almost universally to determine pore size distributions from capillary pressure measurements.

It is apparent, however, that very few real porous media will contain pores of the straight cylindrical capillary shape, and therefore the pore radii calculated from the capillary pressure have to be considered "equivalent pore radii." The equivalent pore radius is the radius of a straight cylindrical capillary that would give rise to the same capillary pressure at the measured value. Since a variety of different capillary shapes may give rise to the same capillary pressure, the pore dimensions calculated from capillary pressure measurements are also somewhat qualitative.

Mercury porosimetry. One of the most popular methods of determining pore size distributions is mercury intrusion porosimetry. There is extensive literature about this technique, and the interested reader is referred to a long list of references given in (*217*). It is widely known that the method consists of measuring simultaneously the capillary pressure and the volume of mercury that has penetrated the sample up to that pressure. This gives the cumulative pore size distribution from which the differential distribution can be derived.

It is evident that mercury intrusion porosimetry will not detect the dimensions of those pores that are accessible only through necks narrower than the pore itself. In addition, the volume of these pores of limited accessibility will be erroneously assigned the dimensions of the necks leading to these pores. To correct this distortion, Meyer (*180*) calculated the probability of a pore of radius r_0 not being connected to pores of radius $r \geq r_0$. He worked out a computational procedure to correct the mercury porosimetry curves for this effect.

Markin (*173*) used a model consisting of a system of randomly arranged intersecting voids with circular cross-section, and continuously varying radius. No more than three branches could converge at any one point. A method was worked out, based on this model, for correcting the experimental capillary pressure curves obtained by mercury porosimetry.

More detailed mathematical analysis of the problem of the true relationship between capillary pressure and the pore sizes has been limited, so far, to the case of uniform-size spheres packed in some regular fashion. Kruyer (*153*) gave an approximate mathematical treatment of mercury retraction from packed spheres. He pointed out that many porous media are better approximated by packed spheres than an assemblage of cylindrical capillary tubes. In such media, after the penetration of mercury, the unpenetrated part of the pack has the shape of toroidal spaces or "pendular rings" (see Figure 5) around the points of contact between neighboring spheres. When the capillary pressure is decreased, the size of these rings will increase, and finally, at the point where these interfere with each other the entire pore space will be empty of mercury. This point is determined by the radius of the sphere inscribed in the interstices. Mercury penetration, on the other hand, will start as soon as a stable surface can pass from one interstice into a neighboring one. The size of the passage is approximately equal to the radius of the circle inscribed in the opening. The radius of this circle is invariably smaller than the radius of the sphere controlling the retraction of mercury from the interstices; therefore, it takes higher pressure to fill the pack than the value under which the pack will empty, resulting in hysteresis. Mercury penetration tests into packs bear out these expectations at least semiquantitatively (Figure 6). If the same experiments are interpreted in terms of the generally used cylindrical model, a peculiarly shaped pore size distribution curve is obtained with a sharp cut-off at the large pore radius end. Kruyer claims that there is complete analogy between mercury retraction from and capillary condensation of vapors in a pack. This analogy is not quite perfect, however. Apart from the complication introduced by the multilayer thickness t in capillary condensation, mercury retraction is complicated by the fact that in parallel channels the retraction of mercury does not always proceed at the same rate; as a result, some mercury may get separated from the bulk and will not be retracted from the pack.

Whereas Kruyer's theoretical treatment dealt with the mercury retraction branch of the hysteresis loop, Frevel and Kressley (*106*) subjected the mercury penetration branch into packs of uniform spheres to theoretical analysis. They also pointed out that the characteristic behavior for a pack of particles is an abrupt threshold for mercury penetration. For porous solids prepared from nonporous particles the "interconnected void" model is preferred to a model based on a system of circular capillaries. Electron micrographs of thin sections should be used for morphological details. The model used by Frevel and Kressley consisted of close-packed layers of spheres. The stacking of the layers was assumed to vary between completely close-packed nesting, and vertical stacking of particles in the layers, one above the other. This model gives rise to a range of porosities varying from 39.54 to 25.95%.

Mayer and Stowe (*177*) considered an alternate model

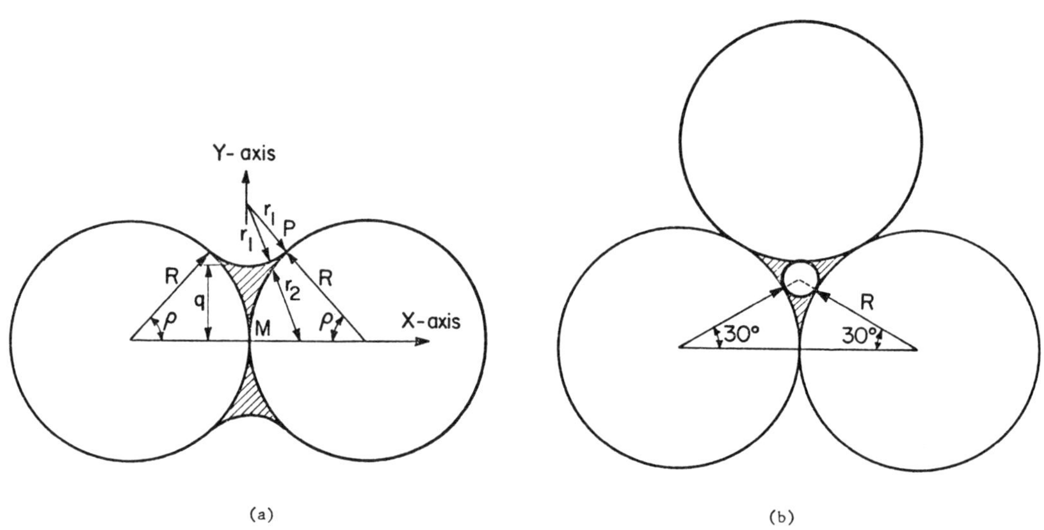

Figure 5. Surface of completely wetting liquid penetrating between touching spheres. (a) side view; (b) top view

*(From Kruyer, Trans. Faraday Soc., **54**, 1758 (1958), by permission)*

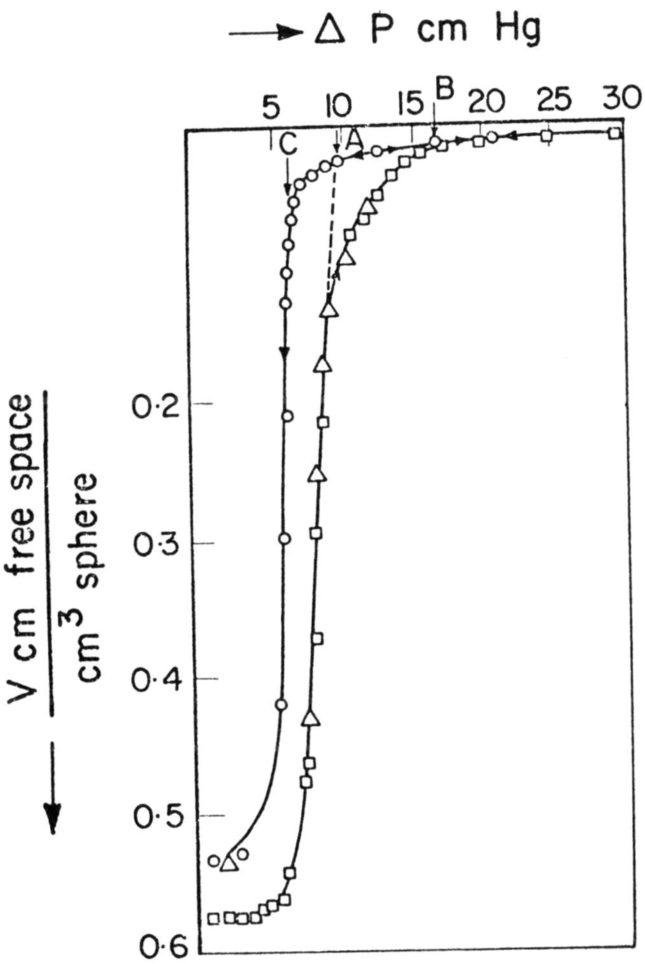

Figure 6. Penetrations and retraction of mercury in glass spheres 0.42–0.35 mm. Penetration, retraction, penetration after retraction

(From Kruyer, Trans. Faraday Soc., ***54****, 1758 (1958), by permission)*

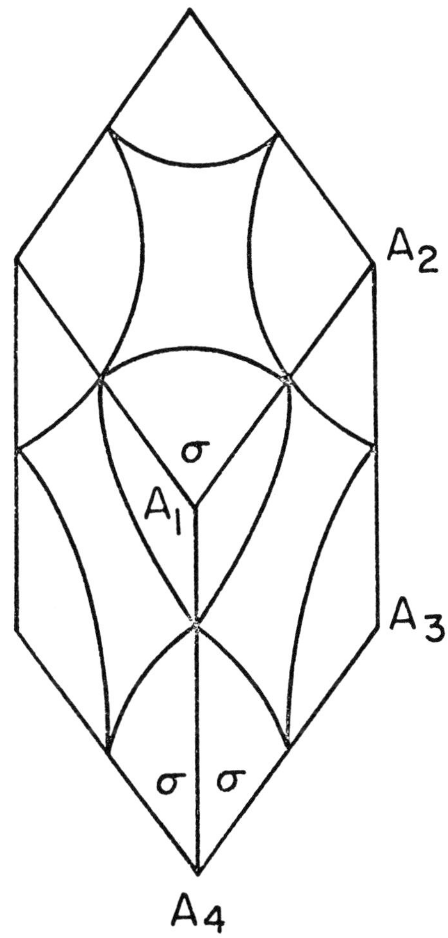

Figure 7. Unit cell of packed spheres

(From Mayer and Stowe, J. Colloid Sci., ***20****, 893 (1965), by permission)*

consisting of uniform spheres varying in packing between the two extremes of three-dimensional close packing and three-dimensional simple cubic packing. The upper porosity limit in this case was 47.64%. The size and shape of the largest openings are identical for both models. In the model of Mayer and Stowe there are 6 square openings, whereas in the model of Frevel and Kressley there are 4 square openings and 4 triangular ones. The model of Mayer and Stowe was defined in terms of a single acute angle, the size of which is determined by the edges of the rhombohedron formed by connecting the centers of a cluster of spheres (see Figure 7). Thus, this model changes packing equally in all three dimensions. At a given value of the angle, the size and shape of the largest opening is identical for either model; the corresponding porosities of the pack, however, differ. Mayer and Stowe tabulated the breakthrough pressure into the pack in terms of the porosity and the contact angle. This model is such that only in complete three-dimensional close packing will the square windows split into two triangular ones. The mercury penetrating through the window (Figures 8 and 9), however, separates into two branches somewhat before complete three-dimensional close-packing is reached, depending on the value of the contact angle. Calculated data covering the porosity range of 47.64 to 25.95%, and contact angles of 90° to 180° were presented. The relationships derived by Mayer and Stowe allow the determination of particle radius from standard mercury penetration and porosity data.

In a second paper (*178*), Mayer and Stowe extended their analysis to the filling of the toroidal void volumes following breakthrough between packed spheres. The solution of their equations gives the toroidal volume as function of the contact angle and the capillary pressure. They pointed out that the assumption of Frevel and Kressley, that the mercury passes through the access openings in the shape of the inscribed circle, leads to a fraction of the space filled following penetration, which is lower than the real value. Mayer and Stowe compared their results on the later part of the penetration branch with Kruyer's results on the early part of the retraction branch of mercury and demonstrated excellent agreement between the results of the two independent methods.

All the theoretical analyses of mercury penetration into and retraction from packed beds have been limited

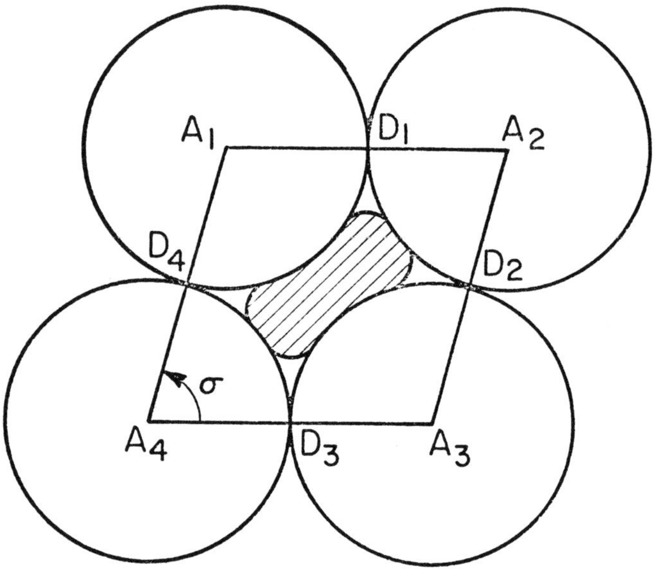

Figure 8. Access opening to unit void

(From Mayer and Stowe, J. Colloid Sci., ***20****, 893 (1965), by permission)*

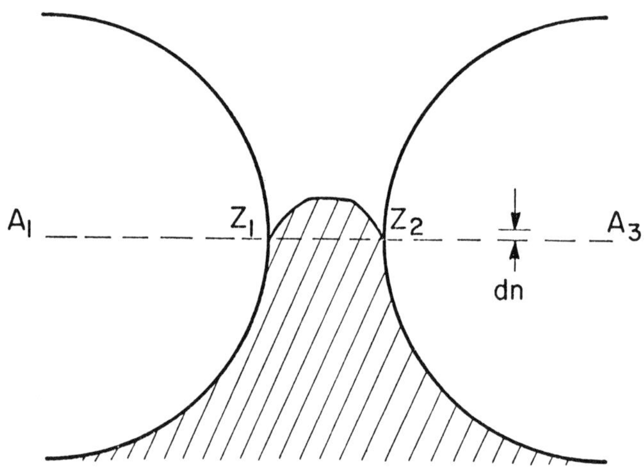

Figure 9. Mercury penetration between spheres

(From Mayer and Stowe, J. Colloid Sci., ***20****, 893 (1965), by permission)*

to the case of uniform spheres. In capillary pressure tests such packs are evidently characterized by two significant values of the capillary pressure, the pressure of penetration, and the pressure of retraction. To each of these pressures one can assign some kind of equivalent diameter characteristic of the pack. On the other hand, penetration of mercury into the toroidal spaces will continue indefinitely if one keeps increasing the pressure. Only at the theoretical limit of infinite pressure will the toroidal spaces be completely filled with mercury. If the cylindrical capillary model were used to interpret the results, the conclusion would be that the sample contains a continuous spectrum of cylindrical pores ranging to molecular diameters. Similar misrepresentation of the experimental results is possible for all pore shapes with the exception of spherical and cylindrical pores.

Notwithstanding the qualitative nature of the pore size distributions derived from capillary pressure measurements, mercury porosimetry has been enjoying an ever-increasing popularity, possibly because of the simplicity and speed characteristic of this technique. A variety of porosimeter designs cover a range of pore sizes from 2 mm down to 15 Å (*49, 161, 188, 200, 209, 265*). At very high pressures, the change in the volume of the lines and vessels, as well as the compression of the sample, are additional factors that may introduce errors into the measurements.

Ortega and Teoule (*193*) designed a porosimeter in which the volume of mercury in the sample is determined by gamma-ray absorption measurement. They claim that their results are unaffected by changes in volume of the lines and vessels under high pressure. Ingles (*127*) considered apparatus corrections applicable in mercury porosimeter measurements.

Mercury porosimetry has been used to determine the pore size distribution of a very large variety of substances. These include petroleum reservoir rocks (*30, 41, 206*), mortars (*151*), nuclear graphite (*28*), aluminum oxides (*56*), membrane filters (*122*), fired clays (*121, 175*), building bricks and limestones (*254, 264*), coal, coke and coal char (*132, 205, 249*), hardened cements (*89*), and others.

Contact angle studies. The value of the contact angle plays an important role in determining the capillary pressure, and therefore its measurement has been subjected to a great deal of research and discussion for many years. It appears that the true value of the contact angle is a rather elusive quantity.

Deryagin *et al.* (*74*) tested a relationship derived earlier by Deryagin between the rate of capillary penetration and the specific surface of a porous medium. This relationship involves the contact angle. In the case of quartz sand they consistently obtained much lower values for the specific surface area as compared with the results of other independent methods. Accepting the specific surface areas obtained from the Kozeny-Carman equation as correct, they back-calculated the value of the contact angle for spreading water on quartz sand and found this to be in the range of 60° to 70°. They blamed this effect on the roughness of the quartz surface. They didn't obtain very different contact angles for a variety of polar and nonpolar liquids in capillary penetration tests of quartz sand and certain soils. The total range of values appears to be 55° to 84°. The fact that they found no deviations from theory in the case of capillary rise experiments conducted in narrow glass capillaries seems to indicate that it was not the dynamic nature of the experiment that gave rise to contact angles that were different from the static value. It would be interesting to test this theory of Deryagin's in capillaries that contain periodic variations in their cross-section. Such a capillary would more closely resemble the capillary path that exists in porous packs. It is easy to show that in a conic capillary section with the apex of the cone pointing downwards the capillary pressure is equal to

$$P_c = \frac{2\sigma \cos (\theta - \alpha)}{R} \tag{19}$$

where α is the half angle of the cone, and R is the distance from the axis of symmetry to the base of the liquid meniscus. If the capillary were cylindrical, then R would be the radius of the capillary. It can be seen from this relationship that, for capillary rise in a cone, the effective contact angle is diminished.

Schwartz and Minor (*231*) calculated capillary pressure and capillary motion of a liquid between two parallel rods, in a yarn, and in a V-shaped notch. The treatment avoids the use of radii of curvature and uses free energy per unit area of the interfaces instead. Problems involving complicated interfaces are much more readily tackled by this technique. In a second paper, Schwartz and Minor (*232*) gave a method of calculating equilibrium contact angles in systems of various shapes such as a drop on a spherical surface, a drop on the hemispherical end of a cylindrical rod, a drop of liquid in a tube, and a drop of liquid on a cylindrical fiber. They discussed contact angle hysteresis in terms of surface roughness.

Braslavskii (*31*) measured rates of capillary rise in capillary tubes of various but uniform cross-sections. The same author (*32*) also obtained various rate of rise curves in real granular porous bodies. The variations observed from case to case could not be explained and were attributed mostly to surface roughness. Braslavskii (*33*) found, in experiments with capillaries of various cross-sections, that if the cross-sectional area is greater, the rate of rise is also greater, but the maximum height of rise is less. If closed and open capillaries (= grooves) of equal cross-sectional areas are used, the height of rise is greater in the closed capillary.

Laughlin and Davies studied the rate of capillary rise in wicking, both experimentally and analytically (*158*).

Rose and Heins (*219*) studied the change of interfacial curvature of Nujol and oleic acid in a capillary by making movies of the moving interface. They found the contact angle to be considerably rate dependent. Whereas the receding contact angle was always zero, the advancing contact angle depended on the rate of advancement, and values up to 40° were reported.

Schwartz *et al.* (*233*) analyzed the question of capillary hysteresis and came to the conclusion that microscopic surface roughness cannot be the sole factor controlling contact angle hysteresis.

Morrow and Harris (*186*) showed that the pendular rings of a wetting liquid left behind after drainage in a packed bed of particles are, for all practical purposes, isolated from the receding bulk liquid and therefore not in mechanical equilibrium with the bulk. They call the state of pendular rings "apparent equilibrium." They provided experimental evidence that the pendular rings are practically independent of the continuous bulk wetting liquid.

Timmons and Zisman (*250*) measured contact angle hysteresis for a variety of liquids on condensed monolayers of 17-(perfluoroheptyl) heptadecanoic acid adsorbed on polished chromium. The hysteresis was simply related to the molecular volume of liquid, and to result from the penetration of liquid molecules into the porous monolayer. It was shown that it is possible to estimate intermolecular pore dimensions of such adsorbed monolayers by contact angle hysteresis measurements with a series of liquids having gradations in molecular volume.

Ingles (*126*) claimed that the advancing contact angle of mercury into coal plugs was considerably less than 140°. He compared the critical penetration pressure of mercury into the plug with the pressure necessary to prevent penetration of completely wetting liquid through the same plug. Kruyer and van Krevelen (*152*), in reply to Ingles, pointed out that the pressure required to force mercury through a plug is determined by the largest pore, but the pressure needed to prevent penetration of a completely wetting liquid in the same plug is determined by the smallest pore. They claimed that Ingles made an error by using the same radius in both experiments. The contact angle, or more accurately the $\sigma \cos \theta$ product, can of course be determined for mercury penetration into a porous plug by performing water capillary pressure measurements in the same plug (provided that the water completely wets the surface of the plug). This was done for example in (*41*).

Capillarity and multiphase flow in porous media. Templeton (*247*) studied oil displacement in microscopic capillaries down to 4 μ. He found Poiseuille's law applied; also, the capillary pressure was independent of the interfacial velocity. Templeton and Rushing (*248*) explained the findings in (*247*) by claiming the existence of an annular water film about 20 Å to 200 Å thick on the walls of cylindrical glass capillaries around the oil slug flowing in the capillary. This claim was later fully confirmed for the case of flow of entrapped air bubbles through a capillary by Marchessault and Mason (*172*). They assumed that the bubble is a circular cylinder surrounded by a film of uniform thickness, and that it moves as a solid plug. Their predictions agreed well with experimental results.

Capillarity plays a very important role in the interpretation of multiphase flow in porous media. These phenomena are of great practical significance in the recovery of petroleum, where the oil very often flows in the presence of water and/or gas in the pores. Capillary and viscous forces compete for the control of multiphase flow in pores. In a discussion of the effect of viscosity in capillarity and the displacement of oil by water, Moore and Slobod (*185*) introduced the capillary doublet model to explain certain phenomena associated with the displacement of oil with water. Engelhardt and Lubben (*93*) investigated the influence of the interfacial tension and the contact angle on the displacement of oil by water in porous bodies. Rockwood *et al.* (*216*) used mercury capillary pressure curves to provide a direct estimate of the oil-occupied pore spaces in the reservoir. They have shown that capillary pressure, connate water, and porosity are interrelated. Perkins (*197*) investigated the role of capillary forces in laboratory water floods and found that the residual oil saturation was independent of the flood-rate or the column length; however, the distribution of oil saturation during the flood depended on these factors.

Pickell *et al.* (*199*) studied capillary hysteresis both by mercury porosimetry and by an "oil-air imbibition system." The residual saturations of the nonwetting phase obtained in these studies correlated very well with the water flood residuals. Even more interestingly, they found that in well-sorted rock the recovery from the higher saturations was more efficient than that from the lower saturations. Poorly sorted rock exhibited uniformly poor recoveries. They discussed these phenomena in terms of "trap hysteresis" caused by restricted openings in the pore structure.

Pacsirszky *et al.* (*196*) used a model of the porous medium consisting of straight cylindrical capillaries of different diameters and lengths. Applying Poiseuille's law, Ohm's law and Laplace's equation to this model they obtained the following equation

$$(Q/\Delta P) = \frac{(2\sigma \cos \theta)^2}{8\mu\gamma} \int_0^c \frac{dc}{P_c^2} \tag{20}$$

where Q equals mercury flow rate through the sample, ΔP equals pressure difference across the sample, σ equals surface tension of mercury, θ equals contact angle of mercury, μ equals viscosity of mercury, γ equals conductivity of mercury, c equals electrical conductivity of the sample, and P_c equals capillary pressure.

The capillary pressures were obtained by mercury porosimetry, the mercury flow rates and the electrical conductivities were measured as functions of the mean capillary pressures. Plotting the experimental

$$Q/\Delta P \quad vs. \quad \int_0^c (dc)/(P_c^2),$$

straight lines passing through the origin were obtained (except for the points at very low and very high saturations). The values obtained for $2\sigma \cos \theta$ from the slopes tended to be higher than the theoretically expected value. Nevertheless, the method can be used to predict relative permeabilities to mercury.

Al'tshuler (*3*) presented an interesting method for the determination of the size distribution of dead-end capillaries in porous bodies. It is based on the entrapment of air in the dead-end capillaries.

Fatt, as well as Dodd and Kiel (*77*) devised network models in which they replaced each pore in a sphere-packed model with a cylinder. The calculations were performed in two dimensions. Different network designs and pore size distributions were used. The displacement of the nonwetting phase by the wetting phase was calculated by the "Monte Carlo" method in a quasi-static fashion, and the results were presented in the form of capillary pressure-fluid saturation curves. The results obtained in the studies are interesting, and they appear to be valid for the assumed models. No allowance was made, however, for the fact that most pores consist of voids that are separated from each other by narrower necks. Therefore, the quantitative aspects of the results obtained by these authors may not be directly applicable to practical situations. It would be of considerable significance to extend this type of calculation to networks that are characterized by a "two-dimensional" pore size distribution $\alpha'(r_e, r_e')\, dr_e dr_e'$, giving the pore volume with radii in the range between r_e and $r_e + dr_e$, and with pore entry radii in the range between r_e' and $r_e' + dr_e'$ (*83*).

The centrifuge method of capillary pressure measurement has been used for pore size distribution determination, using the cylindrical capillary model (*154*). Comparison with the mercury porosimetry curve showed that the centrifuge method gave more weight to the larger pores. This is a surprising result, because if both methods are used properly, they would be expected to give identical pore size distributions.

Experiments on the evaporation of a liquid through simple models of capillary porous solids have demonstrated that in the drying process moisture moves to the surface both as liquid and as vapor (*159*). In the period of constant rate, the transfer of moisture to the surface occurred principally as liquid, and evaporation proceeds at the surface. In the period of falling rate, the moisture is transferred to the surface of the solid as vapor, while the surface of evaporation recedes to the interior. This recession does not proceed frontally, however, since the narrow capillaries are still charged with liquid at the surface.

Stamin and Wagner (*241*) determined the distribution of interstructural openings in wood by using octyl alcohol to displace water. Wood is a highly porous material, with porosity ranging from 45 to 85%. However, it is not very permeable, because the intercommunication between fiber cavities occurs only through openings in the pit membranes that are usually submicroscopic in size. The maximum of the pore size distribution curve was at about 50 mμ.

Microscopic and Electron Microscopic Investigation of Pore Structure

Many theories have been devised that attempt to relate in detail the macroscopic properties of porous materials to the statistical properties of their microscopic structure. Most of these theories attempt to relate "pore size distribution" to the macroscopic properties of the porous materials. Microscopic and electron microscopic investigation of polished sections or thin sections of the porous media is generally considered to be a very useful way of forming a qualitative estimation of the pore size, pore size distribution, or porosity.

Various different methods were developed for the quantitative analysis of ores by mineralogists in the 19th century that are equally applicable for the quantitative determination of porosity of porous materials. These methods have been used for the porosity determination of concrete, and have been reviewed by Mielenz *et al.* (*181*): (1) point count method; (2) areal traverse method; (3) linear traverse method. The point count method ordinarily docs not yield data on the size or spacing of voids; however, by a simple supplement to the point count the specific surface of the voids can be determined. The areal traverse method involves measurement of the area occupied by sections of the voids on plane surfaces of the porous medium. Among these methods, the linear traverse method is the most widely used.

The size distribution of spheres can be calculated from the distribution of chord lengths obtained by piercing the assemblage of spheres with random straight lines. Methods to do this calculation have been developed by a number of authors (*46*, *168*, *226*). The chord length distribution can be determined by traversing polished sections of an isotropic sample under the microscope. Among others, the method was used by Lord and Willis (*168*) to calculate air-bubble size distribution in aerated concrete.

According to another method, the size distribution of spheres can be calculated from the size distribution of circles obtained by intersecting the assemblage of spheres with random planes (*65*, *227*, *272*). The circle diameter distribution can be obtained by measuring the diameters of the pore cross-sections visible in polished sections of the isotropic sample.

Both methods have been critically tested recently by Dullien and coworkers in a computer-simulation study (*86*). They found that relatively small samples permitted accurate back-calculation of the known sphere diameter distributions from the measured cord length or circle diameter distributions.

One basic reason for the lack of widespread use of quantitative microscopic pore structure studies has been that such measurements are extremely time consuming if done visually. With the advent of high speed automatic microscopes, however, the time required for the scanning of the polished sections has become relatively insignificant.

Another problem relating to the use of microscopic methods to determine pore size distribution is that the pores in the vast majority of porous materials are not spherical and, in addition, the pores of greatest interest form interconnecting passages. Nevertheless, the situa-

Figure 10. Photomicrograph of section of sandstone saturated with Wood's metal

tion in this respect is not as serious as it might appear. A very large number of porous materials have a structure consisting of voids that are interconnected through openings or "windows;" also, adjoining voids tend to form irregular tortuous channels rather than straight tubes. As a result, there is relatively little probability that the plane of sectioning will simultaneously cut across several interconnected voids. Inspection of photomicrographs prepared from porous materials proves this contention (see Figure 10). Microscopic methods of pore size distribution determination should be particularly suited for the determination of the distribution of void sizes, rather than the distribution of pore entry radii that is determined by mercury intrusion porosimetry.

The fact that the voids are not spherical necessitates the use of equivalent sphere radii in the statistical calculations. If all or the majority of the pores in the sample belong to a particular shape group, the results of the calculations can be improved by allowing for the statistical shape of the pores, which can also be determined from the polished section of the sample (*227*).

Figure 11. Salt particles suspended in oil

(From Mehta, M.A.Sc. Thesis, University of Waterloo, 1969)

Figure 12. Photomicrograph of section of salt pack saturated with Wood's metal after salt has been dissolved (pores are black)

(From Mehta, M.A.Sc. Thesis, University of Waterloo, 1969)

To form an idea of the reliability of the micrographic method of pore size distribution determination in the case of nonspherical and interconnecting pores, Dullien and Mehta (*85*) have performed the following series of tests. A sample of sodium chloride powder was prepared by grinding, and the particle size distribution of the sample was determined by standard sieve analysis. Part of the sample was then suspended in an oil on a microscope slide in a high enough dispersion so that almost no particles touched one another (Figure 11). The particle size distribution of the salt was determined by one of the above micrographic methods (Scheil's method). A packed bed was also prepared from the same powdered salt, the pores in the pack were saturated with Wood's metal, and the salt was dissolved with water. A porous Wood's metal matrix was obtained (Figure 12). The pore size distribution of this matrix was determined micrographically by the method of Cahn and Fullman (*46, 86*).

The pore size distribution obtained micrographically is shown in Figure 13, along with the particle size distribution of the salt determined by sieve test and the micro-

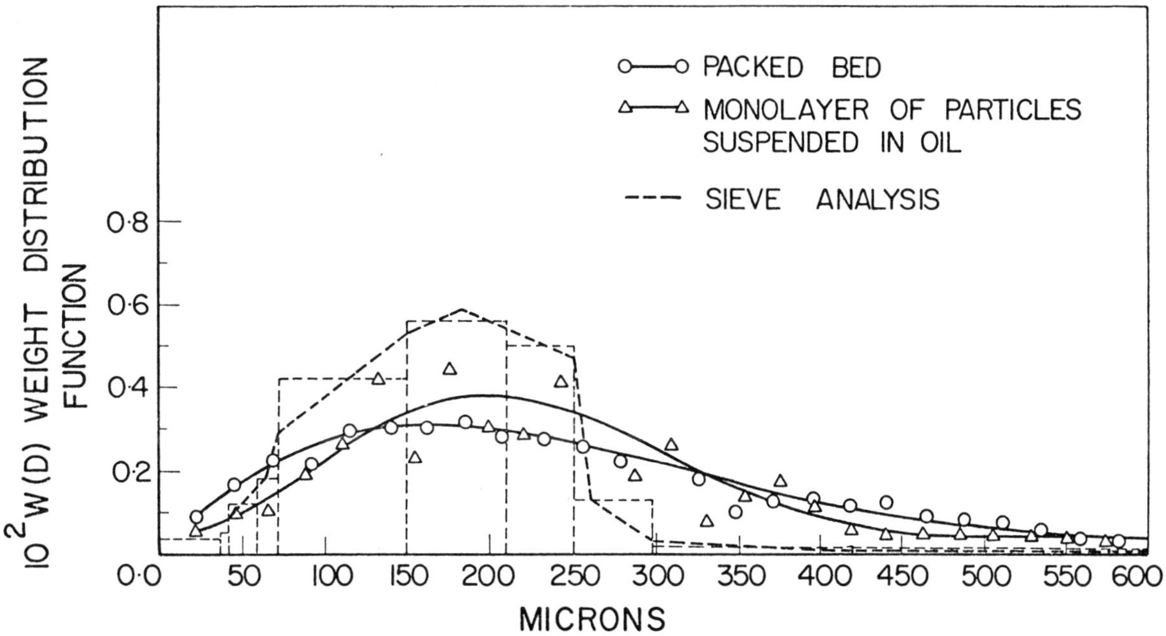

Figure 13. Comparison of results of photomicrographic work with sieve analysis

(From Mehta, M.A.Sc. Thesis, University of Waterloo, 1969)

graphic method. The agreement between the distributions obtained by the various methods is a good one, if one considers that for nonspherical particles (or voids) each method yields a different mean diameter. In particular, the sieve test tends to give lower value than the diameter of the sphere having the same volume as the nonspherical particle.

Mercury porosimetry tests were also performed on samples taken from the leached material. A protective oxide film covering the surface of the Wood's metal held out long enough to permit the penetration tests. It can be seen from Figure 14 that the porosimetry curve gave a very false picture of the pore size distribution of the sample in this case. Since mercury porosimetry measures the size of pore necks, a comparison of the mercury porosimetry distribution curve with the photomicrographic pore size distribution curve results in valuable information on the accessibility of the pores in the sample.

LeFournier (*160*) studied capillary phenomena in porous media by means of thin sections. The sample was first impregnated with red epoxy resin (wetting fluid), desaturated in a centrifuge (air—the nonwetting phase—penetrated the pores), and the void space was impregnated with blue epxoy resin. A study of the sections for a number of unconsolidated and consolidated samples showed that in the unconsolidated media the pores were considerably more accessible than in the consolidated ones. In the latter, many large pores remained uninvaded, probably owing to restrictions in the path of access to these pores.

Bailey (*9*) described the use of petrographic thin sections and polished sections in determining composition, texture, grain size, and cement-silica reaction in steam-cured asbestos portland cement products. The preparation of thin sections and polished sections using epoxy resins as an impregnant, was outlined.

King and Fuchs (*135*) pointed out that when poorly consolidated specimens are polished, the particles on the surface tend to pull out, leaving pits and causing scratches. It is recommended that materials having a higher modulus of elasticity are needed for specimen impregnation. They used lead borosilicate frit, which has the necessary mechanical strength to hold the particles in place, whereas the generally used epoxy resins do not have this property. Dullien and Mehta (*85*) used Wood's metal for specimen impregnation in their study.

Cohen and Sandor described a precision method for determination of pore structure by photographic photometry (*58*). Porosity, homogeneity, average pore size, and average separation can be determined by the technique.

Application of microscopy to the investigation of pore structure of a number of different materials is described in (*55, 104, 204, 212*).

Radiographic Methods of Pore Structure Investigation

Radiography has been only infrequently used for pore structure determination. Ritter and Erich, who applied this technique, wrote (*213*): "The intensity of scattered radiation depends on a difference in scattering power between the two phases and the resultant size determination is a measure of the distance one can travel in the scattering volume without passing a phase boundary in either direction. In the case of a disperse system of solid and air, when the particles are widely separated, the scattering due to their distances apart will be suppressed and one will obtain a particle size distribution. When the particles are squeezed together so closely that the interstices are small compared with particle size, the small angle scattering will be indicative of pore size." "The situation when pore and particle size are about equal is the worst case." Notwithstanding these problems, Ritter and Erich obtained surprisingly good agreement with results of mercury porosimetry for a Fuller's earth sample.

Clark and Liu (*52*) described a method for quantitative determination of porosity by X-ray absorption. After the linear absorption coefficient is determined, the method can be used in a few minutes. Results obtained on commercial rubber-composition storage battery separators were in satisfactory agreement with results of the pycnometric method.

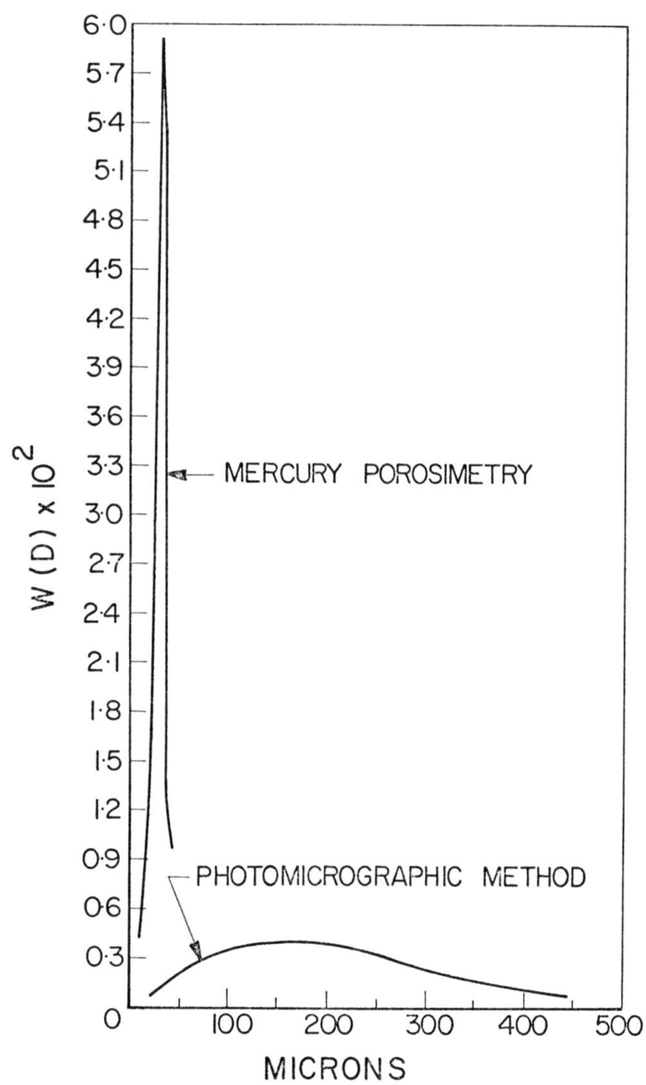

Figure 14. Comparison of results obtained by mercury porosimetry and photomicrographic method

Edwards and Simpson described a method for neutron-derived porosity determination for thin bed boreholes of oil wells (*90*).

Imelik and coworkers reviewed the three existing theories of central X-ray scattering (*266*): the theories of Porod, Soule, and Debye, Anderson and Brumberger. These theories make it possible to obtain the specific surface of porous media. As a result of an improvement made in these theories, the authors could also obtain a mean pore diameter. In another paper, Imelik and coworkers described the original design of a small angle X-ray diffraction spectrometer (*211*).

Clarke used the gamma radiation of Kr^{85} to determine the surface area of graphite (*53*). The amount of gas adsorbed was determined by measuring the gamma radiation intensity. The BET method was used to calculate the surface areas. Excellent agreement was obtained with the results of standard nitrogen adsorption measurements, also using the BET method.

Bonnemay and coworkers discussed possible applications of radiography to the study of structure and operation of porous electrodes (*29*). They envisioned the following applications: apparent density of electrode, distribution of metallic catalyst in the pores, and distribution of gas and liquid phases in the pores.

Plavnik and Dubinin applied small angle X-ray scattering to the investigation of the pore structure of activated charcoals (*201*). Dimensions of the pores and their relative volumes were determined. Micropores of about 6 Å radius were found to have a volume about 20 times that of the transitional pores. They discussed and refuted possible objections to the interpretation of their results.

Integrated Study of the Pore Structure of Various Substances

In this section we are going to review some articles that describe more than one method used to investigate the pore structure of a particular material or a particular class of materials; there was no obvious choice to assign these to any of the previous categories.

Maier and Beutelspacher (*171*) studied the structure and the pore size distribution of filter membranes. Electron microscopy showed that the membranes did not have tubular capillaries, but rather a vacuole structure. Three methods were used to determine the pore size: Poiseuille's law, capillary pressure method, and electron microscopy measurement. In general, the Poiseuille law gave the least mean pore diameter, next came the capillary pressure measurement, and the electron microscopy resulted in the largest. Also, the pore size distribution obtained by capillary pressure measurement appeared as a sharp peak, whereas electron microscopy gave a wider envelope, particularly on the large pore end of the distribution curve.

Corte presented an extensive analysis of cigarette filters (*60*). He came to the conclusion that the suitability of the filter is determined by the pore structure of the filter only; the structure of the filter substance has little influence. The maximum filter pore size was determined by the bubble pressure method. Mercury porosimetry, and gas and vapor flow rate measurements were also used. Grace presented a detailed discussion of the structure and performance of filter media (*108*). He found that the results of permeability, bubble pressure, and mercury intrusion methods were compatible. Akobjanoff and Howe gave a dissertation on the distribution, shape, and size of pores in capillary ultrafilters (*1*). Capillary filters result from arranging fibers of elemental geometrical shape in regular patterns; for example, rods in bundles, sheets in stacks, etc. In Akobjanoff and Howe's paper, bundles of closely packed cylinders were discussed. The largest capillary filter is about 50 μ, while the smallest is close to atomic distances in crystals. Photomicrographs indicated triangular arrangement of the cylinders.

Lohre reviewed various methods to determine pore size distributions, with particular reference to applications in the ceramic industry (*167*). These methods include: mercury porosimetry, microscopy, progressive diminution, capillary suction, gas diffusion, capillary condensation, and capillary rise. He pointed out that in refractory materials the pores appear to be "trumpet-" or "slit-shaped." In his paper, Lohre claimed that the microscopic pore size distribution determination is inapplicable to refractory materials, because such methods yield higher values for the apparent porosities than the accepted values. The various methods used to determine porosities from measurements made on sections of the sample are very well established both thoeretically and experimentally. It may be that in Lohre's work the sections used were not representative of the material investigated, or perhaps there were other inconsistencies.

Lohre made the useful suggestion that in case of hysteresis the results of mercury porosimetry can be corrected by using the results of the capillary rise method, because the latter method reveals the presence of large pores that are accessible only through narrower ones.

Konopicky reported on the pore size distribution of various refractory materials determined by mercury porosimetry and compared these with photomicrographs taken of polished sections (*147*). A large variety of different pore shapes was encountered.

In a paper entitled "Determining the True Magnitude of Open Porosity," Berkman *et al.* (*20*) demonstrated that in ceramic materials the conventional methods of obtaining the open porosity resulted in values that were too low. They compared the results obtained by the different water saturation methods with the result obtained by saturating the sample with mercury under extremely high pressure of 2000 atm. Mercury penetration gave the highest value for the open porosity. It is interesting to compare these findings with the ones obtained by Lohre (*167*).

In a series of two papers, Cameron and Stacy (*47*, *48*) investigated the changes in the pore structure of coke and chars during carbonization and gasification. Mercury porosimetry results were combined with the determination of apparent specific volumes obtained by the displacement of liquids of differing molecular volume. Internal surface area measurements were also performed, using the vapor adsorption method. They found macropores with entrances in the 130,000 to 400 Å diameter range, and micropores with entrances less than 10 Å diameter. The macropores contained many constrictions. The micropores were either cylindrical, or much larger pores with small bottleneck entrances. Spalaris and coworkers studied the surface properties of irradiated graphite (*239*). The BET method was used to calculate surface areas, and the pore size distributions were obtained from nitrogen desorption isotherms, using the B.J.H. method. Heats of adsorption were also obtained from isotherms, using the Clausius-Clapeyron equation. Irradiation appeared to decrease all these quantities. Rostovskii (*221*) used the term "dispersity" for the catalyst, referring to the reciprocal mean pore radius or the surface/volume ratio. Increase in the relative specific surface and the dispersity of the carbon was found to increase the catalytic activity only up to a certain point. Employment of carbons with extremely small pores led to the exclusion of a considerable fraction of the pores from the sphere of reaction.

Harrison and Speakman (*117*) investigated the pore size of keratin, by treating the sample with solutions of iodine in various normal alcohols. Whereas almost all the tyrosine was accessible to iodine when methanol or ethanol was used as the solvent, when a solution of iodine in *n*-propanol was used for iodination, the reaction came to an end before less than half of the tyrosine had been iodinated. With 1-butanol and *n*-pentanol the extent of the reaction was very small. They concluded that the fine structure of wool fibers is inaccessible to molecules larger than those of *n*-propanol.

Batel (*16*) derived a new relationship to correlate flow measurements through powder packs with the properties of the pack. Good correlations were obtained also for nonuniform particle sizes where the Kozeny equation was poor. The height of capillary rise in packs can also be correlated by another equation.

Verbeck gave an interesting review of the pore structure of concrete, and its significance for the properties of this material (*255*).

Freundlich (*105*) gave a brief survey of methods to determine specific surface areas and pore size distributions with particular nitrogen adsorption surface areas and pore size distributions, along with porosities for electrodes prepared from Teflon and various types of platinum black. The structural quantities correlated with electrochemical performance on the basis of an exponential relationship.

Wencke (*269*) reviewed gas adsorption, mercury penetration, and flow test for the investigation of the structure of porous materials. Guyer and Bohlen (*114*) reviewed gas adsorption and mercury porosimetry.

Miscellaneous Methods of Pore Structure Investigation

Yarnton and Simpson (*278*) used an interesting method in determining the size distribution of micropores in powder particles. They used liquids consisting of molecules of different sizes, which gave rise to apparently different densities of the powder particles. From a knowledge of the molecular sizes, the pore size distribution of the micropores could be derived.

Schlogl and Schurig (*228*) used an unusual method to determine the pore sizes in ion exchange resins, by measuring the diffusion coefficient in the resin at various temperatures. The smaller a pore, the smaller the value of the diffusion coefficient, because the water "freezes" in the smaller pores. The size distribution of the pores was calculated, using spheres as models for the pores.

Rosenberg and Shombert (*220*) studied the reaction of oxygen with photo-excited adsorbed dyes. They found that the penetration of oxygen through the pores of the adsorbent was rate determining in some cases. Photochemical observations under these conditions could be used to study the pore characteristics of the adsorbent. Mottlau and Fisher used a titration technique to measure pore volume (*187*).

Oldham and Topol (*192*) claim that the measurement of the current, following the application of a potential step or a potential reversal to a porous material soaked in solution containing electroactive species, offers a possible method of studying pore structure. They considered four different models: (1) the gel model, (2) the diluent model, (3) the uniform-tortuosity model, and (4) the random-tortuosity model. The following pore structure parameters were introduced: retardation factor (diminution of ionic-diffusion coefficient from the bulk solution value), constriction factor (the fraction of cell cross-sectional area occupied by solution), and the tortuosity factor. For each model the parameters were calculated. Krejci (*150*) derived a relationship for porous carriers between the pore volume, pore size distribution, and the change in surface area owing to wetting by a liquid. Experimental verification of the theory was provided.

Kmetko (*145*) studied the compressibility of polycrystalline synthetic graphite, and found that it was about 40 times that of the single graphite crystal. The graphite could be saturated with pentane, and under such conditions its compressibility was small again.

Dullien proposed to determine "two-dimensional" pore size distributions by injecting Wood's metal into the samples and using micrographic analysis (*84*). In this method, the Wood's metal is injected into a series of identical samples under different pressures and the size distribution of pores that were penetrated by the metal is determined in each sample by micrographic methods. From comparison of these distributions, the pore accessibilities can be deduced. The results are expressed as a bivariate distribution function $\alpha'(r_e, r_e')\,dr_e dr_e'$, giving the frequency of voids with radii in the range between r_e and $r_e + dr_e$, and with pore entry radii in the range between r_e' and r_e' and $r_e' + dr_e'$. It is important to note that the pore entry radii are obtained in a generalized sense—*i.e.*, the limiting constriction need not be at the "mouth" of the pore, but may be far removed from it. Hence, this method allows that the domains or voids are often not independent.

Acknowledgments

The authors gratefully acknowledge the assistance of P. N. Mehta in gathering the material for the review. The financial assistance of the Petroleum Research Fund of the ACS, and Imperial Oil Limited, is gratefully acknowledged.

REFERENCES

(1) Akobjanoff, L. and Howe, E. D., *Kolloid-Z.*, **162**, 100 (1959).
(2) Akselrud, G. A., *Zh. Fiz. Khim.*, **34**, 278 (1960).
(3) Al'tshuler, M. A., *Kolloid. Zh.*, **23**, 646 (1961).
(4) Anderson, R. B., *J. Catal.*, **3**, 50 (1964).
(5) Ash, R., Barrer, R. M., and Nicholson, D., *Z. Phys. Chem.*, **37**, 257 (1963).
(6) Atkins, J. H., *Carbon (Oxford)*, **3**, 299 (1965).
(7) Atkins, K. R. and Saki, H., *Phys. Rev.*, **102**, 582 (1956).
(8) Bailey, A., Ph.D. Thesis, Bristol, England, 1965.
(9) Bailey, D., *J. Amer. Ceram. Soc.*, **46**, 348 (1963).
(10) Ballou, E. V. and Barth, R. T., *Advan. Chem. Ser.*, **1**, 133 (1961).
(11) Ballou, E. V., *Anal. Chem.*, **34**, 233 (1962).
(12) Barrer, R. M., "Aspects of Sorption in Porous Crystals," *in* The Structure and Properties of Porous Materials, D. H. Everett and F. S. Stone, Eds., Butterworth, London, 1958, p 6.
(13) Barrer, R. M. and Gabor, T., *Proc. Roy. Soc., Ser. A*, **A251**, 353 (1959).

(14) Barrer, R. M., *Appl. Mater. Res.*, **2**, 129 (1963).
(15) Barrett, E. P., Joyner, L. G., and Halenda, P. P., *J. Amer. Chem. Soc.*, **73**, 373 (1951).
(16) Batel, W., *Chem.-Ing.-Tech.*, **31**, 338 (1959).
(17) Batra, V. K., Fulford, G. D., and Dullien, F. A. L., *Can. J. Chem. Eng.*, in press.
(18) Bauer, W. H., Weber, N., and Wiberley, S. E., *J. Phys. Chem.*, **62**, 1245 (1958).
(19) Bering, B. P., Dubinin, M. M., and Serpinsky, V. V., *J. Colloid Interface Sci.*, **21**, 378 (1966); Dubinin, M. M., *ibid.*, **23**, 487 (1967).
(20) Berkman, A. S., Melnikova, I. G., and Fedotova, E. I., *Steklo Keram.*, **17**, 27 (1960).
(21) Berman, A. S., *J. Appl. Phys.*, **36**, 3356 (1965).
(22) Bernardini, F. and Collepardi, M., *Chim. Ind. (Milan)*, **49**, 136 (1967).
(23) Bernardini, F., Collepardi, M., and Armisi, I., *ibid.*, p 366.
(24) Blackwell, R. J., *Soc. Petrol. Eng. J.*, **2**, 1 (1962).
(25) Blakeney-Edwards, N. J., Ph.D. Thesis, Bristol, England, 1963.
(26) Blokh, T. S., Zaionts, R. M., and Sovetova, L. K., *Steklo Keram.*, **17**, 30 (1960).
(27) Blumenthal, J. L., Sourirajan, S., and Nobe, K., *Can. J. Chem.*, **38**, 783 (1960).
(28) Bocquet, M., Gennison, J., and Sailleau, J., Service de Physics-Chemie Appliquee-Section de Physique de Materiaux Nucleaires, *C.E.A. Rep. 2093*, (1961).
(29) Bonnemay, M., Bronoel, G., Levert, E., and Peslerbe, G., *J. Electrochem. Soc.*, **111**, 265 (1964).
(30) Bourdine, N. T., Gourney, L. S., and Reickertz, P. P., *Petrol. Trans. AIME*, **189**, 195 (1950).
(31) Braslavskii, A. N., *Zh. Prikl. Khim. (Leningrad)*, **34**, 800 (1961).
(32) Braslavskii, A. N., *ibid.*, **35**, 2231 (1962).
(33) Braslavskii, A. N., *ibid.*, **36**, 2648 (1963).
(34) Brigham, W. E., Reed, Ph. W., and Dew, J. N., *Soc. Petrol. Eng. J.*, **1**, 1 (1961).
(35) Broekhoff, J. C. P. and deBoer, J. H., *J. Catal.*, **9**, 8 (1967).
(36) Broekhoff, J. C. P. and deBoer, J. H., *ibid.*, p 15; *Ibid.*, **10**, 153 (1968).
(37) Broekhoff, J. C. P. and deBoer, J. H., *ibid.*, **10**, 153 (1968).
(38) Broekhoff, J. C. P. and deBoer, J. H., *ibid.*, p 377.
(39) Broekhoff, J. G. P. and deBoer, J. H., *ibid.*, p 391.
(40) Broekhoff, J. C. P. and deBoer, J. H., *ibid.*, p 368.
(41) Brown, H., *Petrol. Trans. AIME*, **192**, 67 (1951).
(42) Brunauer, S., Mikhail, R. Sh., and Bodor, E. E., *J. Colloid Interface Sci.*, **24**, 451 (1967).
(43) Brunauer, S., Mikhail, R. Sh., and Bodor, E. E., *ibid.*, **25**, 353 (1967).
(44) Butt, J. B., *J. Catal.*, **4**, 685 (1965).
(45) Bykov, V. T. and Presnyakova, O. E., *Kinet. Katal.*, **3**, 393 (1962).
(46) Cahn, J. W. and Fullman, R. L., *Trans. AIME*, **206**, 610 (1956).
(47) Cameron, A. and Stacy, W. O., *Aust. J. Appl. Sci.*, **9**, 283 (1958).
(48) Cameron, A. and Stacy, W. O., *ibid.*, **10**, 449 (1959).
(49) Cameron, A. and Stacy, W. O., *Chem. Ind. (London)*, 222 (Feb. 27, 1960).
(50) Carman, P. C. and Raal, F. A., *Nature*, **167**, 112 (1951); *Proc. Roy. Soc., London*, **A209**, 59 (1951).
(51) Clampitt, B. H. and German, D. E., *J. Phys. Chem.*, **64**, 284 (1960).
(52) Clark, G. L. and Liu, C. H., *Anal. Chem.*, **29**, 1539 (1957).
(53) Clarke, J. T., *J. Phys. Chem.*, **68**, 884 (1964).
(54) Clement, C., Montarnal, R., and Trambouze, P., *Rev. Inst. Fr. Petrole Ann. Combust. Liquides*, **17**, 558 (1962).
(55) Coble, R. L., *J. Appl. Phys.*, **32**, 787 (1961).
(56) Cockran, C. N. and Cosgrove, L. A., *J. Phys. Chem.*, **61**, 1417 (1957).
(57) Cohan, L. H., *J. Amer. Chem. Soc.*, **66**, 98 (1944).
(58) Cohen, J. and Sandor, A., *J. Opt. Soc. Amer.*, **51**, 1023 (1961).
(59) Cohen, R. M., Marechal, J. E. M., della Faille, M. P., and Fripiat, J. J., *Anal. Chem.*, **37**, 133 (1965).
(60) Corte, H., *Papier (Darmstadt)*, **9**, 290 (1955).
(61) Cranston, R. W. and Inkley, F. A., *in* "Advances in Catalysis," Vol. IX, Academic Press, New York, 1957, p 143.
(62) Cunningham, R. S. and Geankoplis, C. J., *Ind. Eng. Chem., Fundam.*, **7**, 535 (1968).
(63) Danes, V. and Novakova, J., *Collect. Czech. Chem. Commun.*, **24**, 874 (1959).
(64) Danes, V. and Novakova, J., *ibid.*, p 1914.
(65) Debbas, S. and Rumf, H., *Chem. Eng. Sci.*, **21**, 583 (1966).
(66) de Bethune, A. J. and Rowell, R. L., *J. Phys. Chem.*, **67**, 2065 (1963).
(67) deBoer, J. H., "The Shapes of Capillaries," *in* The Structure and Properties of Porous Materials, D. H. Everett and F. S. Stone, Eds., Butterworth, London, 1958, p 68.
(68) deBoer, J. H., Den Heuvel, A., and Linsen, B. G., *J. Catal.*, **3**, 268 (1964).
(69) deBoer, J. H. and Lippens, B. C., *ibid.*, p 38.
(70) deBoer, J. H., Linsen, B. G., and Osinga, Th. J., *ibid.*, **4**, 643 (1965).
(71) deBoer, J. H., van der Plas, Th., and Zondervan, G. J., *ibid.*, p 649.
(72) deBoer, J. H., Linsen, B. G., Broekhoff, J. C. P., and Osinga, Th. J., *ibid.*, **11**, 46 (1968).
(73) Deryagin, B. V., *Acta Physico-Chim. USSR*, **12**, 181 (1940).
(74) Deryagin, B. V., Melnikova, M. K., and Krylova, V. I., *Colloid J., USSR*, **14**, 459 (1952).
(75) Deryagin, B. V., *Proc. Int. Congr. Surface Activ., 2nd*, **2**, London, 1957, p 153.
(76) Deryagin, B. V. and Bakanov, S. P., *Sov. Phys.-Tech. Phys.*, **2**, 1904 (1957).
(77) Dodd, Ch. G. and Kiel, O. G., *J. Phys. Chem.*, **63**, 1646 (1959).
(78) Dollimore, D. and Heal, G. R., *J. Appl. Chem.*, **14**, 109 (1964).
(79) Dubinin, M. M., *Bull. Acad. Sci. USSR, Div. Chem. Sci.*, 549 (1952).
(80) Dubinin, M. M. and Onusaitis, B. A., *Russ. J. Phys. Chem.*, **41**, 1411 (1967).
(81) Dubinin, M. M., Lezin, Y. S., Kadlec, O., and Zukai, A., *ibid.*, **42**, 500 (1968).
(82) Dullien, F. A. L. and Scott, D. S., *Chem. Eng. Sci.*, **17**, 771 (1962).
(83) Dullien, F. A. L., *J. Petrol. Technol.*, **21**, 14 (1969).
(84) Dullien, F. A. L., *ibid.*, p 14.
(85) Dullien, F. A. L. and Mehta, P. N., submitted for publication in *Powder Technol.;* Mehta, P. N., M.A.Sc. Thesis, University of Waterloo, 1969.
(86) Dullien, F. A. L., Rhodes, E., and Schroeter, S. R., *Powder Technol.*, **3**, 124 (1969/70).
(87) Dullien, F. A. L. and Bryson, A. W., in preparation.
(88) Dyer, D. F. and Sunderland, J. E., *Int. J. Heat Mass Transfer*, **9**, 519 (1966).
(89) Edel'man, L. I., Sominskii, D. S., and Kopchikova, N. V., *Kolloid. Zh.*, **23**, 228 (1961).
(90) Edwards, J. M. and Simpson, A. L., *Petrol. Trans. AIME*, **204**, 132 (1955).
(91) Ekedahl, E. and Sillen, L. G., *Acta Chem. Scand.*, **19**, 2323 (1965).
(92) Emig, G. and Hofmann, H., *J. Catal.*, **8**, 303 (1967).
(93) Engelhardt van, W. and Lubben, H., *Erdoel Kohle*, **10**, 747 (1957).
(94) Ergun, S. and Owen, J., *Anal. Chem.*, **25**, 1222 (1953).
(95) Evans, R. B., III, Truitt, J., and Watson, G. M., *At. Energy Comm.*, **ORNL-3067**, 1961.
(96) Evans, R. B., III, Watson, G. M., and Mason, E. A., *J. Chem. Phys.*, **36**, 1894 (1962).
(97) Everett, D. H., "Adsorption Hysteresis," *in* The Solid/Gas Interface, E. A. Flood, Ed., Vol. 2, Marcel Dekker, New York, 1967, Chapter 36.
(98) Eyraud, Ch., Bricout, J., and Grillet, G., *C. R. H. Acad. Sci.*, 2460 (1963).
(99) Fedorov, G. I., Tarasova, G. F., and Izimailov, R. I., *Russ. J. Phys. Chem.* **42**, 545 (1968).
(100) Flood, E. A. and Huber, M., *Can. J. Chem.*, **33**, 203 (1955).
(101) Foster, A. G., *Trans. Faraday Soc.*, **28**, 645 (1932).
(102) Foster, A. G. and Thorp, J. M., "The Preparation and Adsorptive Properties of Silica Gels," *in* The Structure and Properties of Porous Materials, D. H. Everett and F. S. Stone, Eds., Butterworth, London, 1958, p 227.
(103) Foster, R. N. and Butt, J. B., *A.I.Ch.E. J.*, **12**, 180 (1966).
(104) Francois, B., Delmas, R., Caillet, R., and Lacombe, P., *C. R. H. Acad. Sci.*, 925 (1963).
(105) Freundlich, J., *Electrochim. Acta*, **6**, 35 (1962).
(106) Frevel, L. K. and Kressley, L. J., *Anal. Chem.*, **35**, 1492 (1963).
(107) Goodknight, R. C. and Fatt, I., *J. Phys. Chem.*, **65**, 1709 (1961).
(108) Grace, H. P., *A.I.Ch.E. J.*, **2**, 307 (1956).
(109) Greenberg, D. B. and Weger, E., *Chem. Eng. Sci.*, **12**, 8 (1962).
(110) Grove, D. M. and Ford, M. G., *Nature*, **182**, 999 (1958).
(111) Gruber, H. L. and Shipley, C. S., *Vac. Microbalance Tech.*, **3**, 131 (1962).
(112) Grubner, O., *Collect. Czech. Chem. Commun.*, **25**, 180 (1960).
(113) Gunn, R. D. and King, C. J., Paper presented at the Annual Meeting of the A.I.Ch.E., New York, N. Y., November, 1967.
(114) Guyer, A. and Bohlen, B., *Chimia*, **17**, 82 (1963).
(115) Haley, A. J., *J. Appl. Chem.*, **13**, 392 (1963).
(116) Halsey, C. D., *J. Chem. Phys.*, **16**, 931 (1948).
(117) Harrison, D. and Speakman, J. B., *Text. Res. J.*, 1005 (1958).
(118) Henry, J. P., Cunningham, R. S., and Geankoplis, C. J., *Chem. Eng. Sci.*, **22**, 11 (1967).
(119) Hewitt, G. F. and Sharratt, E. W., *J. Nucl. Mater.*, **13**, 197 (1964).
(120) Hewitt, G. F., "Gaseous Mass Transport within Graphite," *Chem. Phys. Carbon*, **1**, 73 (1965), published 1966.
(121) Hill, R. D., *Trans. Brit. Ceram. Soc.*, **59**, 189 (1960).
(122) Honold, E. and Skau, E. L., *Science*, **12**, 805 (1954).
(123) Hoogschagen, J., IND. ENG. CHEM., **47**, 906 (1955).
(124) Houpeurt, A., *Rev. Inst. Fr. Petrole Ann. Combust. Liquides*, **14**, 1468 (1959).
(125) Huber, M. E., Flood, E. A., and Heyding, R. D., *Can. J. Chem.*, **34**, 1300 (1956).
(126) Ingles, O. G., *Fuel*, **36**, 252 (1957).
(127) Ingles, O. G., *ibid.*, **37**, 501 (1958).
(128) Innes, W. B., *Anal. Chem.*, **29**, 1069 (1957).
(129) Izotova, T. I. and Dubinin, M. M., *Russ. J. Phys. Chem.*, **39**, 1492 (1965).
(130) Johansen, R. T. and Dunning, H. N., *J. Colloid. Sci.*, **12**, 68 (1957).
(131) Johnson, M. F. L. and Stewart, W. E., *J. Catal.*, **4**, 248 (1965).
(132) Juntgen, H. and Schwuger, M., *Chem.-Ing.-Tech.*, **38**, 1271 (1966).
(133) Karnaukhov, A. P., *Kinet. Katal.*, **3**, 583 (1962).
(134) Karnaukhov, A. P., *ibid.*, **8**, 172 (1967).
(135) King, A. G. and Fuchs, K., *J. Amer. Ceram. Soc.*, **50**, 328 (1967).
(136) Kington, G. L. and Smith, P. S., *Trans. Faraday Soc.*, **60**, 705 (1964).
(137) Kipling, J. J. and Wilson, R. B., *ibid.*, **56**, 557 (1960).
(138) Kipling, J. J. and Wilson, R. B., *ibid.*, p 562.
(139) Kipling, J. J., Sherwood, J. N., Shooter, P. V., and Thomson, N. R., *Carbon (Oxford)*, **1**, 321 (1964).
(140) Kiselev, A. V., *Usp. Khim.*, **14**, 367 (1945).
(141) Kiselev, A. V. and Lopatkin, A. A., *Kinet. Katal.*, **4**, 786 (1963).
(142) Kiselev, A. V. and Lopatkin, A. A., *ibid.*, **5**, 183 (1964).
(143) Klinkenberg, L. J., *Trans. AIME*, **210**, 366 (1957).
(144) Klyunkina, N. G., *Zh. Prikl. Khim. (Leningrad)*, **33**, 76 (1960).
(145) Kmetko, E. A., Morgan, J. R., and Andrew, J. F., *Carbon (Oxford)*, **6**, 571 (1968).
(146) Kohlschutter, H. W. and Kampf, G., *Z. Anorg. Allg. Chem.*, **292**, 298 (1957).

(147) Konopicky, K., *Tonind-Ztg. Keram. Rundsch.*, **83**, 535 (1959).
(148) Kraemer, E. O., *in* Treatise on Physical Chemistry, H. S. Taylor, Ed., Van Nostrand–Reinhold, New York, 1931, p 1661.
(149) Kreith, F. and Eisenstadt, R., *Trans. ASME*, **79**, 1070 (1957).
(150) Krejci, M., *Collect. Czech. Chem. Commun.*, **32**, 1152 (1967).
(151) Kroone, B. and Crook, D. N., *Mag. Concr. Res.*, **14**, 43 (1962).
(152) Kruyer, S. and van Krevelen, D. W., *Fuel*, **37**, 118 (1958).
(153) Kruyer, S., *Trans. Faraday Soc.*, **54**, 1758 (1958).
(154) Ksenzhek, O. S., Kalinovskii, Petrova, S. A., and Litvinova, V. I., *Russ. J. Phys. Chem.*, **41**, 856 (1967).
(155) Kuchinskii, E. M., Lipikhin, N. P., and Flisskii, M. M., *Zh. Prikl. Khim.*, **37**, 460 (1964).
(156) Kuhn, I., *J. Colloid. Sci.*, **19**, 685 (1964); *J. Colloid Interface Sci.*, **23**, 563 (1967).
(157) Kuzin, I. A. and Andronow, E. A., *Zh. Prikl. Khim.*, **36**, 2600 (1963).
(158) Laughlin, R. D. and Davies, J. E., *Text. Res. J.*, 904 (1961).
(159) Lebedeva, V. K., *Sov. Phys.-Tech. Phys.*, **1**, 1042 (1957).
(160) LeFournier, M. J., *Rev. Ass. Fr. Tech. Petrole*, 17, (1966).
(161) Leppard, C. J. and Spencer, D. H. T., *J. Sci. Instrum., Ser. 7*, **1**, 573 (1968).
(162) Lester, G. R., *J. Catal.*, **8**, 283 (1967).
(163) Lifson, S., *J. Polym. Sci.*, **20**, 1 (1956).
(164) Lippens, B. C., Linsen, B. G., and deBoer, J. H., *J. Catal.* **3**, 32 (1964).
(165) Lippens, B. C. and deBoer, J. H., *ibid.*, p 44.
(166) Lippens, B. C. and deBoer, J. H., *ibid.*, **4**, 319 (1965).
(167) Lohre, W., *Tonind.-Ztg. Keram. Rundsch.*, **83**, 334 (1959).
(168) Lord, G. W. and Willis, T. F., *ASTM Bull.*, **177**, 56 (1951).
(169) Lund, L. M. and Berman, A. S., *J. Chem. Phys.*, **28**, 363 (1958).
(170) Maidanik, G., *Trans. Faraday Soc.*, **56**, 118 (1960).
(171) Maier, K. H. and Beutelspacher, H., *Naturwissenschaften*, **40**, 605 (1953).
(172) Marchessault, R. N. and Mason, S. G., IND. ENG. CHEM., **52**, 79 (1960).
(173) Markin, V. S., *Izv. Akad. Nauk SSSR, Ser. Khim.*, **11**, 1923 (1963).
(174) Marshall, T. J., *J. Soil Sci.*, **9**, 1 (1958).
(175) May, J. O. and Butterworth, B., *Sci. Ceram.*, **1**, 201 (1962).
(176) May, P., Nallet, J. P., and Daneyrolle, C., *RILEM (Reunion Int. Lab. Essais Rech. Mater. Constr.) Bull.*, **1966** (33), 413.
(177) Mayer, R. P. and Stowe, R. A., *J. Colloid Sci.*, **20**, 893 (1965).
(178) Mayer, R. P. and Stowe, R. A., *J. Phys. Chem.*, **70**, (1966).
(179) McBain, J. W., *J. Amer. Chem. Soc.*, **57**, 699 (1935).
(180) Meyer, H. I., *J. Appl. Phys.*, **24**, 510 (1953).
(181) Mielenz, R. C., Wolkodoff, V. E., Backstrom, J. E., and Burrows, R. W., *J. Amer. Concr. Inst.*, 507 (1958).
(182) Mikhail, R. Sh., Copeland, L. E., and Brunauer, S., *Can. J. Chem.*, **42**, 426 (1964).
(183) Mikhail, R. Sh., Brunauer, S., and Bodor, E. E., *J. Colloid Interface Sci.*, **26**, 45 (1968); *ibid.*, **26**, 54 (1968).
(184) Millington, R. J. and Quirk, J. P., *Trans. Faraday Soc.*, **57**, 1200 (1961).
(185) Moore, T. F. and Slobod, R. L., *Prod. Mon.*, **20**, 20 (1956).
(186) Morrow, N. R. and Harris, C. C., *Soc. Petrol. Eng. J.*, **5**, 15 (1965).
(187) Mottlau, A. Y. and Fisher, N. E., *Anal. Chem.*, **34**, 714 (1962).
(188) Neher, M. B., *ibid.*, **33**, 1133 (1961).
(189) Nelsen, F. M. and Eggertsen, F. T., *ibid.*, **30**, 1387 (1958).
(190) Nicholson, D. and Petropoulos, J. H., *Brit. J. Appl. Phys.*, **1**, 1379 (1968).
(191) Nikitin, Y. S., *Russ. J. Phys. Chem.*, **33**, 499 (1957).
(192) Oldham, K. B., and Topol, L. E., *J. Phys. Chem.*, **71**, 3007 (1967).
(193) Ortega, M. and Teoule, R., *Bull. Soc. Chim. Fr.*, 130 (1968).
(194) Otani, S., Wakao, N., and Smith, J. M., *A.I.Ch.E. J.*, **11**, 446 (1965).
(195) Otani, S., Wakao, N., and Smith, J. M., *ibid.*, p 439.
(196) Pacsirszky, J., Gaulier, C., and Marle, C., *Colloq. Ass. Rech. Tech. Forage Prod., 2nd, Comm. 14*, Paris, June 3, 1965.
(197) Perkins, F. M., Jr., *Petrol. Trans. AIME*, **210**, 409 (1957).
(198) Petersen, E. E., *A.I.Ch.E. J.*, **4**, 343 (1958).
(199) Pickell, J. J., Swanson, B. F., and Hickman, W. B., *Soc. Petrol. Eng. J.*, **6**, 55 (1966).
(200) Plachenov, T. G., *J. Appl. Chem. USSR*, **28**, 223 (1955).
(201) Plavnik, G. M. and Dubinin, M. M., *Izv. Akad. Nauk SSSR, Ser. Khim.*, 628 (1966).
(202) Pope, C. G., *Trans. Faraday Soc.*, **64**, 566 (1968).
(203) Pouchly, J., *Collect. Czech. Chem. Commun.*, **25**, 1397 (1960).
(204) Presland, A. E. B., Greenwood, A. D., and Block, M., *Nature*, **212**, 394 (1966).
(205) Pruss, W., *Brennst.-Chem.*, **42**, 43 (1961).
(206) Purcell, W. R., *Trans. AIME*, **186**, 39 (1949).
(207) Quinn, H. W. and McIntosh, R., *Can. J. Chem.*, **35**, 745 (1957).
(208) Ramser, J. H. and Hill, P. B., IND. ENG. CHEM., **50**, 117 (1958).
(209) Reich, B., *Chem.-Ing.-Tech.*, **39**, 1275 (1967).
(210) Renon, J., Francois-Rossetti, J., and Imelik, B., *Bull. Soc. Chim. Fr.*, 446 (1960).
(211) Renouprez, A., Bottazzi, H., Weigel, D., and Imelik, B., *J. Chem. Phys.*, **62**, 131 (1965).
(212) Ridge, M. J. and Crook, D. N., *Aust. J. Appl. Sci.*, **10**, 497 (1959).
(213) Ritter, H. L. and Erich, L. C., *Anal. Chem.*, **20**, 665 (1948).
(214) Roberts, B. F., *J. Colloid Interface Sci.*, **23**, 266 (1967).
(215) Robertson, J. L. and Smith, J. M., *A.I.Ch.E. J.*, **9**, 342 (1963).
(216) Rockwood, S. H., Lair, G. H., and Langford, B. J., *Petrol. Trans. AIME*, **210**, 252 (1957).
(217) Rootare, H. M., Aminco Laboratory News, **24**, (3), (1968).
(218) Rose, W. D., *J. Geophys. Res.*, **64**, 103 (1959).
(219) Rose, W. and Heins, R. W., *J. Colloid Sci.*, **17**, 39 (1962).
(220) Rosenberg, J. L. and Shombert, D. J., *J. Phys. Chem.*, **65**, 2103 (1961).
(221) Rostovskii, E. N., *Zh. Prikl. Khim.*, **32**, 2691 (1959).
(222) Rothfield, L. B., *A.I.Ch.E. J.*, **9**, 19 (1963).
(223) Rouquerol, F. and Imelik, B., *Bull. Soc. Chim. Fr.*, 2802 (1966).
(224) Salman, T. and Robertson, R. F., *Can. Mining Met. Bull.*, **58**, 1088 (1965).
(225) Sandstede, G. and Robens, E., *Chem.-Ing.-Tech.*, **34**, 708 (1962).
(226) Scarlett, B. and Todd, A. C., Paper presented at the Materials Handling Conference of the American Society of Mechanical Engineers, Boston, Mass., October 21–23, 1968.
(227) Scheil, E., *Z. Metallk.*, **27**, 199 (1935).
(228) Schlogl, R. and Schurig, H., *Z. Electrochem.*, **65**, 863 (1961).
(229) Schlosser, J., *Nucl. Sci. Eng.*, **24**, 123 (1966).
(230) Schneider, P. and Smith, J. M., *A.I.Ch.E. J.*, **14**, 762 (1968).
(231) Schwartz, A. M. and Minor, F. M., *J. Colloid Sci.*, **14**, 572 (1959).
(232) Schwartz, A. M. and Minor, F. M., *ibid.*, p 584.
(233) Schwartz, A. M., Radar, Ch. E., and Huey, E., *Advan. Chem. Ser.*, **43**, 250 (1964).
(234) Scott, D. S. and Dullien, F. A. L., *A.I.Ch.E. J.*, **8**, 293 (1962).
(235) Scott, D. S. and Dullien, F. A. L., *ibid.*, p 113.
(236) Shull, C. G., *J. Amer. Chem. Soc.*, **70**, 1405 (1948).
(237) Sinyakova, I. B., *Kinet. Katal.*, **6**, 155 (1965).
(238) Skibitzke, H. E., Chapman, J. I., Robinson, G. M., and McCullough, R. A., *Int. J. Appl. Radiat. Isotop.*, **10**, 38 (1961).
(239) Spalaris, C. N., Bupp, L. P., and Gilbert, E. C., *J. Phys. Chem.*, **61**, 350 (1957).
(240) Spencer, D. H. T., and Fereday, F., *Chem. Ind. (London)*, 847 (1968).
(241) Stamin, A. J. and Wagner, E., *Forest Prod. J.*, 141 (1961).
(242) Steggerda, J. J., Thesis, University of Delft, Delft, The Netherlands, 1955.
(243) Stevenson, D. G., Hill, T., and Anderson, A., *J. Sci. Instrum.*, **44**, 922 (1967).
(244) Suetin, P. E. and Volobuev, P. V., *Sov. Phys.-Tech. Phys.*, **9**, 859 (1964).
(245) Sutherland, K. L. and Winfield, M. E., *Aust. J. Chem.*, **6**, 244 (1954).
(246) Talaev, M. V., Deryagin, B. V., and Zakhavaeva, N. N., Surface Forces Conference, Moscow, 1960, p 151.
(247) Templeton, Ch. C., *Petrol. Trans. AIME*, **201**, 161 (1954).
(248) Templeton, Ch. C. and Rushing, S. S., *Amer. Inst. Mining Met. Petrol. Eng., Tech. Publ. 4340*, in *J. Petrol. Technol.*, **8**, 211 (1956).
(249) Teoule, R., Ortega, M., and Bochirol, L., *Bull. Soc. Chim. Fr.*, 134 (1968).
(250) Timmons, C. O. and Zisman, W. A., *J. Colloid Interface Sci.*, **22**, 165 (1966).
(251) Turner, G. A., *Chem. Eng. Sci.*, **7**, 156 (1958).
(252) Turner, G. A., *ibid.*, **10**, 14 (1959).
(253) Tzuruizumi, A., *Bull. Chem. Soc. Jap.*, **34**, 1457 (1961).
(254) Ulmer, G. C. and Smothers, W. J., *Ceram. Bull.*, **46**, 649 (1967).
(255) Verbeck, G., *ASTM Spec. Tech. Publ. 169-A*, 211 (1966); Reprinted by *Portland Cem. Ass. Res. Develop. Lab., Res. Dep. Bull. 197.*
(256) Viswanathan, B. and Sastri, M. V. C., *J. Catal.*, **8**, 312 (1967).
(257) Voigt, E. M. and Tomlinson, R. H., *Can. J. Chem.*, **33**, 215 (1955).
(258) Volarovich, M. P. and Churaev, N. V., *Int. J. Appl. Radiat. Isotop.*, **13**, 179 (1962).
(259) Volobuev, P. V. and Suetin, P. E., *Sov. Phys.-Tech. Phys.*, **10**, 269 (1965).
(260) Vysotskii, Z. Z., Dionich, L. F., and Polyakov, M. V., *Kolloid. Zh.*, **23**, 248 (1961).
(261) Wakao, N. and Smith, J. M., *Chem. Eng. Sci.*, **17**, 825 (1962).
(262) Wakao, N., Otani, S., and Smith, J. M., *A.I.Ch.E. J.*, **11**, 435 (1965).
(263) Waldmann, L. and Schmitt, K. H., *Z. Naturforsch., A*, **16a**, 1343 (1961).
(264) Watson, A., May, J. O., and Butterworth, B., *Trans. Brit. Ceram. Soc.*, **56**, 37 (1957).
(265) Watson, A., May, J. O., and Butterworth, B., *Sci. Ceram.*, **1**, 187 (1962).
(266) Weigel, D., Renouprez, A., and Imelik, B., *J. Chem. Phys.*, **62**, 125 (1965).
(267) Weissman, E. Y., *J. Electrochem. Soc.*, **114**, 658 (1967).
(268) Weisz, P. B., *Z. Phys. Chem.*, **11**, 1 (1957).
(269) Wencke, K., *Wiss. Ann. V/Suppl.*, 223 (1956).
(270) Wheeler, A., Presentations at Catalysis Symposia, *Amer. Ass. Advan. Sci. Conf.*, Gibson Island, June 1945, 1946.
(271) Wicke, E., and Hugo, P., *Z. Phys. Chem.*, **28**, 401 (1961).
(272) Wicksell, S. D., *Biometrike*, **17**, 84 (1925).
(273) Wittman, F. and Englert, G., *Mater. Sci. Eng.*, **2**, 14 (1967).
(274) Wolff, W. F., *J. Phys. Chem.*, **62**, 829 (1958).
(275) Wyllie, M. R. J., and Gregory, A. R., *Petrol. Trans. AIME*, **198**, 103 (1953).
(276) Wyllie, M. R. J. and Gregory, A. R., IND. ENG. CHEM., **47**, 1379 (1955).
(277) Wynne-Jones, W. F., "The Surface Characteristics of Carbons," *in* The Structure and Properties of Porous Materials, D. H. Everett and F. S. Stone, Eds., Butterworth, London, 1958, p 35.
(278) Yarnton, D., and Simpson, G. R., *Powder Met.*, 42 (1961).
(279) Zhdanov, S. P., "Coll. Methods of Investigating the Structure of Highly Disperse and Porous Solids" (in Russian), *Izv. Akad. Nauk SSSR*, (1), (1953); (2), p 117 (1958).
(280) Zsigmondy, R., *Z. Anorg. Allg. Chem.*, **71**, 356 (1911).
(281) Zukal, A., Dubinin, M. M., Kadler, O., Plachenov, T. D., and Polak, R. *Russ. J. Phys. Chem.*, **39**, 634 (1965).

2

Advances in Theory of Fluid Motion in Porous Media

General theory of creep flow is proposed which indicates permeability tensor can be represented in terms of only four scalar constants

Advances in Theory of

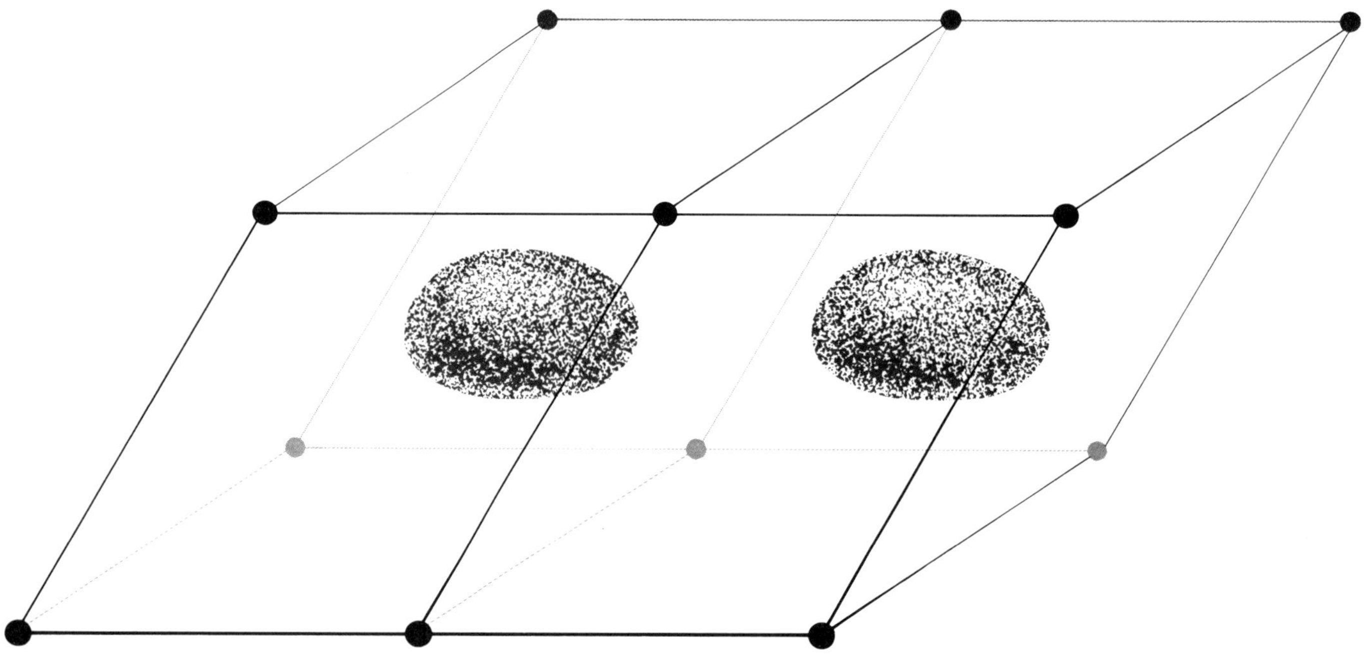

Figure 1. Spatially periodic porous media

The problem of flow in porous media is of interest to an enormous range of engineers and scientists—from the chemical engineer who quite likely is concerned with multicomponent reacting systems, to the civil engineer who is often faced with a two-phase (air-water) flow process, to the seismologist who must deal with a deformable porous media in a transient state. Even with this vast array of important problems awaiting our attention, this paper will consider only the simplest of all possible cases; the steady, incompressible flow of a Newtonian fluid under conditions such that inertial effects can be neglected. A careful survey of this problem will hopefully provide a sound basis for the study of some of the more important aspects of flow in porous media.

The ideas presented in this paper are drawn primarily from the early work of Hubbert (*15*) who has given a lucid account of the famous work of Darcy (*7*) and has presented, on an intuitive basis, a logical route to a theoretical treatment of this problem; from the exhaustive study by Brenner (*5*) on creep flow in a spatially periodic porous media; from the work of Slattery (*30*, *31*) on the flow of viscoelastic fluids in porous media; and from the author's own attempt at a theoretical treatment of this problem (*38*).

There are perhaps three fairly distinct, although certainly not mutually exclusive, approaches to the problem of flow in porous media. Since a porous medium is generally not an ordered structure, the idea of developing statistical models is appealing, and a number of attempts have been made in this area [*1*, (*3*, Chap. 6), (*28*, Sec. 6.6), (*29*)]. Another approach is the method of

Fluid Motion in Porous Media

STEPHEN WHITAKER

geometric modeling in which one postulates a geometry which hopefully bears some resemblance to the porous media, yet is sufficiently simple to allow the governing differential equations to be solved. Only two geometric models will be referred to in this work; the spatially periodic model of Brenner (Figure 1), and a skewed capillary model (Figure 2).

In Brenner's model the porous media are made up of an infinite array of unit cells (two shown in Figure 1) having boundaries of arbitrary shape and containing one or more solid particles of arbitary shape. So that the flow in a unit cell will be amenable to analysis, it is required that the boundary of the unit cell be a fluid surface. Thus one must accept the existence of rigid supports of negligible hydrodynamic resistance which hold the matrix of unit cells in place.

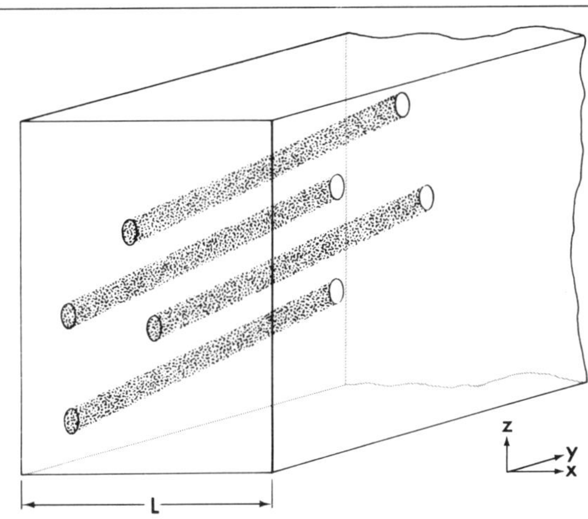

Figure 2. Skewed capillary model of porous media

The skewed capillary model shown in Figure 2 is a minor variation of the well-known straight capillary model (*28*, page 114), but it cannot be obtained from that model by a simple transformation of coordinates because the boundary conditions, which may be imposed on the flow, also change under a coordinate transformation. A third approach, which lies somewhere between statistical models and geometric models, is the development of correct averaged forms of the governing differential equations. These equations should be valid for any geometry, thus the results obtained for specific geometric models must satisfy the averaged equations. In addition, any statistical model should be in accord with the averaged equations. All three methods lead to unspecified parameters which must be determined experimentally, and the primary objective of theoretical work in this general area is to aid in the interpretation of experimental data.

In attacking the problem of incompressible flow in porous media, one is confronted with the fact that the final result is pretty well established—*i.e.*, Darcy's law gives an accurate description of the flow. Because of this, it is easy to proceed along a variety of approaches, some of which might well be erroneous or wholly intuitive, to the correct final result. We will try to avoid this pitfall in the present study and establish as carefully as possible a logical, correct route to the final result.

1. The Problem

We are concerned here with the incompressible, creep flow of a constant viscosity Newtonian fluid in a rigid porous media. The equations governing this process are:

1. The continuity equation

$$\nabla \cdot \mathbf{v} = 0 \tag{1-1}$$

The velocity, $\mathbf{v}$, and position vector, $\mathbf{r}$, are measured relative to an inertial frame imbedded in the porous media; thus the porous media may move with a constant velocity, $\mathbf{u}$, relative to a frame that is stationary with respect to the fixed stars.

2. The equations of motion

$$0 = -\nabla p + \rho \mathbf{g} + \mu \nabla^2 \mathbf{v} \tag{1-2}$$

The restriction of constant viscosity is imposed by the form of Equation 1-2.

3. The void volume distribution function

$$\alpha(\mathbf{r}) = \begin{cases} 1, \text{ if } \mathbf{r} \text{ lies in the fluid region} \\ 0, \text{ if } \mathbf{r} \text{ lies in the solid region} \end{cases} \tag{1-3}$$

If the function $\alpha(\mathbf{r})$ were known it would be possible, in principle, to solve the governing equations and thus determine completely the pressure and velocity fields. However, $\alpha(\mathbf{r})$ is never known and we are obviously forced into a different type of analysis—*i.e.*, a derivation of the volume averaged equations. In this approach we associate with every point in space an averaging volume V which contains a volume of fluid V_f. The volume porosity is given by

$$\Phi = \frac{V_f}{V} \tag{1-4}$$

or more explicitly in terms of the void volume distribution function

$$\Phi = \frac{1}{V}\int_V \alpha(\mathbf{r}) dV \tag{1-5}$$

Throughout this work we will require that the averaging volume be constant in time and space; however, the magnitude of the fluid volume V_f may vary with position subject to certain restrictions to be discussed later.

Before attacking the problem at hand—*i.e.*, obtaining the averaged forms of the governing equations—we must establish some ideas about area and volume averages and what is meant by a "meaningful average," and we must develop an averaging theorem to relate averages of derivatives to the derivatives of averages.

2. Averages

In treating problems of flow in porous media, we assume that some microscopic characteristic length, d, exists which is representative of the distance over which significant variations in the point velocity, $\mathbf{v}$, take place. Similarly, we designate the macroscopic characteristic length as L, and assume it is representative of the distance over which significant variations in the volume averaged velocity, $\langle \mathbf{v} \rangle$, take place. In general, one associates d with the poorly defined but intuitively appealing mean pore diameter, and L with some macroscopic dimension representative of the process under consideration. In Darcy's original work one would consider d to be on the order of magnitude of the "diameter" of the sand grains used in the filter, while L would be associated with the diameter of the filter bed.

Letting ψ be some point quantity (tensor of any order) associated with the fluid (velocity, density), we define the volume average of ψ as

$$\langle \psi \rangle = \frac{1}{V}\int_{V_f} \psi dV \tag{2-1}$$

One can now ask whether the average $\langle \psi \rangle$ is a function that is suitable to use in analyzing flow in porous media. If we were to plot $\langle \psi \rangle$ versus V we might obtain a function similar to that in Figure 3. Such a curve would be obtained if the point associated with V was in the solid region; thus V_f would be zero for small values of V. As V becomes larger, portions of the fluid are contained within V and the average increases from zero going through some fluctuations representative of the variations in the point value of ψ. For values of V larger than V^* the microscopic variations in ψ are essentially smoothed out, but the value of $\langle \psi \rangle$ need not become constant. It should be clear that $\langle \psi \rangle$ is a continuous function for any value of V, but for values of V larger than V^* the volume average $\langle \psi \rangle$ becomes "smooth," and thus amenable to the type of analysis we have in mind.

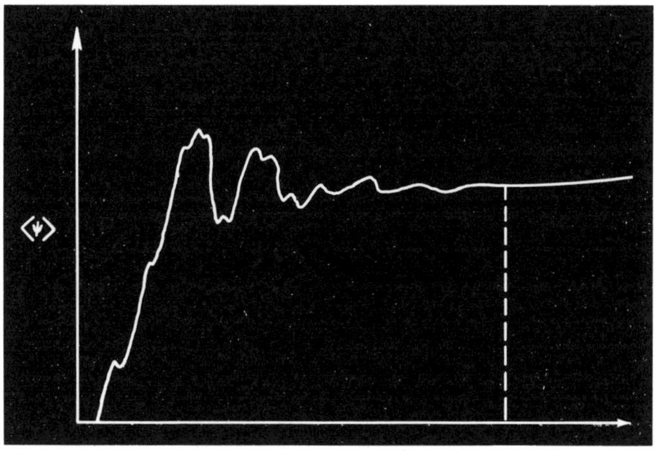

Figure 3. Dependence of average on averaging volume

If we let l be a characteristic length for the averaging volume, we can impose our first restriction (throughout this paper restrictions will be denoted by R.1, R.2, and assumptions by A.1, A.2, and so forth).

$$d \ll l \qquad \text{R.1}$$

In addition to this restriction we will require that the average of the average be equal to the average, a concept perhaps more clearly expressed by

$$\langle\langle\psi\rangle\rangle = \langle\psi\rangle \tag{2-2}$$

This is an important restriction, for it must be true if the average is to be a useful quantity with which to work. A similar situation occurs in the time averaging of turbulent fields (*39*, pp 187, 209). In considering $\langle\psi\rangle$, we note that it is defined everywhere in space, not just in the space occupied by the fluid. Thus the proper average of this function is given by

$$\langle\langle\psi\rangle\rangle = \frac{1}{V}\int_V \langle\psi\rangle dV \tag{2-3}$$

For convenience only, we choose the point with which we associate $\langle\psi\rangle$ to be the centroid of V and expand $\langle\psi\rangle$ in a Taylor series (32, p 228) choosing the centroid as the origin of our coordinate system. (Here x_i is the position vector in index notation. Repeated indices are summed from 1 to 3.)

$$\langle\psi\rangle = \langle\psi\rangle_0 + x_i\left(\frac{\partial\langle\psi\rangle}{\partial x_i}\right)_0 + \frac{x_ix_j}{2!}\left(\frac{\partial^2\langle\psi\rangle}{\partial x_i\partial x_j}\right)_0 + \frac{x_ix_jx_k}{3!}\left(\frac{\partial^3\langle\psi\rangle}{\partial x_i\partial x_j\partial x_k}\right)_0 + \ldots \tag{2-4}$$

Here the subscript zero indicates that the function is evaluated at $x_i = 0$. Forming the average of $\langle\psi\rangle$ indicated by Equation 2-3 yields

$$\langle\langle\psi\rangle\rangle = \langle\psi\rangle_0 + \frac{1}{2!}\left(\frac{\partial^2\langle\psi\rangle}{\partial x_i\partial x_j}\right)_0 \frac{1}{V}\int_V x_ix_j dV + \frac{1}{3!}\left(\frac{\partial^3\langle\psi\rangle}{\partial x_i\partial x_j\partial x_k}\right)_0 \frac{1}{V}\int_V x_ix_jx_k dV + \ldots \tag{2-5}$$

The linear term in Equation 2-4 has dropped out since x_i is measured from the centroid of V, thus

$$\int_V x_i dV = 0 \tag{2-6}$$

If $\langle\psi\rangle$ is a linear function of x_i, Equation 2-2 is immediately satisfied; however, we wish to consider a somewhat more general case.

An order of magnitude estimate of the first integral in Equation 2-5 can be expressed as,

$$\frac{1}{V}\int_V x_ix_j dV = \mathbf{0}\,(l^2) \tag{2-7}$$

and the order of magnitude estimate for the second derivative of $\langle\psi\rangle$ is given by

$$\left(\frac{\partial^2\langle\psi\rangle}{\partial x_i\partial x_j}\right)_0 = \mathbf{0}\,(\langle\psi\rangle/L^2) \tag{2-8}$$

Neglecting the higher order terms in Equation 2-5, we express the average of the average as

$$\langle\langle\psi\rangle\rangle = \langle\psi\rangle + \mathbf{0}[\langle\psi\rangle(l/L)^2] \tag{2-9}$$

Clearly our analysis must be bound by the restriction

$$l \ll L \qquad \text{R.2}$$

in order that Equation 2-2 be satisfied.

AREA AVERAGES

Subject to restriction R.2, it will be of interest to compare area averages with volume averages to determine any possible differences between the area and volume averaged forms of the basic equations. To make a comparison, it is necessary to choose an averaging volume V which can bound a plane averaging area $A_{(j,k)}$ such that the area is always constant. As an example, we choose V to be a cube of volume l^3 and $A_{(j,k)}$ to be an area of magnitude l^2 parallel to the (j, k) plane. We let the center of the cube be the origin of our coordinate system, the sides of the cube being orthogonal to the coordinate axes. The area average of ψ at the origin is given by

$$\bar{\psi}_0^{(j,k)} = \frac{1}{A_{(j,k)}}\int_{A_f} \psi dA \tag{2-10}$$

Here A_f represents the fluid area contained within $A_{(j,k)}$, the dependence of A_f on the particular coordinate surface under consideration being understood. The volume average can now be expressed as

$$\langle\psi\rangle_0 = \frac{1}{l}\int_{x_{(i)} = -l/2}^{x_{(i)} = +l/2} \bar{\psi}^{(j,k)} dx_{(i)} \tag{2-11}$$

where (i), (j), and (k) are distinct. If we expand $\bar{\psi}^{(j,k)}$ in a Taylor series in $x_{(i)}$, we obtain

$$\bar{\psi}^{(j,k)} = \bar{\psi}_0^{(j,k)} + x_{(i)}\left(\frac{\partial\bar{\psi}^{(j,k)}}{\partial x_{(i)}}\right)_0 + \frac{x_{(i)}^2}{2!}\left(\frac{\partial^2\bar{\psi}^{(j,k)}}{\partial x_{(i)}^2}\right)_0 + \ldots \tag{2-12}$$

where the summation convention does not apply to indices in parentheses. Substituting Equation 2-12 into Equation 2-11 and evaluating the integrals yield

$$\langle\psi\rangle_0 = \bar{\psi}_0^{(j,k)} + \frac{l^2}{24}\left(\frac{\partial^2\bar{\psi}^{(j,k)}}{\partial x_{(i)}^2}\right)_0 + \ldots \qquad (2\text{-}13)$$

Estimating the order of magnitude of the second derivative as

$$\left(\frac{\partial^2\bar{\psi}^{(j,k)}}{\partial x^2_{(i)}}\right)_0 = 0(\bar{\psi}^{(j,k)}/L^2) \qquad (2\text{-}14)$$

we can express Equation 2-13 as

$$\langle\psi\rangle_0 = \bar{\psi}_0^{(j,k)} + \left(\frac{l}{L}\right)^2 0[\bar{\psi}^{(j,k)}(l/L)^2] \qquad (2\text{-}15)$$

Thus if our analysis is to be restricted by $l \ll L$, it follows that volume and area averages are essentially equivalent.

It is of some interest to apply this result to the void volume distribution function, α, which leads us to

$$\langle\alpha\rangle \approx \bar{\alpha}^{(j,k)} \qquad (2\text{-}16)$$

Since the volume average of α is the volume porosity, Φ, while the area average is the plane porosity, we find that our restriction $l \ll L$ requires the volume porosity be essentially equal to the three plane porosities

$$\Phi \approx \bar{\alpha}^{(1,2)} \approx \bar{\alpha}^{(2,3)} \approx \bar{\alpha}^{(3,1)} \qquad (2\text{-}17)$$

This relation between the porosities applies in general to anisotropic porous media; however, it is not necessarily valid for a highly structured porous media in which the order of magnitude estimate given by Equation 2-14 might not apply to $\bar{\alpha}^{(j,k)}$. Note that if porous media are homogeneous or uniform, Equation 2-17 takes the form

$$\Phi = \bar{\alpha}^{(1,2)} = \bar{\alpha}^{(2,3)} = \bar{\alpha}^{(3,1)} \qquad (2\text{-}18)$$

regardless of whether or not the medium is isotropic.

3. Slattery's Averaging Theorem

Up to this point we have been discussing the averaging of some point function ψ; however, in developing the averaged form of the basic equations we will be dealing with gradients of point functions, and we need to explore the problem of averaging gradients of functions for we are interested in obtaining gradients of averages rather than averages of gradients.

A general relationship between gradients of averages and averages of gradients for functions defined in both the solid and the liquid phase and suffering a jump discontinuity at the phase interface has been presented by the author (*36*); however, the result (*37*) is not available in the readily accessible literature, and furthermore the development is a rather heavy-handed one. A similar result has been obtained by Slattery (*30*) for functions defined in the fluid phase by a rather ingenious application of the general transport theorem [(*35*, p 347), (*39*, p 88)]. Since the original presentation given by Slattery is rather terse, and since the development has value in the analysis of more complex systems than the one treated here, we will present a detailed version of the theoretical development.

The general transport theorem can be written as

$$\frac{d}{dt}\int_{V(t)} \psi dV = \int_{V(t)} \frac{\partial\psi}{\partial t} dV + \int_{A(t)} \psi \boldsymbol{w}\cdot\boldsymbol{n} dA \qquad (3\text{-}1)$$

where $V(t)$ is a volume bounded by the surface $A(t)$ moving at a velocity $\boldsymbol{w}$ which may be different from the velocity of the fluid, $\boldsymbol{v}$. If $\boldsymbol{w}$ is taken to be equal to $\boldsymbol{v}$ the volume $V(t)$ is the region occupied by a body and Equation 3-1 is referred to as the Reynolds' transport theorem (*2*, p 84). To apply the general transport theorem to the problem of averaging gradients, we consider a point in the porous media (in either the fluid or solid) located on an arbitrary, continuous curve, the arc length measured along this curve being s. With each point on the curve, we associate an averaging volume $V(s)$ bounded by a surface $A(s)$. Of this volume, a portion is occupied by the fluid; we designate this portion by $V_f(s)$ and its bounding surface by $A_f(s)$. The local volume porosity is then given by

$$\Phi(s) = \frac{V_f(s)}{V(s)} \qquad (3\text{-}2)$$

We consider now how the integral over $V_f(s)$ of some quantity ψ changes as a function of s. Although the transport theorem is generally viewed as a rule for taking the time derivative of an integral having limits which are functions of time, the analysis can be carried over directly to the problem of obtaining the directional derivative (*16*, p 107). Thus, from Equation 3-1, we may proceed directly to

$$\frac{d}{ds}\int_{V_f(s)} \psi dV = \int_{V_f(s)} \left(\frac{\partial\psi}{\partial s}\right) dV + \int_{A_f(s)} \psi \left(\frac{d\boldsymbol{r}}{ds}\right)\cdot \boldsymbol{n} dA \qquad (3\text{-}3)$$

where $\boldsymbol{n}$ is the outwardly directed unit normal for $A_f(s)$.

AUTHOR *Stephen Whitaker is Associate Professor in the Department of Chemical Engineering, University of California, Davis, Calif.* *This work was supported by the National Science Foundation, Grant GK-477. The author also acknowledges the courtesy of the hand-written notes made available to him by Professor Howard Brenner, Carnegie-Mellon University, and the helpful discussions with Professor J. C. Slattery, Northwestern University.*

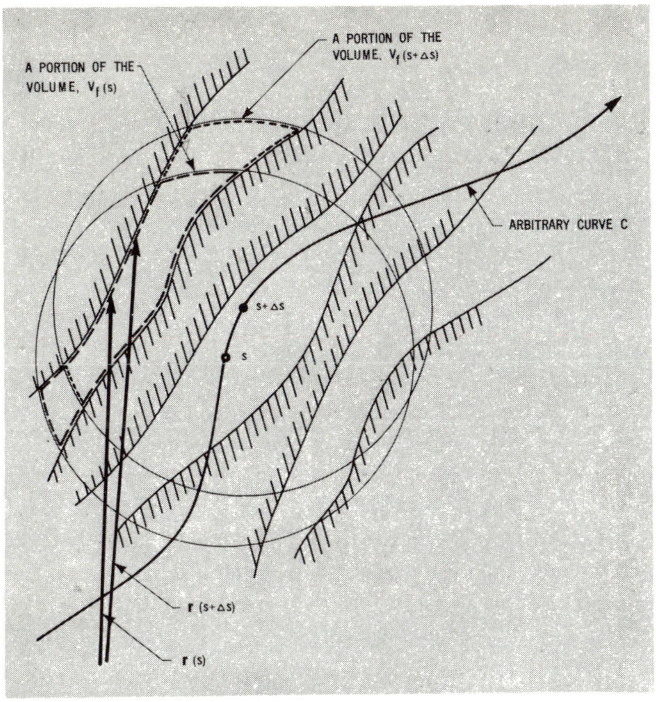

Figure 4. Physical significance of derivative of volume integral with respect to s

The formal mathematical development of Equation 3-3 is obtained by replacing t in the derivation of the transport theorem with the arc length, s. This simply means that we assume a continuous and invertible mapping exists

$$r = r[\xi, s]$$

$$\xi = r(s = 0)$$

which maps $V_f(s = 0)$ into $V_f(s)$. Here r and ξ are comparable to spatial and material position vectors (*39*, p 76). We will concern ourselves with quantities which are explicit functions of time and the spatial coordinates only, thus $\partial\psi/\partial s = 0$ giving us

$$\frac{d}{ds}\int_{V_f(s)} \psi ds = \int_{A_f(s)} \psi\left(\frac{dr}{ds}\right)\cdot n dA \qquad (3\text{-}4)$$

To clarify the dependence of ψ upon s, we note that ψ is, in general, a function of the spatial coordinates and time, $\psi = \psi(x_i, t)$, and the spatial dependence may be represented in terms of the arc-length s along an arbitrary curve to give, $\psi = \psi[x_i(s), t]$. It should be clear that the total derivative of ψ with respect to s is generally nonzero and given by

$$\left(\frac{d\psi}{ds}\right) = \left(\frac{\partial\psi}{\partial x_i}\right)\left(\frac{dx_i}{ds}\right) \qquad (3\text{-}5)$$

while the partial derivative is zero since ψ does not depend explicitly upon s. The physical significance of the derivative of the volume integral with respect to s is shown in Figure 4. It is clear from that figure that dr/ds over the solid-fluid interface is a tangent vector to the interface, thus

$$\left(\frac{dr}{ds}\right)\cdot n = 0, \text{ on the solid-fluid interface} \qquad (3\text{-}6)$$

We may express the area $A_f(s)$ as

$$A_f(s) = A_i(s) + A_e(s) \qquad (3\text{-}7)$$

where $A_e(s)$ represents the area of the entrances and exits on the surface $A(s)$, and $A_i(s)$ represents the area of the solid-fluid interface contained within $V(s)$. In view of Equations 3-6 and 3-7, we may write Equation 3-4 as

$$\frac{d}{ds}\int_{V_f(s)} \psi dV = \int_{A_e(s)} \psi\left(\frac{dr}{ds}\right)\cdot n dA \qquad (3\text{-}8)$$

Letting $r_0(s)$ be the position vector locating the reference point on the arbitrary curve and $p(s)$ the position vector locating points on $A_f(s)$ relative to the reference point (which is at the center of a sphere for the case illustrated in Figure 4), we can write

$$r(s) = r_0(s) + p(s) \qquad (3\text{-}9)$$

The directional derivative (*39*, p 232) can be expressed in terms of $r_0(s)$ as

$$\frac{d}{ds} = \left(\frac{dr_0}{ds}\right)\cdot \nabla \qquad (3\text{-}10)$$

Substituting Equations 3-9 and 3-10 into Equation 3-8 yields

$$\left(\frac{dr_0}{ds}\right)\cdot \nabla \int_{V_f(s)} \psi dV = \int_{A_e(s)} \psi\left(\frac{dr_0}{ds}\right)\cdot n dA + \int_{A_e(s)} \psi\left(\frac{dp}{ds}\right)\cdot n dA \qquad (3\text{-}11)$$

Since dr_0/ds is not a function of the limits of integration for a fixed value of s, we may remove dr_0/ds from the integral sign to obtain

$$\frac{dr_0}{ds}\cdot\left(\nabla \int_{V_f(s)} \psi dV - \int_{A_e(s)} \psi n dA\right) = \int_{A_e(s)} \psi\left(\frac{dp}{ds}\right)\cdot n dA \qquad (3\text{-}12)$$

Provided the volume $V(s)$ is translated without rotation along the arbitary curve C, any differential variation in p is a tangent vector to the surface of the volume (*21*, p 168)—*i.e.*,

$$\left(\frac{dp}{ds}\right) = \begin{Bmatrix}\text{a unit tangent vector}\\ \text{to the surface } A_e(s)\end{Bmatrix} \qquad (3\text{-}13)$$

Thus dp/ds and n are orthogonal and the right-hand

side of Equation 3-12 is zero. Since $d\mathbf{r}_0/ds$ is an arbitary vector, we obtain

$$\nabla \int_{V_f} \psi dV = \int_{A_e} \psi \mathbf{n} dA \tag{3-14}$$

Here the dependence of V_f and A_e upon s has been omitted; however, we must remember that the fluid volume V_f contained within the averaging volume V very definitely depends upon the spatial coordinates. The same may also be said of A_e.

The above result, due to Slattery (*30*), can be used to provide a useful expression for the average of a derivative in terms of the derivative of the average. Making use of the divergence theorem,

$$\int_{V_f} \nabla \psi dV = \int_{A_i} \psi \mathbf{n} dA + \int_{A_e} \psi \mathbf{n} dA \tag{3-15}$$

we may write Equation 3-14 as

$$\int_{V_f} \nabla \psi dV = \nabla \int_{V_f} \psi dV + \int_{A_i} \psi \mathbf{n} dA \tag{3-16}$$

Remembering that the volume average is defined as

$$\langle \psi \rangle = \frac{1}{V} \int_V \psi dV$$

we may divide Equation 3-16 by V (remember that V is a constant) in order to obtain

$$\langle \nabla \psi \rangle = \nabla \langle \psi \rangle + \frac{1}{V} \int_{A_i} \psi \mathbf{n} dA \tag{3-17}$$

A more general form of this result, relating the average of the derivative to the derivative of the average, has been presented elsewhere by the author (*37*). We must remember that there was one important restriction made in the derivation of Equation 3-17 namely

> The averaging volume V is constant, and its orientation relative to some inertial frame remains unchanged R.3

Having discussed averages and averages of derivatives, we are now in a position to proceed with our analysis of flow in porous media.

4. Darcy's Law

The governing equations that we wish to analyze are given by Equations 1-1 and 1-2. The volume averaged form of the continuity equation may be written as

$$\frac{1}{V} \int_{V_f} \nabla \cdot \mathbf{v} dV = 0 \tag{4-1}$$

and by our averaging theorem, Equation 3-17, we obtain

$$\nabla \cdot \langle \mathbf{v} \rangle + \frac{1}{V} \int_{A_i} \mathbf{v} \cdot \mathbf{n} dA = 0 \tag{4-2}$$

if the porous media is fixed in space $\mathbf{v} = 0$ over A_i and the second term in Equation 4-2 is zero. If the porous media is moving at a constant velocity $\mathbf{u}$, we must remember that $\mathbf{v}$ is to be interpreted as the relative velocity which is also zero over A_i. With this in mind we write Equation 4-2 as

$$\nabla \cdot \langle \mathbf{v} \rangle = 0 \tag{4-3}$$

where it is understood that $\langle \mathbf{v} \rangle$ is the average fluid velocity relative to the constant porous media velocity, $\mathbf{u}$.

Before attacking the equations of motion we define a piezometric pressure, P, as

$$P = p - p_0 + \rho \varphi \tag{4-4}$$

where p_0 is some reference pressure and φ is the gravitational potential function which must satisfy the condition

$$\mathbf{g} = -\nabla \varphi \tag{4-5}$$

Equation 4-5 specifies φ to within an arbitrary constant, and we will choose this constant so that $\varphi = 0$ when $p = p_0$. In terms of Equation 4-4, the equations of motion can be written as

$$0 = -\nabla P + \mu \nabla^2 \mathbf{v} \tag{4-6}$$

Again, it should be kept in mind that $\mathbf{v}$ is the velocity relative to $\mathbf{u}$. Taking the divergence of Equation 4-6 and making use of the continuity equation show that P must satisfy the Laplace equation

$$\nabla^2 P = 0 \tag{4-7}$$

If we now operate on Equation 4-6 with the Laplacian, the velocity must satisfy the biharmonic equation

$$\nabla^4 \mathbf{v} = 0 \tag{4-8}$$

In addition, the velocity must satisfy the "no slip" condition at the solid-fluid interface.

$$\mathbf{v} = 0 \text{ on } A_i \text{ of } V \tag{4-9}$$

If we were to specify $\mathbf{v}$ over the entrances and exits of any averaging volume V—*i.e.*,

$$\mathbf{v} = \mathbf{f}(\mathbf{x}) \text{ on } A_e \text{ of } V \tag{4-10}$$

a unique solution for $\mathbf{v}$ would result (*17*, Chap. 2) in the region V. It follows that Equations 4-8, 4-9, and 4-10 are sufficient to determine $\langle \mathbf{v} \rangle$ for the particular region in question. Thus, these three equations lead to

$$\langle \mathbf{v} \rangle = \mathbf{g}(\mathbf{r}_0) \tag{4-11}$$

where $\mathbf{r}_0$ is the point with which the averaging volume is associated.

Now we are going to assume that this problem can be, in effect, turned around and stated in the following way:

> Given the governing differential equation (Eq. 4-8), the no-slip condition (Equation 4-9), and the average velocity (Equation 4-11), the velocity field is uniquely determined A.1

One can prove this for a number of flows. For example, the one-dimensional flow in a capillary tube is uniquely determined by the equations of motion, the no-slip condition at the walls of the capillary, and the average velocity. Flow past a sphere is uniquely determined by the equations of motion, the no-slip condition, and the average velocity, provided the velocity is uniform at infinity. Brenner (*5*) has been able to prove assumption A.1, provided the flow is spatially periodic. It seems doubtful that a general proof of assumption A.1 will ever be given, for in each case where we know it holds, there is always one additional requirement placed on the flow—*i.e.*, the flow is one-dimensional, or uniform at infinity, or spatially periodic. Still, assumption A.1 has some intuitive appeal, and, one way or another, it appears in all previous theoretical treatments of this problem.

Since $\mathbf{v}$ and $\langle \mathbf{v} \rangle$ are continuous and defined everywhere in the region under consideration, we can map $\langle \mathbf{v} \rangle$ into $\mathbf{v}$—*i.e.*,

$$\mathbf{v} = \mathbf{M} \cdot \langle \mathbf{v} \rangle \tag{4-12}$$

where $\mathbf{M}$ is the transformation which maps the average velocity into the point velocity. (Tensors are indicated by capital letters in boldface type, while vectors are represented by lower case letters in boldface type.) By the assumption A.1, the mapping is unique although not necessarily linear. If we substitute the above expression for $\mathbf{v}$ into Equation 4-8 we obtain

$$(\nabla^4 \mathbf{M}) \cdot \langle \mathbf{v} \rangle + \mathbf{M} \cdot (\nabla^4 \langle \mathbf{v} \rangle) = 0 \tag{4-13}$$

From previous order of magnitude arguments, it is clear that $\nabla^4 \langle \mathbf{v} \rangle$ is negligible and the transformation matrix is determined by

$$\nabla^4 \mathbf{M} = 0 \tag{4-14a}$$

We must keep in mind here that Equation 4-14a follows rigorously from Equation 4-13 if $\langle \mathbf{v} \rangle$ is a constant vector or depends on the spatial coordinates to the third order or less. For the more general case, we need only remember that significant variations in $\langle \mathbf{v} \rangle$ take place over a distance L and significant variations in $\mathbf{M}$ take place over a distance d where $d <<< L$. The boundary conditions for $\mathbf{M}$ are

$$\mathbf{M} = 0 \text{ on } A_i \tag{4-14b}$$

$$\langle \mathbf{M} \rangle = \mathbf{U} \text{ at } \mathbf{r}_0 \tag{4-14c}$$

where Equation 4-14c is a logical extension of assumption A.1. From Equations 4-14 we conclude that $\mathbf{M}$ is independent of $\langle \mathbf{v} \rangle$, thus the point velocity is a linear vector function of the average velocity. From Halmos (*13*, p 62) we know that a linear transformation is invertible if, and only if, $\mathbf{M} \cdot \langle \mathbf{v} \rangle = 0$ implies that $\langle \mathbf{v} \rangle = 0$; however, $\mathbf{v} = 0$ everywhere in the solid regardless of the value of $\langle \mathbf{v} \rangle$. Thus we cannot say that $\mathbf{M}$ has an inverse.

Returning now to Equation 4-6, we form the scalar product with $\boldsymbol{\lambda}$

$$0 = -\boldsymbol{\lambda} \cdot \nabla P + \mu \boldsymbol{\lambda} \cdot \nabla^2 \mathbf{v} \tag{4-15}$$

where $\boldsymbol{\lambda}$ is a tangent vector to some arbitrary curve lying wholly within the fluid region. Letting s be the arc length measured along this curve and substituting Equation 4-12 for $\mathbf{v}$ allow us to write

$$\frac{dP}{ds} = \mu \boldsymbol{\lambda} \cdot (\nabla^2 \mathbf{M}) \cdot \langle \mathbf{v} \rangle \tag{4-16}$$

Here we have used the relation $\boldsymbol{\lambda} \cdot \nabla = d/ds$ and used the same order of magnitude arguments that were imposed between Equation 4-13 and Equation 4-14. (This step and several subsequent ones are all rigorous if $\langle \mathbf{v} \rangle$ is a constant vector or is a linear function of the spatial coordinates. For the general case, we constantly invoke the argument that the spatial variations in $\langle \mathbf{v} \rangle$ are small compared to those of $\mathbf{M}$.) Integrating from $s = 0$ where $P = P_0$ to any point on the arbitrary curve yields

$$P(\mathbf{r}) = P_0 + \mu \left\{ \int_{\eta=0}^{\eta=s(\mathbf{r})} [\boldsymbol{\lambda} \cdot (\nabla^2 \mathbf{M})] d\eta \right\} \cdot \langle \mathbf{v} \rangle \tag{4-17}$$

The term in braces depends only on the geometry through Equation 4-14 and on the position vector $\mathbf{r}$ through the upper limit of integration. Somewhat more simply we can state that P is determined to within an arbitrary constant by the expression

$$P = -\mu \mathbf{m} \cdot \langle \mathbf{v} \rangle \tag{4-18}$$

where $\mathbf{m}$ depends on the position vector $\mathbf{r}$ in addition to the detailed structure of the porous media. (The minus sign is included in Equation 4-18 so that the permeability for isotropic porous media will be positive.)

Returning now to Equation 3-14, we can substitute P for ψ on the left-hand side and $-\mu \mathbf{m} \cdot \langle \mathbf{v} \rangle$ for ψ on the right-hand side and divide by sides by V to obtain

$$\nabla \left[\frac{1}{V} \int_{V_f} P dV \right] = -\frac{1}{V} \int_{A_e} \mu \mathbf{m} \cdot \langle \mathbf{v} \rangle \mathbf{n} dA \tag{4-19}$$

This can be expressed as

$$\nabla \langle P \rangle = -\mu \mathbf{K}^{-1} \cdot \langle \mathbf{v} \rangle \tag{4-20}$$

where $\mathbf{K}^{-1}$ is given by

$$\mathbf{K}^{-1} = \frac{1}{V} \int_{A_e} \mathbf{n} \mathbf{m} dA = -\frac{1}{V} \int_{A_e} \mathbf{n} \int_{\eta=0}^{\eta=s(\mathbf{r})} [\boldsymbol{\lambda} \cdot (\nabla^2 \mathbf{M})] d\eta dA \tag{4-21}$$

We are now confronted with the question as to whether $\mathbf{K}^{-1}$ has an inverse. We know that if P is constant, the pressure variation is hydrostatic, the point velocity is zero, and $\langle \mathbf{v} \rangle = 0$. The same line of reasoning does not necessarily follow if $\langle P \rangle$ is constant; however, it is intuitively appealing, and we will make the assumption

$$\nabla \langle P \rangle = 0 \text{ implies } \langle \mathbf{v} \rangle = 0 \tag{A.2}$$

The theorem of Halmos (*13*, p 62) then indicates that $\mathbf{K}^{-1}$ has an inverse which we designate by $\mathbf{K}$ so that Equation 4-19 takes the form

$$\langle \mathbf{v} \rangle = -\frac{1}{\mu} [\mathbf{K} \cdot \nabla \langle P \rangle] \tag{4-22}$$

Up to this point we have dealt with the equations of motion in a form which kept the analysis as neat as possible. Now we would like to express the result in a form which is more directly applicable to the interpretation of experimental data. Substitution of Equation 4-4 yields

$$\langle v \rangle = -\frac{1}{\mu}\mathbf{K}\cdot\{\nabla[\langle p - p_0\rangle + \rho\langle\varphi\rangle]\} \tag{4-23}$$

Here we have removed the density from the average since we are considering only incompressible flows. To within an arbitrary constant, we can express the gravitational potential function as

$$\varphi = -r \cdot g \tag{4-24}$$

and the average becomes

$$\langle\varphi\rangle = -\frac{1}{V}\int_{V_f}(r \cdot g)dV \tag{4-25}$$

We note that g is constant which reduces to

$$\langle\varphi\rangle = -\Phi\bar{r}\cdot g \tag{4-26}$$

where $\bar{r}$ is the position vector locating the centroid of the fluid volume, V_f. Because of the restrictions we have placed on the variation of the porosity, $\bar{r}$ can be considered essentially identical to r_0, the position vector locating the centroid of the averaging volume, V.

An experimental measurement of the fluid pressure in a porous medium would probably determine the quantity $\langle p - p_0\rangle_f$ which is defined

$$\langle p - p_0\rangle_f = \frac{1}{V_f}\int_{V_f}(p - p_0)dV \tag{4-27}$$

Thus the average pressure in Equation 4-23 can be written as

$$\langle p - p_0\rangle = \Phi\langle p - p_0\rangle_f \tag{4-28}$$

Substitution of Equations 4-26 and 4-28 into Equation 4-23 yields

$$\langle v \rangle = -\frac{1}{\mu}\mathbf{K}\cdot\{\nabla[\Phi(\langle p - p_0\rangle_f - \rho\bar{r}\cdot g)]\} \tag{4-29}$$

The general interpretation of the quantity ∇r is

$$\nabla r = \mathbf{U} \tag{4-30}$$

where $\mathbf{U}$ is the unit tensor. In taking the gradient of averaged functions, we think of these functions as being defined at points in space, and if we associate the centroid of the averaging volume it appears consistent to write

$$\nabla r_0 = \mathbf{U} \tag{4-31}$$

In as much as $\bar{r}$ is essentially identical to r_0 we can express the body force term in Equation 4-29 as

$$\nabla[-\Phi\rho\bar{r}\cdot g] = -\Phi\rho g - \rho\bar{r}\cdot g\nabla\Phi \tag{4-32}$$

In general then, Equation 4-29 takes the form

$$\langle v \rangle = -\frac{1}{\mu}\mathbf{K}\cdot\{\Phi[\nabla\langle p - p_0\rangle_f - \rho g] + [\langle p - p_0\rangle_f - \rho\bar{r}\cdot g]\nabla\Phi\} \tag{4-33}$$

We have already restricted the analysis to porous media having only slowly changing physical properties, thus the practical form of Equation 4-32 is

$$\langle \mathbf{v} \rangle = -\frac{\Phi\mathbf{K}}{\mu}[\nabla\langle p - p_0\rangle_f - \rho g] \tag{4-34}$$

Note that the average form used for the velocity has been chosen with the idea that volumetric flow rates can be readily calculated by the relation

$$Q = \int_S \langle v\rangle \cdot n dS \tag{4-35}$$

while the pressure and body force terms are given in a form most suitable for interpretation by the experimentalist. (Here the surface area S must be large compared to l^2 so that the integrals accurately represent the volumetric flow rate.) Massarani (*20*) has experimentally verified Equation 4-34 for one-dimensional flow in a nonhomogeneous porous media.

We can carry our analysis just a bit further by examining Equation 4-21 and noting that an order of magnitude estimate gives

$$\nabla^2\mathbf{M} = 0(d^{-2})$$

thus

$$\mathbf{K}^{-1} = 0(d^{-2})$$

and the inverse of $\mathbf{K}^{-1}$ has an order of magnitude given by

$$\mathbf{K} = 0(d^2)$$

We can define a dimensionless permeability tensor $\tilde{\mathbf{K}}$ by the equation

$$\mathbf{K} = d^2\tilde{\mathbf{K}}$$

and express Equation 4-34 as

$$\langle v \rangle = -\left(\frac{\Phi d^2}{\mu}\right)\tilde{\mathbf{K}}\cdot[\nabla\langle p - p_0\rangle_f - \rho g] \tag{4-36}$$

At this point we can say little about $\tilde{\mathbf{K}}$ except that it depends on the structure of the porous media. In general, the determination of the scalar components of $\tilde{\mathbf{K}}$ represents an enormous experimental task, although geometric models can be helpful in shedding some light on this problem. In particular, one can argue from the capillary model of Kozeny (*4*, p 196) that $\tilde{\mathbf{K}}$ should be proportional to $\Phi^2/(1 - \Phi)^2$, thus the difficulty in correlating data is reduced. The spatially periodic model of Brenner (*5*) indicates that $\tilde{\mathbf{K}}$ should be symmetric; thus the results of those investigators who have tacitly assumed this to be self-evident become more credulous, and the experimentalist is faced with six undetermined components instead of nine. The experimental determination of the scalar components of $\tilde{\mathbf{K}}$ indeed represents a formidable task and in the next section we would like to address this problem.

5. The Permeability Tensor

In discussing the problem of the experimental determination of the scalar components of $\tilde{\mathbf{K}}$ or $\mathbf{K}$, we will consider two geometric models: Brenner's spatially periodic model shown in Figure 1 and the skewed capillary model shown in Figure 2. For the former, Brenner has proved that $\mathbf{K}$ is symmetric even if the porous media is anisotropic. The skewed capillary model shown in Figure 2 has apparently never been analyzed, presumably because it appears to be nothing more than a minor variation of the straight capillary tube model. This is quite true—still there is something to be learned from the analysis.

The skewed capillary model consists of a slab of thickness L, infinite in the y- and z-directions, and embedded with capillary tubes of diameter d. We require that the tube diameter be much smaller than the slab thickness, $d <<< L$, so that our requirements for meaningful averages are satisfied. The flow occurring in the capillaries will be one-dimensional, laminar, and will satisfy assumptions A.1 and A.2. Thus the average pressure gradient will be related by Darcy's law, Equation 4-22, to the average velocity.

In analyzing the flow that takes place in the capillaries, we imagine that either the pressure or the pressure gradient can be maintained at some constant value over y-z planes at $x = 0$ and $x = L$. Under these circumstances we can consider three distinct flow processes given by

$$\text{I.}\quad \frac{\partial\langle P\rangle}{\partial x} = -C_1, \quad \frac{\partial\langle P\rangle}{\partial y} = \frac{\partial\langle P\rangle}{\partial z} = 0$$

$$\langle \mathbf{v}\rangle^{(1)} = u_0^{(1)}(\mathbf{i}\alpha + \mathbf{j}\beta + \mathbf{k}\gamma)$$

$$u_0^{(1)} = C_1(\alpha d^2/32\mu)$$

$$\text{II.}\quad \frac{\partial\langle P\rangle}{\partial x} = \frac{\partial\langle P\rangle}{\partial z} = 0, \quad \frac{\partial\langle P\rangle}{\partial y} = -C_2$$

$$\langle \mathbf{v}\rangle^{(2)} = u_0^{(2)}(\mathbf{i}\alpha + \mathbf{j}\beta + \mathbf{k}\gamma)$$

$$u_0^{(2)} = C_2(\beta d^2/32\mu)$$

$$\text{III.}\quad \frac{\partial\langle P\rangle}{\partial x} = \frac{\partial\langle P\rangle}{\partial y} = 0, \quad \frac{\partial\langle P\rangle}{\partial z} = -C_3$$

$$\langle \mathbf{v}\rangle^{(3)} = u_0^{(3)}(\mathbf{i}\alpha + \mathbf{j}\beta + \mathbf{k}\gamma)$$

$$u_0^{(3)} = C_3(\gamma d^2/32\mu)$$

Here α, β, and γ represent the direction cosines for the orientation of the capillary tube, and u_0 represents the magnitude of the average velocity determined by the Hagen-Poiseuille law. Substituting these expressions for the pressure gradients and average velocities into Equation 4-22 allows us to determine the scalar components of $\mathbf{K}$ given by

$$K_{ij} = \left(\frac{d^2}{32}\right)\lambda_i\lambda_j \tag{5-1}$$

where λ_i are the scalar components of the unit vector pointing in the direction of the capillary tubes—*i.e.*, $\lambda_1 = \alpha$, $\lambda_2 = \beta$, and $\lambda_3 = \gamma$. Here we have found that the permeability tensor is symmetric, an interesting result especially in light of Brenner's analysis of spatially periodic porous media. In that structure, the anisotropy results from the geometry and orientation of a solid particle within a unit cell; thus the orientation occurs on a length scale comparable to d. The situation is quite different in the skewed capillary model where the orientation occurs over a length scale on the order of L. Since these two rather different types of porous media have symmetric permeability tensors, it is appealing to assume that $\mathbf{K}$ is always symmetric, and a number of investigators have presented arguments in support of this point of view [*11*, *25*, (*27*, p 625)].

Our next step is to try to put forth some general ideas about the analysis of anisotropic porous media and to offer a proof, suggested by Slattery (*31*), that the permeability tensor is always symmetric.

Following the work of Ericksen (*8*, *9*), we will assume at the outset that there is a unique orientation vector $\boldsymbol{\omega}$, representing a preferred direction, which describes the anisotropic nature of the porous media:

$$\mathbf{K} = \mathbf{K}(\boldsymbol{\omega}) \tag{5-2}$$

Subject to the constraints on the porous media listed in Sec. 2, the vector, $\boldsymbol{\omega}$, may depend on the spatial coordinates. For isotropic porous media, $\boldsymbol{\omega}$ will clearly be zero, and for the skewed capillary model, $\boldsymbol{\omega}$ is obviously parallel to the capillaries. If the particles in Brenner's spatially periodic porous media are ellipsoids, $\boldsymbol{\omega}$ would be parallel to the major axis, but what can we say about the orientation vector for the more general porous media? Very little, it seems, at this point; however, we will overlook for the present the fact that the three scalar components of $\boldsymbol{\omega}$ may have to be determined experimentally and proceed.

If a porous media is symmetric about one plane, then a change of coordinates of the type

$$\bar{x}_1 = -x_1, \quad \bar{x}_2 = x_2, \quad \bar{x}_3 = x_3 \tag{5-3}$$

should leave the permeability tensor unchanged, thus

$$\bar{K}_{ij} = K_{ij}$$

for the transformation given by Equation 5-3. This requires that

$$K_{12} = K_{21} = K_{31} = K_{13} = 0 \tag{5-4}$$

and the array of coefficients becomes

$$K_{ij} = \begin{pmatrix} K_{11} & 0 & 0 \\ 0 & K_{22} & K_{23} \\ 0 & K_{32} & K_{33} \end{pmatrix} \tag{5-5}$$

If the media are symmetric about two orthogonal planes, a coordinate transformation of the form,

$$\bar{x}_1 = x_1, \quad \bar{x}_2 = -x_2, \quad \bar{x}_3 = x_3 \tag{5-6}$$

in addition to the transformation given by Equation 5-3, leaves the permeability tensor unchanged. In addition to the restrictions given by Equation 5-4, we now require that

$$K_{23} = K_{32} = 0 \tag{5-7}$$

and the array of coefficients becomes

$$K_{ij} = \begin{pmatrix} K_{11} & 0 & 0 \\ 0 & K_{22} & 0 \\ 0 & 0 & K_{33} \end{pmatrix} \tag{5-8}$$

Clearly porous media symmetric about two orthogonal planes must be symmetric about a third orthogonal plane so that only the diagonal elements are nonzero in the principal axis coordinate system established by the three orthogonal planes of symmetry. Such materials are called orthotropic (*12*, p 159) and **K** is completely specified by the principal values K_{11}, K_{22}, and K_{33}. One can easily show that if **K** is symmetric in one coordinate system it is symmetric in all coordinate systems; thus the permeability tensor for orthotropic materials is symmetric. The skewed capillary model represents an orthotropic material—one plane of symmetry being perpendicular to the direction of the capillaries and the other two planes being orthogonal to the first.

If, in addition to the transformations given by Equations 5-3 and 5-6, the permeability tensor is unchanged by a transformation of the type

$$\begin{aligned} \bar{x}_1 &= x_1 \cos\theta + x_2 \sin\theta \\ \bar{x}_2 &= -x_1 \sin\theta + x_2 \cos\theta \\ \bar{x}_3 &= x_3 \end{aligned} \tag{5-9}$$

the porous media are said to be transversely isotropic (*12*, p 159) and we find that

$$K_{22} = K_{33}$$

At this point we have learned nothing about how one might expect the orientation vector, $\boldsymbol{\omega}$, to depend on the nature of the porous media. For orthotropic, transversely isotropic, and isotropic media, one might guess the form of $\boldsymbol{\omega}$ to be

$$\boldsymbol{\omega} = \boldsymbol{e}_{(1)}\left[\frac{K_{11}\sqrt{3}}{\sqrt{K_{ij}K_{ij}}} - 1\right] + \boldsymbol{e}_{(2)}\left[\frac{K_{22}\sqrt{3}}{\sqrt{K_{ij}K_{ij}}} - 1\right] + \boldsymbol{e}_{(3)}\left[\frac{K_{33}\sqrt{3}}{\sqrt{K_{ij}K_{ij}}} - 1\right] \tag{5-10}$$

to within a constant multiplier. Here K_{11}, K_{22}, K_{33} represent the principal values, and $\boldsymbol{e}_{(1)}$, $\boldsymbol{e}_{(2)}$, and $\boldsymbol{e}_{(3)}$ are the orthogonal base vectors for the principal directions. For isotropic porous media, $K_{11} = K_{22} = K_{33}$, and $\boldsymbol{\omega}$ equals zero as required. For transversely isotropic media, the angle between $\boldsymbol{\omega}$ and $\boldsymbol{e}_{(3)}$ (in our example) would depend on the ratio of K_{11} to K_{22}.

For the skewed capillary model, $\boldsymbol{\omega}$ would be parallel to the capillary tubes as required. The general problem seems to indicate that $\boldsymbol{\omega}$ must be measured experimentally; however, it would seem that good intuition and some ingenious modeling might lead to some general rules regarding the nature of $\boldsymbol{\omega}$.

To put our analysis of the dependence of **K** upon $\boldsymbol{\omega}$ on a sound basis, we first note that the vector $\boldsymbol{\omega}$ can be represented by the skew-symmetric tensor $\boldsymbol{\Omega}$ by the relation

$$\omega_i = \frac{1}{2} e_{ijk}\Omega_{jk} \quad \text{or} \quad \Omega_{jk} = e_{jki}\omega_i \tag{5-11}$$

The functional dependence of the scalar components of **K** can therefore be expressed as

$$K_{ij} = K_{ij}(\omega_m) = K_{ij}(\Omega_{np}) \tag{5-12}$$

We now assume that K_{ij} can be expressed as a polynomial in Ω_{np}, and apply the Caley-Hamilton theorem (*14*, p 64) to obtain a closed-form expression.

$$K_{ij} = A_{ij}{}^{I} + A_{ijmn}{}^{II}\Omega_{mn} + A_{ijmn}{}^{III}\Omega_{mp}\Omega_{pn} \tag{5-13}$$

Here the A-tensors are polynomial functions of the single scalar invariant of Ω_{mn}—*i.e.*, the magnitude of $\boldsymbol{\omega}$ Substituting Equation 5-11 into Equation 5-13 yields

$$K_{ij} = A_{ij}{}^{I} + A_{ijmn}{}^{II}e_{mnp}\omega_p + A_{ijmn}e_{mpq}{}^{III}e_{pnr}\omega_q\omega_r \tag{5-14}$$

With a little algebraic effort we can show that

$$e_{mpq}e_{pnr}\omega_q\omega_r = -\omega_m\omega_n + \delta_{mn}\omega_k\omega_k \tag{5-15}$$

Since the A-tensors are unspecified we can redefine and regroup these terms to obtain

$$K_{ij} = B_{ij}{}^{\mathrm{I}} + B_{ijk}{}^{\mathrm{II}}\lambda_k + B_{ijmn}{}^{\mathrm{III}}\lambda_m\lambda_n \tag{5-16}$$

where once again the B-tensors are polynomial functions of the magnitude of $\boldsymbol{\omega}$ In going from Equations 5-14 and 5-15 we have replaced $\boldsymbol{\omega}$ with $\omega\boldsymbol{\lambda}$ where $\boldsymbol{\lambda}$ is the unit orientation vector. No generality is lost in this step since the B-tensors are already functions of ω. If $\boldsymbol{\lambda}$ is indistinguishable from $-\boldsymbol{\lambda}$, as it will be for any orthotropic material, the second term must vanish to yield

$$K_{ij} = B_{ij}{}^{I} + B_{ijmn}{}^{III}\lambda_m\lambda_n \tag{5-17}$$

for orthotropic porous media. If the porous media is isotropic, $\boldsymbol{\lambda} = 0$ and $B_{ij}{}^{II}$ must be an isotropic second-order tensor and we obtain

$$K_{ij} = k\delta_{ij} \tag{5-18}$$

for isotropic porous media.

At this point we proceed with a development suggested by Slattery (*31*) and require that Darcy's law satisfy the principle of material frame–indifference [(*22*), (*34*, p 41)]. This follows from the fact that Darcy's law is nothing more than a balance of forces, and forces are regarded as being frame-indifferent (*33*, p 27). Thus the equation

$$\langle v_i \rangle = -\frac{1}{\mu}\left[B_{ij}{}^{I}\frac{\partial\langle P\rangle}{\partial x_j} + B_{ijk}{}^{II}\lambda_k \cdot \frac{\partial\langle P\rangle}{\partial x_j} + B_{ijmn}{}^{III}\lambda_m\lambda_n\frac{\partial\langle P\rangle}{\partial x_j}\right] \tag{5-19}$$

must be frame-indifferent. The most general change of frame is given by (*34*, p 41)

$$x^* = c(t) + Q(t) \cdot [x - x_0] \tag{5-20}$$

$$t^* = t - a \tag{5-21}$$

where $c(t)$ is a time-dependent point, $Q(t)$ is a time-dependent orthogonal tensor, x_0 is a fixed point that is mapped into $c(t)$. The time-dependent orthogonal transformation, $Q(t)$, represents a rotation and, possibly, a reflection from a right-(left)-handed coordinate system to a left-(right)-handed coordinate system. A vector b is said to be frame-indifferent if it satisfies the condition

$$b^* = Q(t) \cdot b \tag{5-22}$$

In considering Equations 5-19 we express the functional dependence of $\langle v\rangle$ upon λ and $\nabla\langle P\rangle$ as

$$\langle v\rangle = F(\lambda, \nabla\langle P\rangle) \tag{5-23}$$

and require that (*33*, p 34)

$$\langle v\rangle^* = F(\lambda^*, \nabla\langle P\rangle^*) \tag{5-24}$$

This expression means that the material properties of the porous media—*i.e.*, the relationships among velocity, orientation, and pressure gradient are indifferent to the choice of observer. In writing Equation 5-24 we have used the fact that the relative velocity is frame-indifferent. We also regard forces—*i.e.*, $\nabla\langle P\rangle$—as being frame-indifferent (*33*, p 27), and we assume that the orientation vector λ is frame-indifferent. Thus λ must be thought of as the gradient of some scalar property of the media.

Equation 5-24 requires that the B-tensors satisfy the conditions

$$B_{ij}{}^{I*} = B_{ij}{}^{I}$$
$$B_{ijk}{}^{II*} = B_{ijk}{}^{II} \tag{5-25}$$
$$B_{ijmn}{}^{III*} = B_{ijmn}{}^{III}$$

which leads to the restriction that the B-tensors must be isotropic (not that the porous media are isotropic) (*34*, p 23), and we conclude

$$B_{ij}{}^{I} = B^{I}\delta_{ij} \tag{5-26a}$$

$$B_{ijk}{}^{II} = 0 \tag{5-26b}$$

$$B_{ijmn}{}^{III} = B_{(1)}{}^{III}\delta_{ij}\delta_{mn} + B_{(2)}{}^{III}(\delta_{ij}\delta_{mn} + \delta_{im}\delta_{jn}) + B_{(3)}{}^{III}(\delta_{ij}\delta_{mn} - \delta_{in}\delta_{jm}) \tag{5-26c}$$

We note here that the third-order tensor e_{ijk} is isotropic only to right-(left)-handed transformations and must be considered a hemitropic tensor-valued function, if reflections are to be allowed (*34*, p 24).

Subject to Equations 5-26, we write Equation 5-19 as

$$\langle v_i\rangle = -\frac{1}{\mu}\,[B_{(1)}\delta_{ij} + B_{(2)}\lambda_i\lambda_j]\,\frac{\partial\langle P\rangle}{\partial x_j} \tag{5-27}$$

Putting this result in the form expressed by Equation 4-36, we obtain our final form of Darcy's law for anisotropic porous media

$$\langle v\rangle = -\left(\frac{\Phi d^2}{\mu}\right)(\tilde{B}_{(1)}\mathbf{U} + \tilde{B}_{(2)}\lambda\lambda) \cdot [\nabla\langle p - p_0\rangle_f - \rho\mathbf{g}] \tag{5-28}$$

This result is consistent with that given by Slattery (*31*) and indicates that the permeability tensor is symmetric. If $\tilde{B}_{(1)} = 0$, we obtain the proper form of Darcy's law for the skewed capillary model, and the symmetry requirement is in accord with the spatially periodic model. From the experimentalist's point of view, the problem has been reduced to measuring four scalars ($\tilde{B}_{(1)}$, $\tilde{B}_{(2)}$, λ_1, λ_2) although the problem could be simplified if a method of making a good guess about the nature of λ could be developed. Our guess, given by Equation 5-10 is not entirely out of line with the result obtained here. Comparing Equation 5-28 and Equation 4-36, we find

$$\tilde{B}_{(1)}\delta_{ij} + \tilde{B}_{(2)}\lambda_i\lambda_j = \tilde{K}_{ij} \tag{5-29}$$

Taking the trace of this expression allows us to solve for $\tilde{B}_{(2)}$

$$\tilde{B}_{(2)} = (\tilde{K}_{ij}\tilde{K}_{ij} - 3\tilde{B}_{(1)}) \tag{5-30}$$

For orthotropic materials, we can put Equation 5-29 in the principal axis coordinate system and solve for λ

$$\lambda = e_{(1)}\sqrt{\frac{\tilde{K}_{11}/\tilde{B}_{(1)} - 1}{(\tilde{K}_{ij}\tilde{K}_{ij}/\tilde{B}_{(1)}) - 3}} + e_{(2)}\sqrt{\frac{(\tilde{K}_{22}/\tilde{B}_{(1)}) - 1}{(\tilde{K}_{ij}\tilde{K}_{ij}/\tilde{B}_{(1)}) - 3}} + e_{(3)}\sqrt{\frac{(\tilde{K}_{33}/\tilde{B}_{(1)}) - 1}{(\tilde{K}_{ij}\tilde{K}_{ij}/\tilde{B}_{(1)}) - 3}}$$

We must keep in mind that two assumptions were made in the course of this development:

> The anisotropic nature of a porous media can be uniquely described by a single orientation vector, ω A.3
>
> The principle of material frame-indifference applies to Darcy's law A.4

6. Transient Creep Flow

For transient creep flow our governing equations take the form

$$\rho\frac{\partial v}{\partial t} = -\nabla P + \mu\nabla^2 v \tag{6-1}$$

$$\nabla \cdot v = 0 \tag{6-2}$$

We can follow the analysis of Sections 4 and 5, provided assumption A.1 can still be applied; a reasonable procedure only if the flow is quasi-steady. This means that the left-hand side of Equation 6-1 is negligible and the time dependence of $\langle v\rangle$ and $\langle P\rangle$ enters the problem only through the boundary conditions. We can examine the problem of transient flow in a tube (*4*, p 129) to gain some insight as to when the quasi-steady assumption might be valid. If a tube filled with fluid is subjected to a sudden change in the pressure drop, essentially steady flow occurs for times greater than t_0 where

$$t_0 \approx d^2/4\nu \tag{6-3}$$

Here d is the tube diameter and ν is the kinematic viscosity. For the purpose of estimating microscopic transient times in porous media, a practical lower bound on ν is 10^{-2} cm²/sec, and an upper bound on d might be on the order of 10^{-1} cm. This gives a microscopic transient time on the order of 1 sec, and if the transient time for the macroscopic process is much larger (say on the order of minutes), we should be able to treat the flow as quasi-steady and the analysis presented in Sections 4 and 5 will be valid. This means that Equation 5-28 and Equation 4-3 are to be solved subject to boundary conditions of the type

$$\langle \mathbf{v} \rangle = \mathbf{g}(\mathbf{r}, t) \text{ over } S \tag{6-4}$$

where S is the surface of some macroscopic region.

If the left-hand side of Equation 6-1 is not negligible, we reformulate the problem so the governing equation for **v** is

$$\left(\frac{\partial}{\partial t} - \nu\nabla^2\right)\nabla^2\mathbf{v} = 0 \tag{6-5}$$

and the boundary conditions are

$$\mathbf{v} = 0 \quad \text{on } A_i \text{ of } V \tag{6-6}$$

$$\mathbf{v} = \mathbf{f}(\mathbf{r}, t) \quad \text{on } A_e \text{ of } V \tag{6-7}$$

The solution of Equation 6-5, subject to these boundary conditions, is unique (*17*, Chap. 4), thus Equations 6-5, 6-6, and 6-7 are sufficient to determine $\langle \mathbf{v} \rangle$ for the region in question, and we write

$$\langle \mathbf{v} \rangle = \mathbf{g}(\mathbf{r}_0, t) \tag{6-8}$$

Following the development in Section 4, we assume

> Given the governing differential equation (Equation 6-5), the no slip condition (Equation 6-6), and the average velocity (Equation 6-8), the velocity field is uniquely determined A.1′

This assumption has less appeal than assumption A.1, primarily because we do not have a set of special cases for which we know it holds true. Nevertheless, once the assumption is made we can express $\mathbf{v}$ in terms of $\langle \mathbf{v} \rangle$ by the unique mapping, **M**

$$\mathbf{v} = \mathbf{M} \cdot \langle \mathbf{v} \rangle \tag{6-9}$$

Substitution of Equation 6-9 into the governing differential equation yields

$$\left[\left(\frac{\partial}{\partial t} - \nu\nabla^2\right)\nabla^2\mathbf{M}\right] \cdot \langle \mathbf{v} \rangle + \mathbf{M} \cdot \left[\left(\frac{\partial}{\partial t} - \nu\nabla^2\right)\nabla^2\langle \mathbf{v} \rangle\right] = 0 \tag{6-10}$$

Following the arguments given in Section 4, we may conclude that $\nabla^2\langle \mathbf{v} \rangle$ is negligible and **M** must satisfy the equation

$$\left(\frac{\partial}{\partial t} - \nu\nabla^2\right)\nabla^2\mathbf{M} = 0 \tag{6-11}$$

subject to the boundary conditions of

$$\mathbf{M} = 0 \text{ on } A_i \tag{6-12}$$

$$\langle \mathbf{M} \rangle = \mathbf{U} \text{ at } \mathbf{r}_0 \tag{6-13}$$

Since the boundary conditions are independent of time, we conclude that **M** cannot depend on time, but only on the structure of the porous media. This immediately leads us to the conclusion that the time dependence of the point velocity is completely determined by the time dependence of the average velocity—*i.e.*,

$$\mathbf{v} = \mathbf{M} \cdot \mathbf{g}(\mathbf{r}_0, t) \tag{6-14}$$

This, of course, is the quasi-steady flow condition, and it is not reasonable at all for the conditions we are now considering. This unhappy condition is brought about by the assumption A.1′, for in making that assumption the time dependence of Equation 6-7 is replaced by the time dependence of Equation 6-8. We might consider the simplification in going from Equation 6-10 to Equation 6-11 as a possible error in our analysis; however, this simplification is quite reasonable, and is strictly satisfied if $\langle \mathbf{v} \rangle$ is a linear function of the spatial coordinates. As an example, consider the case of one-dimensional, transient, incompressible flow in homogeneous porous media. For such a flow, the average velocity $\langle \mathbf{v} \rangle$ is always independent of the spatial coordinates, being a function of time alone, and Equation 6-11 logically follows from Equation 6-10.

If we were to throw caution to the winds, we might discard assumption A.1′ but retain Equation 6-9, allowing **M** to be a function of time. Under these circumstances we can modify the development given in Section 4 as follows. Beginning at Equation 4-15 we would write

$$\boldsymbol{\lambda} \cdot \nabla P = \mu\boldsymbol{\lambda} \cdot \nabla^2\mathbf{v} - \rho\boldsymbol{\lambda} \cdot \left(\frac{\partial \mathbf{v}}{\partial t}\right) \tag{6-15}$$

Substitution of Equation 6-9, keeping in mind that **M** is a function of time leads to

$$\frac{dP}{ds} = \mu\boldsymbol{\lambda} \cdot (\nabla^2\mathbf{M}) \cdot \langle \mathbf{v} \rangle - \rho\left[\boldsymbol{\lambda} \cdot \left(\frac{\partial \mathbf{M}}{\partial t}\right)\right] \cdot \langle \mathbf{v} \rangle - \rho(\boldsymbol{\lambda} \cdot \mathbf{M}) \cdot \frac{\partial\langle \mathbf{v} \rangle}{\partial t} \tag{6-16}$$

As usual, we assume negligible spatial variation of $\langle \mathbf{v} \rangle$ and $\partial\langle \mathbf{v} \rangle/\partial t$ to obtain

$$P(\mathbf{r}) = P_0 + \mu\left\{\int_{\eta=0}^{\eta=s(\mathbf{r})} \left[\boldsymbol{\lambda} \cdot (\nabla^2\mathbf{M}) - \left(\frac{\boldsymbol{\lambda}}{\nu}\right) \cdot \left(\frac{\partial \mathbf{M}}{\partial t}\right)\right] d\eta\right\} \cdot \langle \mathbf{v} \rangle - \rho\left\{\int_{\eta=0}^{\eta=s(\mathbf{r})} (\boldsymbol{\lambda} \cdot \mathbf{M}) d\eta\right\} \cdot \frac{\partial\langle \mathbf{v} \rangle}{\partial t} \tag{6-17}$$

Here we see that, to within an arbitrary constant, P is given by

$$P = -\mu\mathbf{m} \cdot \langle \mathbf{v} \rangle - \rho\mathbf{b} \cdot \frac{\partial\langle \mathbf{v} \rangle}{\partial t} \tag{6-18}$$

Here we must remember that $\mathbf{m}$ is a function of ν and t, and that $\mathbf{b}$ is a function of t. Rearranging and substituting into Equation 3-14 yield

$$\nabla\left[\frac{1}{V}\int_{V_f}\left(P - b \cdot \rho\frac{\partial\langle v\rangle}{\partial t}\right)dV\right] = -\frac{1}{V}\int_{A_e}\mu m \cdot \langle v\rangle n dA \tag{6-19}$$

This can be expressed as

$$\nabla\langle P\rangle - \nabla\left[\langle b\rangle \cdot \rho\frac{\partial\langle v\rangle}{\partial t}\right] = -\mu\mathbf{K}^{-1} \cdot \langle v\rangle \tag{6-20}$$

If the spatial variation of $\langle b\rangle$ is large compared to the spatial variation of $\partial\langle v\rangle/\partial t$, we could express Equation 6-20 as

$$\nabla\langle P\rangle - \mathbf{B} \cdot \rho\frac{\partial\langle v\rangle}{\partial t} = -\mu\mathbf{K}^{-1} \cdot \langle v\rangle \tag{6-21}$$

The existence of an inverse for $\mathbf{K}^{-1}$ is not so easy to argue in favor of as it was in the case of steady flow, nevertheless we will assume it exists and write Equation 6-21 as

$$\langle v\rangle = -\frac{1}{\mu}\left[\mathbf{K} \cdot \nabla\langle P\rangle - (\mathbf{K}^t) \cdot \rho\frac{\partial\langle v\rangle}{\partial t}\right] \tag{6-22}$$

where $\mathbf{K} \cdot \mathbf{B}$ has been replaced by $\mathbf{K}^t$. It must be remembered that both $\mathbf{K}$ and $\mathbf{K}^t$ are time dependent. Several workers (*19, 23, 24, 34*) have suggested forms similar to this, but have assumed that $\mathbf{K}^t = \mathbf{K}$ which requires that $\mathbf{B}$ be equivalent to the unit tensor. It should be remembered that two assumptions were required to arrive at Equation 6-22.

A unique time-dependent mapping $\mathbf{M}$ exists which maps $\langle v\rangle$ into v A.5

The time-dependent tensor $\mathbf{K}^{-1}$ has an inverse A.6

Further theoretical work and experimental studies are certainly needed to confirm or reject the intuitive development outlined here. If assumptions A.5 and A.6 prove to be satisfactory, the principle of material frame-indifference can again be invoked to put $\mathbf{K}$ and $\mathbf{K}^t$ in the simplified form suggested by Equation 5-28.

7. Inertial Effects

In considering inertial effects, we restrict our analysis to incompressible flows so that the governing equations are

$$\rho\left(\frac{\partial v}{\partial t} + v \cdot \nabla v\right) = -\nabla P + \mu\nabla^2 v \tag{7-1}$$

$$\nabla \cdot v = 0 \tag{7-2}$$

The volume averaged continuity equation quickly becomes

$$\nabla \cdot \langle v\rangle = 0 \tag{7-3}$$

while the volume averaged Navier-Stokes equations may be written as

$$\left(\frac{\partial\langle v\rangle}{\partial t} + \langle v \cdot \nabla v\rangle\right) = -\nabla\langle P\rangle + \frac{1}{V}\int_{A_i} PndA + \mu\langle\nabla^2 v\rangle \tag{7-4}$$

We can define a new velocity $\tilde{v}$ which represents the deviations of the local point velocity from the local average velocity

$$v = \langle v\rangle + \tilde{v} \tag{7-5}$$

Substituting this expression into Equation 7-4 and noting $\langle\langle v\rangle\rangle = \langle v\rangle$ and therefore $\langle\tilde{v}\rangle = 0$, we find

$$\rho\left(\frac{\partial\langle v\rangle}{\partial t} + \langle v\rangle \cdot \nabla\langle v\rangle\right) = -\nabla\langle P\rangle + \mu\langle\nabla^2 \mathbf{v}\rangle + \frac{1}{V}\int_{A_i} PndA - \rho\nabla \cdot \langle\tilde{v}\tilde{v}\rangle \tag{7-6}$$

Assuming that $\langle v\rangle = 0$ if, and only if, v is everywhere equal to zero, a mapping of $\langle v\rangle$ into v exists which cannot be linear nor is it likely to be unique, and we can write

$$v = \mathbf{M} \cdot \langle v\rangle \tag{7-7}$$

Expressing the last two terms in Equation 7-6 as a force f, we can write

$$\rho\left(\frac{\partial\langle v\rangle}{\partial t} + \langle v\rangle \cdot \nabla\langle v\rangle\right) = -\nabla\langle P\rangle + \mu(\nabla^2\mathbf{M}) \cdot \langle v\rangle + f \tag{7-8}$$

If $\langle v\rangle = 0$ implies $f = 0$ we can map $\langle v\rangle$ into f by the expression

$$f = \mathbf{A} \cdot \langle v\rangle \tag{7-9}$$

and rewrite Equation 7-8 in the form

$$\rho\left(\frac{\partial\langle v\rangle}{\partial t} + \langle v\rangle \cdot \nabla\langle v\rangle\right) = -\nabla\langle P\rangle + \mathbf{R} \cdot \langle v\rangle \tag{7-10}$$

Here we have written $\mathbf{R} = \mu(\nabla^2\mathbf{M}) + \mathbf{A}$ where $\mathbf{R}$ is a total resistance tensor. It will depend upon the structure of the porous media, the local velocity, and perhaps velocity gradients, in addition to the viscosity and density of the fluid. In general we would expect the components of $\mathbf{R}$ to be proportional to the viscosity at low Reynolds numbers and proportional to the density and velocity at high Reynolds. This generalized type of approach to the analysis of high Reynolds number flows does not appear to have been considered previously, probably because such flows are usually one-dimensional and one need only determine a friction factor or drag coefficient to characterize the macroscopic flow field.

Conclusions

The development presented in this paper has as the primary objective a careful listing of the assumptions and restrictions that must be imposed if the point equations describing steady, incompressible, creep flow in a rigid porous media can be integrated over an averaging volume to produce Darcy's law. The functions to be averaged must behave in a manner such that the following restriction is satisfied

$$d \ll l \ll L$$

and the averaging must be performed in such a way that

The averaging volume is constant and maintains a fixed orientation relative to an inertial frame

Here d, l, and L represent characteristic lengths for: (1) the structure of the porous media, (2) the averaging volume, and (3) the macroscopic process. With these restrictions placed on the averaging process, Darcy's law can be derived from the point equations on the basis of two assumptions

$$\mathbf{v} \text{ is a unique function of } \langle \mathbf{v} \rangle$$
$$\nabla \langle P \rangle = 0 \text{ implies } \langle \mathbf{v} \rangle = 0$$

On the basis of two additional assumptions

The anisotropic nature of a porous media can be uniquely represented by a single orientation vector, $\boldsymbol{\omega}$

Darcy's law must satisfy the principle of material indifference

one can prove that the permeability tensor is symmetric and takes the form

$$K_{ij} = B_{(1)}\delta_{ij} + B_{(2)}\omega_i\omega_j$$

The analysis of transient flow indicates that the relationship between velocity, pressure gradient, and local acceleration is an extremely complex one, and further theoretical study is obviously needed.

Nomenclature

Roman Letters

A = surface of the averaging volume, cm²
A_f = surface of the fluid volume contained within the averaging volume, cm²
A_e = area of entrances and exits of A_f, cm²
A_i = area of solid-fluid interface of A_f, cm²
$\mathrm{A}_{(j,k)}$ = an averaging area parallel to the (j, k) plane, cm²
A_f = fluid area contained within $\mathrm{A}_{(j,k)}$, cm²
d = characteristic length for the structure of the porous media, cm
$\mathbf{g}$ = gravity vector, cm/sec²
$\mathbf{K}$ = permeability tensor, cm²
$\tilde{\mathbf{K}}$ = dimensionless permeability tensor
$\mathbf{K}^{-1}$ = inverse of the permeability tensor, cm⁻²
k = isotropic permeability, cm²
K_{ij} = permeability tensor in index notation, cm²
L = characteristic length for the macroscopic process, cm
l = characteristic length for the averaging volume, cm
$\mathbf{M}$ = the tensor mapping $\langle \mathbf{v} \rangle$ into $\mathbf{v}$
$\mathbf{m}$ = a vector relating the average pressure to the average velocity, cm⁻¹
$\mathbf{n}$ = outwordly directed unit normal for V and V_f
p = pressure, dyne/cm²
p_0 = reference pressure, dyne/cm²
$\mathbf{p}$ = position vector relative to the centroid of the averaging volume, cm
P = piezometric pressure, dyne/cm²
$\mathbf{r}$ = position vector, cm
$\mathbf{r}_0$ = position vector locating the centroid of the averaging volume, cm
$\bar{\mathbf{r}}$ = position vector locating the centroid of the fluid volume contained within the averaging volume, cm
$\mathbf{R}$ = resistance tensor, dyne-sec/cm⁴
s = arc length, cm
S = surface area of a macroscopic region, cm²
$\mathbf{U}$ = unit tensor
$\mathbf{u}$ = constant porous media velocity, cm/sec.
V = averaging volume, cc
V_f = fluid volume contained within the averaging volume, cc
$\mathbf{v}$ = velocity vector, cm/sec
$\langle \mathbf{v} \rangle$ = volume averaged velocity vector, cm/sec
x_i = position vector for rectangular Cartesian coordinates in index notation, cm

Greek Letters

$\alpha(\mathbf{x})$ = void volume distribution function
$\langle \alpha \rangle$ = volume porosity
$\bar{\alpha}^{(j,k)}$ = plane porosity for the (j, k) plane
ρ = fluid density, g/cc
φ = gravitational potential function, cm²/sec²
Φ = volume porosity $(= \langle \alpha \rangle)$
$\boldsymbol{\psi}$ = any tensor valued function associated with the fluid
$\boldsymbol{\lambda}$ = unit tangent vector to an arbitrary curve, unit orientation vector
ν = kinematic viscosity, cm²/sec
μ = coefficient of viscosity, dyne sec/cm²
$\boldsymbol{\omega}$ = orientation vector
$\boldsymbol{\Omega}$ = skew-symmetric orientation tensor
$\boldsymbol{\xi}$ = reference position vector, cm

REFERENCES

(1) Aranow, R. H., "Statistical approach to flow through porous media," *Phys. Fluids*, **9**, 1721 (1966).
(2) Aris, Rutherford, "Vectors, Tensors, and the Basic Equations of Fluids Mechanics," 1962, Prentice-Hall, Englewood Cliffs, N. J.
(3) Beran, M. J., "Statistical Continuum Theories," 1968, Interscience, New York, N. Y.
(4) Bird, R. B., Stewart, W. E., and Lightfoot, E. N., "Transport Phenomena," 1960, Wiley & Sons, Inc., New York, N. Y.
(5) Brenner, Howard, "Elements of transport processes in porous media," to be published as a monograph by Springer-Verlag.
(6) Brenner, Howard, "The Stokes resistance of an arbitrary particle," *Chem. Engr. Sci.*, **18**, 1 (1963).
(7) Darcy, Henry, "Les Fontaines Publiques de la Ville de Dijon," 1856, Victor Dalmont, Paris.
(8) Ericksen, J. L., "Anisotropic fluids," *Archive Rat. Mech. Anal.*, **4**, 231 (1959–60).
(9) Ericksen, J. L., "General solutions in the hydrostatic theory of liquid crystals," *Trans. Soc. Rheol.*, **11**, 5 (1967).
(10) Ericksen, J. L., "Tensor fields," "Handbuch der Physik," 1960, Vol. III, Pt. 1, Springer-Verlag, Berlin.
(11) Ferrandon, J., *Le Genie Civil*, **125**, 24 (1948).
(12) Green, A. E., and Zerna, W., "Theoretical Elasticity," 1963, Oxford Press.
(13) Halmos, P. R., "Finite-Dimensional Vector Spaces," 2nd ed., 1958, D. van Nostrand Co., Princeton, N. J.
(14) Hildebrand, F. B., "Methods of Applied Mathematics," 1952, Prentice-Hall, Englewood Cliffs, N. J.
(15) Hubbert, M. K., "Darcy's law and the field equations of the flow of underground fluids," *A.I.M.E. Petrol. Trans.*, **207**, 222 (1956).
(16) Kaplan, Wilfred, "Advanced Calculus," 1952, Addison-Wesley, Reading, Mass.
(17) Ladyzhenskraya, O. A., "The Mathematical Theory of Viscous Incompressible Flow," 1963, Gordon and Breach, New York, N. Y.
(18) Landau, L. D., and Lifshitz, E. M., "Fluid Mechanics," 1959, Addison-Wesley, Reading, Mass.
(19) Lapwood, E. R., *Proc. Camb. Phil. Soc.*, **44**, 508 (1948).
(20) Massarani, Giulio, "Criticism on the validity of the law of Darcy in heterogeneous media," *Compt. Rend.*, **265**, 19, 587a (1967).
(21) McConnell, A. J., "Applications of Tensor Analysis," 1957, Dover Pub., New York, N. Y.
(22) Noll, W., "On the continuity of the solid and fluid states," *J. Rat. Mech. Anal.*, **4**, 3 (1955).
(23) Orovy Ann, T. and Sil'vyann, E., *Rev. Mecan. Appl.*, **5**, 215 (1960).
(24) Paria, G., *Appl. Mech. Rev.*, **16**, 421 (1963).
(25) Polubarinova-Kochin P. Ya., "Theory of Ground Water Movement," 1960, Princeton University Press, Princeton, N. J.
(26) Poreh, M. and C. Elata, "Analytical Derivation of Darcy's Law," *Israel J. Tech.*, **4**, 214 (1966).
(27) Scheidegger, A. E., "Handbuch der Physik," 1960, Vol 3, Part 1, Springer-Verlag, Berlin.
(28) Scheidegger, A. E., "The Physics of Flow Through Porous Media," 1960, Macmillan, New York, N. Y.
(29) Scheidegger, A. E., "Statistical theory of flow through porous media," *Trans. Soc. Rheol.*, **9**, 313 (1965).
(30) Slattery, J. C., "Flow of viscoelastic fluids through porous media," *A.I.Ch.E. J.*, **13**, 1066 (1967).
(31) Slattery, J. C., "Single-phase flow through porous media," in press, *A.I.Ch.E. J.*
(32) Taylor, A. E., "Advanced Calculus," 1955, Ginn and Co., New York, N. Y.
(33) Truesdell, C., "The Elements of Continuum Mechanics," 1966, Springer-Verlag, New York, N. Y.
(34) Truesdell, C., and Noll, W., "The non-linear field theories of mechanics," "Handbuch der Physik," 1965, Vol. 3, Pt. 3, Springer-Verlag, Berlin.
(35) Truesdell, C., and Toupin, R., "The classical field theories," *ibid.*, 1960, Vol. 3, Pt. 1.
(36) Whitaker, S., "Diffusion and dispersion in porous media," *A.I.Ch.E. J.*, **13**, 420 (1967).
(37) Whitaker, S., Document 9234 deposited with the American Documentation Institute, Photoduplication Service, Library of Congress, Washington 25, D. C.
(38) Whitaker, S., "The equations of motion in porous media," *Chem. Engr. Sci.*, **21**, 291 (1966).
(39) Whitaker, S., "Introduction to Fluid Mechanics," 1968, Prentice-Hall, Englewood Cliffs, N. J.

3

Anisotropic Permeability in Porous Media

Anisotropic Permeability in Porous Media

Measurements of resistivity are clues to anisotropy

PHILIP A. RICE DANIEL J. FONTUGNE
RAIMONDO G. LATINI ALLEN J. BARDUHN

Methods of measurement of directional permeabilities in anisotropic porous beds have been reviewed for both consolidated and unconsolidated media. Anisotropic permeability data on consolidated beds and unconsolidated beds are summarized here, and the relationships among the bed formation process, the bed structure, and the observed anisotropic permeabilities are discussed. Anisotropy is generally a result of the orientation and shape of the asymmetric grains making up the porous bed. Measurements indicate that the extent of anisotropy can be approximately predicted by measurements of resistivity in different directions. Several methods for predicting the potential and stream functions in anisotropic porous media are also cataloged.

If the permeability of a porous medium varies with the direction of fluid flow or pressure gradient, the permeability is anisotropic. This variation in permeability with direction reflects the differences in path length through which a fluid element moves and the forces it experiences in moving through the porous media in a given direction. On the other hand, the permeability of a purely anisotropic media will not vary with position—*i.e.*, the media is homogeneous. The directional nature of the permeability can be expressed as a symmetric second-order tensor with six independent components.

$$\bar{\bar{K}} = \begin{bmatrix} K_{11} & K_{12} & K_{13} \\ K_{12} & K_{22} & K_{23} \\ K_{13} & K_{23} & K_{33} \end{bmatrix} \quad (1)$$

This term can be reduced to three principal components by choosing a coordinate system which corresponds to the principal axes of the bed. The permeability in any direction is then given by the relation

$$\frac{1}{K} = \frac{\cos^2 \delta_1}{K_1} + \frac{\cos^2 \delta_2}{K_2} + \frac{\cos^2 \delta_3}{K_3} \quad (2)$$

where K_1, K_2, and K_3 are the three principal components of permeability and δ_1, δ_2, and δ_3 are the angles of the

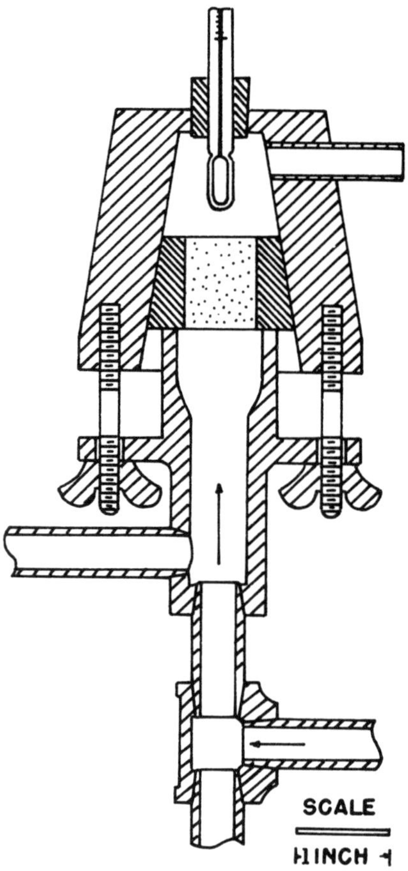

Figure 1. Apparatus of Fancher et al.

AUTHORS *Philip A. Rice is an Assistant Professor, and Allen J. Barduhn is a Professor in the Department of Chemical Engineering, Syracuse University, Syracuse, N. Y. 13210. Daniel Fontugne and Raimondo Latini are former graduate students of the above department.*

direction cosines of the direction of flow.

If the inertial effects of the flow are negligible, the velocity vector for the flow is given by Darcy's Law.

$$\bar{v} = \frac{\bar{\bar{K}}}{\mu}\,\overline{\text{grad}\ \phi} \tag{3}$$

where $\bar{v}$ is the velocity vector for flow
μ is the fluid viscosity
ϕ is the hydrodynamic potential

This relation was first derived by Ferrandon (*7*) for the capillary model of a porous medium.

The differences in directional permeabilities depend on the shape, orientation, and concentration of the particles which form the bed or, for a consolidated medium, were consolidated into a porous bed. Beds in which the orientation, size, and shape of the flow channels are randomly oriented would be expected to be isotropic—*i.e.*, show no variation in permeability with direction. Beds in which the size and shape of the flow channels are different on the average in different directions would be expected to be anisotropic. In unconsolidated particulate beds or in naturally occurring sedimentary structures, the orientation of the flow channels often results from the orientation of the particles making up the bed. The orientation of particles reflects in turn the circumstances surrounding the formation of the bed.

In this paper we will discuss methods of measuring anisotropic permeability in consolidated and unconsolidated porous media and will summarize the anisotropic permeability data of several investigators. In addition, work relating the circumstances of bed formation to anisotropic permeability of the porous medium will be described and the prediction of anisotropic permeabilities from knowledge of the structural parameters discussed. Finally, a short summary of mathematical methods useful for the prediction of flow in anisotropic porous media is included.

Measurement of Anisotropy

General. Measurements of anisotropic permeabilities are made by simultaneously measuring the flow rate and pressure drop for one-, two-, or three-dimensional flow. Usually the measurements are assumed to be made along the principal axes of the specimen—*i.e.*, the directions which characterize the three principal components of the permeability tensor. In practice this generally corresponds to making the measurements perpendicular to the bedding plane and at various angles parallel to the bedding plane.

In general, the directional permeability could be measured in any direction, but Scheidegger (*37*) has pointed out that these measurements can be made in two ways with two different results. If the flow is confined to the direction of interest and the component of the pressure gradient measured in that direction, the measured permeability is $K_n = 1/\bar{n}\bar{\bar{K}}^{-1}\bar{n}$, where $\bar{n}$ is the vector representing the direction of flow. On the other hand, if the pressure gradient is fixed and the component of flow in the direction of the pressure gradient is measured, then the permeability measured is $K_n = \bar{n}\bar{\bar{K}}\bar{n}$, where $\bar{n}$ is the vector representing the direction of the pressure gradient. These measured permeabilities are not the same. Only when the measurements are made in the direction of the principal axes of the media are the pressure gradient vector and the flow vector in the same direction and the measurement of permeability unambiguous. Fortunately the actual difference between the permeabilities observed by either method is often small.

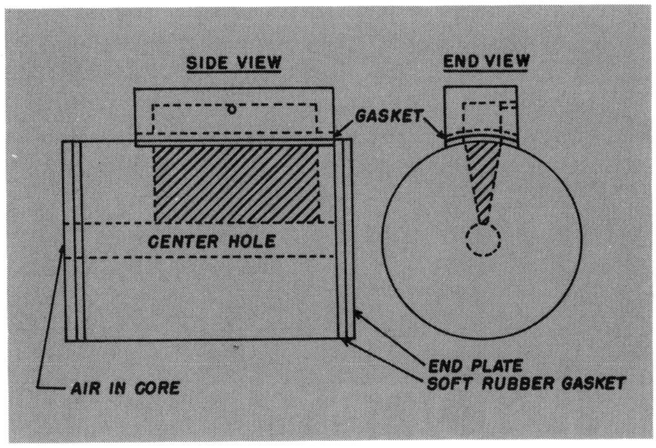

Figure 2. Apparatus of Johnson and Hughes

Marcus (*22*) and Liakopoulos (*20*) have derived methods interpreting two- and three-dimensional permeabilities measured in directions other than the principal direction. In general, depending on the length to width ratio of the measurement cell, neither the pressure gradient nor the flow vector is in the direction over which the measurements are made. Nonetheless, Marcus shows that if the angles the principal axes make with the angle of measurement and the geometry of the sample are known, the principal components of the permeability can be determined. Both authors discuss ways of determining the direction of the velocity vector if the direction of pressure gradient vector is known and *vice versa*. To avoid these difficulties, most measurements are made parallel or perpendicular to the bedding plane with the direction perpendicular to the bedding plane being assumed to be one of the principal directions. Measurements are made at all angles parallel to the bedding plane and the maximum and minimum permeabilities in this plane, if perpendicular, are assumed to be the principal permeability components in that plane. The direction of the bedding plane, however, does not necessarily have to correspond to a principal direction of the specimen, a situation which complicates the interpretation of the measurements.

Measurements in consolidated media. The most direct way of measuring directional permeability in consolidated media is to cut a sample in such a way that the flow rate and pressure drop in the desired direction can be measured. One of the first apparatus of this type (Figure 1) was built and used by Fancher (*6*). Cylindrical samples were cut parallel or perpendicular to the bedding plane and inserted into a bored rubber stopper which constituted a holder (see Figure 1). Fluid entered from the bottom, passed through the sample and left through a flowmeter near the top. Manometers near the top and bottom were used to measure the pressure drop. The permeability is given by Darcy's Law to be

$$K = \frac{\mu v L}{\Delta P}$$

Johnson and Hughes (*16*) designed an apparatus for measuring permeabilities as a function of angle in the bedding plane. In their apparatus (Figure 2) a hole was bored axially through the center of the sample perpendicular to the bedding plane to make an annular cylinder. The sample was then clamped in position, and gas was forced into the center hole to give radial flow. A gas collecting head, which fitted tightly over the sample, was used to collect the gas which left the sample at a given angle. Pressure drop and flow rates were measured with manometers and a flowmeter, respectively. Measurements made in both these apparatus prove to be reproducible, and unchanging with time.

Maasland and Kirkham (*21*) have described an *in situ* test which can be used for determining the anisotropic permeabilities in soil clods. In this method, the time necessary for the pressure to fall in a pressurized air tank is recorded after an inlet tube from the tank is connected to a soil clod in two or three mutually perpendicular positions. From the time lag measurements and the form factors derived by Maasland and Kirkham (which depend on the anisotropic nature of the sample), the principal components of the permeability tensor can be determined.

Snow (*33*) has extended Maasland and Kirkham's work and developed a three-hole pressure test for determining anisotropic permeabilities. Discharge rates are measured from the three holes oriented orthogonally, and from the discharge rates from these holes, the three principal components of the permeability tensor are calculated, assuming that the principal directions parallel to the bedding plane can be determined.

Weeks (*34*) also has derived a relationship between the pressure drawdown curve of a discharge hole and the ratio of radial to vertical permeabilities for aquifers of different thicknesses. Accurate determination of the permeability ratio by this method requires pressure measurements of an even higher order of accuracy.

Measurements in unconsolidated media. Because of the lack of structural rigidity in unconsolidated porous media, the permeability measurement methods used for consolidated media, which require core samples to be taken, cannot be used with unconsolidated media. In unconsolidated media the permeability measurements must be taken on the bed *in situ*, since samples of the bed at various orientations with respect to the bedding plane cannot be taken without altering significantly the structure of the bed. The *in situ* methods discussed above (Maasland and Kirkham, Snow, and Weeks) could be used for large unconsolidated porous media, but to study small samples of unconsolidated media requires that the bed be formed in the same device in which the directional permeability measurements are made.

Two types of two-dimensional permeability apparatus

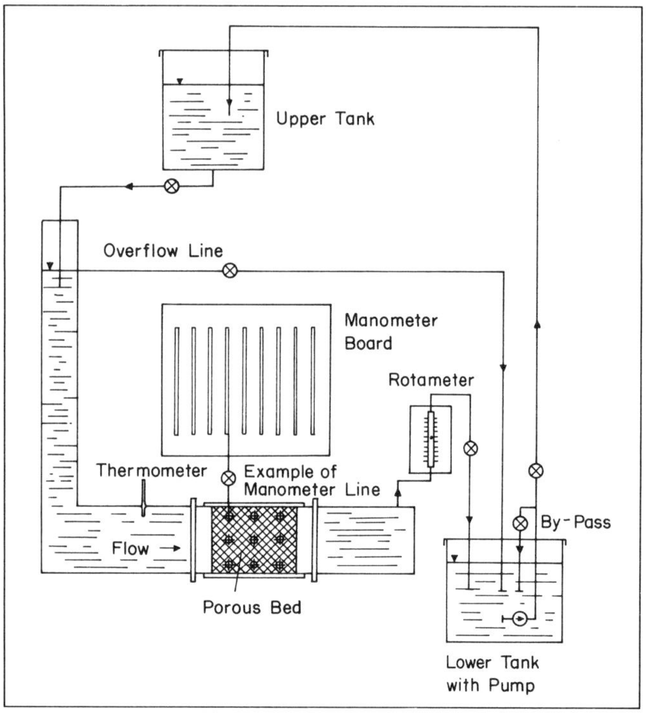

Figure 3. Latini's permeability apparatus—horizontal flow

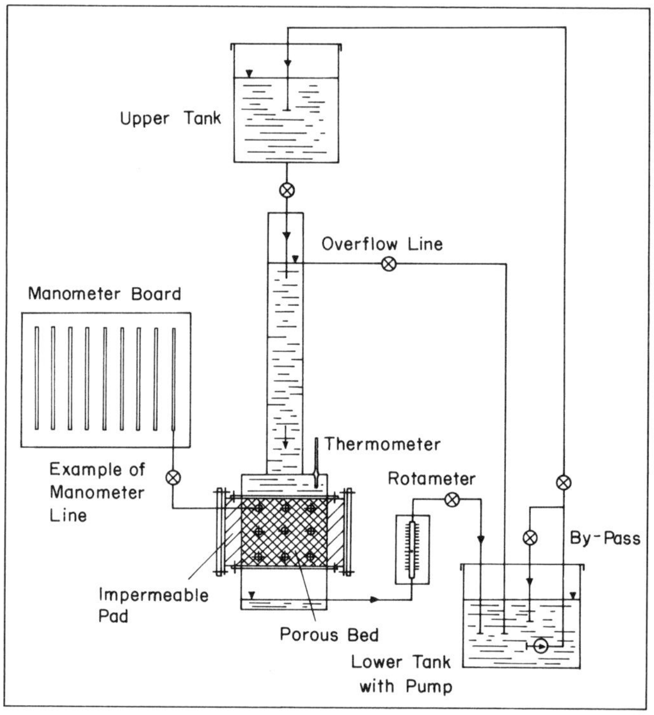

Figure 4. Latini's permeability apparatus—vertical flow

for unconsolidated media have been constructed in the Department of Chemical Engineering at Syracuse University. The first, built by Latini (*19*) and shown in Figures 3 and 4, allows flow and pressure in two orthogonal directions to be measured separately. It consists of a rectangular box 4″ × 4″ × 5″ with four walls made of porous stainless steel, one of which can be removed to form the bed. The two other walls are made of solid stainless steel and have fittings for manometers. Either pair of opposite porous walls can be blocked off with soft, flat, impermeable pads. It is thus possible to obtain one-dimensional flow in either the vertical direction in which the bed was formed or in a horizontal direction.

The crucial step in the design of the box is the choice of the porous walls. They are available in a wide range of thicknesses and permeabilities, and both these quantities have an influence on the regularity of the flow that will be attained. Since the impermeable pads are pressed against the outer surfaces of the walls, part of the total flow rate will flow in the porous walls instead of flowing through the bed. This anomaly not only causes an error in the experimental value of the flow rate, but it also gives origin to a nonlinear flow in the region of the bed close to the bed entrance and exit. Therefore, the porous plates should have the smallest possible thicknesses, within the requirements for mechanical resistance. For this apparatus the thickness was 3/16 in.

Secondly, it would be desirable to make the walls very permeable to reduce the pressure drop normal to the face. Unfortunately, the more permeable the walls, the larger will be the fraction of flow rate passing through them parallel to the faces. Thus, a compromise is necessary. For walls 3/16 in. thick, the flow was sufficiently uniform if the permeability of the walls does not exceed the permeability of the bed. Boxes with porous walls of different permeabilities are necessary for measurements on beds of different permeabilities. To use this apparatus, the box is first completely filled, usually by natural settling of the particles through the fluid. The box is covered with the removable porous wall or Plexiglass plate avoiding air pockets between the porous plate and the solid particles. The permeability is then measured by determining the flow rates and pressure drops for flow in both the horizontal and vertical directions.

Latini's apparatus has several disadvantages. For accurate measurements, corrections were made for the parallel flow through the permeable walls.
A rather complicated procedure for switching the flow direction is required, and the allowable range of permeabilities which can be measured in a single apparatus is limited to about 1/3 to 3 times the permeability of the walls.

The second Syracuse apparatus shown in Figure 5 was built by Fontugne (*8*; Rouher *29*) and employs simultaneous two-dimensional flow. It eliminates the necessity of porous walls and enlarges the range of permeability which can be measured. This cell has a rectangular outlet on either side and on the bottom. The related mathematical problem for flow in this system is two-dimensional. Horizontal and vertical flows are collected separately from two overflow tanks. The difference of liquid heads between the inlet and the two outlets is the same. Then, the ratio between horizontal and vertical flow rates, together with the solution of the flow equation, yields the ratio between horizontal and vertical permeabilities. The procedure for forming the bed is essentially the same as Latini's.

The flow equations for the cell were solved numerically, and the flow pattern in the cell was obtained for various degrees of anisotropy. An example of a flow pattern is given in Figure 6. Equipotential lines and streamlines are represented for one-half of the cell in the case where the horizontal permeability is twice the vertical permeability. The ratio between horizontal and vertical flow rates can be deduced from the location of the particular streamline, which separates the flow to the two outlets. The ratio between horizontal and vertical permeabilities is plotted as a function of the ratio between horizontal and vertical flow rates in Figure 7. The calculated values were checked for a bed of isotropic glass beads by comparing the observed and calculated flow ratios. The experimental points for glass beads are shown on Figure 7.

During a run the vertical permeability is first mea-

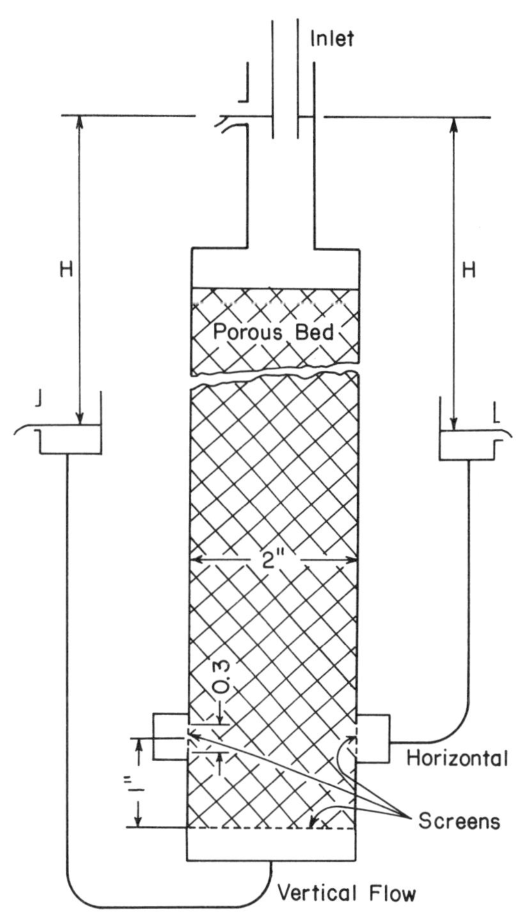

Figure 5. Fontugne's permeability apparatus

sured with the outlets clamped to the horizontal screen. Then the side outlets are unclamped and the ratio of flow rates is measured. This ratio of flow rates is then related to the ratio of permeabilities by Figure 7.

Fontugne's apparatus can be extended to three-dimensional permeability measurements by placing the outlets at right angles instead of across from one another. To measure the permeability in one horizontal direction, the other horizontal screen outlet is closed and *vice versa.*

Directional Permeability Measurements

Directional permeability measurements most often have been made on oil sands because of the interest in recovering oil from these sands and in sand, salt, and clay aquifers because of their water content. Directional permeabilities in unconsolidated porous media have been measured for soils and for beds of crushed limestone and sand. Knowledge of the anisotropy in unconsolidated media is useful for predicting aeration and irrigation rates in soils and more recently for the prediction of washing rates in ice crystal counterwasher columns in the freezing process for desalting sea water (*2*).

Measurements on consolidated media. An early extensive set of measurements on oil sands has been published by Fancher, Lewis, and Barnes (*6*). They include 65 pairs of data in which the permeability, both perpendicular and parallel to the bedding plane, were measured. In about two thirds of these samples, the permeability parallel to the bedding plane was higher than the permeability normal to it; the ratio of the two values ranging as high as 42. In the cases where the normal permeability was higher than the parallel permeability, the maximum ratio was 7.3 These measurements were made by cutting and measuring the permeabilities of cores cut parallel and perpendicular to the bedding plane.

Johnson and Hughes (*16*) have measured permeability as a function of the angular direction in some consolidated sands. The values of the permeability parallel to the bedding plane varied 25% to 30% depending on the angular direction. Scheidegger (*32*) has plotted $\sqrt{1/K}$ as a function of angle in the bedding from Johnson and Hughes data, and has shown that this plot forms an ellipse as would be expected on the basis of Ferrandon's theory.

More recently anisotropic permeability measurements have been made on water bearing deposits in Central California by Johnson, Moston, and Morris (*14*). These deposits were generally fine-textured materials consisting of mixtures of sand, salt, and clay. Samples taken in three different areas had permeabilities varying from 3.8×10^{-14} cm^2 to 3.5×10^{-7} cm^2 with average of

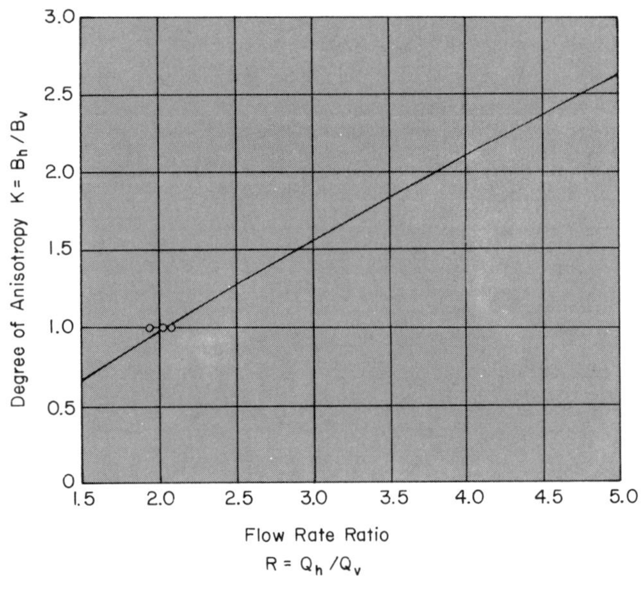

Figure 6. Fontugne's calibration curve

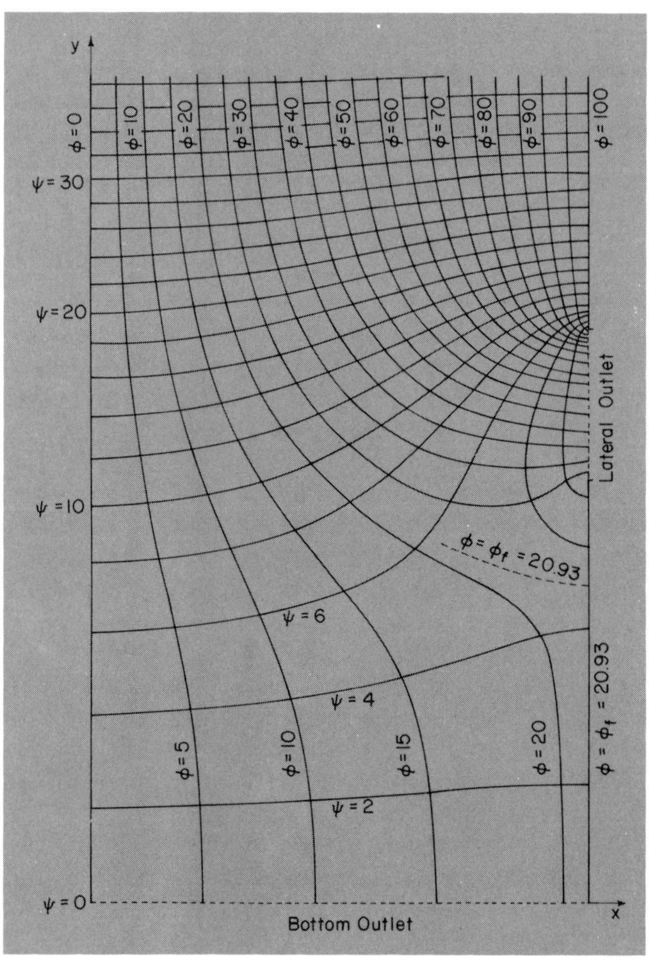

Figure 7. Fontugne's flow pattern ($K_x = 2\,K_y$)

the ratio of the horizontal to vertical permeabilities varying from 1.4 in the Tulane-Wasco area to 2.7 in the Los Banos-Kettleman City area to 7.0 in the Santa Clara Valley.

Greenkorn, Johnson, and Shallenberger (*9*) measured the 142 samples of a sandstone reservoir. They found that 81 of the samples showed directional differences in major and minor axis permeabilities of more than 5%. Twenty-nine of these samples had directional differences in permeability of more than 20%.

Other measurements of anisotropic permeability on oil sandstones have been made by Griffiths (*10*), Hutta and Griffiths (*13*), Hutta (*12*), and Johnson and Breston (*15*). They found differences in permeability in different directions parallel to the bedding plane in many samples, as well as differences in permeability between flow perpendicular and parallel to the bedding plane.

Measurements on unconsolidated media. Two different sets of directional permeability measurements on two different pieces of apparatus have been made by Latini (*19*) and Fontugne (*8*) at Syracuse. Latini measured the permeabilities of samples of sand and crushed limestone normal and perpendicular to the bedding plane (horizontal) after natural settling and after shaking the bed until it attained a minimum volume. Permeabilities were measured separately with the entire flow being either horizontal or vertical. His data (Table I) show no significant anisotropy in the beds of naturally settled sand particles (particles settled in still water), but the vertical permeabilities were about 20% less than the horizontal permeabilities for naturally settled crushed limestone samples. Mechanically shaking the bed always decreased the permeability in each direction, and generally reduced the degree of anisotropy.

Fontugne measured two-dimensional permeabilities on samples similar to Latini's in an apparatus in which flow in two directions took place simultaneously. His data are shown in Table II. Fontugne found a horizontal permeability in the sand sample with smaller particle size to be about 20% larger than the vertical permeability. Also for the larger diameter limestone bed, Fontugne observed a vertical permeability higher than the horizontal permeability by about 20%. Fontugne (*8*) also measured horizontal to vertical permeability ratios of mica particles and of activated carbon particles. The large permeability ratio was not surprising for the mica particles because they are very flat and flake-like, leading to marked directional differences in the bed structure. For the activated carbon particles this large anisotropy is probably due to the planar structure of the graphitic carbon particles.

The equivalent diameter calculated from the Carman-Kozeny formula using the vertical permeability is of the same order of magnitude as the volume surface mean diameter for the sand and limestone samples, but is more than the order of magnitude less for the mica and activated carbon samples. The volume surface mean diameter has little meaning for these samples because of the high degree of anisotropy.

Relationship of Anisotropic Permeability to Bed Structure

The anisotropy of a porous medium is undoubtedly related to internal structure of the medium and the structure is related to the circumstances under which the bed was formed. A porous medium may be expected to be anisotropic if the elementary particles from which the bed was formed are asymmetric in shape and have, on the average, a particular orientation.

Marked anisotropy can occur also if a sedimentary rock contains thin breaks parallel to the bedding plane. In these cases, the permeability perpendicular to the bedding plane may be close to zero and consequently many times less than the permeability parallel to the bedding plane. This type of bed, with the permeability perpendicular to the bedding plane, is determined largely by the shale while the permeability parallel to the bedding plane is determined by the sediment. Even beds of symmetric particles can be anisotropic if they are packed in certain regular patterns. In naturally occurring beds, however, the packing pattern is sufficiently random so that anisotropy does not occur in beds made up of symmetric particles. The size distribution of the particles making up the bed also can affect the magnitude of the permeability, but ordinarily does not contribute to the anisotropy.

For beds made up of randomly oriented symmetric particles, the Kozeny-Carman equation (*3*) based on the capillary model of a porous media

$$K = \frac{\epsilon^3}{(1 - \epsilon)^2} \frac{d_e^2}{36\, k}$$

with k equal to 5, predicts quite well the permeability if the particles are approximately the same size. The hydraulic radius of the bed, d_e, can be estimated from measurements of pore-size distribution [*e.g.*, Wyllie and Gregory (*35*)]. The porosity which must be used in this equation is the "effective porosity"—*i.e.*, the porosity which reflects only the interconnecting flow channels.

For flat asymmetric-shaped particles such as disks and cylinders, Wyllie and Gregory (*35*) have found that the Kozeny constant, k, is higher than 5. Fontugne also calculated this constant for the vertical flow in his beds of sand, limestone, mica, and activated carbon using the Sauter mean diameter determined by sieve analysis. For the anisotropic limestone and sand particles, the k for the vertical permeability varied from about 14 to 18. For the mica and activated carbon particles, the k for the vertical permeability was much higher on the order of 10^5 for mica and 2000 for carbon particles. When the Sauter mean diameter is used, changes in k reflect changes in the tortuosity, but particularly for mica and activated carbon particles, the reduction in permeability probably reflects a reduction in the mean pore diameter as well. In fact, the mini-

Table I. Directional Permeabilities–Unidirectional Flow (*19*)

Sample	Porosity ε	Vertical permeability $B_v \times 10^7$ $(cm^2) \times 10^7$	Horizontal permeability $B_h \times 10^7$ $(cm^2) \times 10^7$	Permeability ratio B_h/B_v	Equivalent diameter d_e, μ[a]	Volume surface mean diameter d_p, μ[b]
Sand, sample 1	0.383[c]	5.1	5.4	1.057	254	320
	0.441[d]	9.6	10.1	1.052	254	
Sand, sample 2	0.381[d]	0.760	0.793	1.042	100	230
Limestone, sample A	0.463[c]	9.9	11.0	1.111	230	376
	0.508[d]	14.4	17.4	1.209	220	
Limestone, sample B	0.451[c]	17.8	19.7	1.105	326	547
	0.499[d]	30.5	36.5	1.196	334	
Limestone, sample C	0.460[c]	25.1	28.1	1.120	369	711[e]
	0.501[d]	41.0	50.5	1.230	383	

[a] d_e, calculated by Carman-Kozeny formula using vertical permeability.
[b] d_p, obtained from sieve analysis.
[c] Bed formed by natural settling followed by shaking to minimum volume.
[d] Bed formed by natural settling only.
[e] All values measured by Fontugne except this one.

Table II. Directional Permeabilities–Two-Dimensional Flow (*8*)

Sample	Porosity ϵ	Vertical permeability $B_v \times 10^7$ $(cm^2) \times 10^7$	Horizontal permeability $B_h \times 10^7$ $(cm^2) \times 10^7$	Permeability ratio B_h/B_v	Equivalent diameter d_e, μ[a]	Volume surface mean diameter d_p, μ[b]
Sand, sample 1	0.422[c]	5.21	4.85	0.93	204	320
	0.477[d]	7.80	8.54	1.10	220	
Sand, sample 2	0.422[c]	1.98	2.43	1.23	126	230
	0.438[d]	2.61	3.54	1.36	133	
	0.451[d]	3.15	4.01	1.28	136	
Limestone, sample A	0.485[c]	9.26	11.3	1.22	191	376
	0.516[d]	14.3	16.7	1.17	210	
Limestone, sample B	0.475[c]	20.0	22.2	1.11	304	547
	0.516[d]	32.4	32.5	1.01	315	
Limestone, sample C	0.502[c]	50.7	40.8	0.81	424	711[e]
	0.531[d]	70.4	56.3	0.80	432	
MICA	0.866[d]	0.088	0.17	1.93	2	334
Activated carbon	0.766[c]	1.38	2.98	2.16	17	322
	0.758[d]	0.97	2.37	2.46	15	

[a] d_e, calculated by Carman-Kozeny formula using vertical permeability.
[b] d_p, obtained from sieve analysis.
[c] Bed formed by natural settling, followed by settling overnight.
[d] Bed formed by natural settling alone.
[e] Obtained from Latini, other values measured by Fontugne.

mum dimensions of flat particles or the internal pore size of the activated carbon may be much better estimates of the hydraulic diameter in anisotropic beds than the mean diameter determined from sieve analysis. All of these measurements indicate that the asymmetry of these particles always tends to decrease the permeability. The extent of anisotropy also seems to increase with the asymmetry as indicated by the much higher anisotropy of the beds of flake-like mica particles compared to the beds of limestone or sand.

In sandstone, several studies of the relationship between sand fabrics and anisotropy have been made. Mast and Potter (*23*) found that imbibition and polar permeability plots are nearly isotropic parallel to bedding in sandstones even though the fabrics are slightly anisotropic. In sections perpendicular to the bedding plane, the anisotropic sand fabrics and elliptical inhibition patterns, indicating anisotropic permeabilities, are much better defined.

Greenkorn, Johnson, and Shallenberger (*9*) found a strong correlation between bedding and directional permeability. In 75% of the plugs with bedding that they tested, the permeability perpendicular to the bedding plane was lower than the permeability parallel to the bedding plane. Fancher, Lewis, and Barnes (*6*) also found that samples of anisotropic sands usually had horizontal permeabilities higher than the vertical permeabilities.

Griffiths (*10*), Hutta and Griffiths (*13*), and Hutta (*12*) found no strong relation between the direction of the sand fabric parallel to the bedding plane and the maximum permeability, but in measurements made perpendicular to the bedding plane the largest permeability correlated with the bedding direction. Mast and Potter point out, however, that the results from many samples can be confusing because the surface structure of fabric does not necessarily reflect the fabric throughout the sample. Heterogeneity of this type was found in many of the samples tested by Hutta also. Thus, many permeability measurements, supposed to reflect the anisotropy of the surface fabric of the sample tested, reflect the anisotropy and heterogeneity in the internal volume of sample.

The structure of a porous medium depends on the circumstances of its formation and these can be very complex, especially in naturally formed beds. The simplest case is that of a particle with three axes of symmetry settling in an infinite quiescent medium. Milne-Thompson (*24*) [generalizing an earlier derivation by Lamb (*18*)] showed that the only stable mode of settling at low Reynolds number is with the large axis of the particles in the horizontal position, although positions with the other axes parallel to the horizontal also represent unstable equilibrium positions. Heiss and Coull (*11*) confirmed experimentally these modes of settling for several different types of particles. For beds formed by natural settling, this tendency leads to beds made up of particles oriented with their flat side downward, giving the bed a layered appearance. When the Reynolds is above 200 to 500 depending on the shape of the particles, Christiansen and Barker (*4*) observed that the particles oscillated about their largest axis in a somewhat helical path as they fell. This additional motion leads to beds made of particles with a more random orientation, although the general tendency still is for the particles to orient themselves flatly.

During the formation of many naturally formed beds, a flow current is also present. Anczurowski and Mason (*1*) have observed rods and disks under couette (shear) flow. They found that rods oriented themselves with their long side perpendicular to the direction of flow but with their axis parallel to the shear planes. Disks oriented themselves with their flat sides in the planes of shear. They also noted complex orbiting behavior in both types of particles. Other types of complex flow deposition effects, such as scouring, have been discussed by Jopling (*17*) and the interaction of grains during settling at high concentrations by Rees (*28*). These works suggest that asymmetric particles such as disks and rods tend to settle on the average in laminations with particular orientations, but that considerable randomness is possible because of their complex motions. The shear forces from the flow and interaction of particles can lead to imbrication of grains, while scouring can lead to cross-bedding. The events after the initial sedimentation are also important. Compaction from the weight of the overlaying sediment and cementation from the hydrolysis of some of the bed material can occur. Upheavals resulting from hydrostatic pressure and the elasticity of the medium can also affect the bed structure.

It would be desirable to predict the directional permeabilities from fundamental knowledge about the pore size distribution, the porosity, and the structure of the porous bed. These relationships are not easy to develop, however, especially with naturally occurring porous media, because of the large pore-size distribution, the differences in grain shape, the presence of two separate fluid phases, cementation, and the heterogeneity of the formation. More success has been realized in predicting porosities and permeabilities in porous media, not directly from knowledge of the bed structure, but from some more easily measured property *in situ* such as the electrical resistivity. Ordinarily in oil formations, the resistivity is expressed [Pirson (*26*)] in terms of a formation factor to account for the relative saturation with oil. The formation factor can be used to predict the effective porosity by using Archie's or Humble's equation and the tortuosity of the flow path. From the tortuosity and porosity and a knowledge of the pore size derived from capillary pressure measurements, the permeability can be predicted [Wyllie and Spangler (*36*) and Pirson (*26*)]. These methods have been applied to anisotropic media by Wyllie and Spangler (*36*) who found that the ratio of permeabilities perpendicular and parallel to the bedding plane is inversely proportional to the square of inverse ratio of the formation factors. This use of the resistivity to predict anisotropy in porous media probably can be extended to three dimensional anisotropic media since the resistivity of the media is a tensor of the same order as the permeability.

Predicting Potential and Stream Functions

Many methods available for the prediction of the potential and stream functions for flow of fluids in isotropic media are available for anisotropic media through a simple transformation of coordinates

$$\alpha = \frac{x}{\sqrt{K_1}}, \beta = \frac{y}{\sqrt{K_2}}, \gamma = \frac{z}{\sqrt{K_3}}$$

For a homogeneous medium this variable change reduces the steady-state potential and stream function equations for anisotropic media

$$K_1 \frac{\partial^2 \phi}{\partial x^2} + K_2 \frac{\partial^2 \phi}{\partial y^2} + K_3 \frac{\partial^2 \phi}{\partial z^2} = 0$$

$$K_1 \frac{\partial^2 \psi}{\partial x^2} + K_2 \frac{\partial^2 \psi}{\partial y^2} + K_3 \frac{\partial^2 \psi}{\partial z^2} = 0$$

$$K_1 \frac{\partial^2 \chi}{\partial x^2} + K_2 \frac{\partial^2 \chi}{\partial y^2} + K_3 \frac{\partial^2 \chi}{\partial z^2} = 0$$

to the following equations for isotropic media

$$\frac{\partial^2\phi}{\partial\alpha^2}+\frac{\partial^2\phi}{\partial\beta^2}+\frac{\partial^2\phi}{\partial\gamma^2}=0$$

$$\frac{\partial^2\psi}{\partial\alpha^2}+\frac{\partial^2\psi}{\partial\beta^2}+\frac{\partial^2\psi}{\partial\gamma^2}=0$$

$$\frac{\partial^2\chi}{\partial\alpha^2}+\frac{\partial^2\chi}{\partial\beta^2}+\frac{\partial^2\chi}{\partial\gamma^2}=0$$

For the case of incompressible fluids in a homogeneous medium, many different methods of solution have been reviewed by Muskat (*25*) and Scheidegger (*32*), including separation of variables, the Laplace transform, and conformal mapping. For the more general problems, integral approximation such as those of Ritz [Collatz (*5*)] and numerical methods are available.

One of the most powerful numerical methods is the alternating direction implicit matrix procedure (ADIP) developed by Peaceman and Rachford (*25a*) and the alternating direction explicit procedures developed recently by Quon, Dranchuk, Allada, and Leung (*27*). For the elliptic equations, which result when the fluid density is constant, these schemes generally require less computational time than the older numerical procedure of overrelaxation, but relaxation generally requires less computer storage. The ADIP and the ADEP do have the advantage of also being applicable to gas reservoir problems where the density is not constant.

The stream function equations are more difficult to solve in the three dimensions than in two, since at the boundaries the two stream functions are coupled in their relationships to the velocity [Rouse (*30*)]—*i.e.*,

$$v_x=\lambda\left(\frac{\partial\psi}{\partial y}\frac{\partial\chi}{\partial z}-\frac{\partial\psi}{\partial z}\frac{\partial\chi}{\partial y}\right)$$

$$v_y=\lambda\left(\frac{\partial\psi}{\partial z}\frac{\partial\chi}{\partial x}-\frac{\partial\psi}{\partial x}\frac{\partial\chi}{\partial z}\right)$$

$$v_z=\lambda\left(\frac{\partial\psi}{\partial x}\frac{\partial\chi}{\partial y}-\frac{\partial\psi}{\partial y}\frac{\partial\chi}{\partial x}\right)$$

where λ is a constant. For flow in two dimensions, there is only one stream function equation, and the velocities are simply related to the stream function by

$$v_x=-\frac{\partial\psi}{\partial y}$$

$$v_y=+\frac{\partial\psi}{\partial x}$$

Those in three dimensions, the ψ and χ equations must be solved simultaneously to determine the velocity profile.

Nomenclature

d_e = the equivalent diameter
d_p = the volume surface mean diameter
$\bar{\bar{K}}$ = the permeability tensor
K_1, K_2, K_3 = the principal values of the permeability tensor
K_{ij} = the ij^{th} component of the permeability tensor
K_n = the component of the permeability direction $\bar{n}$
$\bar{n}$ = unit vector designating a given direction
v_x, v_y, v_z = the components of the velocity
$\bar{v}$ = the velocity vector
x, y, z = coordinate variables
α, β, γ = coordinate variables
$\delta_1, \delta_2, \delta_3$ = the angles of the direction cosines
ϕ = the potential function
ψ, χ = the stream functions
μ = the fluid
ρ = the fluid density
ϵ = the porosity

BIBLIOGRAPHY

(1) Anczurowski, E., and Mason, S. G., "The Kinetics of Flowing Dispersions, *J. Colloid Interface Sci.*, **23**, 522, 533 (1967).

(2) Barduhn, A. J., "The Freezing Process for Water Conversion in the U.S.," *Proc. Int. Symp. Water Desalination, 1st*, Wash., D.C., **2** (1967).

(3) Carman, P. C., "Flow of Gases Through Porous Media," 1–33, Academic Press, Inc., New York, N.Y., 1956.

(4) Christiansen, E. B., and Barker, D. H., "The Effect of Shape and Density on the Free Settling of Particles at High Reynolds Numbers," *A.I.Ch.E. J.*, **11** (1), 145 (1965).

(5) Collatz, L., "Numerical Treatment of Differential Equations," 425–438, Springer-Verlag, New York, N.Y., 1966.

(6) Fancher, G. H., Lewis, J. A., and Barnes, K. B., "Some Physical Characteristics of Oil Sands," *Bull. Miner. Ind. Exp. Sta., Pa. State Coll.*, **12** (1933).

(7) Ferrandon, J., Le Genie Civil, "Les Lois de L'Ecoulement de Filtration," **125** (2), 24 (1948).

(8) Fontugne, D., "Permeability Measurement in Anisotropic Media," M.S. Thesis, Dept. of Chemical Engineering and Metallurgy, Syracuse University, 1969.

(9) Greenkorn, R. A., Johnson, C. R., and Shallenberger, L. K., "Directional Permeability of Heterogeneous Anisotropic Porous Media," *Soc. Petrol. Eng. J.*, **4**, 124 (June 1964).

(10) Griffiths, J. C., "A Review of Dimensional Orientation of Quartz Grains in Sediments," *Prod. Mon.*, **17** (3), 14 (1952).

(11) Heiss, J., and Coull, J., "The Effect of Orientation and Shape on the Settling Velocity of Non-isometric Particles in a Viscous Medium," *Chem. Eng. Progr.*, **48** (3), 133 (1952).

(12) Hutta, J. J., "Relation of Dimensional Orientation of Quartz Grains to Directional Permeability in Sandstones," M.S. Thesis, Pa. State Univ., Coll. Min. Industries, 1952.

(13) Hutta, J. J., and Griffiths, J. C., "Directional Permeabilities of Sandstone, a Test of Techniques," *Bull. Miner. Ind. Exp. Sta., Pa. State Univ.*, **64**, 75 (1955).

(14) Johnson, A. I., Moston, R. P., and Morris, D. A., "Physical and Hydrologic Properties of Water Bearing Deposits in Subsiding Areas in Central California," Geological Survey, Prof. Paper 497A, 1968.

(15) Johnson, W. E., and Breston, J. N., "Directional Permeability on Sandstones from Various States," *Prod. Mon.*, **15**, 10–19 (1951).

(16) Johnson, W. E., and Hughes, R. V., "Directional Permeability Measurements and Their Significance," *Bull. Miner. Ind. Exp. Sta., Pa. State Coll.*, **52**, 180 (1948).

(17) Jopling, Alan V., "Origin of Laminae Deposited by the Movement of Ripples Along a Streambed," *J. Geol.*, **75** (3), 287–305 (1967).

(18) Lamb, H., "Hydrodynamics," 6th ed., 177–8, Dover Publications, 1945.

(19) Latini, R. G., "Measurem nt of Directional Permeabilities," M.S. Thesis Dept. of Chemical Engineering and Metallurgy, Syracuse University, 1967.

(20) Liakopoulos, A. C., "Variation of the Permeability Tensor Ellipsoid in Homogeneous Anisotropic Soils," *Water Resour. Res.*, **1** (1), 135 (1965).

(21) Maasland, M., and Kirkham, D., "Theory and Measurement of Anisotropic Air Permeability in Soil," *Soil Sci. Soc. Amer., Proc.*, **19** (4), 395 (1955).

(22) Marcus, Henri, "The Permeability of a Sample of Anisotropic Porous Media," *J. Geophys. Res.*, **67** (13), 5215–25 (Dec. 1962).

(23) Mast, Richard F., and Potter, P. E., "Sedimentary Structure, Sand Shape Fabrics and Permeability," *J. Geol.*, **71** (4), 441–71, 548–65 (1952).

(24) Milne-Thompson, L. M., "Theoretical Hydrodynamics," pp 504–5, The Macmillan Co., New York, N.Y., 1950.

(25) Muskat, M., "The Flow of Homogeneous Fluids Through Porous Media," 103–9, J. W. Edwards, Ann Arbor, Mich. (1946).

(25a) Peaceman, D. W., and Rachford, H. H., Jr., "The Numerical Solution of Parabolic and Elliptic Differential Equations," *J. Soc Ind. Appl. Math.*, **3**, 22–41 (1955).

(26) Pirson, S. J., "Oil Reservoir Engineering," Chap. 2 and 3, McGraw-Hill, 1958.

(27) Quon, D., Dranchuk, P. M., Allada, S. R., and Leung, P. K., "Application of the Alternating Direction Explicit Procedure to Two Dimensional Natural Gas Reservoirs," *Soc. Petrol. Engr. J.*, **6** (2), 137 (1966).

(28) Rees, A. I., "The Production of Preferred Orientation in a Concentrated Dispersion of Elongated and Flattened Grains," *J. Geol.*, **76** (4), 457–65 (1968).

(29) Rouher, O. S., Fontugne, D. S., and Barduhn, A. J., "The Freezing Processes: Butane Hydrates and Anisotropic Porous Media," *Prog. Rept. No.* **366**, Office of Saline Water, U.S. Dept. of Interior, June 1968.

(30) Rouse, H., "Advanced Mechanics of Fluids," 42–5, Wiley, 1959.

(31) Scheidegger, A. E., "On Directional Permeability," *Geofis. Pura Appl.*, **33** (111) (1956).

(32) Scheidegger, A. E., "Physics of Flow in Porous Media," Macmillan, 1960.

(33) Snow, David T., "Three Hoie Pressure Test for Anisotropic Foundation Permeability," *Felsmechanik Ingenieurgeol.*, **4** (4), 298–316 (1966).

(34) Weeks, Edwin P., "Field Methods for Determining Vertical Permeability and Aquifer Anisotropy," U.S. Geol. Survey, Prof. Paper 501-D, 193–8 (1964).

(35) Wyllie, M. R. J., and Gregory, A. R., "Fluid Flow Through Unconsolidated Porous Aggregates—Effect of Porosity and Particle Shape on Kozeny-Carman Constants," IND. ENG. CHEM., **47**, 1379 (1955).

(36) Wyllie, M. R. J., and Spangler, M. B., "Application of Electrical Resistivity Measurements to Problems of Fluid Flow in Porous Media," *Amer. Assoc. Petrol. Geol., Bull.*, **36**, 359–403 (1952).

4

Diffusion and Flow of Gases in Porous Solids

Diffusion and Flow of Gases in Porous Solids

Diffusion and flow fluxes for gases through porous solids are discussed

Gordon R. Youngquist

The prediction and correlation of fluxes for diffusion and flow of gases through porous solids, such as catalysts and adsorbents, are of considerable importance. Reactions catalyzed by porous solids are limited in many cases by the rate of transport of reactant molecules through the catalyst pores to the active catalyst surface. Similarly, the rate of approach to adsorption equilibrium is frequently limited by transport through adsorbent pores. Additional examples are found in cases of non-catalytic reactions where one reactant is located in the porous solid and a second must be transported into the solid. Such reactions may include, for example, the combustion of carbon deposits involved with the regeneration of catalysts and the oxidation or reduction of ores.

Mass transport in a porous solid is complex. Several mechanisms of transport, including molecular diffusion, Knudsen diffusion, viscous and slip flow, as well as surface migration, may contribute to the flux of gas through the pores of the solid. In addition, the pore structure of solids is not well understood, so that the relationship between structure and fluxes is not easily delineated. As a consequence, most models for prediction of gaseous diffusion and flow in porous solids are based on reasonably well developed theories for diffusion and flow in capillaries. These capillary theories have been adapted for use with porous solids through the use of relatively simple structural models for the solid. Most such models contain at least one adjustable parameter, which must be determined by experiment, and in this sense are not completely predictive. Given the structural complexity of porous solids, and the limited ability to determine the nature of the pore structure, it is quite possible that a completely predictive approach may never be developed. Nevertheless, the models formulated to date, when used with a minimum of experimental data, appear reasonably useful for extrapolation and interpolation of data, and give at least a qualitative indication of the effects of pore structure.

The developments leading to the formulation of various models for the description of mass fluxes in porous materials are reviewed. Experimental methods are discussed briefly, and comparison of the model predictions to experimental results is also described.

DIFFUSION AND FLOW IN CAPILLARIES

Consider a single cylindrical capillary of radius r and length L through which steady-state transport of components A and B of a binary gas mixture occurs. The gas mixture at ends of the capillary are maintained at constant pressures P_0 and P_L, and constant compositions y_{A_0} and y_{AL}, respectively, as shown in Figure 1.

Diffusion

When the system pressure is uniform ($P_0 = P_L$), the mass flux is diffusive in nature and may involve Knudsen diffusion, ordinary molecular diffusion, and surface migration. The latter will only occur when the diffusing gases are adsorbed in a mobile layer; here it will be considered negligible. If the radius of the capillary and the gas pressure is such that the mean free path is large compared to the diameter of the pores, the rate of transport of the molecules A and B is governed only by collisions with the capillary wall. This type of transport is usually referred to as Knudsen diffusion, after Knudsen who first investigated this type of behavior for capillaries (*15*). Qualitatively, Knudsen diffusion appears to dominate for values of r/λ less than 0.1. Knudsen showed that the diffusive flux is given by

$$N_A = -D_{KA}\frac{dC_A}{dx} \tag{1}$$

where

$$D_{KA} = 2/3\, r\, \bar{v}_A \left(\frac{2-f}{f}\right)$$

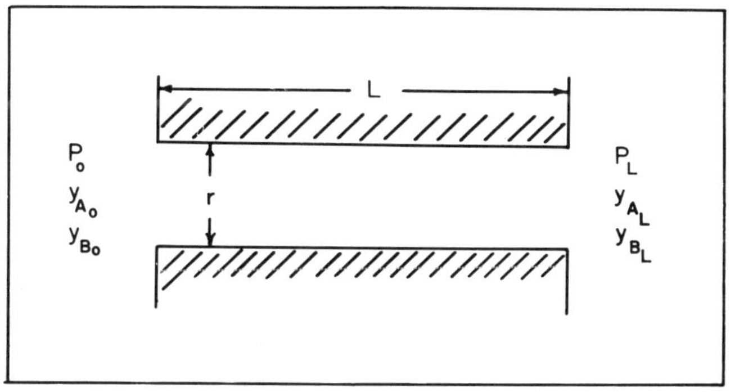

Figure 1. Capillary schematic

The mean molecular velocity $\bar{v}_A$ is given by the kinetic theory of gases as

$$\bar{v}_A = \left(\frac{8\,RT}{\pi M_A}\right)^{1/2}$$

and the fraction f of molecules, which undergo diffuse reflections at the walls, is usually taken to be one. Thus, the Knudsen diffusion coefficient is

$$D_{KA} = 2/3\, r \left(\frac{8\,RT}{\pi M_A}\right)^{1/2} \tag{2}$$

and is seen to be proportional to the capillary radius r, to $T^{1/2}$ and inversely proportional to the square root of molecular weight of the diffusing specie. Since the rate of transport is governed by collisions with the capillary wall, the gas A diffuses independent of the presence of other species, and Equation 1 applies regardless of the existence of a total pressure gradient. At steady state, Equation 1 may be integrated to give

$$N_A = -\frac{D_{KA}}{RTL}(P_{AL} - P_{A0}) \tag{3}$$

Similarly, for component B,

$$N_B = -\frac{D_{KB}}{RTL}(P_{BL} - P_{B0}) \tag{4}$$

At constant pressure, Equations 3 and 4 add to give

$$\frac{N_A}{D_{KA}} + \frac{N_B}{D_{KB}} = 0$$

By use of Equation 2, the ratio of fluxes is given by

$$N_A/N_B = -\left(\frac{M_B}{M_A}\right)^{1/2} \tag{5}$$

When the ratio r/λ is greater than about 10, ordinary molecular diffusion predominates, and the flux of A may be described by

$$N_A = -D_{AB}\frac{dC_A}{dx} + y_A(N_A + N_B) \tag{6}$$

The molecular diffusion coefficient D_{AB} is, of course, independent of the size of the capillary. From kinetic theory, D_{AB} is proportional to $T^{3/2}$ and inversely proportional to the total pressure. Integrating Equation 6 for the steady-state case and at constant total pressure gives

$$N_A = \frac{D_{AB}P}{RTL\alpha} \ln\left[\frac{1 - \alpha y_{AL}}{1 - \alpha y_{A0}}\right] \tag{7}$$

where

$$\alpha = 1 + \frac{N_B}{N_A}$$

The transition region between Knudsen diffusion and ordinary molecular diffusion has received considerable attention in recent years, principally because many cases of diffusion in porous solids that are of practical interest lie in this region. Early attempts at describing the diffusion flux in the transition region were more or less intuitive—*e.g.*, Wheeler (*33*). Bosanquet (*3*), using a random walk argument, suggested that the flux be described by

$$N_A = -D_N\frac{dC_A}{dx} \tag{8}$$

where

$$\frac{1}{D_N} = \frac{1}{D_{KA}} + \frac{1}{D_{AB}} \tag{9}$$

Equation 9 is a special case derived from a more general flux relationship. More recently, Rothfeld (*19*) and Scott and Dullien (*24*), using a momentum balance, and Spiegler (*27*), using a friction factor analysis, have shown that the flux in the transition region is more properly described by

$$N_A = \frac{-D_{AB}\dfrac{dC_A}{dx} + (N_A + N_B)y_A}{1 + \dfrac{D_{AB}}{D_{KA}}} \tag{10}$$

or

$$N_A = -\frac{1}{\left[\dfrac{1 - \alpha y_A}{D_{AB}} + \dfrac{1}{D_{KA}}\right]}\frac{dC_A}{dx} \tag{11}$$

When the fluxes are equal, $\alpha = 0$ and the Bosanquet expression results. Integrating at constant pressure and temperature results in

$$N_A = \frac{D_{AB}P}{RTL\alpha} \ln\left[\frac{1 - \alpha y_{AL} + \dfrac{D_{AB}}{D_{KA}}}{1 - \alpha y_{A0} + \dfrac{D_{AB}}{D_{KA}}}\right] \tag{12}$$

When molecular diffusion dominates, $D_{KA} \gg D_{AB}$ and Equations 10 and 11 become identical to Equations 6 and 7, respectively. Similarly, when $D_{AB} \gg D_{KA}$, the equations reduce to Equations 1 and 3. To use Equation 12, some knowledge of the flux ratio N_B/N_A is required. At constant pressure the flux ratio for Knudsen diffusion follows the inverse square root molecular weight relationship, Equation 5. This relationship has further been shown (*12, 26*) to apply to all open systems involving the diffusion of a binary gas at uniform pressure, regardless of the diffusion mechanism—Knudsen, molecular, or transition. Thus, at constant pressure

$$\alpha = 1 - \left(\frac{M_A}{M_B}\right)^{1/2} \tag{13}$$

Combined Diffusion and Flow

In the presence of a pressure gradient ($P_0 \pm P_L$) both diffusion and flow contribute to the net flux through the capillary. Evans *et al.* (*8*) have shown that the flux relationship given by Equation 10 for pure diffusion is also appropriate to combined diffusion and flow. Account must be taken of the pressure dependence of the molecular diffusion coefficient, and of the fact that the flux ratio no longer obeys the inverse square root molecular weight relationship. To integrate Equation 10, the variation of pressure with x must be known. The total molar flux is given by the sum of contributions from diffusion and forced flow:

$$N_T^\circ = N_T + F = N_A + N_B + F = N_A^\circ + N_B^\circ \tag{14}$$

where N_A and N_B are pure diffusion fluxes given by Equation 11, and F is the forced flow term. Combining Equations 14, 11, and 13 we obtain

$$N_A^\circ = \frac{-D_{AB}\dfrac{dC_A}{dx}}{1 + \dfrac{D_{AB}}{D_{KA}} - (\alpha)\, y_A} + \frac{Fy_A}{1 + \dfrac{D_{AB}}{D_{KA}}} \tag{15}$$

For component B,

$$N_B^\circ = \frac{-D_{AB}\,(1 - \alpha)\dfrac{dC_B}{dx}}{1 + \dfrac{D_{AB}}{D_{KA}} - (\alpha)\, y_A} + \frac{Fy_B}{1 + \dfrac{D_{AB}}{D_{KB}}} \tag{16}$$

In terms of the overall pressure drop,

$$(1 - \alpha)\, N_A^\circ + N_B^\circ = \frac{-\dfrac{D_{AB}}{RT}\,(1 - \alpha)\dfrac{dP}{dx}}{1 + \dfrac{D_{AB}}{D_{KA}} - (\alpha)\, y_A} + F\left[\frac{(1 - \alpha)y_A}{1 + \dfrac{D_{AB}}{D_{KA}}} + \frac{y_B}{1 + \dfrac{D_{AB}}{D_{KB}}}\right] \tag{17}$$

AUTHOR *Gordon R. Youngquist is an Associate Professor of Chemical Engineering at the Clarkson College of Technology, Potsdam, N. Y. 13676.*

The nature of the forced flow term F has been the subject of some discussion. Clearly, where $r/\lambda \ll 1$, Knudsen diffusion exists and F is negligible. Also, when $r/\lambda \gg 1$, Poiseuille flow is observed. In the transition region, $r/\lambda \approx 1$, slip flow results. Most treatments of this transition flow region use a modification of the slip flow term, which corrects for the nonzero velocity at the wall when slip occurs. Various methods of establishing the correction term have been proposed. Recently, Scott and Dullien (*25*) have used modern kinetic theory arguments which show that, for a pure gas, F is given by

$$F = -\left[\frac{r^2P}{8\,\mu} + \frac{4}{3}\frac{rRT}{M\bar{v}}\right]\frac{1}{RT}\frac{dP}{dx} \qquad (18)$$

Wakao *et al.* (*30*) suggest that for a binary gas,

$$M\bar{v} = M_A\,\bar{v}_A\,y_A + M_B\,\bar{v}_B y_B \qquad (19)$$

so that Equation 18 becomes

$$F = -\left[\frac{r^2P}{8\,\mu} + \frac{4\,rRT}{3\,(M_A\bar{v}_A y_A + M_B y_B \bar{v}_B)}\right]\frac{1}{RT}\frac{dP}{dx}$$

$$= -\left[\frac{r^2P}{8\,\mu} + \frac{\pi D_{KA}}{4\left(y_A + \dfrac{y_B}{1-\alpha}\right)}\right]\frac{1}{RT}\frac{dP}{dx} \qquad (20)$$

Equation 17 thus may be written

$$(1-\alpha)\,\overset{\circ}{N}_A + \overset{\circ}{N}_B = -\frac{C}{RT}\frac{dP}{dx} \qquad (21)$$

where

$$C = \frac{(1-\alpha)D_{KA}}{[1+(1-\alpha)y_A]\dfrac{D_{KA}}{D_{AB}}} + \left[\frac{r^2P}{8\,\mu} + \frac{D_{KA}\,\pi}{4\left(y_A + \dfrac{y_B}{1-\alpha}\right)}\right]\left[\frac{1-\alpha y_A}{1+\dfrac{D_{AB}}{D_{KA}}} + \frac{1-y_A}{1+\dfrac{D_{AB}}{D_{KA}}}\right] \qquad (22)$$

Equations 15 and 21 must be integrated simultaneously over the length of the capillary to predict the fluxes for specified end conditions. This may be done by trial and error using numerical integration. However, it appears that little error is introduced if arithmetic averages for the compositions and pressure are used to evaluate the concentration and temperature dependent terms in C. Thus, the assumption of a linear pressure relationship for use with Equation 15 seems quite reasonable.

APPLICATION TO POROUS SOLIDS

The pore structure of solids is exceedingly complex and for the most part unknown. Gross properties of the solid are easily measured—*e.g.*, the porosity—but the details of the structure defy direct observation. Indeed, if a packed bed is visualized as a large-scale representation of a porous solid, with the interstices between the packing as pores, one gets some appreciation of the complex structure of the solid. The size of pores is in most cases determined by either or both of two techniques: mercury porosimetry (*18*) for relatively large pores (usually $r > 100$ Å) and low temperature nitrogen adsorption (*2*) for small pores (<100 Å). Neither method provides the means for determining the shape or length of the pores, and in each case the pores are normally represented as interconnected cylindrical capillaries having some effective radius. While the pore sizes for some solids are more or less uniform, many solids have a fairly broad pore size distribution. The data of Figure 2 are typical.

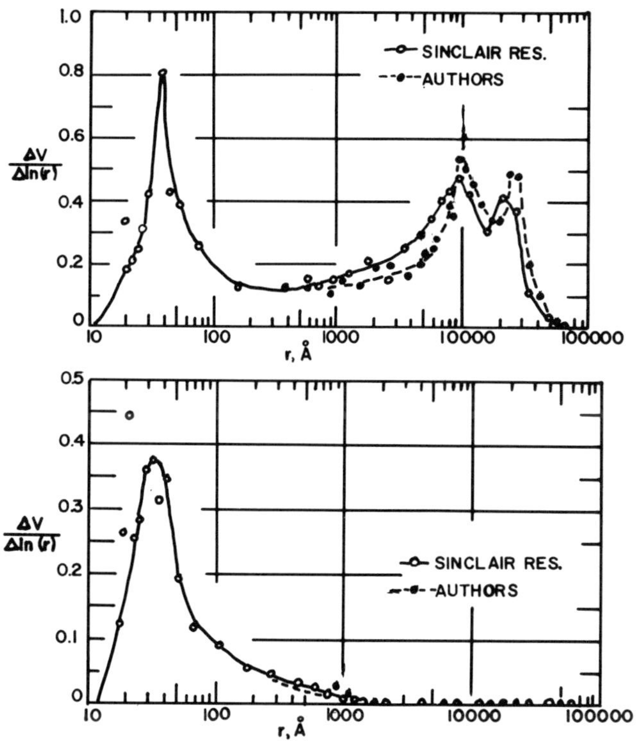

Figure 2. Typical pore size distributions for compressed alumina powder [Cunningham and Geankoplis (5)]

In treating diffusion and flow through porous solids, frequent use has been made of the capillary equations describing these phenomena and discussed earlier. Relatively simple structural models for the solids based on capillaries have been proposed. To account for the shape and orientation of pores, and for the effective length of the pores, an adjustable constant (the "tortuosity") is often included. Various attempts have been made to deduce theoretically the nature of the tortuosity, but with only limited success, so this constant must be determined experimentally. More detailed structural models have also been suggested, but often these are of such complexity that they become both difficult to validate and to use conveniently.

Parallel Pore Models

Perhaps the simplest model for diffusion and flow in a porous solid pictures transport as occurring through a number of parallel capillaries all of which are the same size. As proposed by Wheeler (*33*), the model suggests that the flux be given by

$$N_{AS} = n\pi r^2 \frac{L}{L_e} N_A \qquad (23)$$

where N_{AS} represents the superficial molar flux (mol/unit area of solid $\times$ time), N_A is the molar flux in each capillary of radius r, $n\pi r^2$ is the total pore area per unit solid area, n being the number of pores and r the radius of the pores, and L_e/L is the ratio of the actual diffusion path length to the straight line distance through the solid.

The porosity of the solid, ϵ, equals $n\pi r^2 L_e/L$, so Equation 23 becomes

$$N_{AS} = \epsilon \left(\frac{L}{L_e}\right)^2 N_A \qquad (24)$$

Wheeler takes the length ratio $L_e/L = \sqrt{2}$. However, some account must also be taken of the shape and orientation of the pores. Petersen (*16*), for example, by examining diffusion in a pore of varying cross section shows that the shape of the pore may markedly reduce the flux from that obtained in an equivalent cylindrical capillary. The form of Equation 24 in many cases has been retained, and the flux expressed as

$$N_{AS} = \frac{\epsilon}{\Upsilon} N_A \qquad (25)$$

where Υ is the tortuosity. According to Wheeler's model, $\Upsilon = (L_e/L)^2$. Obviously, Υ must include shape factors as well. In most instances, Υ is determined from experimental data because of the lack of detailed information about pore geometry. In principle, Υ could be determined from a single flux measurement using Equation 25 and the diffusion flux N_A predicted by Equation 12 for a single capillary. To use Equation 12, a single average pore size must be chosen in order to evaluate the Knudsen diffusion coefficient. Given the pore size distribution, the average pore size may be calculated from

$$\bar{r} = \frac{1}{V_t} \int_0^{V_t} r\, dV \qquad (26)$$

where V represents the pore volume as determined by mercury porosimetry or nitrogen adsorption. Alternatively, an estimate of the average pore size may be obtained from the void volume–surface area ratio:

$$\bar{r} = \frac{2\, V_t}{S} \qquad (27)$$

Equation 14 assumes the pores to be equal cylinders of radius $\bar{r}$.

The use of Equation 25 is likely to be satisfactory only for solids having a relatively narrow pore size distribution unless ordinary molecular diffusion dominates so that the diffusion flux becomes independent of pore size. The latter is unlikely, since this would be true only for large pores and/or very high pressures, neither of which is probable in most cases of interest. The parallel pore model may be adapted (*14*) to consider the variation in pore size as follows. A pore size distribution function $f(r)$ is defined such that $f(r)dr$ is the void fraction of pores with radii between r and $r + dr$. Using Equation 25, the contribution to the flux from pores in this size range is

$$N_{AR} = \frac{1}{\Upsilon} N_{A(r)}\, f(r)dr \qquad (28)$$

Assuming constant Υ and integrating over all possible pore sizes,

$$N_{AS} = \frac{1}{\Upsilon} \int_0^{\infty} N_{A(r)}\, f(r)dr \qquad (29)$$

For pure diffusion at constant temperature and pressure, $N_{A(r)}$ may be replaced by Equation 12 to give

$$N_{AS} = \frac{D_{AB}\, P}{\Upsilon RTL\alpha} \int_0^{\infty} \ln \left[\frac{1 - \alpha y_{AL} + \frac{D_{AB}}{D_{KA}}}{1 - \alpha y_{A_0} + \frac{D_{AB}}{D_{KA}}} \right] f(r)dr \qquad (30)$$

The integration with respect to r may be performed numerically. If the pore size distribution function $f(r)$ may be assumed to consist of one or more sharp peaks (δ functions), the integral may be evaluated directly. Indeed, for a single peak Equation 25 is obtained. Alternatively, the integral of Equation 30 may be evaluated incrementally as indicated by Satterfield and Cadle (*21*).

$$N_A = \frac{D_{AB}\, P}{\Upsilon RTL\alpha} \sum_{r\,\max}^{r\,\max} \ln \left[\frac{1 - \alpha y_{AL} + \frac{D_{AB}}{D_{KA}}}{1 - \alpha y_{A_0} + \frac{D_{AB}}{D_{KA}}} \right] \Delta\epsilon_{(r)} \qquad (31)$$

where $\Delta\epsilon_{(r)}$ is the void fraction for pores of average size r.

Random Pore Models

One of the few completely predictive models is that proposed by Wakao and Smith (*29*). These authors noted that many porous solids are pelleted materials prepared by compressing powder particles that are themselves porous. Thus, the resulting pellet may have a bidisperse pore structure consisting of micropores (pores within the powder particles) and macropores (spaces between the powder particles). They represented the solid, as shown in Figure 3, and noted that diffusion through the solid could occur by three parallel paths:

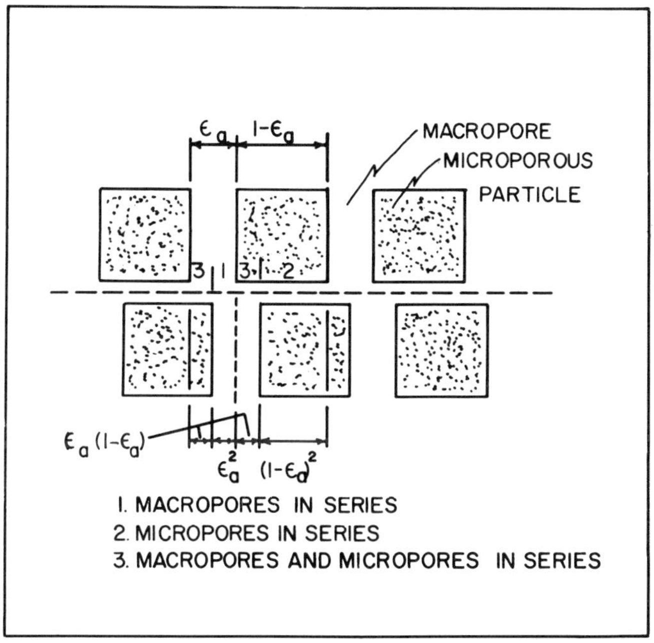

Figure 3. Random pore model for bidisperse solids [Wakao and Smith (29)]

(1) through the macropores of average radius $\bar{r}_a$
(2) through the micropores of average radius $\bar{r}_i$
(3) through macropores and micropores in series

Associated with the macropores is a void fraction ϵ_a, and with the micropores a void fraction ϵ_i. To predict the contributions to the flux due to each path, a sample is pictured as being cut and rejoined at random. The flux area for each mechanism at the rejoined plane is taken as the probability of lining up macropores with macropores, micropores with micropores, and macropores with micropores. The flux per unit area of pellet is the sum of the three contributions. For diffusion at constant pressure, Wakao and Smith (*29*) find

$$N_{AS} = -\left[\epsilon_a^2 D_a + (1-\epsilon_a)^2 D_i + 2\,\epsilon_a(1-\epsilon_a)\left(\frac{2}{\dfrac{1}{D_a}+\dfrac{1}{D_i}}\right)\right]\frac{P}{RT}\frac{dy_A}{dx} \qquad (32)$$

where

$$D_a = \frac{D_{AB}}{(1-\alpha y_A) + \dfrac{D_{AB}}{D_{KA_a}}}$$

$$D_i = \frac{D_{AB}\,\epsilon_i^2/(1-\epsilon_a)^2}{(1-\alpha y_A) + \dfrac{D_{AB}}{D_{KA_i}}}$$

If either the macropores in the micropores do not exist—*i.e.*, $\epsilon_a = 0$ or $\epsilon_i = 0$—the pore size distribution is monodispersed and Equation 32 reduces to

$$N_{AS} = -\epsilon^2\frac{D_{AB}}{(1-\alpha y_A) + \dfrac{D_{AB}}{D_{KA}}}\frac{P}{RT}\frac{dy_A}{dx} \qquad (33)$$

thus implying that the tortuosity of Equation 25 equals the reciprocal of the porosity, $\Upsilon = 1/\epsilon$. Cunningham and Geankoplis (*5*) have extended the Wakao and Smith treatment to tridispersed pore size distributions.

Other Models

Recently, Foster and Butt (*9*) have proposed a somewhat more detailed computational model for the prediction of diffusion fluxes in porous solids. The entire pore volume is represented by two approximately conical ducts, shown in Figure 4, which are themselves made up of straight cylindrical capillary segments. One of the ducts is centrally convergent and the other centrally divergent, each being the inverse of the other. The exact shape of these ducts is determined from the pore size distribution of the solid by assigning the pore volume corresponding to a given pore radius to capillary segments of that radius in the two ducts. Given the gas composition at opposite ends of a solid sample, the diffusional flux may be calculated by an iterative procedure. By using an assumed value for the flux, the capillary equation (Equation 12) is rearranged to give the change in composition across each capillary segment. For the *j*th segment,

$$y_{A_{j+1}} = \frac{1}{\alpha}\left\{\left[(1-y_{A_j}) + \frac{D_{AB}}{D_{KA}}\right]\exp\left[\frac{N_A RTL_j\alpha}{P\,D_{AB}}\right] - \left(\frac{D_{AB}}{D_{KA}}+1\right)\right\} \qquad (34)$$

Starting at one end, Equation 34 is applied consecutively to each capillary segment. Mixing between the two ducts may be allowed at various points, and the extent to which it is allowed provides an adjustable parameter for data correlation. The computations are carried through the total length of the ducts, and correctness of the assumed flux is determined by comparing the calculated and given end compositions.

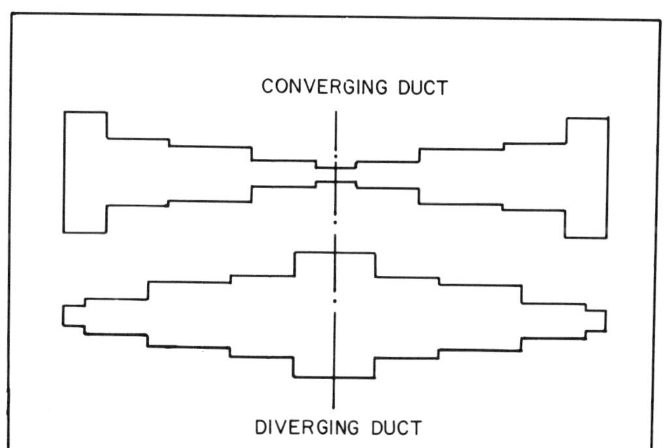

Figure 4. Converging–diverging pore model [Foster and Butt (9)]

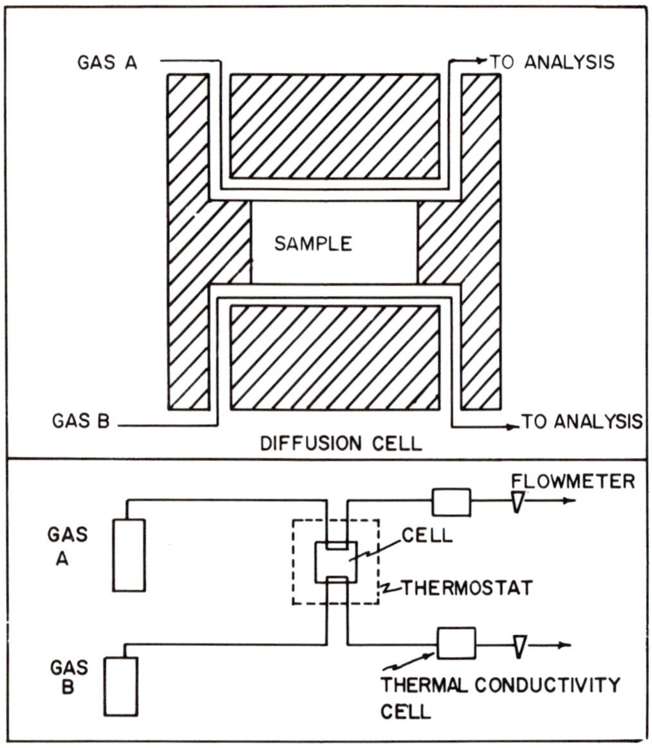

Figure 5. Wicke-Kallenbach diffusion apparatus

EXPERIMENTAL METHODS

Perhaps the most common method presently in use for the study of diffusion and flow in porous solids is that first developed by Wicke and Kallenbach (*34*) and later modified by Weisz (*31*). In this method, a sample of the test solid is placed in a diffusion cell between gas streams of different composition. For pure diffusion measurements the pressure and temperature are maintained uniform. The gases on either side of the solid are normally pure, although this is not a necessary restriction. A typical apparatus is shown schematically in Figure 5. Care must be taken in the design of the diffusion cell to ensure that boundary layer effects are not present at the cell ends. This is normally done by using a high gas velocity and contacting the plug ends uniformly with the gas stream. Concentrations of the effluent streams often have been determined by either thermal conductivity measurement or gas chromatography. From the flow rates and compositions of the feed and exit streams, the fluxes may be readily determined. Brief descriptions of apparatus of this type are given in several papers (*5, 10, 21, 31*).

Although very useful and perhaps the best of available techniques, the Wicke-Kallenbach technique has its disadvantages. Single samples must be used and these must be cylindrical or shaped into cylinders. Since the axial flux is measured, no indication of the effects of the possible anisotropy of the solid or of dead end pores is possible. If the solid has a wide distribution of the pore sizes, most of the measured flux is likely to be due to the large pores. Since small pores contribute greatly to the active surface involved with catalysis or adsorption, fluxes measured in this way may not reveal the importance of the small pores in these applications. Moreover, the temperatures and pressures conveniently used with this type of apparatus are somewhat limited, so measurements at reaction conditions cannot easily be made.

An unsteady-state method, which overcomes some of these objections, has been discussed by Davis and Scott (*6*). Based on an elementary theory of gas chromatography of van Deemter *et al.* (*28*), the method allows the determination of an effective diffusion coefficient for particular gas–solid systems. A carrier gas is passed continuously through a packed bed of the particles of interest, and the response of the system to an injected pulse of test gas is measured, much as in gas chromatography, although the gases used here must be nonadsorbing. By measuring the HETP (height equivalent to a theoretical plate) for the bed as a function of gas velocity, the effective diffusion coefficient may be determined. Davis and Scott give the details. The method appears adaptable to a wide range of operating conditions, gives results reflecting the behavior of a large number of particles, and since it is a transient technique should reflect the presence of small and dead-end pores. However, the theory on which the method is based is only approximate, and some difficulties associated with wall effects and external mass transfer effects arise.

Improvements to the chromatography theory as well as experimental results have been given by Schneider and Smith (*23*). Their results indicate that the chromatographic technique should develop into a very useful experimental tool.

An experimental approach that is particularly useful for cases where Knudsen diffusion dominates is the "time-lag" method developed by Barrer (*1*). In this method, the transient flow of pure gas through a porous plug is observed. The pressure on the upstream side of the plug is increased suddenly, and the downstream pressure maintained constant. The total quantity Q of gas that has passed through the plug is measured as a function of time. The transient condition that exists in the medium is described by

$$\frac{\partial C}{\partial t} = D_e \frac{\partial^2 C}{\partial x^2} \tag{35}$$

where D_e is the effective diffusivity for Knudsen flow in the medium. Subject to the conditions of the experiment, Q approaches the asymptote for large t

$$Q = \frac{D_e C_0}{L}\left[t - \frac{L^2}{6\,D_e}\right] \tag{36}$$

Thus D_e may be determined conveniently from the intercept of the asymptote with t axis:

$$I = \frac{L^2}{6\,D_e} \tag{37}$$

I is referred to as the "time lag" and corresponds essentially to the time required to reach steady-state flow. Goodknight and Fatt (*11*) have shown that the magnitude of the time lag depends on the dead-end pore volume so that values of D_e determined from the time lag are likely to be somewhat different from those obtained by steady-state methods.

EXPERIMENTAL RESULTS AND COMPARISON WITH THEORY

Although a wealth of experimental data for diffusion and flow of gases in porous solids do not exist, there is sufficient data available for at least a preliminary evaluation of the models that have been proposed to describe these phenomena. Most diffusion and flow measurements have been made with such gases as helium, nitrogen, hydrogen, and argon. These gases have been chosen partly because of their ready availability, partly because of the high molecular weight ratios between pairs such as helium–nitrogen, and also because surface diffusion effects are often negligible. Experiments have been carried out at pressures as low as a few mm Hg, and less, and as high as 65 atm. The effect of temperature has not been as widely investigated, with most measurements made around room temperature. A reasonable spectra of solids have been tested, ranging from commercial catalysts to materials such as alumina, which have been pelleted especially so as to control porosity and macropore size.

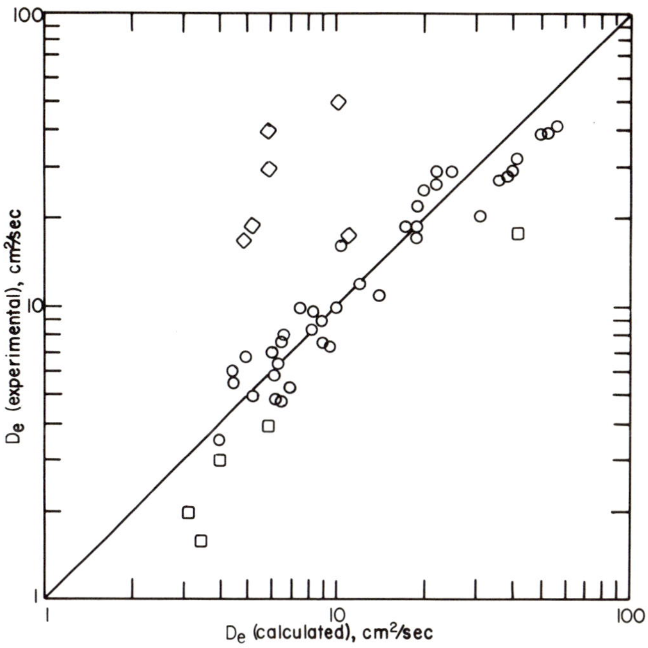

Figure 6. Comparison of experimental to calculated diffusivities [Weisz and Schwartz (32)]

Diffusion

Diffusion measurements in the Knudsen region have been reported for some 59 different samples of porous catalysts by Weisz and Schwartz (*32*). The Wicke-Kallenbach technique was employed, using apparatus described by Weisz (*31*), and hydrogen and nitrogen at atmospheric pressure as the test gases. The solids were various silica-alumina, silica-magnesia, and chromia-alumina catalysts with pore sizes usually well below 200 Å. Because the mean free path at atmospheric pressure is of the order of 1000 Å, Knudsen diffusion was assumed to dominate. Effective diffusivities, defined as

$$D_e = \frac{N_{AS}RTL}{P(y_{AL} - y_{A_0})} \qquad (38)$$

were reported for hydrogen, but no data are given for nitrogen. The data were compared initially to values calculated using the method proposed by Wheeler (*33*). Knudsen diffusivities were evaluated using an average pore radius defined as

$$\bar{r} = \frac{2\,V_t}{S} \qquad (27)$$

The experimental effective diffusivities were generally found to be somewhat lower than those calculated. From the structure of gel-type solids, Weisz and Schwartz argue that the pore structure is cellular rather than tubular as suggested by Wheeler. They subsequently take the effective diffusivity to be

$$D_e = \frac{\epsilon^2}{\sqrt{3}} D_K \qquad (39)$$

The porosity appears as ϵ^2 because of an area correction, and also from a random walk argument for progress from cell to cell. $\sqrt{3}$ appears as a correction to the diffusion length. D_K is calculated using $\bar{r} = 3\,V_t/S$, which represents the pores as spherical cells. With this model, the agreement between experimental and calculated results is excellent, as indicated by Figure 6. The few data points that deviate widely have bimodal structures or high densities. For the bimodal solids, the deviations are possibly caused by the broad pore size distribution, with high experimental values for D_e arising from large fluxes through large pores, which are not properly accounted for in the calculated data. The high density solids give low values for D_e, perhaps because of dead-end pores which contribute to the surface area and hence $\bar{r}$, but not to the flux. Thus it appears that the simple model given by Equation 39 will give good predictions for D_e if the pores are small with a narrow size distribution and if there are few dead-end pores.

Similar measurements have been made by Johnson and Stewart (*14*) for pelleted alumina catalysts having somewhat broader pore size distributions. The distributions were either bimodal or trimodal with pore sizes from 20 to 4500 Å. Some 26 samples having different macropore and micropore volumes were tested using nitrogen and hydrogen at atmospheric pressure and room temperature. Because of the pore size range, the data obtained were in the transition region. Effective diffusivities for H_2 were calculated from Equation 38 and compared with values predicted by using Equation 30 and a tortuosity $\Upsilon = 3$. No diffusivity data are given for N_2 and the observed flux ratio is not mentioned. Figure 7 shows the result. In most cases the experimental diffusivities agree with predicted

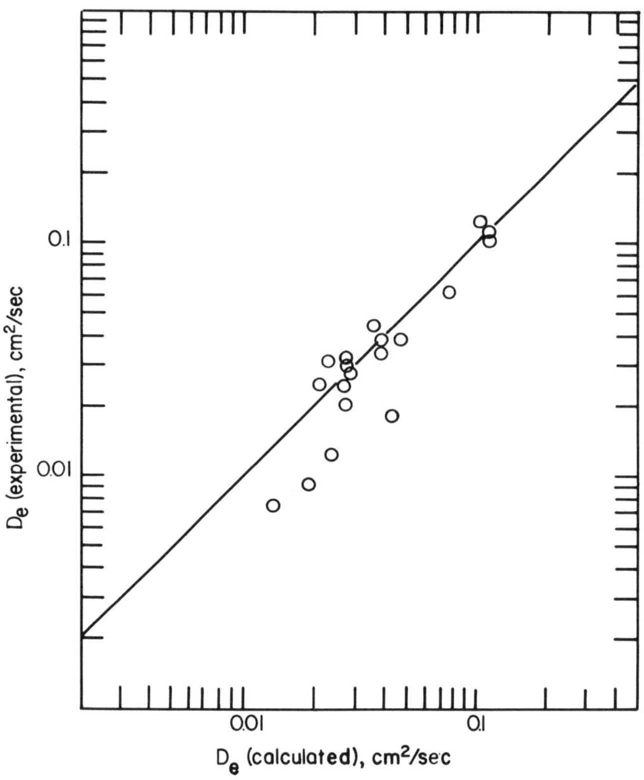

Figure 7. Comparison of experimental results to parallel pore model, $\Upsilon = 3$ [Johnson and Stewart (14)]

values within about $\pm 20\%$, although some of the data show considerably greater variation.

Satterfield and Cadle (*21*) made diffusion measurements with hydrogen and nitrogen for seventeen commercial pelleted catalysts and catalyst supports. Measurements were made at atmospheric pressure and 20° to 30°C. With the exception of two methanol synthesis catalysts, the experimental flux ratio for H_2/N_2 was -3.75 ± 0.11, which agrees very well with the theoretical value of -3.73. The flux ratios for the methanol synthesis catalysts were -3.46 and -3.53, and these low values were attributed to surface diffusion of nitrogen. The catalysts generally had wide pore size distributions, so Equation 31 was used to calculate tortuosities. The tortuosities were between 2.8 and 7.3 except for two materials, which were calcined at very high temperatures, that showed abnormally large values (11.1 and 79). These tortuosities agree well with those observed by Johnson and Stewart for similar solids. For comparison, Satterfield and Cadle also predicted the fluxes using the random pore model of Wakao and Smith (*29*). The agreement between predicted and experimental values was generally not very good, with deviations ranging from 8 to 160%.

In a second study, Satterfield and Cadle (*20*) made pure diffusion measurements for helium and nitrogen on five commercial catalysts over the pressure range 1 to 65 atm and at ambient temperature. Combined diffusion and flow measurements were also made for a single catalyst under similar conditions. The results of the diffusion studies indicated that surface diffusion of nitrogen was important for four of the five catalysts tested. The observed dependence of the helium flux on pressure was as predicted by Equation 31. At low pressures, the flux increases rapidly and at high pressures, when bulk diffusion dominates, the flux becomes essentially independent of the pressure. With the exception of one catalyst, the nitrogen flux continued to increase with pressures, thus indicating the presence of surface diffusion for N_2. Correspondingly, the flux ratios $-N_{He}/N_{N_2}$ decreased markedly as a function of pressure to values much below the theoretical value of 2.65. For the one catalyst where no surface diffusion was evident, the flux ratio was very close to the theoretical at all pressure levels. Tortuosities were calculated from the diffusion data using Equation 31 and as with the previous H_2–N_2 experiments all fall in the range 2.8–7.5. The tortuosities were largely independent of pressure, so that in principle a single diffusion measurement would suffice for extrapolation. The random pore model of Wakao and Smith (*29*) was found inadequate for extrapolation purposes, giving poor agreement with experiment and predicting trends with pressure incorrectly.

In contrast to the generally poor agreement between experimental and predicted results from the random pore model obtained by Satterfield and Cadle, several studies that show excellent agreement have been reported. Wakao and Smith (*29*) did nitrogen–helium counterdiffusion experiments in the pressure range 1–12 atmospheres using Boehmite pellets of various densities and

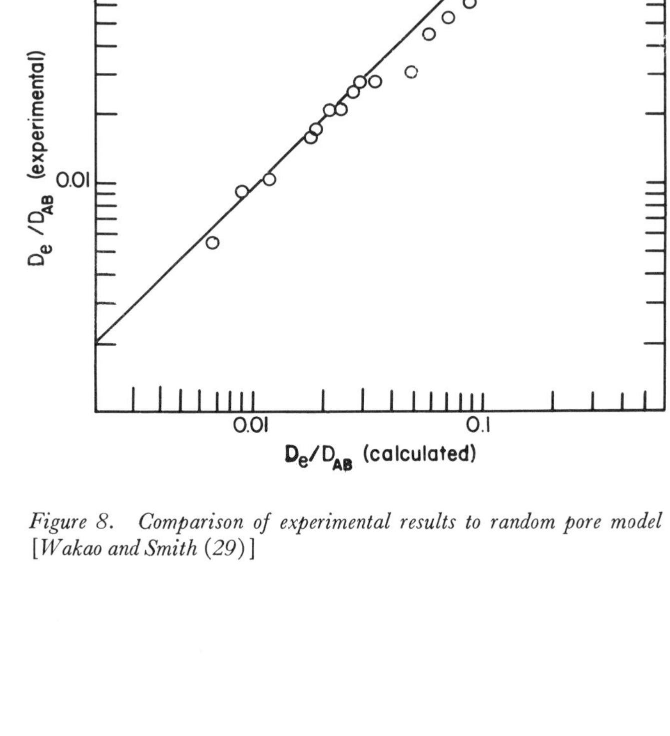

Figure 8. Comparison of experimental results to random pore model [Wakao and Smith (29)]

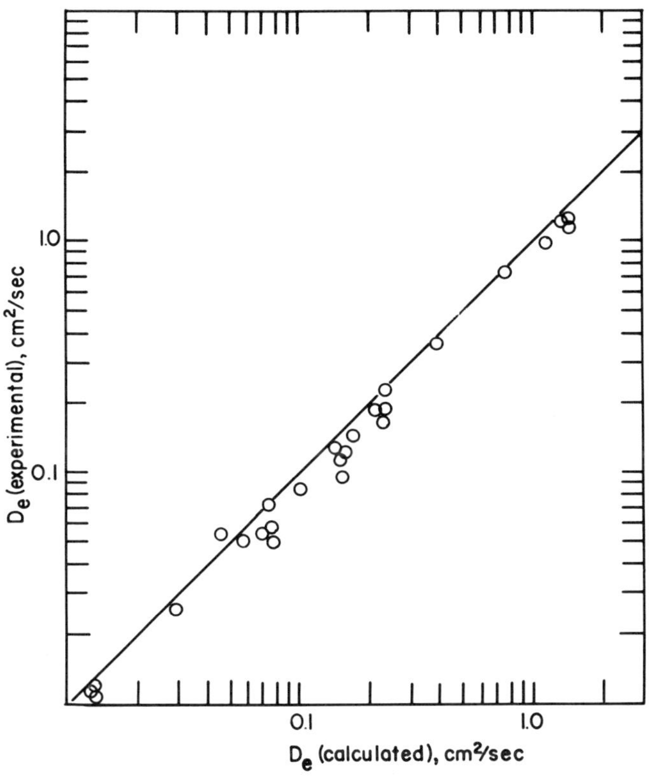

Figure 9. Comparison of experimental results to random pore model [Cunningham and Geankoplis (5)]

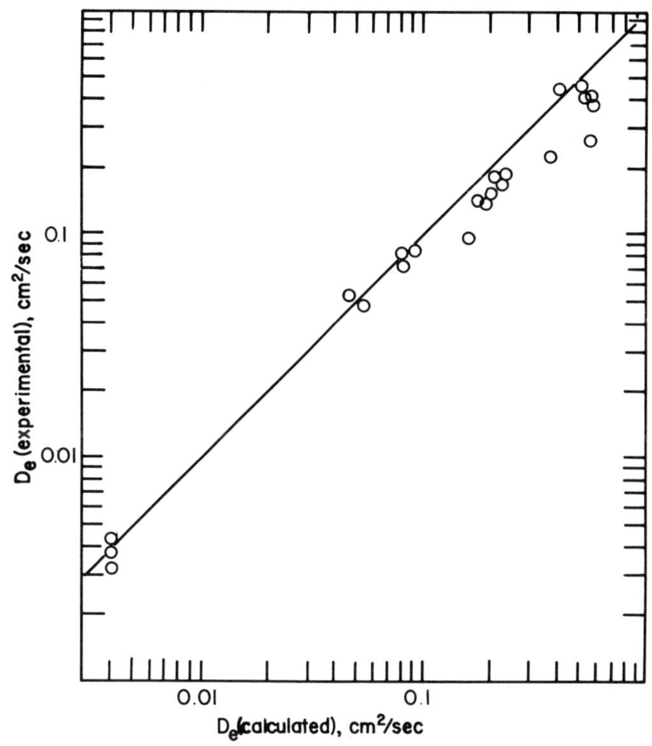

Figure 10. Comparison of experimental results to random pore model [Henry et al. (12)]

porosities. Henry *et al.* (*12*) did similar experiments for a much lower pressure range (0.5 to 600 mmHg) using alumina and Vycor, as did Cunningham and Geankoplis (*5*) using alumina. In all three studies, good agreement between experimental and predicted results was obtained as is indicated by Figures 8, 9, and 10. The flux ratios also agreed well with theoretical values. The test pellets were made from powder in a pellet press, so that the structure of the pellets closely approximates that used as the basis for the random pore model. The powder particles were themselves microporous (radii generally less than 100 Å). The size of the macropores, which consist largely of spaces between the particles, could be controlled over rather wide limits by the pressure applied in making the pellets. As is intuitively expected, the random pore model predicts that where the number and size of macropores is large—*i.e.*, low pellet density and large macropore radius—most of the diffusion occurs through the macropores with relatively small contributions from micropores and macropores in series with micropores. Cunningham and Geankoplis also used a random pore model for tridispersed pore size distributions to predict fluxes for two samples that had quite broad distributions. Agreement between experimental and predicted values was only slightly better using the tridispersed model rather than the bidispersed model.

Foster *et al.* (*10*) measured diffusion fluxes for the argon–helium system at pressures from 1 to 14.6 atm and temperatures from 0° to 69°C. Pelleted test samples from powders of iron and nickel oxide were supported on kieselguhr. The average of 102 observations of the flux ratio $-N_{He}/N_{Ar}$ was 3.04 ± 0.13 compared to the theoretical value of 3.16. Several low values were obtained, especially at high pressures, indicating that surface diffusion of argon may have occurred. The dependence of the measured fluxes with pressure agrees with model predictions. At low pressures the flux increases rapidly with pressure and at high pressures the flux becomes independent of the pressure. The effect of temperature on the fluxes is also that expected from theoretical predictions. For low pressures and/or small pores where Knudsen diffusion dominates, the flux is expected to be inversely proportional to $T^{1/2}$ at constant pressure. In large pores and/or high pressures where molecular diffusion dominates, the flux is directly proportional to $T^{1/2}$. In the intermediate region, fluxes will be more or less independent of temperature. These expected trends were confirmed by the experiments. At 1 atm the measured fluxes were essentially independent of temperature, while at 14.6 atm a modest increase in the fluxes was observed over the range tested. The flux data correlated well using the convergent–divergent pore model devised by Foster and Butt (*9*) and the mixing efficiency as an adjustable parameter. The predictive Wakao and Smith (*29*) random pore model, in contrast, gave very poor agreement. It was determined that the mixing efficiency could be correlated with the properties of the porous structures tested by a relationship of the form

$$(L/\bar{d})\ (\epsilon_s/\eta_E)^{1/2} = \beta \qquad (40)$$

where L = particle thickness, $\bar{d}$ the average pore diameter, ϵ_s the surface porosity taken as 2/3 the volume porosity, η_E the mixing efficiency, and β = a constant determined from experiment. For five materials examined, β was constant within $\pm 14\%$. The form of Equation 40 was justified by considering the solid as an Emmenthal structure.

Combined Diffusion and Flow

Very few experimental measurements for combined diffusion and flow have been reported. Such experiments were conducted by Satterfield and Cadle (*20*) for one of the catalysts noted above. They used three methods to correlate their data: a parallel pore model similar to that used for interpretation of diffusion measurements, a capillary tube model, and a dusty gas model. The parallel pore model is given by Equation 31. The experimental data were fit to the model using an iterative technique after correction of measured fluxes for surface diffusion. Various values of the tortuosity were assumed until values of end compositions determined by numerical integration matched the experimental values. The pressure was assumed to be a linear function of distance through the pellet. Sufficient cross passages between pores to assure uniform pressure and composition at any axial position were assumed to exist. Tortuosities determined in this way were remarkably constant, showing an 8% variation from an average value of 7.1 in excellent agreement with the tortuosity obtained from pore diffusion measurements, which was 6.9. The capillary tube model was not nearly as satisfactory for correlation of the data. Here, the flux is taken as the sum of the pure diffusion flux and the forced flow flux

$$N_T^{\circ} = N_T + F$$

The diffusion flux N_T was evaluated using the pure diffusion equation at the mean pressure. The forced flow contribution was evaluated using the developments of Wakao *et al.* (*30*). If the same tortuosity factor was applied to both terms, wide variations of Υ resulted, including several negative values. This deficiency was overcome somewhat by using different values for the diffusion and flow terms, but these results are difficult to reconcile with the results from the parallel pore model above since the two should be equivalent. In the dusty gas model, fluxes are described by

$$N_T^{\circ} = \frac{C_D}{RTL}\,\Delta P + \alpha N_A^{\circ} \qquad (41)$$

where C_D is the dusty gas constant. Evans *et al.* (*8*) neglecting Knudsen and slip flow terms, suggest that C_D should be linear with pressure. Satterfield and Cadle show that their data for C_D agree well with this prediction over the entire pressure range used.

Wakao *et al.* (*30*) obtained data for pure gas flow and combined diffusion and flow for a binary gas system using pellets of Boehmite and Vycor. The data, for nitrogen and hydrogen, were used to test their random pore model for diffusion and flow. For pure gas flow, average permeabilities were calculated from

$$N_{AS} = \frac{\bar{K}}{RT}\frac{\Delta P}{L} \qquad (42)$$

using measured values of the flux. For a single capillary, the permeability for a pure gas is given by

$$K = \frac{D_{KA}}{1 + \frac{D_{KA}}{D_{AA}}} + \frac{1}{1 + \frac{D_{AA}}{D_{KA}}}\left(\frac{\pi}{4}\,D_{KA} + \frac{r^2P}{8\,\mu}\right) \qquad (43)$$

When we use the random pore model, the permeability for the porous solid is

$$K = \epsilon_a^2 K_a + \frac{\epsilon_i^2(1 + 3\epsilon_a)}{1 - \epsilon_a}\,K_i \qquad (44)$$

For comparison with the experimental results, average values were calculated from this relationship using the arithmetic average pressure to evaluate the pressure dependent terms. As Wakao *et al.* note, little error is introduced by doing this. In most cases the predicted permeabilities were somewhat higher (by a few % to about 100%) than experimental values, but showed the same trends with pressure and pressure drop. For combined diffusion and flow an effective flow parameter C_e was introduced and evaluated from the experimental data:

$$C_e = \frac{RTL}{\Delta P}\,(m\,N_{AS} + N_{BS}) \qquad (45)$$

Values of C_e were predicted from the random pore model using arithmetic average values of y and P to evaluate the macropore, micropore, and series contributions to the flux from Equation 43. These contributions were combined according to the random pore model by

$$C_e = \epsilon_a^2 C_a + \frac{\epsilon_i^2(1 + 3\epsilon_a)}{1 - \epsilon_a}\,C_i$$

Because of the differences in pore size and porosity, as well as pressure, C_e varied by a factor of several thousand. Agreement between experimental and predicted values of C_e was quite good and generally better than for the pure gas experiments.

Conclusions

No *a priori* method for predicting diffusion and flow fluxes for gases through porous solids is yet available. However, on the basis of the relatively meager experimental data available, it appears that simple capillary models coupled with elementary knowledge of the pore size distribution and a few flux measurements may be used with reasonable success to extrapolate data. For solids with rather broad pore size distributions, parallel pore models have given good correlations of data obtained under a wide range of pressure conditions. Random pore models in some cases have been used with considerable success in predicting fluxes through pellets made from pressed microporous powders.

Whether these methods may be used to predict the

fluxes appropriate to diffusion limited reaction or adsorption conditions remains largely unanswered. For solids with broad pore size distributions, the models predict that most of the flux occurs through the large pores. No effect of dead-end pores is in the models. The surface area to volume ratio is greatest for small pores. Thus, the small pores, as well as dead-end pores, which may be of great importance to reaction or adsorption, are largely ignored by the models. In addition, most experimental flux measurements have been made using low-molecular weight and inert gases at relatively low pressures and temperatures. Little is known of the applicability of the models for large molecules or at more extreme conditions.

A further point of considerable importance should be made. Depending on the method of manufacture, the structure of porous solids may exhibit considerable heterogeneity that can cause the local diffusivity to vary with position. As Satterfield *et al.* and (*4*, *22*) have shown this effect may be quite large with the local diffusivity varying by about a factor of three. This suggests that agreement between theory and experiment in many cases may be at least partly fortuitous.

These facts point up the need for better experimental testing. Ideally, diffusivity or flux measurements obtained from reaction or adsorption rate measurements should be compared with model predictions and/or measurements calculated by other methods. Results obtained by transient methods such as the time-lag or chromatographic methods, when compared with steady-state results, may also provide some indication of the relative importance of small or dead-end pores. Experiments with a wider variety of diffusion gases and operating conditions are also indicated.

Nomenclature

C = defined by Equation 22, sq cm/sec
C_A = concentration of A, mol/cc
C_B = concentration of B, mol/cc
C_D = dusty gas constant, sq cm/sec
C_e = effective diffusion and flow parameter, sq cm/sec
C_a = diffusion and flow parameter for macropores, sq cm/sec
C_i = diffusion and flow parameter for micropores, sq cm/sec
C_0 = upstream gas concentration, mol/cc
D_{AA} = molecular self-diffusion coefficient, sq cm/sec
D_{AB} = molecular diffusion coefficient, sq cm/sec
D_{KA} = Knudsen diffusion coefficient for A, sq cm/sec
D_{KB} = Knudsen diffusion coefficient for B, sq cm/sec
D_{KA_a} = Knudsen diffusion coefficient based on macropore radius, sq cm/sec
D_{KA_i} = Knudsen diffusion coefficient based on micropore radius, sq cm/sec
D_N = effective diffusion coefficient defined by Equation 9, sq cm/sec
D_a = effective diffusion coefficient for macropores defined by Equation 32, sq cm/sec
D_i = effective diffusion coefficient for micropores defined by Equation 32, sq cm/sec
D_e = effective diffusivity defined by Equation 38, sq cm/sec
$\bar{d}$ = average pore diameter, cm
F = forced flow flux, mol/sq cm sec
f = fraction of molecules diffusely reflected at pore walls
$f(r)$ = pore size distribution function
$\bar{K}$ = average permeability, sq cm/sec
K_a = macropore permeability, sq cm/sec
K_i = micropore permeability, sq cm/sec
L = capillary length or pellet thickness, cm
L_e = pore length, cm
L_j = pore segment length, cm
M_A = molecular weight of A
M_B = molecular weight of B
N_A = diffusion flux of A, mol/sq cm sec
N_B = diffusion flux of B, mol/sq cm sec
N_T = $N_A + N_B$
N_T° = combined diffusion and flow flux, mol/sq cm sec
N_{AS} = flux based on total area of solid, mol/sq cm sec
n = number of pores per unit area of solids
P = total pressure, atm
P_0 = total pressure at $x = 0$, atm
P_L = total pressure at $x = L$, atm
P_{AL} = partial pressure of A at $x = L$, atm
P_{A0} = partial pressure of A at $x = 0$, atm
P_{BL} = partial pressure of B at $x = L$, atm
P_{B0} = partial pressure of B at $x = 0$, atm
Q = total quantity of gas passed through cell in time-lag apparatus, mol
R = gas constant, cc atm/°K
r = pore or capillary radius, cm
$\bar{r}$ = average pore radius, cm
S = surface area of solid, cm²/g solid
V = pore volume, cc/g solid
V_t = total pore volume, cc/g solid
$\bar{v}_A$ = mean molecular speed of A, cm/sec
$\bar{v}_B$ = mean molecular speed of B, cm/sec
x = axial distance through capillary or pellet, cm
y_{A0} = mole fraction A at $x = 0$
y_{AL} = mole fraction A at $x = L$
α = $1 + N_A/N_B$
β = constant defined by Equation 40
η_E = mixing efficiency
ϵ = porosity
ϵ_a = void fraction of macropores
ϵ_i = void fraction of micropores
ϵ_s = void area fraction
μ = viscosity
λ = mean free path
Υ = tortuosity

LITERATURE CITED

(1) Barrer, R. M., *J. Phys. Chem.*, **57**, 35 (1953).
(2) Barrett, R. P., Joyner, L. G., and Halenda, P. P., *J. Amer. Chem. Soc.*, **73**, 373 (1951).
(3) Bosanquet, C. H., cited in Ref. 16.
(4) Cadle, P. J., and Satterfield, C. N., Ind. Eng. Chem., Fundam., **7**, 192 (1968).
(5) Cunningham, R. S., and Geankoplis, C. J., Ind. Eng. Chem., Fundam., **7**, 535 (1968).
(6) Davis, B. R., and Scott, D. S., Symposium on Fundamentals of Heat and Mass Transfer, 58th Annual Meeting, American Institute of Chemical Engineers, Philadelphia, Pa., 1965.
(7) Evans, R. M., Watson, G. M., and Mason, E. A., *J. Chem. Phys.*, **35**, 2076 (1961).
(8) Evans, R. M., Watson, G. M., and Mason, E. A., *ibid.*, **36**, 1894 (1962).
(9) Foster, R. N., and Butt, J. B., *A.I.Ch.E. J.*, **12**, 182 (1966).
(10) Foster, R. N., Butt, J. B., and Bliss, H., *J. Catal.*, **7**, 179, 191 (1967).
(11) Goodknight, R. C., and Fatt, I., *J. Phys. Chem.*, **65**, 1709 (1961).
(12) Henry, J. P., Cunningham, R. S., and Geankoplis, C. J., *Chem. Eng. Sci.*, **22**, 11 (1967).
(13) Hewitt, G. F., and Sharratt, E. W., *Nature*, **198**, 952 (1963).
(14) Johnson, M. F. L., and Stewart, W. E., *J. Catal.*, **4**, 248 (1965).
(15) Knudsen, M., *Ann. Phys.*, **28**, 75 (1909).
(16) Petersen, E. E., *A.I.Ch.E. J.*, **4**, 343 (1958).
(17) Pollard, W. G., and Present, R. D., *Phys. Rev.*, **73**, 762 (1948).
(18) Ritter, H. L., and Drake, R. C., Ind. Eng. Chem., Anal. Ed., **17**, 787 (1945).
(19) Rothfeld, L. B., *A.I.Ch.E. J.*, **9**, 19 (1963).
(20) Satterfield, C. N., and Cadle, P. J., Ind. Eng. Chem., Fundam., **7**, 202 (1968).
(21) Satterfield, C. N., and Cadle, P. J., Ind. Eng. Chem., Process Des. Develop., **7**, 256 (1968).
(22) Satterfield, C. N., and Saraf, S. K., Ind. Eng. Chem., Fundam., **4**, 451 (1965).
(23) Schneider, P., and Smith, J. M., *A.I.Ch.E. J.*, **14**, 762 (1968).
(24) Scott, D. S., and Dullien, F. A. L., *ibid.*, **8**, 113 (1962).
(25) Scott, D. S., and Dullien, F. A. L., *ibid.*, p 293.
(26) Scott, D. S., and Dullien, F. A. L., *Chem. Eng. Sci.*, **17**, 77 (1962).
(27) Spiegler, K. S., Ind. Eng. Chem., Fundam., **5**, 529 (1966).
(28) van Deemter, J. J., Zuidereg, F. J., and Klinkenberg, A., *Chem. Eng. Sci.*, **5**, 271 (1956).
(29) Wakao, N., and Smith, J. M., *Chem. Eng. Sci.*, **17**, 825 (1962).
(30) Wakao, N., Otani, S., and Smith, J. M., *A.I.Ch.E. J.*, **11**, 435 (1965).
(31) Weisz, P. B., *Z. Phys. Chem.*, **11**, 1 (1957).
(32) Weisz, P. B., and Schwartz, A. B., *J. Catal.*, **1**, 399 (1962).
(33) Wheeler, A., *Catalysis*, **2**, 105 (1955).
(34) Wicke, E., and Kallenbach, R., *Kolloid Z.*, **97**, 135 (1941).

5

Non-Newtonian Flow Through Porous Media

Non-Newtonian Flow Through Porous Media

J. G. SAVINS

A review of rheologically complex flow phenomena likely to have a role in mechanisms responsible for behavior observed in non-Newtonian flow through porous media

Contemporary literature relates interesting descriptions of how rheologically complex fluids behave in flow through a variety of porous media. From these studies, several correlations of flow phenomena with the rheology of grossly "non-Newtonian" and "Newtonian-like" fluids have been described or attempted. In the present paper, we will be mainly concerned with the phenomenological aspects of these effects. These occur under different flow conditions in porous structures, and seem to be coupled with the unusual rheological behavior of complex fluids.

We will begin with a necessarily restricted overview of non-Newtonian fluids, focusing our attention on those rheological characteristics which appear to describe, at least under conditions of viscometric flow, the fluid systems of contemporary interest in studies of non-Newtonian flow through porous media. Phenomena likely to have a role in mechanisms responsible for the behavior observed in non-Newtonian flow through porous media will also be recognized. Some of these non-Newtonian effects have long been recognized and treated as the principal modes of expression of "non-Newtonian behavior." Other effects have not been considered such in the past. Ironically, as chemically distinct non-Newtonian systems are investigated, these equally important and different manifestations of complex flow behavior have been "rediscovered," so to speak.

We will make some brief remarks concerning the diverse kinds of porous media which have been used in experiments involving complex fluids. No attempt will be made to discuss the structure and properties of porous materials *per se*, or to consider topics related to pore size distribution, local voidage characteristics, etc. These topics are thoroughly reviewed elsewhere in this symposium. A brief summary of the characteristics of known fluid systems relating to non-Newtonian flow through porous media will be presented. We will deal with the generally accepted techniques for reduction and analysis of data from flow of non-Newtonian fluids in porous media. We will consider, next, examples out of the literature which illustrate the fascinating, and sometimes bizarre, flow behavior of complex fluids in different models of porous media. Different descriptions are found in the literature. Several mechanisms explaining the different phenomena will be illustrated and discussed. Finally we will consider some applications to show the relevance of non-Newtonian flow through porous media to processes of technological importance. Applications in such diverse areas as oil recovery and naval architecture will be noted, and examples from the patent art will be cited.

Rheologically Complex Flow Behavior

As indicated earlier, we will deliberately restrict ourselves to an overview of a few examples or categories of rheologically complex flow behavior. Mainly those properties which have been shown to be characteristic in some significant way of the behavior of fluid systems of interest in studies of non-Newtonian flow through porous media will be considered. Inasmuch as it is the most obvious flow property, and since most published studies involve this function in correlations of these systems, we will direct major attention to the gradient-dependent viscous resistance characteristics of complex fluids. Certain viscoelastic phenomena will also be recognized. Interestingly, the shear rate range of interest in non-Newtonian flow through porous media may range from values on the order of one sec^{-1} in technological applications related to oil recovery, to 10^5 sec^{-1} in problems related to naval architecture and bearing applications. There are a variety of other categories of complex flow behavior which are beyond the scope of this review. Some of these other properties may very well be significant in a more general discussion of non-Newtonian flow through porous media. These other rheological properties may warrant closer scrutiny as other candidate fluid systems are considered in such studies. Hierarchies of rheologically complex flow behavior are discussed by several authors (*1*, *2*).

Gradient dependent viscosity. The characteristic most frequently cited in discussions of the behavior of complex fluids is that of the shear rate dependent viscosity function $\eta(\Gamma)$ derived from viscometric experiments (symbols are defined in the Nomenclature section). The gradient dependent viscosity function takes on a variety of forms in real fluids, *e.g.* (Figures 3–7 in Reference *2*). The apparent viscosity may appear to decrease monotonically with increasing Γ over a finite but limited range of experimental conditions, in which case the descriptor "shear thinning" may be applied. Alternatively, $\eta(\Gamma)$ may appear to increase monotonically with increasing Γ, *i.e.*, exhibit a "shear thickening" behavior. Over wider ranges of shearing conditions, one usually observes such behavior coupled with a "Newtonian-like" response, although combinations of shear thinning and shear thickening behavior are not uncommon (*3*). Of course, other combinations of gradient dependent behavior cannot be excluded either (*2*).

Shear thinning. A monotonic decrease in apparent viscosity with increasing shear rate is readily observed over easily accessible ranges of flow conditions in polymeric melts and their solutions, moderately concentrated suspensions, colloidal dispersions, biological fluid systems, etc. Actually, as shown in Figure 4 of Reference *2*, the $\eta(\Gamma)$ function for the shear thinning fluid has three more or less distinct regimes: a Newtonian behavior at low Γ characterized by a limiting viscosity η_0; a nonlinear region characterized by a spectrum of apparent viscosities—*i.e.*, the non-Newtonian region; and a

Table I. Functions for Describing Gradient Dependent Viscosity Behavior Under Steady Simple Shearing Conditions

Descriptor	Function	Reference
Ostwald-de-Waele (power law)	$\eta(\Gamma) = K(\Gamma)^{n-1}$	(14)
Ellis	$\frac{1}{\eta(\Gamma)} = \frac{1}{\eta_0}\left[1 + \frac{\tau}{\tau_{1/2}}\right]^{\alpha-1}$	(14, 15)
Oldroyd (3 Constant)	$\eta(\Gamma) = \eta_0\left[\frac{1 + 2/3\lambda_1\lambda_2\Gamma^2}{1 + 2/3\lambda_1^2\Gamma^2}\right]$	(16, 17)
Bernstein-Kersley-Zapas	$\eta(\Gamma) = -2\int_0^\infty \alpha'(\xi)\Gamma^2\xi^3 d\xi - 2\int_0^\infty \frac{\beta'(\xi)\xi d\xi}{1 + 2/9\Gamma^2\xi} - 2\int_0^\infty C'(\xi)\xi d$	(18)
Bogue	$\eta(\Gamma) = \frac{G_1\lambda_1}{1 + \alpha\lvert\Gamma\rvert\lambda_1} + \frac{G_2\lambda_2}{1 + \alpha\lvert\Gamma\rvert\lambda_2} + \ldots.$	(19)
Pao	$\eta(\Gamma) = \int_{-\infty}^{\infty} \frac{H(\lambda)d\ 1n\ \lambda}{1 + \Gamma^2\lambda^2} + \Gamma^2 \frac{\left\{\int_{-\infty}^{\infty} \frac{\lambda^2 H(\lambda)d\ 1n\ \lambda}{1 + \Gamma^2\lambda^2}\right\}^2}{\left\{\int_{-\infty}^{\infty} \frac{H(\lambda)d\ 1n\ \lambda}{1 + \Gamma^2\lambda^2}\right\}}$	(19, 20)

Newtonian behavior at high Γ, characterized by a limiting viscosity η_∞. In the case of polymeric systems, such details of the $\eta(\Gamma)$, Γ curve as the ratio of η_0/η_∞, the extent of the different flow regimes, and the coordinates of the shear rates at the inflection points separating the Newtonian and non-Newtonian regions, are affected by a number of morphological parameters, the concentration of solute, and the characteristics of the solute-solvent system (*4–7*). A very large number of examples of shear thinning behavior have been documented in various reviews (*2, 5, 6*). A number of molecular theories of the gradient dependence of polymeric melts and dilute and concentrated solutions have been developed (*5, 6, 8–13*). A variety of empirical viscosity functions and continuum-theory derived viscosity functions have been described. A very few examples of these functions are given in Table I. Functions such as these and others have been compiled by Reiner (*14*), White and Metzner (*21*), Peterlin (*6, 8*), Spriggs (*22*), and Spriggs *et al.* (*23*). Some of these functions are useful only for approximately curve fitting the $\eta(\Gamma)$, Γ behavior. The derived parameters have no mechanistic significance. Others are so complicated as to be useless from a practical point of view. The Ostwald-de-Waele or "power law" model (*14*) and the Ellis model, as originally formulated (*14, 15*), are empirical functions. The deficiencies of such models have been discussed by numerous authors. However, their inherent simplicity plus the documented fact that such functions approximate the $\eta(\Gamma)$ behavior of shear thinning fluids over useful or accessible ranges of flow conditions have led to their use in a number of special applications (*24–26*). The power law and Ellis models are of interest here because, as will be shown later, the "capillary model" for flow of Newtonian fluids in porous media has been extended to these empirical models as devices to approximate the flow behavior of certain fluids which exhibit a shear thinning behavior both in viscometric flows and in flow experiments involving porous media. The other viscosity functions shown in Table I have been derived from rheological models which exhibit varyng degrees of success in simulating the $\eta(\Gamma)$, Γ response and other behavior observed in real shear thinning fluids—*e.g.*, normal stresses, recoil, and relaxation. All of these viscosity functions readily simulate a gradient-dependent viscosity function of the shear thinning type. In the case of the power law model, $n < 1$ generates this behavior. Setting $\eta_\alpha(\Gamma) = 0$ in the *BKZ* model also leads to a shear thinning response. For critiques of the advantages and limitations of the rheological models from which the functions listed in Table I have been derived, that of Spriggs (*22*), and discussions by Spriggs and Bird (*27*), Bogue and Doughty (*28*), Pao (*20*), White and Metzner (*21*), and Metzner *et al.* (*29*) are recommended. As noted by Bogue and Doughty (*28*), the evidence of similarities between the steady state $\eta(\Gamma)$ and the dynamic viscosity function, and the observed correlations between pronounced non-Newtonian viscosity and strong normal stress effects, for example, illustrate the interdependence between the diverse phenomena characteristic of complex fluids.

Shear thickening. Perhaps the most controversial and least understood gradient-dependent viscosity behavior is that in which an increase in viscous resistance with increase in shear rate is observed over some part of the shear rate spectrum. Moderately concentrated to highly concentrated solutions of polymeric or micellar materials, proteins, and protoplasm, oftentimes with less than 1 wt % solute, have been observed to exhibit this viscosity behavior. Numerous investigators have reported the occurrence of "shear thickening behavior," but other descriptors such as "negative thixotropy," "antithixotropy," and "rheopexy," have also been applied, depending upon whether the observed thickening action is reversible or accompanied by time-dependent effects. In the contemporary literature, one still occasionally finds shear thickening equated with the "dilatant" behavior observed in suspensions and slurries

Table II. Mechanisms Advanced to Explain Gradient- and Time-Dependent Shear Thickening Phenomena

Mechanism	
Large-scale network associations	(*31*)
Inter- and intramolecular bonding and cross-linking of hydrogen bonds	(*32*)
Chain unwinding and intensification of intermolecular interactions	(*33, 34*)
Formation of complete hydrophobic regions	(*35*)
Strong structures built up from quasicrystallites in solution	(*36, 37*)
Inter- and intramolecular bridges	(*38*)
Intramolecular effects due to the isolated molecule	(*39, 40*)
Competition between the formation of rheological units under shear and stress relaxation	(*41*)
Inability of the flowing volume element to respond to the shear field	(*42, 43*)
Temporary network formation during shearing	(*44, 45*)
Intermolecular effects	(*46*)
Nonlinear behavior analogous to solid deformation	(*47*)

of particulate materials at high volume concentrations. Shear thickening behavior is of particular interest in connection with non-Newtonian flow through porous media because, as will be shown later, certain dilute polymeric solutions which exhibit shear thinning behavior in a viscometric flow seem to exhibit a shear thickening response under appropriate conditions of flow through porous media. Combinations of shear thinning and shear thickening behavior, as well as Newtonian behavior and shear thickening behavior, accompanied by elastic effects, have been documented (*3*). This review of prior work reveals that clear instances of shear thickening have been documented for a large number of diverse systems, including micellar or detergent systems, aqueous solutions of poly(methacrylic acid), dilute solutions of poly(methylmethacrylate) in a variety of solvents, and aqueous solutions of Dextran prepared with a highly purified enzyme of the strain *Leuconostoc mesenteriodes* B512F. Peterlin and coworkers (*30*) have ample experimental confirmation of a theory that with the appropriate solute molecular weight (preferably high) and appropriate solvent viscosity (preferably high), a variety of shear thickening phenomena can occur. As noted in (*3*), several mechanisms have been advanced to explain specific occurrences of gradient and time-dependent shear thickening phenomena, as illustrated by the summary given in Table II. Some of the viscosity functions shown in Table I will simulate a gradient-dependent viscosity of the shear thickening type. In the case of the power law model, $n > 1$ generates a monotonically increasing shear thickening response. Varying degrees of shear thickening may be simulated by the *BKZ* model when $\eta_\alpha(\Gamma)$, $\eta_\beta(\Gamma)$, and η_c all contribute to the viscosity.

Viscoelastic effects. Included here are such well-known and thoroughly documented effects as normal stress behavior, *e.g.* the "Weissenberg climb" effect, "jet swell" effect, and a variety of memory effects, including stress relaxation, elastic recoil, stress recovery, etc. Some of these effects are observable under steady flow (viscometric) conditions. Other distinctly elastic-like effects become detectable in nonsteady and non-viscometric flow situations. The detection of elastic effects in very dilute aqueous solutions, *circa* a few ppm, is very difficult. Often the presence of these elastic effects is inferred indirectly, as by the occurrence of drag reduction phenomena under nonlaminar flow conditions (*48*). As in Figure 1 of Reference *49*, it is not unusual for the normal stress difference to become considerably larger than the shear stress. This behavior is characteristic of a variety of polymer melts and their solutions, and micellar or detergent systems. Normal stress behavior cannot be predicted or interpreted by the power law or Ellis models. Various theories have been devised to describe this behavior and other examples of linear and nonlinear viscoelastic phenomena. The continuum or phenomenological theories of incompressible memory fluids developed by Green *et al.* (*50*), and by Coleman and Noll (*51, 52*), the general theory of Oldroyd (*53, 54*), and the theories of Pao (*20*), Bogue *et al.* (*19*), Bernstein *et al.* (*18, 55*) and White and Metzner (*21*), are examples. For example, in steady shearing motion of an incompressible second-order memory fluid, the stress response can be described by a zero shear viscosity coefficient η_0 and two parameters (β, γ), related to normal stress and viscoelastic effects. For a Cartesian coordinate system x_1, x_2, and x_3, and velocity components,

$$\begin{aligned} v_1 &= \Gamma x_2 \\ v_2 &= v_3 = 0 \end{aligned} \tag{1}$$

the stresses are given within terms of order of two in the shear rate as (*56*).

$$\frac{\tau_{12}}{\Gamma} = \eta_0 \tag{2a}$$

$$\frac{\tau_{11} - \tau_{22}}{\Gamma^2} = -2\gamma \tag{2b}$$

$$\frac{\tau_{22} - \tau_{33}}{\tau_{11} - \tau_{22}} = -\left(1 + \frac{\beta}{2\gamma}\right) \tag{2c}$$

With regard to the behavior predicted by Equations 2, some dilute polymeric solutions seem to be characterized by Newtonian-like viscosity behavior at low but accessible shear rates. A quadratic dependence of normal stress differences on shear rate over a modest range of flow conditions has also been documented (*57*). However, at higher shear rates a pronounced gradient-dependent viscosity effect is generally observed. Under these conditions experimental evidence suggests the primary normal stress difference is approximately proportional to the first power of the shear rate (*58*). Also, there is evidence that the point on the $\eta(\Gamma)$, Γ curve at which a fluid begins to deviate from Newtonian behavior, and the resultant rate of change of apparent viscosity with shear rate depend, in large measure, on the molecular weight distribution (*5, 59*). Considerable controversy exists regarding the relative magnitudes of the normal stress differences given by Equations 2b and 2c. There is evidence that $(\tau_{22} - \tau_{33})$ is considerably smaller than the primary normal stress difference. Both positive and negative values for the coefficient of the secondary normal stress difference have been reported (*49*). For a summary and critique of some of the predictions of these phenomenological theories, refer to References *9, 19, 21, 23, 28,* and *29*. Most of these theories have been partially successful in describing some of the observed features of viscoelastic phenomena. Naturally none of them seem to be able to simulate, without ambiguity, all features of viscoelastic phenomena and certain other effects observed in complex fluids. Most of these theories and special cases of the more general theories are naturally subject to certain assumptions. Typical assumptions include: (i) an absence of peculiarities within the fluid or between the fluid and the confining boundaries which would cause the no-slip criterion to be violated; (ii) no preferred orientation by the fluid with respect to the confining boundaries; (iii) absence of a yield stress; (iv) no time-dependent behavior of the type referred to as thixotropy, rheopexy, etc.; and (v) the fluid is a continuum rather than a collection of discrete particles. Certain of these complications have been considered by Giesekus (*60*), Ericksen (*61*), Eringen (*62*), Leslie (*63*), and Stokes (*64*), among others. Molecular theories have also been developed which attempt to portray certain viscoelastic phenomena and non-Newtonian viscosity behavior in terms of inter- and intramolecular forces, polydispersity, coil rigidity, network and entanglement effects, and other morphological parameters. For comparisons of these different theories, the reviews of van Oene (*9*), Peterlin (*6, 8*), Graessley (*4*), and Kotaka and Osaki (*65*) may be consulted. As noted (*65*), some of the molecular models also satisfy certain of the phenomenological equations, while others fail in predicting a realistic dependence of stresses on shear rate.

Coupling of Rheological Behavior and Boundary Effects

There is a voluminous literature which illustrates that under appropriate conditions the steady flow behavior of some non-Newtonian materials may become coupled with the type and dimensions of the flow channel, *e.g.*, capillary diameter in Poiseuille flow, and spacing between inner and outer cylinders in Couette flow. Very recently, theories have been devised by Eringen (*62*), Leslie (*63*), Stokes (*64*), and Ericksen (*66*), which predict the occurrences of macroscopic effects in certain flow geometries and to degrees which depend on the scale of the geometry. Scott-Blair (*67*), Mooney (*68*), and Oldroyd (*69*) were among the first to investigate these coupling effects. In a careful documentation, Scott-Blair (*70*) distinguishes between positive and negative wall effects. An example of a positive wall effect is the case where a solute-deficient layer exists adjacent to the boundary. An example of a negative wall effect is the case where the solute adsorbs onto the solid boundary. The latter is a serious problem in intrinsic viscosity measurements. Wall effects are of interest here for several reasons. Briefly, we note that Kozicki *et al.* (*71*) claim the analysis of some experimental data pertaining to non-Newtonian flow through porous media reveals the existence of an anisotropic anomalous layer on the surface of the media. Burcik (*72*) claims that the unusual behavior observed on flowing slightly shear thinning fluids through porous media results from polymer adsorption or retention within the pore structure. Processes for enhancing the recovery of oil from subterranean formations have been disclosed which embody complex fluids which exhibit a coupling between rheological behavior and the type and scale of the flow geometry (*73–84*). Lastly, the fact that rheological behavior and flow geometry can become coupled under certain conditions gives renewed warnings that subtle complications, if not accounted for in the analysis of viscometric data, can lead to an erroneous description of rheological behavior because of an inadequate interpretation of viscometric data.

There are a variety of ways that a boundary or "wall" effect can develop in a flow field. Numerous real fluids exhibit such behavior. Lucid treatments of wall effects in tube flow are given by Oldroyd (*69*) and Reiner (*14*), among others. In the presence of anomalous behavior, in a region close to the solid boundary, the well-known relationship between pressure loss and flow rate for the steady flow called Poiseuille or capillary flow becomes

$$\frac{V' - V_s}{R} = \frac{1}{\tau_w{}^3} \int_{0,\phi}^{\tau_w} \tau^2 \Gamma(\tau) d\tau \qquad (3)$$

Here V' is the observed bulk velocity when the effective velocity of slip V_s is finite and

$$\tau_w = \frac{R \Delta P}{2L} \qquad (4)$$

is the shear stress at the boundary expressed in terms of the radius of the tube R and the pressure drop ΔP over length L. Methods for evaluating V_s and correcting the data to regain some measure of the homogeneous rheological behavior of the material in the region remote from the boundary are described in (*2, 24*). In the analysis leading up to Equation 3, it is assumed that V_s is not a function of radius, but only a function of τ_w.

Most published studies relating to anomalous behavior near the solid boundary in capillary flow indicate that V_s is positive so that within the anomalous layer, some normal distance δ from the wall, the material has a viscosity lower than that of the material remote from the boundary. By contrast, when V_s is negative it can be interpreted to mean that the anomalous layer is more viscous than the bulk of the material. Wall effects have also been observed in Couette flow (*68*). Thus, by the occurrences of such phenomena we have situations where the gradient-dependent viscosity function can depend upon the dimensions or scale of a flow channel. Rheological behavior which has apparently been interpreted as resulting from a positive wall effect has been reported for a variety of systems, *e.g.* dilute solutions of a moderately high-molecular-weight grade of poly(methylmethacrylate) in monochlorobenzene (*85*), GRS lattices (*86*), aqueous solutions of the sodium salt of carboxymethylcellulose (*87*), blood (*88*), and association colloids (*89*). Adsorption effects have been reported from studies of non-Newtonian flow through porous media (*71, 72, 90–93*). Hence the possibility of negative wall effects arises. In this regard, it is of interest to recall the studies of Rowland *et al.* (*94*) and Little (*95*). An interesting observation of a solute deficient wall layer, induced by applying an oscillatory motion to a suspension of rigid spheres, has been reported by Shizgal *et al.* (*96*). A pronounced radial migration at very small Reynolds numbers of liquid drops away from the boundaries, and the absence of such effects with rigid particles has been observed in several studies. At Reynolds numbers on the order of 10^{-4}, particles have been observed to migrate inward from the boundaries and outward from the axis of the flow channel (*97*). Another important observation is radial migration of rigid spheres suspended in a viscoelastic carrier at vanishingly small Reynolds numbers (*97*). Obviously, these phenomena create additional opportunities for a coupling to occur between rheological behavior and the type and scale of the flow geometry.

As might be anticipated, a variety of mechanisms seem to explain, qualitatively at least, specific occurrences of boundary or wall effects. Oldroyd (*69*) argues that the presence of a boundary could exert an orientation effect on a flowing solution containing extended macromolecular segments. The result could be a region of anisotropic rheological behavior extending over a considerable fraction of the flow cross section. A summary of some other mechanisms is given in Table III. For a critical analysis of some of these theories, the interested reader is referred to Brenner (*113*) and Cox and Brenner (*114*).

Table III. Mechanisms Advanced to Explain Specific Occurrences of Boundary or Wall Effects

Mechanism	References
Least action forces	(*98, 99, 100*)
Preferred particle orbit	(*101*)
Fluid layers arising from boundary constraint	(*102, 103*)
Negative concentration gradient	(*99, 104*)
Magnus effect	(*105, 106*)
Interactions between walls and outer layers of particles	(*107*)
Radially directed forces arising from shear of deformable particles	(*108*)
Finite migration velocity	(*109, 110*)
Transverse forces associated with nonlinear terms in Navier-Stokes equations	(*111, 112*)

A coupling between rheological behavior and the type and scale of the flow geometry seems to occur naturally in theories devised by Eringen (*62*), Leslie (*63*), Stokes (*64*), and Ericksen (*66*). In a contemporary review of the continuum theories of liquid crystals, Ericksen (*115*) remarks that when the channel width is sufficiently small, anomalous values of apparent viscosity result, "....the effect being as if the gap were smaller than it is in fact...." There is some feeling that adsorption phenomena are involved in the observed boundary layer effect. Ericksen also notes it is not unlikely that size effects may occur in systems whose microstructures are quite unlike those of liquid crystals (*116*). Leslie (*63*) has developed a theory of the viscometry of liquid crystals in which the apparent viscosity is a complicated function of shear rate and the gap between cylinders in the Couette geometry. Stokes' theory of couple stress fluids (*64*) admits macroscopic effects in certain geometries and to degrees which depend on the scale of the geometry. In the case of Couette flow, the couple stresses are absent so the theory predicts no coupling between observed rheological behavior and flow geometry. However, a dependence of apparent viscosity on channel dimensions is predicted to occur in Poiseuille and plane Poiseuille flows of the couple stress fluid. Similarly, Eringen (*62*), and Ariman and Cakmak (*117*) have shown that a coupling between rheological behavior and flow geometry is predicted to arise quite naturally in some basic viscous flows of micropolar fluids.

Porous Materials of Interest

A variety of "porous media" have been used in flow studies involving rheologically complex fluids. In-

AUTHOR *J. G. Savins is a Research Associate with the Mobil Research and Development Corp., Field Research Laboratory, Dallas, Tex.* *Mr. Savins thanks R. F. Burdyn, W. R. Foster, and A. B. Metzner University of Delaware, S. A. Williams of Mobil, and Eugene Wissler University of Texas for their encouragement, suggestions, and data contributions. He is also indebted to the Mobil Research and Development Corp. for allowing publication of this material.*

Table IV. Characteristics of Porous Media Used in Flow Studies Involving Rheologically Complex Fluids

Material	Geometry	Av. Particle Diam. Range (cm)	Permeability Range (darcies)	Porosity Range (%)	Remarks	References
Unconsolidated Porous Media						
Glass	Spheres	0.0053–0.50	0.522–532	27–40	Mostly homogeneous packs, but some binary mixtures	(*71, 90, 118, 119, 121–133*)
Plastic	Spheres	0.348		38–41		(*134*)
Steel or lead	Spheres	0.1124–0.3175				(*71, 90, 121, 130*)
Glass wool	Compressed slab		2–4		Radial flow	(*135*)
Sand and sand-fire-clay mixtures			0.40–179		Mostly linear flow; some packs up to 40 ft in length	(*125, 126, 135, 136*)
Aluminum oxide	Irregular	0.147		37		(*134*)
York mat	Mesh					(*124*)
Crushed sandstone			0.622	36.3		(*137*)
Consolidated Porous Media						
Sandstone	Cylindrical core, disk, or slab		5.74×10^{-4} to 1.92	16–27	Mostly linear but some radial and stratified flow	(*91–93, 123, 135–143*)
Alundum	Cylindrical core		0.028–4.16	23–35	Linear	(*92, 141, 142, 147*)
Sintered glass	Disk				Nominal pore size range from 1 to 77 μ	(*140, 143*)
Sintered bronze	Disk		46.4	48.6	Av. pore diam. of 0.01266 cm	(*120*)

cluded are models of porous media such as sand packs and matrices of uniformly packed spheres and woven screen, as well as alundum plugs, sandstone cores, porous metal disks, sintered glasses, and compressed glass wool. Each example of a model porous medium is in itself somewhat unique in geometric morphology, and there are formidable problems in precisely defining the flow conditions existent within any particular porous structure. Several of the published investigations concerned with more fundamental aspects of the flow of non-Newtonian fluids in porous media have involved the use of unconsolidated models, such as packed columns of spheres, fabricated from such diverse materials as lead, glass, plastic, and steel. Some consolidated models have also been used in these studies, including powdered bronze metal (compressed, sintered, and processed into a porous structure), sandstone plugs, and fritted glass disks. The models of porous media used in these particular studies have generally been incorporated into small flow cells, and the permeabilities are quite varied, usually somewhat high in the case of the unconsolidated models. For example, the unconsolidated models used by Christopher (*118*) consisted of flow cells with a nominal internal diameter of 1 in. and lengths ranging from 3 to 10 in. The diameters of the glass beads used as packing materials were in the 710- to 840-μ range, and the bed permeability was on the order of 445 darcies. In the glass bead packs used by Dauben (*119*), bead diameters ranged from 53 to 300 μ and bed permeabilities varied between 2 and 18 darcies. The characteristics of the consolidated porous media used in these studies have been equally diverse. For example, McKinley *et al.* (*91*) used a 15-cm long sandstone core characterized by a permeability of 0.305 darcies, while Gogarty (*93*) used sandstone cores, 1 in. in diameter by 3 in. in length, characterized by permeabilities in the range from 0.066 to 0.395 darcy. In contrast, the consolidated model used by Marshall and Metzner (*120*) consisted of a sintered bronze porous disk, 0.50 in. in diameter and 0.25 in. in length; the spherical particles comprising the porous structure had a number average diameter of 126.6 μ and the structure was characterized by a permeability on the order of 46 darcies. The characteristics of the porous materials which have been employed in engineering-type studies involving rheologically complex fluids are also quite varied. A summary of the characteristics of some of the models and materials which have been used in studies

Table V. Fluid Systems of Interest in Studies Related to Flow of Rheologically Complex Fluids in Porous Media

Chemical Identification	Mol. Wt. Range ($\times 10^{-6}$)	Solvent	Conc. Range (%)	Remarks	References
Poly(vinyl) alcohol	0.15	Water	3.9–6.13	[a]	(*121*)
Poly(acrylamide)	3 to 6	Water or NaCl solns	0.02–0.05	30% hydrolyzed [b,c,d]	(*123, 137, 140, 141, 143, 144*)
Poly(acrylamide)	2 to 3	Water or NaCl solns	0.005–1.0	25% hydrolyzed [b,c,d]	(*92, 139, 144, 145*)
	0.25	Water or NaCl solns	0.05	10% hydrolyzed	(*123, 143*)
Poly(acrylamide)	3 to 5	Water	0.02–0.20	2–4% hydrolyzed [b,c,d]	(*144*)
Poly(acrylamide)	3 to 10	Water, NaOH, or NaCl solns	0.05–0.75	Partially hydrolyzed [b,c,d]	(*93, 120, 131, 133, 135, 136, 142, 146*)
Poly(acrylamide)	1.0	Water and NaCl solns	0.05–1.0	[b,c,d]	(*92*)
Poly(acrylamide)	2.0	Water and NaCl solns	0.05–1.0		
Poly(acrylamide)	...	Water	0.001–0.05	[b,c,d]	(*128, 129*)
Poly(acrylamide)	14.0	Water	0.02–0.08	[b,c,d]	(*144*)
	10.0		0.05–0.25		
	5–7		0.02–0.05		
	6.5–10		0.01–0.075		(*172*)
Hydroxyethylcellulose		Water	0.04–1.0	[c,e]	(*121, 132*)
Carboxymethylcellulose		Water	0.005–2.85	[c,e]	(*71, 118, 122, 128, 129, 134*)
Polyethylene glycol	0.006	Water	39.0	[f]	(*121*)
	0.02	Water	10–18.5	[a]	
Poly(ethylene oxide)	0.2	Water	0.05–0.585	[b,c,d]	(*92, 119, 127*)
	0.6		0.05–0.304		
	4.0		0.05–1.0		
	>5.0		0.025–0.486		
Poly(ethylene oxide)	...	Water	0.04	[b,c,d]	(*134*)
Poly(isobutylene)	...	Toluene or decalin	5.0–9.9	[c,d]	(*118, 120, 122, 125, 126*)
Poly(acrylic acid)	...	Water	0.2–0.6	[a,c]	(*71, 120, 134*)
Poly(ethylene)	...	Melt	...		(*130*)
-glucosyl polysaccharide	...	Brine	1.5		(*91*)
Modified xanthan gum	...	Water	0.04–0.1		(*144*)
Sec-butylamine oleate				Micellar system[e,g]	(*82, 147*)
Rubber		Light hydrocarbon			(*124*)
Ethylene oxide polymer-petroleum sulfonate mixtures		Water			(*148*)

[a] Usually exhibits a purely viscous (inelastic) response in low concentrations.

[b] At lowest concentration, solutions generally exhibit a Newtonian-like (constant) viscosity behavior in viscometric flow. A markedly rheologically complex response occurs under nonviscometric or nonsteady flow conditions.

[c] Higher concentrations exhibit markedly non-Newtonian behavior and a shear-thinning gradient dependent-viscosity behavior.

[d] Usually very susceptible to shear, auto, or biodegradation.

[e] Usually resistant to shear degradation.

[f] Newtonian behavior at this concentration.

[g] Markedly viscoelastic behavior. Gradient-dependent viscosity behavior may be of shear-thinning or shear-thickening types. Not uncommon for flow behavior to be coupled with type and dimensions of flow geometry.

[h] Shear-thickening and viscoelastic behavior above critical shear rate.

related to the flow of rheologically complex fluids in porous media is given in Table IV.

Fluid Systems of Interest

The published literature relating to non-Newtonian flow through porous media indicates that relatively few chemically distinct fluid systems have been employed in the more fundamental investigations. This is somewhat surprising considering the bewildering variety of systems which exhibit grossly different types of rheologically complex behavior. A summary of some of these fluid systems has been compiled in Table V. With few exceptions, the bulk of the published investigations of non-Newtonian flow through porous media have involved systems derived from so-called "high polymer" additives. Typical examples include aqueous solutions of synthetic polymers such as poly(acrylamides), poly(ethyleneoxides), copolymer derivatives such as partially hydrolyzed poly(acrylamides), cellulose derivatives, polyethylene glycols, poly(vinyl) alcohol, and poly(acrylic) acid. Aqueous solutions of naturally derived polymeric substances, including polysaccharides and gums, have been utilized in these studies. In general, the majority of these solutions exhibit a rheological behavior which can be loosely described as "Newtonian-like" or shear thinning in a viscometric flow. However, their behavior is usually more complex rheologically under nonviscometric or nonsteady flow conditions. As an example, a very dilute solution of a high-molecular-weight grade of poly(ethylene oxide) usually exhibits a Newtonian-like (constant) viscosity over a finite range of shear rates in Poiseuille or Couette flow. This same solution is spectactularly non-Newtonian—*i.e.*, drag reducing, under fully developed turbulent flow conditions. A few studies have involved the use of a polymer melt, hydrocarbon solutions of a medium-molecular-weight grade of high polymers, and a suite of more complicated, fluid systems containing water, hydrocarbons, and other materials. The patent literature discloses several processes for enhancing the recovery of petroleum from subterranean formations which exploit the unusual rheological properties of a variety of complex systems—*e.g.* (*73–84*, *148*). Here, the gradient-dependent viscosity behavior may be either shear thinning, shear thickening, or combinations of these properties and other rheological characteristics. Apparently it is not uncommon for the flow behavior of these micellar systems to be coupled with "wall effects" and other aberrations related to the type and dimensions of the flow geometry.

Scale-up Methods

Some effort has been expended to establish methods for predicting a non-Newtonian flow behavior in porous media and for correlating the results of porous media experiments with viscometric data. At the present time there is no universally acceptable scale-up method for flow of rheologically complex fluids in porous media. The various methods currently employed, or suggested, can be arbitrarily divided into at least four major categories. The method which seems to have received the most attention is based on the coupling of a particular model for a porous medium—*i.e.*, the so-called capillary or hydraulic radius model, with an assumed functional relationship between shear rate and shear stress to describe the rheological behavior of a non-Newtonian fluid. The approaches of Sadowski and Bird (*90*, *121*), Christopher, Middleman, and Gaitonde (*118*, *122*, *125*, *126*), and Gregory and Griskey (*130*) are typical. These have involved correlations of experimental data from one-dimensional flow experiments in unconsolidated porous media—*i.e.*, mostly bead packs, with the appropriate rheological parameters derived from viscometric experiments on the fluid of interest. The power law and Ellis models have been used to describe the purely viscous behavior of the non-Newtonian fluids used in these studies. Another category involves generalized scale-up methods which adapt Darcy's law to non-Newtonian fluids without invoking a particular rheological model of purely viscous behavior. The appropriate rheological description can, in principle, be derived from viscometric and porous media flow experiments. McKinley *et al.* (*91*) and Kozicky *et al.* (*71*) have explored this category. A third approach, Slattery's suggested method (*149*), based on the concept of the simple fluid, involves the application of dimensional analysis to the scale-up of porous media flow data for an arbitrary viscoelastic fluid. Within the fourth category are other correlation methods, such as that described by Hassell and Bondi (*124*). Variations on these different categories and other methods are illustrated by the investigations of Gogarty (*93*), Dauben, Menzie, and Harvey (*119*, *127*, *144*), Jones and Maddock (*128*, *129*), Benis (*150*), and White (*132*, *151*). Roughly speaking, the rheological complexity of the fluid of interest, the morphological complexity of the porous medium, and interactions between the fluid and porous medium, *e.g.*, plugging, adsorption, chemical reaction, etc., determine the general applicability of any given scale-up method.

A generalized scale-up method based on the capillary model-hydraulic radius concept. Evidently Bird (*152*) was the first to suggest that the capillary or hydraulic radius concept could be extended to describe flow of rheologically complex fluids in porous media. These early ideas were the precursor to the generalization of Darcy's law for non-Newtonian fluids in terms of the Ellis and power law models. We will review the state-of-the-art relating how the capillary model of a porous medium has been coupled with these particular rheological models to develop equations similar to the integrated forms for capillary viscometric flow. However, the capillary model of a porous medium can be used in a generalized scale-up method without invoking the use of a particular rheological model. It cannot be overemphasized that this generalized method encompasses these earlier methods as special cases. We make this observation at this point to prepare the reader for the need to discriminate between the different contributions reviewed. Hopefully, this review will provide a sharp perspective not usually available from a study of the fragmentary literature alone. We will show that in all situations where a particular rheological model of purely viscous behavior has been coupled with the capillary or hydraulic radius model to describe the flow of a non-Newtonian fluid through porous media, it naturally follows that the generalized scale-up method is also applicable.

The integrated form of any rheological model describing the relationship between pressure loss and flow rate for a complex fluid in the steady flow called Poiseuille or capillary flow is derived from the following well-known equation relating macroscopic variables (*51*, *153*).

$$F(\tau_w) \equiv \frac{V}{R} = \frac{1}{\tau_w^3}\int_0^{\tau_w} \tau^2\Gamma(\tau)d\tau \tag{5}$$

where

$$\eta_{\text{cap}} = \frac{\tau_w}{4F(\tau_w)} \tag{6}$$

In extending the Poiseuille flow equation to a porous medium, it is tacitly assumed the flow is still viscometric—*i.e.*, steady and isochoric, so that both the overall geometry of flow and the scale of the deformation process are unchanged. Of course, it is anticipated from the results of numerous tests of Darcy's law with Newtonian fluids that these conditions are only approximately achieved. Deviations from Darcy's law readily occur in the case of Newtonian fluids without the flow becoming "turbulent" because of the presence of inertial forces generated by accelerations and decelerations within the continuous but tortuous flow channels interspersed within the porous medium. One now assumes the average pore velocity $\langle V\rangle$ can be related to the superficial velocity V_0 by:

$$V_0 = \epsilon\langle V\rangle \tag{7}$$

where ϵ is the porosity or void fraction. This is the Dupuit-Forcheimer assumption (*154*). Following Bird (*152*), the concept of hydraulic radius is introduced into Equation 5 so that the generalized flow equation is scaled to the particle diameter. This is done by expressing $R = 2R_H$ in terms of the porosity and the wetted surface S per unit volume of porous material, namely:

$$R = \frac{2\epsilon}{S} \tag{8}$$

Noting that S is related to S_0, the total particle surface per volume of particles, in the following way,

$$\frac{S}{S_0} = (1 - \epsilon) \tag{9}$$

and defining a mean particle diameter in terms of the specific surface of the packing

$$D_P = \frac{6}{S_0} \tag{10}$$

Equations 9 and 10 can now be combined to replace the capillary radius R in Equation 8 by an equivalent radius expressed in terms of ϵ and D_P

$$R_H = \frac{D_P\epsilon}{6(1 - \epsilon)} \tag{11}$$

It is also customary to account for the actual length of the flow path or "tortuosity" of the capillaries. One method is to correct the linear length of the porous medium by a numerical factor. Following Christopher and Middleman (*118*, *122*), one choice of correction is to set $L = C'L$ and accept the value of C' found to be valid for Newtonian fluids. By introducing the above equations, the generalized flow equation becomes

$$F(\tau_\epsilon) \equiv \frac{3(1 - \epsilon)V_0}{D_P\epsilon^2} = \frac{1}{\tau_\epsilon^3}\int_0^{\tau_\epsilon} \tau^2\Gamma(\tau)d\tau \tag{12}$$

where

$$\tau_\epsilon = \frac{\omega\Delta P}{6C'L} \tag{13}$$

$$\omega = \frac{D_P\epsilon}{(1 - \epsilon)} \tag{14}$$

and a "Darcy viscosity" may be defined as

$$\eta_\epsilon = \frac{\tau_\epsilon}{4F(\tau_\epsilon)} = \left(\frac{\omega^2\epsilon}{72C'}\right)\frac{\Delta P}{V_0L} \tag{15}$$

It is important to note the motivation leading to the generalized method. In principle, there is no discrimination between results from porous media and classical viscometry experiments in the instance of a given fluid. One assumes superposition of data between these experiments, thereby allowing, for example, the prediction of flow behavior in a porous structure from experiments in a capillary viscometer. Implicit in this approach is the assumption that conditions in the capillary and porous structure are such as to result in constancy of the Reynolds, Weissenberg, and Deborah numbers, defined as

Reynolds number = inertial/viscous forces
Weissenberg number = elastic/viscous forces
Deborah number = duration of fluid memory/duration of deformation process

In the case of a viscometric flow, $N_{\text{DEB}} = 0$. However, as major viscoelastic effects are encountered under

Lagrangian unsteady flow conditions, either N_{DEB} or N_{WS} may become significant. For these conditions, the generalized method and Equations 12 or 15 may no longer suffice.

Here τ_ϵ is an averaged shear stress scaled to the porous medium. Similarly, $4F(\tau_\epsilon)$ is an averaged nominal shear rate, equivalent to the usual capillary flow expression, with the capillary radius replaced by its scaled value for a porous medium. If the correction or "shift factor" C' is introduced following the suggestion of Bird (*152*), a different form is obtained, namely:

$$F(\tau_{RH}) \equiv \frac{3(1-\epsilon)V_0}{D_P\epsilon^2} = \frac{1}{C'\tau_{RH}^3}\int_0^{\tau_{RH}} \tau^2\Gamma(\tau)d\tau \quad (16)$$

where

$$\tau_{RH} = \frac{\omega J}{6} \quad (17)$$

Alternatively, Equation 12 can be expressed in terms of the permeability of the porous medium. Here, a variety of approaches is possible. A commonly used correlation between permeability, porosity, and particle diameter is Equation 18:

$$k = \frac{D_P^2\epsilon^3}{72C'(1-\epsilon)^2} \quad (18)$$

The theoretical and experimental basis for Equation 18 is discussed in a number of places (*152, 154–156*). By use of this relation, the generalized flow equation is made dependent on permeability, and one obtains the results

$$F(\tau_\epsilon) \equiv \frac{V_0}{2^{3/2}\sqrt{C'k\epsilon}} = \left(\frac{L}{\Delta P\sqrt{2k/C'\epsilon}}\right)^3 \times \int_0^{\sqrt{(2k/C'\epsilon)}(\Delta P/L)} \tau^2\Gamma(\tau)d\tau \quad (19)$$

$$\tau_\epsilon = \sqrt{\frac{2k}{C'\epsilon}}\frac{\Delta P}{L} \quad (20)$$

$$\eta_\epsilon = \frac{k\Delta P}{V_0L} \quad (21)$$

Similarly, by use of Equation 18, the generalized flow Equation 16 is made dependent on k and one obtains

$$F(\tau_{RH}) = \frac{V_0\sqrt{C'}}{2^{3/2}\sqrt{k\epsilon}} = \frac{1}{\tau_{RH}^3}\int_0^{\tau_{RH}} \tau^2\Gamma(\tau)d\tau \quad (22)$$

where

$$\tau_{RH} = J\sqrt{\frac{2C'k}{\epsilon}} \quad (23)$$

As will be shown later, inserting the power law model in Equations 19 and 22, we recover the expressions for one-dimensional non-Newtonian flow through porous media derived by Christopher and Middleman (*118, 122*), and Bird (*152*), respectively. Equation 21 is equivalent to the well-known one-dimensional form of Darcy's law for Newtonian fluid (*154*) flowing through an isotropic homogeneous medium.

With the exception of relating hydraulic radius to equivalent pore radius differently *via* permeability, porosity, tortuosity, and pore size distribution, the generalized scale-up method is analogous to the modification of Darcy's law described by McKinley *et al.* (*91*). For example, an equivalent pore size for a particular kind of porous medium can be arbitrarily defined as

$$R_{eq} = 2T\alpha_0\sqrt{\frac{k}{\epsilon}} \quad (24)$$

Here T is some measure of tortuosity and α_0 is some measure of both tortuosity and pore size distribution. Again, replacing the average pore velocity by V_0, and replacing R by R_{eq}/T, the generalized flow equation becomes

$$F(\tau_\epsilon) = \frac{V_0}{2\alpha_0\sqrt{k\epsilon}} = \left(\frac{L}{\Delta P\alpha_0\sqrt{k/\epsilon}}\right)^3 \times \int_0^{\alpha_0\sqrt{k/\epsilon}\,\Delta P/L} \tau^2\Gamma(\tau)d\tau \quad (25)$$

The results (*91*) suggest that α_0 should increase with tortuosity. Gogarty (*93*) also considered how the shear rate function is related to $\langle V\rangle$ for consolidated porous media, and arrived at an expression of the form

$$\overline{F(\tau_\epsilon)} = \left[\frac{B}{f(k)\sqrt{k\epsilon}}\right]^y \quad (26)$$

where

$$f(k) = A_1 \log R(k) + A_2 \quad (27)$$

The parameter y accounts for possible deviation between the slope of the apparent viscosity-shear rate curve from a capillary viscometer experiment, and the slope of the corresponding curve, for the same fluid, but determined from an experiment in the porous medium. Also, for a given "kind of rock," the constants A_1 and A_2 depend on fluid type, and $R = k/k_r$, where k_r is some reference permeability. Gogarty observed considerable differences in both A_1 and A_2. In this regard, his experiments pertain to shear thinning surfactant stabilized dispersions of water in hydrocarbon. Flow experiments were performed in sandstone cores characterized by permeabilities in the range from 0.069 darcy through 0.425 darcy, and porosities in the range from 17% through 21.7%. Viscometric data were obtained *via* capillary and cone/plate experiments.

Darcy's law for one-dimensional flow is recovered by requiring that $\alpha_0 = \sqrt{2}$. Through the use of Equations 12, 16, 19, 22, or 25, we have a method for applying the capillary or hydraulic radius model of a porous

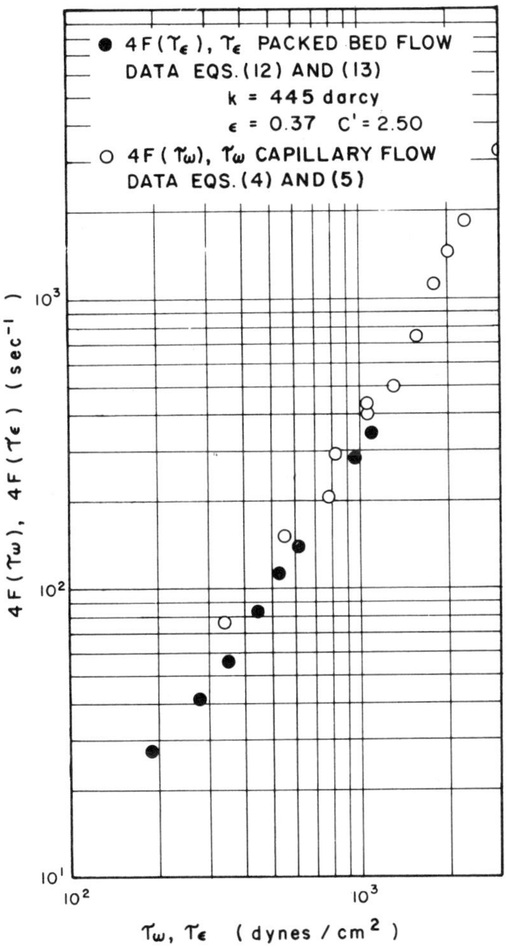

Figure 1. Flow behavior of 2 wt % carboxymethylcellulose solution in bead pack and comparison with capillary flow behavior (118)

medium to non-Newtonian fluids without coupling the macroscopic variables to a particular rheological model of purely viscous behavior. It follows from the form of Equation 5 that, irrespective of the dimensions of the capillary, the viscometric flow of a complex fluid can be represented by a unique curve in which, by using data from capillary rheometer experiments, $F(\tau_w)$ is graphed as a function of τ_w. Similarly, from Equation 12 or 16, the flow of a complex fluid can be represented by a unique curve in which, by using data from flow experiments in porous media, $F(\tau_\epsilon)$ is graphed as a function of τ_ϵ. Provided the appropriate choice of the correction or shift factor, C' or α_0 is made, it follows that for the same fluid, the $F(\tau_\epsilon)$, τ_ϵ curve should map directly onto the $F(\tau_w)$, τ_w curve. Some work with unconsolidated porous media (*90*, *118*, *121*, *122*) indicates that for certain complex fluids it is acceptable to apply the value of C' determined from experiments with Newtonian fluids. That the shift factor is even the same for Newtonian and different non-Newtonian fluids is surprising, particularly in the instance of flow through a porous medium considering the hydraulic radius distribution of the pore system. Even in the case of power law flow in a single capillary, possibly the simplest description of a rheologically complex flow behavior, there exists a close coupling between flow geometry and rheological parameters. Thus, it might be anticipated that the "averaged" flow configuration seen in non-Newtonian flow through a porous medium would not be the same as that average geometry seen by the Newtonian fluid.

An illustrative experimental flow curve, comparing capillary viscometry data with packed bed data on a shear thinning solution of a cellulose derivative (*118*), is shown in Figure 1. A factor of $C' = 2.50$ has been applied to these packed bed data. However, some experimental work related to flow of non-Newtonian fluids through consolidated porous media indicates it may be more appropriate in certain instances to determine the correction factor needed to achieve superposition of the flow curves by experimentation with the fluid of interest. An illustrative experimental flow curve for this case is shown in Figure 3 of Reference *91* for an aqueous solution of a polysaccharide exhibiting shear thinning behavior. Other viscometric data may be compared with results from flow experiments in porous media. If a basic description of rheological data is available in the form of a tabulation of τ, $\Gamma(\tau)$ data, a capillary flow curve can be "synthesized" by a scheme of numerical evaluation of Equation 12. Illustrative examples for this case are shown in Figures 2 and 3, respectively, for aqueous solutions of a vinyl polymer and a cellulose derivative (*121*). By way of comparison, numerical factors of $C' = 25/12$ and 2.5 have been applied to each set of packed bed data. The solid

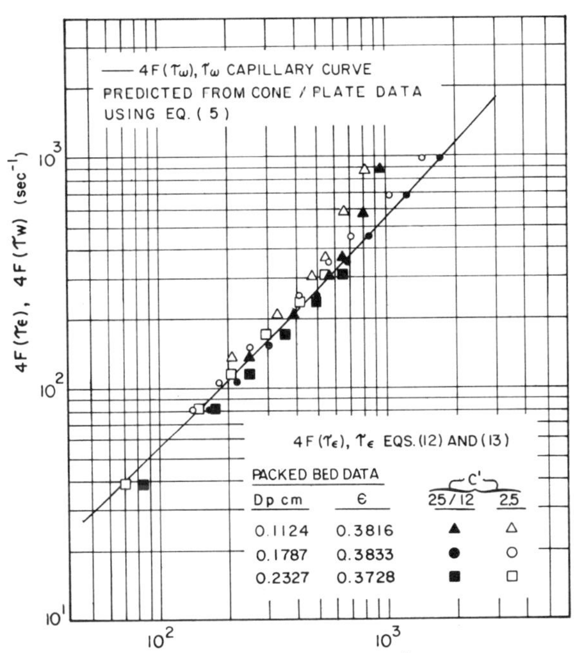

Figure 2. Flow behavior of 6 wt % poly(vinyl) alcohol solution in bead packs—comparison with behavior predicted for capillary flow from cone/plate experiments (121)

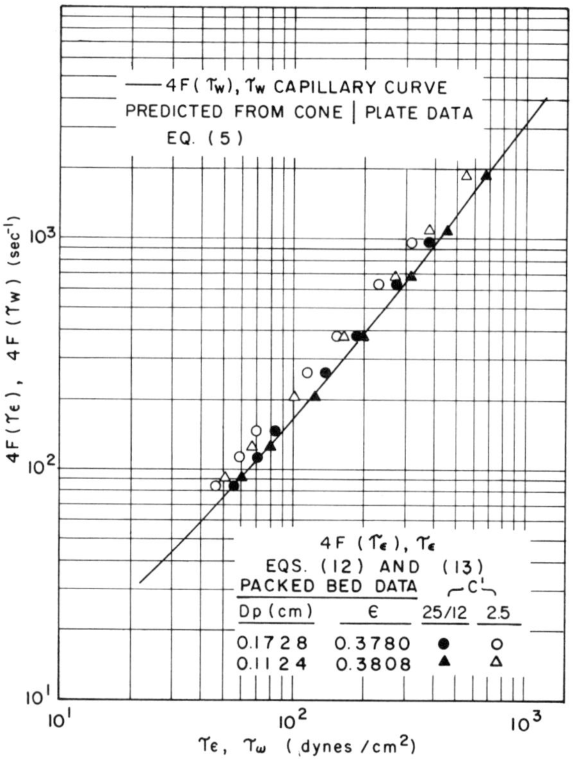

Figure 3. Flow behavior of 1.4 wt % hydroxyethyl cellulose solution in bead packs—comparison with behavior predicted for capillary flow from cone/plate experiments (121)

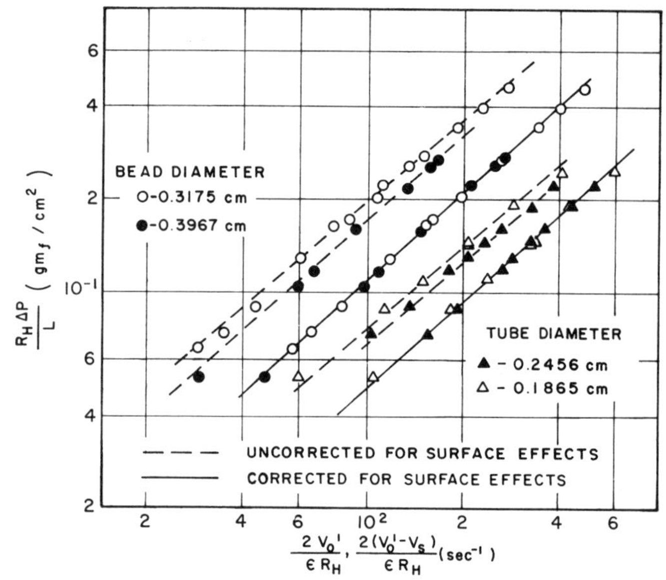

Figure 4. Examples of anomalous surface effects in flow of 2.85% carboxymethylcellulose solution through bead packs and capillaries (71)

line is the synthesized capillary curve derived from Equation 12, using published viscometric data from cone/plate experiments (*121*).

Alternatively, graphs of apparent viscosity and Darcy viscosity *vs.* nominal shear rate, should lead to superposition if the generalized scale-up method is applicable. Illustrative examples are shown later in this paper. Varying degrees of agreement between the theory of the generalized scale-up method and experiments are apparent in the results presented in Figures 1–3. These findings by no means confirm or disprove the applicability of the capillary or hydraulic radius model of a porous medium to the description of the one-dimensional flow of non-Newtonian fluids through porous media. Rather, the generalized scale-up method should probably be viewed as a device for avoiding complications due to uncertainty about how well a particular choice of rheological model of purely viscous behavior simulates the flow behavior of a particular fluid.

In general, in addition to the assumptions discussed earlier, departures from the behavior predicted by the generalized scale-up method can also be ascribed to several other implied assumptions. By way of recapitulation, these include the following assumptions: (i) the correct description of purely viscous behavior has been obtained from basic viscometric flow experiments; (ii) the porous medium is unaltered by the flowing fluid; (iii) absence of viscoelastic effects; (iv) absence of significant nonlaminar flow conditions in the porous medium; and (v) absence of a coupling between the geometry of the porous medium and the rheological properties of the flowing fluid. The complications related to items (ii) to (iv) are discussed later. An illustrative example of item (ii) is worth noting at this point. The results shown in Figure 4 were obtained by Kozicki *et al.* (*71*). Represented are some of their results from flow experiments on an aqueous solution of a cellulose derivative in both capillaries and in bead packs. Distinct flow curves are obtained for each packed bed and each capillary. Following a procedure to be discussed later, these data were corrected for an anomalous layer of thickness estimated to be on the order of 2.6×10^{-3} cm on the particles of the packed beads, and on the order of 6.8×10^{-3} cm in the case of the capillary data. The two sets of data now superimpose onto single master curves. Superposition of these corrected master curves onto a single curve, as predicted by the generalized scale-up method, would result on applying a shift factor $C' = 2.4$ (see Equation 17) to the ordinate variable of the packed bed data.

Scale-up methods utilizing rheological models. Admittedly, there are occasionally good reasons for choosing a specific scale-up method. Simplicity, for example, may provide for insights otherwise obscured by a general technique. Similarly, one may wish to choose a specific model as a basis for the analysis, say, of wall effects or some similar added complication. The adoption of a specific model at this point may be useful in this context. However, noting the generality of Equations 12 to 15, these more restrictive methods cannot be commended without such justification. Admittedly, some readers may question the wisdom of emphasizing these methods in the detail which follows. In defense, an in-depth review seems justified inasmuch as the scale-up methods for one-dimensional flow of non-Newtonian fluids in porous media which have received the most attention in the literature are those in which the power law and Ellis models of purely viscous behavior have been coupled with the capillary or hydraulic radius model of a porous medium. These studies have led to the formulation of modified equations for describing laminar flow of rheologically complex fluids through packed beds and porous media. Discussions of these methods are provided (*90*, *118*, *121*, *122*, and *152*). These derivations have taken as their starting point the integrated forms of the power law and Ellis models for Poiseuille flow, Equations 28 and 29, respectively

$$\frac{V}{R} = \left(\frac{n}{3n+1}\right)\left[\frac{R\Delta P}{2LK}\right]^{1/n} \tag{28}$$

$$\frac{V}{R} = \frac{\tau_w}{4\eta_0}\left[1 + \frac{4}{\alpha+3}\left(\frac{\tau_w}{\tau_{1/2}}\right)^{\alpha-1}\right] \tag{29}$$

The analogs for one-dimensional flow through porous media have been developed from these expressions using Equations 7 to 10 and 18, and the desired correlation or shift factor C'. However, as noted earlier, a more direct approach to the same result is to insert the functional relationship between τ and $\Gamma(\tau)$ in the integral Equation 12 or 16, for example.

Power law model. Thus, the result for one-dimensional flow of a power model fluid in porous media is, from Equation 12,

$$\tau_\epsilon = K'[4F(\tau_\epsilon)]^n \tag{30}$$

where $4F(\tau_\epsilon)$ and (τ_ϵ) are given by Equations 12 and 13, respectively, and

$$K' = K\left(\frac{1+3n}{4n}\right)^n \tag{31}$$

This result can also be expressed (*118*, *122*) as:

$$V_0^n = \frac{k\Delta P}{\psi L} \tag{32}$$

Here ψ is referred to as the "viscosity level parameter" (*120*) and, expressed in terms of permeability porosity, and rheological parameters is

$$\psi = \frac{K}{12}\left[3\left(\frac{3n+1}{n}\right)\right]^n [72C'k\epsilon]^{(1-n)/2} \tag{33}$$

Equation 33 is identical to the result derived by Christopher and Middleman (*118*, *122*) and applied by Gregory and Griskey (*130*) when $C' = 25/12$. Alternatively, using Equation 18, expressed in terms of particle diameter, ψ is given as

$$\psi = \frac{K}{12}\left[3\left(\frac{3n+1}{n}\right)\right]^n \left[\frac{\epsilon^2 D_P}{(1-\epsilon)}\right]^{(1-n)/2} \tag{34}$$

A different form for the power model analog is presented in Reference *152*. The difference results from the manner in which the correction term C' is introduced into Equation 28. Christopher and Middleman (*118*, *122*) argue that the length term should be corrected and accordingly have inserted the correction in the term within brackets, replacing L by $C'L$. In Reference *152*, the correction term is applied to the velocity term (Equation 16). For this alternate form, the viscosity level parameter, or "effective viscosity" (*152*) is

$$\psi = \frac{K}{12}\left[3\left(\frac{3n+1}{n}\right)\right]^n [72C'k\epsilon]^{(1-n)/2} (C')^{n-1} \tag{35}$$

Thus, as noted by Christopher and Middleman (*118*, *122*), their expression for flow of a power law fluid in a porous medium differs from that presented in Bird *et al.* (*152*) by the factor $(C')^{n-1}$. Expressing Equation 32 in the form

$$V_0^n = (\Phi)^n \frac{D_P{}^{n+1}\Delta P}{K'L} \tag{36}$$

where

$$\Phi_{CM} = \left\{\left(\frac{1}{2C'}\right)^{1/n}\left[\frac{\epsilon}{3(1-\epsilon)}\right]^{(1+1)/n} \frac{\epsilon}{4}\right\} \tag{37}$$

$$\Phi_{BSL} = \Phi_{CM}(C')^{(1-n)/n} \tag{38}$$

is particularly useful as the function Φ may also be used to infer how well the assumed coupling between the capillary-hydraulic radius model and the power law model represents the dependence of non-Newtonian behavior on the geometry of the porous medium. Here Φ_{CM} and Φ_{BSL} refer to the different approaches outlined in References *122* and *90*, respectively.

Figures 5 and 6 compare experimentally determined (averaged) values of Φ with values predicted by Equations 37 and 38. A single porosity ($\epsilon = 0.37$) characterized the bead packs used by Christopher (*118*), and for the computation of Φ_{CM} the correction factor $C' = 25/12$ recommended by the authors (*118*, *122*) was used. Porosities in the range ($0.3653 \leq \epsilon \leq 0.3883$) characterized the bead packs used by Sadowski (*121*). This accounts for the pairs of curves computed from Equations 37 and 38 using the extremal values of ϵ. Here, the correction factor $C' = 2.5$ recommended by the authors (*90*, *121*) was used. The power law parameters for Christopher's data were obtained from capillary experiments. Sadowski's data were fit to the power law expression, using the parameters cited (*118*). Using the capillary model for power law flow through porous media, Equation 30, Gaitonde and Middleman (*126*) correlated their data on flow of poly(isobutylene) solutions through bead packs with an average error of 9.8%. Similarly, from experiments related to flow of molten poly(ethylene) through bead packs, Gregory and Griskey (*130*) concluded that measured superficial velocities and values predicted from Equation 32 agreed to within 10%. Figures 7 and 8 compare experimentally determined pressure gradients with values predicted

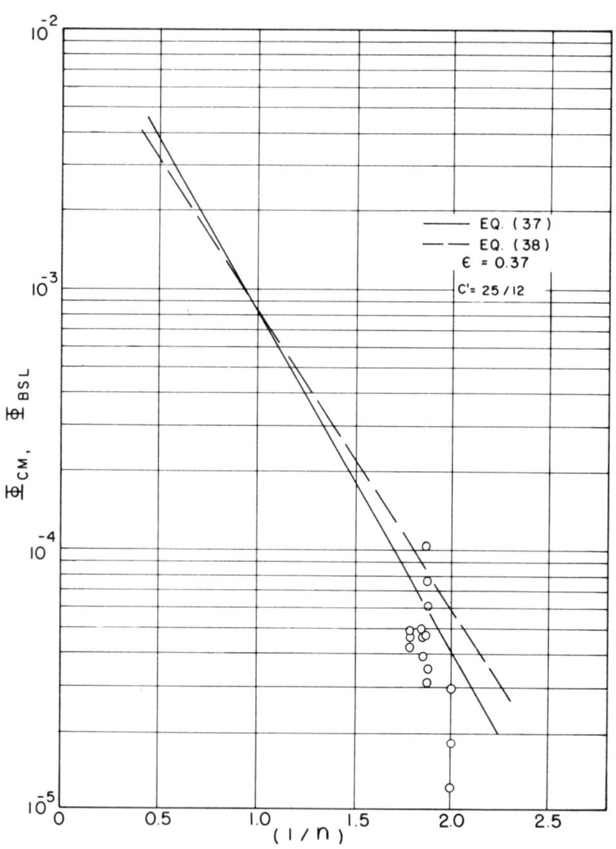

Figure 5. Test of modified capillary models for non-Newtonian flow through porous media using data of Christopher (118)

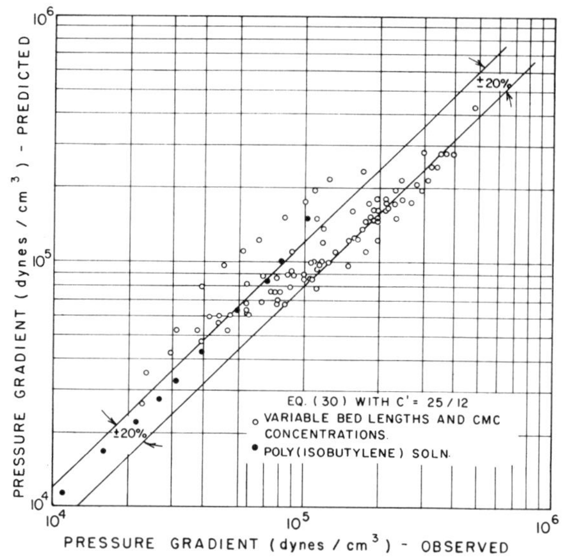

Figure 7. Test of Christopher-Middleman capillary model for power law flow through porous media (118)

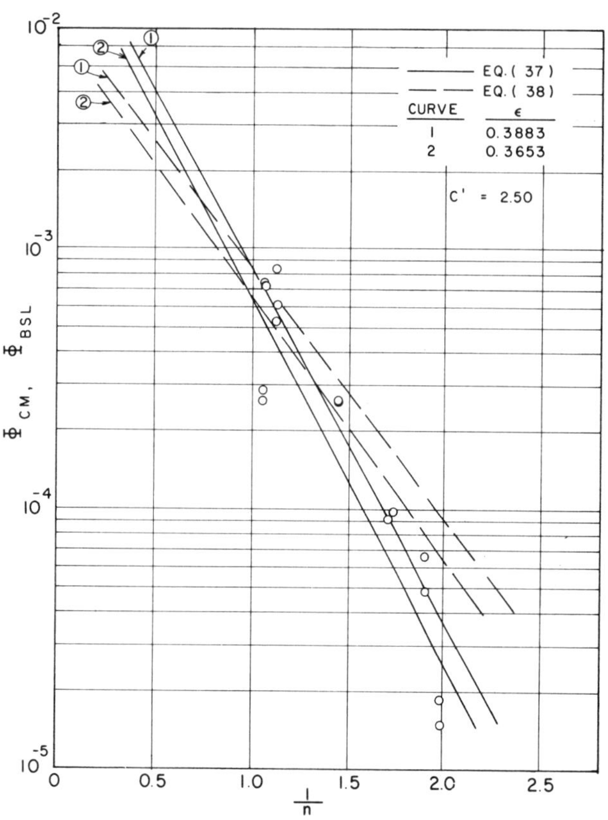

Figure 6. Test of modified capillary models for non-Newtonian flow through porous media using data of Sadowski (121)

from Equation 32. Again, Christopher's data (*118*) and Sadowski's constant flow rate data (*121*) are represented. No attempt was made to eliminate from the results presented in Figures 7 and 8 those data which might be questioned because of possible length effect errors. Kozicki *et al.* (*71*) claim some of these data show an anomalous wall layer effect. There is always the question as to how well the power law parameters, determined from viscometric experiments, describe the rheological behavior of the fluids used in the porous media flow experiments.

The significance of inertial effects also requires consideration. A reasonable starting point for determining if inertial effects are negligible in porous media experiments is to develop a friction factor *vs.* Reynolds number relationship. Precise values have not been established for a critical Reynolds number above which the "laminar" flow condition breaks down in porous media. Values in the range from 0.1 to 75 have been reported (*154*). In the case of Newtonian fluids, unlike the noticeable breakdown effect usually observed in tube flow, a gradual departure from laminar-like behavior seems to characterize some packed bed data (see Figure 6.4-1 of Reference *152*). Several choices of friction factor–Reynolds number relationships may be made. Following Christopher and Middleman (*122*), Ergun's (*157*) definition of the friction factor leads to the result

$$f_\epsilon = \frac{\beta J \epsilon^2}{\rho V_0^2} \qquad (39)$$

Arbitrarily defining the Reynolds number such that

$$f_\epsilon N'_{\mathrm{Re}} = 1 \qquad (40)$$

we obtain

$$N'_{\mathrm{Re}} = \frac{D_P V_0^{2-n} \rho}{72C'(1 - \epsilon)\psi} \qquad (41)$$

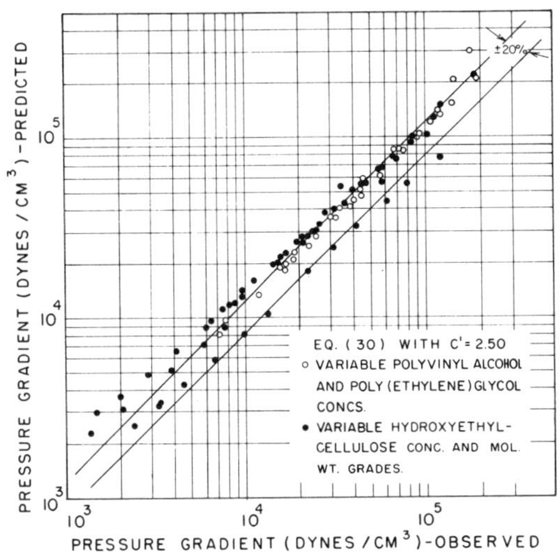

Figure 8. Test of Christopher-Middleman capillary model for power law flow through porous media using data of Sadowski (121)

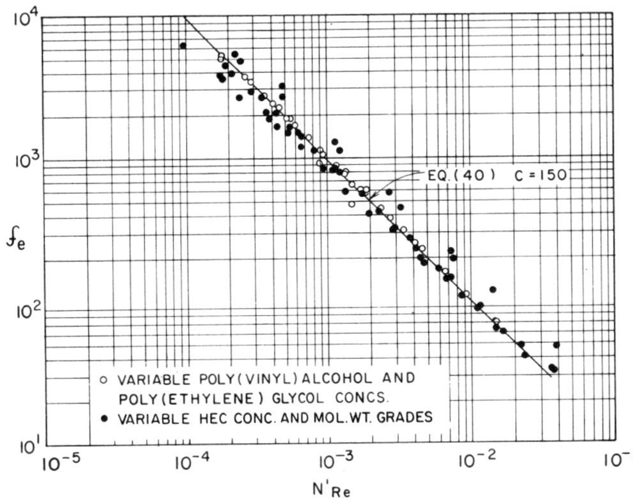

Figure 9. Reynolds number vs. friction factor for power law flow through bead packs using data of Sadowski (121)

When $n = 1$, Equation 41 reduces to the familiar bed Reynolds number (*152*). In Figure 9 the results from Sadowski's experiments in bead packs (*121*) are replotted according to Equation 40 using the power law parameters cited in Reference *118*, with f_ϵ and N'_{Re} defined by Equations 39 and 41, respectively; here ψ is defined by Equation 33 and $C' = 25/12$ has been used. We observe the majority of the experimental data lie between Reynolds numbers from 10^{-4} to 10^{-2}, well within the range considered to be the slow flow region, where the inertia term in the equation of motion can be neglected. No significant trends are apparent.

Ellis model. A development similar to that described above has been outlined by Sadowski and Bird (*90, 121*) using Equation 29, the integrated form of the Ellis model for capillary flow. Their development leads to

$$f_\epsilon = \frac{72C'}{\mathrm{Re}} \tag{42}$$

Here f_ϵ is defined by Equation 39 and the Reynolds number is given as

$$\mathrm{Re} = \frac{D_P V_0 \rho}{(1 - \epsilon)\eta_{\mathrm{eff}}} \tag{43}$$

where

$$\frac{1}{\eta_{\mathrm{eff}}} = \frac{1}{\eta_0}\left[1 + \frac{4}{\alpha + 3}\left(\frac{\tau_{RH}}{\tau_{1/2}}\right)^{\alpha - 1}\right] \tag{44}$$

Equation 42 is identical to the result given (*90*) when $C' = 2.5$. Figure 10 shows some of Sadowski's constant flow rate results from bead pack experiments plotted according to Equation 42 with f_ϵ and Re defined by Equations 39 and 43, respectively. The Ellis model parameters, η_0, $\tau_{1/2}$, $(\alpha - 1)$ were determined from

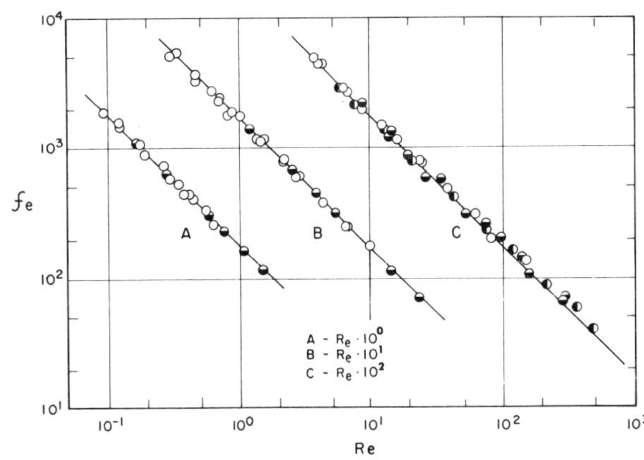

Figure 10. Reynolds number vs. friction factor for Ellis fluid flow through bead packs, using data of Sadowski (121) (Equation 42). Fluid systems containing different concentrations of (A) poly(ethylene) glycol, (B) poly(vinyl) alcohol, and (C) hydroxyethylcellulose

experiments with cone/plate and falling sphere viscometers. The results indicate a maximum error in the calculated value of f_ϵ ranging from +15.8% to −19.0%. Other data obtained on a moderately concentrated solution of a cellulose derivative exhibited larger deviations which have been attributed to viscoelastic effects. As will be discussed later, better agreement with Equation 42 was obtained for these data by addition of another dimensionless term to Equation 44. Sadowski found experiments on non-Newtonian flow through porous media gave results which depended upon whether constant flow or constant pressure drop conditions were employed. Constant flow rate conditions produced results which were steady and reversible, *e.g.*, the data summarized in Figure 10. By contrast, unsteady and irreversible results were obtained when the pressure drop across the packed bed was held constant. The tentative explanation offered by the authors was that during the latter experimental conditions polymer adsorption and gel formation occurred.

A method based on dimensional analysis. Slattery (*149*) obtains equations of motion for steady isochoric flow of Newtonian and rheologically complex fluids through rigid porous media by "volume averaging" the stress equations of motion applied to fluid filling the pore space. A "resistance transformation" K acting on the average velocity vector of the fluid is also introduced. The form of K is determined from the Buckingham-Pi theorem for the Newtonian fluid, for two simple models of non-Newtonian fluid behavior, *i.e.*, the power law and Ellis models, and for the Noll simple fluid. Methods for correlating data for steady rectilinear flow of incompressible viscoelastic fluids through a uniformly packed bed are described for each example. Slattery (*149*) shows how the correlations presented by Sadowski and Bird (*90, 121*) and Christopher and Middleman (*118, 122*) represent particular cases of his treatment for the simpler models of fluid behavior discussed above. Since the concept of the simple fluid does not require adoption of a specific constitutive equation to represent material behavior, we will focus attention on Slattery's suggested scheme for correlating data for the simple fluid. We will follow closely the treatment outlined (*149*) and illustrate the correlations with examples. In Reference *51*, an incompressible simple fluid is defined as "...material with the property that all of its local configurations are intrinsically equivalent in response, with all observable differences in response being due to definite differences in history." The incompressible simple fluid is characterized by the property that the stress is essentially determined, given a description of the strain of each past configuration relative to the present configuration:

$$\mathbf{t} + p\,\mathbf{I} = \frac{\mu_0}{t_0} \mathop{\mathcal{H}}_{\sigma=-\infty}^{0} [\mathbf{c}(t + t_0\sigma)] \tag{45}$$

The constants μ_0 and t_0 are a characteristic viscosity and a characteristic time, respectively, and $\mathop{\mathcal{H}}_{\sigma=-\infty}^{0}$ is a dimensionally invariant tensor-valued functional. For the particular case of steady state flow of an incompressible fluid through an isotropic porous medium, the generalized Darcy's law (*149*) can be specialized to the form

$$\nabla[\overline{\mathcal{P} - p_0}] + K\mathbf{v} = 0 \tag{46}$$

Slattery suggests a reasonable approach, that is to say that K should be a function of a characteristic length ($\mathcal{L}$) of the system, the magnitude of a characteristic velocity, say $\overline{V}$, and the material constants in Equation 45, namely:

$$K = H_1(\mathcal{L}, \overline{V}, \mu_0, t_0) \tag{47}$$

Other dimensionless parameters in terms of which any particular functional could be included in Equation 47 have been omitted for simplicity. Slattery observes that by the Buckingham-Pi theorem (*158*), Equation 47 can be written in terms of a dimensionless permeability given by

$$K = \frac{\mu_0}{\mathcal{L}^2 k^*} \tag{48}$$

where the dimensionless permeability k^* has the following functional dependence:

$$k^* = h_1\left(\frac{\overline{V}t_0}{\mathcal{L}}\right) \tag{49}$$

Alternatively, one might assume K is a function of the magnitude of the local averaged pressure gradient, namely:

$$K = H_2\,(\mathcal{L}, |\nabla(\overline{\mathcal{P} - p_0})|, \mu_0, t_0) \tag{50}$$

By the Buckingham-Pi theorem, Equation 48 is again obtained where now

$$k^* = h_2\left(|\nabla(\overline{\mathcal{P} - p_0})|\,\frac{\mathcal{L}t_0}{\mu_0}\right) \tag{51}$$

If a correlation of data from packed bed experiments in geometrically similar situations is desired for all fluids of a given class, it would be necessary to select a specific functional for that class. However, assume one wants a correlational scheme which will allow predicting the behavior of a particular fluid, *i.e.*, holding fixed the material constants μ_0 and t_0 and all dimensionless parameters needed to describe $\mathop{\mathcal{H}}_{\sigma=-\infty}^{0}$, in a given geometry using results from flow experiments in geometrically similar situations. For a given fluid, μ_0 and t_0 need not be specified, and a correlation could result by regarding $\mathcal{L}^2K$ as a function of a single parameter. Thus correlations in the form of graphs of $\mathcal{L}^2K$ against $\overline{V}/\mathcal{L}$ or $\mathcal{L}^2K$ against $|\nabla(\overline{\mathcal{P} - p_0})|$ are possibilities from (say) Equations 48 and 49, or 48 and 51, respectively, namely:

$$\mathcal{L}^2K = k_1(\overline{V}/\mathcal{L}) \tag{52}$$

$$\mathcal{L}^2K = k_2[\mathcal{L}|\nabla(\overline{\mathcal{P} - p_0})|] \tag{53}$$

Furthermore, consider the kinds of experiments discussed, for example, in References *90* and *122* and in the preceding scale-up methods. Slattery gives an interpretation of rectilinear flow appropriate to these experiments, and argues such examples can be reasonably described as steady-state rectilinear flow through a uniformly packed bed. We know that superficial velocity is proportional to the volume averaged velocity $\overline{V}$ (Equation 7), and the volume averaged pressure gradient corresponds to the pressure drop measured across a packed bed—*i.e.*, $\Delta P/L = |\nabla(\overline{\mathcal{P} - p_0})|$. Furthermore, we arbitrarily choose

$$\mathcal{L}^2 = \frac{k}{\epsilon} \tag{54}$$

where k is permeability. From Equations 52 and 53, and introducing the above substitutions for $\bar{V}$ and $|\nabla(\overline{\mathcal{P} - p_0})|$, we find that

$$\mathcal{L}^2 K = \frac{\Delta P k}{V_0 L} \tag{55}$$

$$\frac{\bar{V}}{\mathcal{L}} = \frac{V_0}{\sqrt{k\epsilon}} \tag{56}$$

$$|\nabla(\overline{\mathcal{P} - p_0})| = \frac{\Delta P}{L}\sqrt{\frac{k}{\epsilon}} \tag{57}$$

Comparing Equations 21 and 55, we see for the experiments considered here that $\mathcal{L}^2K$ can be identified with the "Darcy" viscosity. Likewise, comparing the L.H.S. of Equation 19 with Equation 56, $\bar{V}/\mathcal{L}$ is proportional to the average nominal shear rate parameter used in the capillary model of a porous medium. Apparently, the only difference between the coordinates employed here and those of the generalized design method for purely viscous fluids is that the former includes a pressure drop term on both coordinates. This reduces the sensitivity of the plot, thus lowering the discriminatory powers desired. Figures 11 and 12 compare the correlations suggested by Equations 55 and 56, respectively. The points are the unsmoothed values obtained from the raw data of Dauben (*119*), pertaining to packed bed experiments. By contrast, viscometric experiments indicate the behavior of this solution is that of a Newtonian fluid with constant viscosity equal to 0.042 poise. Finally choosing the characteristic length to be proportional to particle diameter, and going through a treatment similar to that outlined above, we find that the friction factor Reynolds number correlation given in (*90*) is a modification of the form of Equation 53

$$\frac{\mathcal{L}K}{\bar{V}} = f[\mathcal{L}|\nabla(\overline{\mathcal{P} - p_0})|]/\mathcal{L}\bar{V} \tag{58}$$

In the case of a particular fluid, Slattery claims Equation 58 should also lead to the same degree of correlation as that observed in correlations of the kind described in (*90*).

However superficially similar they appear to be, the development leading separately to the generalized method and the analysis applied by Slattery are motivated by different concepts. The concepts embodied in the former method have already been discussed (page 27). By contrast, as noted above, Slattery's method does not attempt to directly couple results from viscometric and porous media experiments. Instead, as indicated by the results shown in Figures 11 and 12, scale-up for a given non-Newtonian fluid in a porous medium is accomplished by performing flow experiments in geometrically similar porous media.

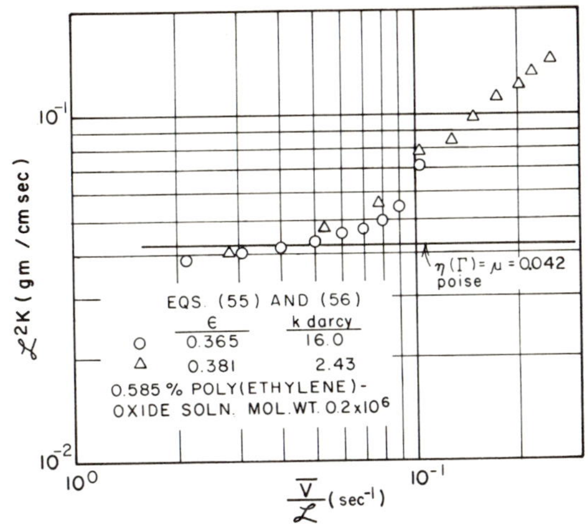

Figure 11. Slattery-type correlation of non-Newtonian flow through bead packs using data of Dauben (119)

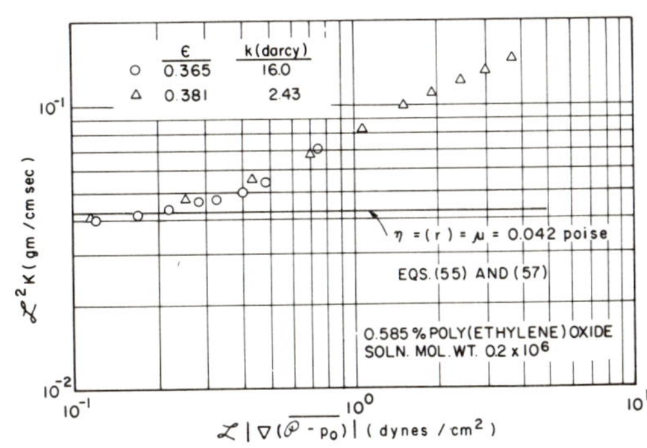

Figure 12. Slattery-type correlation of Darcy viscosity of non-Newtonian flow through bead packs using data of Dauben (119)

Other Examples of Non-Newtonian Flow Behavior in Porous Media

In the preceding discussion we were primarily concerned with the type of non-Newtonian flow behavior in porous media illustrated, for example, in Figures 1–4, 9, 10. More recent investigations reveal a wider variety of rheologically complex flow behavior can occur. One may observe, apparently under flow conditions through porous media not much different from previous studies, abnormal increases in flow resistance which resemble a "shear thickening" response. This general type of behavior has been observed in porous media flow experiments involving a variety of dilute to moderately concentrated solutions of high-molecular-weight polymers. Flow destabilization or premature departure from "laminar-like" behavior in dilute polymeric solutions have also been observed in flow experiments with porous media. Geometry-dependent flow behavior in porous media has been detected and seems to be characteristic of certain micellar systems.

A variety of descriptors are given in the literature relating to these and other phenomena observed in flow of non-Newtonian fluids through porous media. Included are "viscoelastic effects," "adsorption," "pore blockage," etc. Here we outline the salient features of some of these phenomena, summarize the various mechanisms proposed to explain the observed behavior, and illustrate these different effects with examples.

Viscoelastic effects. Several mechanisms have been considered in discussions of specific occurrences of viscoelastic effects involving rheologically complex fluids in porous media. Sadowski and Bird (*90*) recognized their coupling of the capillary-hydraulic radius model of a porous medium with a viscosity function ignores time-dependent elastic phenomena. They reasoned that no elastic effects would be observed provided the fluid relaxation time is small with respect to the transit time through a contraction or expansion in a tortuous channel of the porous medium—*i.e.*, the fluid can readjust to the changing flow conditions. By the same reasoning, major elastic effects can be expected when the fluid relaxation time is large compared to the transit time.

Thus the concept of a ratio of characteristic times also emerges as an ordering parameter in non-Newtonian flow through porous media. Astarita and Metzner (*159*) and Metzner *et al.* (*29*), among others, have discussed the importance of the ratio of the natural time of a fluid Θ_f to the duration time of a process Θ_p—*i.e.*, residence time of the material in a process, and have shown this to be a most useful concept. This Deborah number, defined as

$$N_{\mathrm{DEB}} = \frac{\Theta_f}{\Theta_p} = \Theta_f \left[\frac{D|\mathrm{II}_d|^{1/2}}{Dt} \right]^{1/2} \tag{59}$$

where II_d, the second invariant of the deformation tensor, represents the magnitude or intensity of the deformation rate process. If N_{DEB} is small and Θ_p is large, fluid-like behavior is implied in that the response may be indistinguishable from a purely viscous liquid. By contrast, solid-like behavior is implied by small Θ_p and large N_{DEB}. As noted (*29, 120*), flow studies in porous media are particularly useful for evaluating the significance of N_{DEB} and N_{WS}. A reasonable starting point is to develop suitable measures of these characteristic times. Truesdell (*160*) has indicated that by defining a natural time for a material in terms of measurable parameters, in such a way as to make its magnitude reflect the relative importance of normal stress effects, such a number could be used as an ordering parameter by which viscoelastic fluids could be classified. However, a specific value of this parameter is not important. Rather, the dimensionless function formed by the ratio of the natural time of the fluid to that of the flow field or macroscopic process is of interest. Measures of the natural time of a fluid can be arrived at in several ways. By performing a dimensional analysis of the flow model used in their investigation of non-Newtonian flow through porous media, Sadowski and Bird (*90*) defined a natural time for the Ellis fluid in terms of material parameters derived from the shear dependency of viscosity $\eta(\Gamma)$, namely

$$\Theta_f = \eta_0/\tau_{1/2} \tag{60}$$

By contrast, in their investigation, Marshall and Metzner (*120*) used the contravariant convected "Maxwell" model (Equations 70) to define Θ_f in terms of the viscosity function $\eta(\Gamma)$ and normal stress effects, namely

$$\Theta_f = \frac{1}{2\Gamma^2} \left(\frac{\tau_{11} - \tau_{22}}{\eta(\Gamma)} \right) \tag{61}$$

in which τ_{11} and τ_{22} denote the deviatoric normal stresses observed in viscometric flow. However, a variety of other possible definitions of Θ_f are discussed, for example, by Walters and Kemp (*161*), Truesdell (*160*), and Astarita and Metzner (*159*). Depending upon the fluid class studied in a program related to non-Newtonian flow through porous media, other measures might be useful. As an example, the parameters $\int_0^\infty \Theta^m N(\Theta) d\Theta$ (where m is a positive integer and $N(\Theta)$ is the distribution function of relaxation times Θ) occur in the equations of state for simple fluids with fading memory in slow flow (*161, 162*). Assuming the fluid of interest can be described by a lower limiting viscosity η_0 and one relaxation time at $\Theta = \lambda_1^*$, $\int_0^\infty N(\Theta) d\Theta$ and $\int_0^\infty \Theta N(\Theta) d\Theta$ may be used to define a natural time, namely (*161*)

$$\Theta_f = \lambda_1^* = \frac{\int_0^\infty \Theta N(\Theta) d\Theta}{\int_0^\infty N(\Theta) d\Theta} \tag{62}$$

Following Truesdell (*163*), a natural time may be evaluated from normal stress data as

$$\Theta_f = \frac{1}{2\mu(0)} \lim_{\Gamma \to 0} \left\{ \frac{1}{\Gamma^2} \left[(\tau_{11} + \tau_{22} - \tau_{33})^2 + (\tau_{11} - \tau_{22})^2 \right]^{1/2} \right\} \tag{63}$$

Comparing Equations 61, 62, and 63, for example, we observe that Θ_f may be finite in a constant viscosity liquid. As noted by Astarita and Metzner (*159*), there are major problems related to the evaluation of Θ_f for real fluids which exhibit elastic effects but are characterized by constant viscosity behavior. In this regard, there has been some discussion of the evaluation of Θ_f from the shear dependent viscosity function $\eta(\Gamma)$. As an example, Metzner *et al.* (*29*) have commented on Θ_f defined by Equation 60, with the observation that this parameter appears to be related only to $\eta(\Gamma)$ and not to viscoelastic properties. However, Sadowski and Bird (*75*) emphatically state that their Θ_f is approximately twice the time constant for the Spriggs-Bird Model (*27*). A natural time can also be developed from molecular theories. For example, in their study of non-Newtonian flow through porous media, Gaitonde and Middleman (*126*) used Θ_f as defined by Bueche (*164*), namely

$$\Theta_f = \frac{12\eta_0 M}{\pi^2 cRT} \tag{64}$$

Estimating some measure of the process time, Θ_p, is more uncertain in the case of a porous medium. Metzner *et al.* (*29*) argue that an estimate of Θ_p is meaningful provided the flow field is nonsteady in the Lagrangian sense. That is, the time function of interest is that representing the rate of change of the local fluid deformation rate given by the material derivative. One intuitively visualizes the porous matrix as consisting of pore openings interspersed with diverging and converging flow channel boundaries within which a nonsteady flow field may exist. The smoothness of these transitions is perhaps some function of the mode of packing or layering of the particles which constitute the porous matrix. Recalling Equation 59, it is more difficult to approximate II_d for complicated flows. Thus, in their study of viscoelastic effects in non-Newtonian flow through porous media, Marshall and Metzner (*120*) assumed that conditions within a porous medium could be approximated as a converging flow accompanied by shearing of material elements and stretching or elongation in the direction of flow. They modeled a porous medium in terms of flows in the frusta of cylindrical cones with alternating converging and diverging sections. For a porous medium composed of spherical particles of diameter D_P, Θ_P is evaluated as

$$\Theta_P = \alpha \left(\frac{\epsilon D_P}{V_0}\right) \tag{65}$$

where α is a numerical factor. The Θ_P definition used by Marshall and Metzner is equivalent to that given by Equation 65 with $\alpha = 1$. Combining Equations 59 and 65, the ratio of characteristic times used by Marshall and Metzner in their study of viscoelastic effects in non-Newtonian flow through porous media is given by the following Deborah number group:

$$N_{\text{DEB}} = \frac{\Theta_f V_0}{\epsilon D_P} \tag{66}$$

By contrast, Sadowski and Bird (*90*) evaluated Θ_P as

$$\Theta_P = \frac{D_P}{V_0} \tag{67}$$

so that combining Equations 60 and 67, the ratio of characteristic times used in their study is given by the Ellis number, defined as

$$\text{El} = \left(\frac{\eta_0}{\tau_{1/2}}\right)\left(\frac{V_0}{D_P}\right) \tag{68}$$

Bird (*165*) provides an interesting discussion of how the dimensionless group given by Equation 68 is a measure of the importance of elasticity of the fluid, particularly relating this group to flow through porous media. The Θ_P definition given by Equation 67 was also used by Gaitonde and Middleman (*126*), so that combining Equations 64 and 67 their ratio of characteristic times is given as

$$\frac{\Theta_f}{\Theta_P} = \left(\frac{12\eta_0 M}{\pi^2 cRT}\right)\left(\frac{V_0}{D_P}\right) \tag{69}$$

Admittedly, a fluid element is probably not subjected to a pure elongational flow in a porous medium. However, for flow fields of this type, those rheological models which admit an exponential rate of stress relaxation—*e.g.*, the contravariant convected Maxwell model (*29, 166–168*), as defined by the equations

$$\boldsymbol{\tau}' + \Theta_f \frac{\delta \boldsymbol{\tau}'}{\delta t} = 2\mu \mathbf{d} \tag{70a}$$

$$\boldsymbol{\tau} = -p\,\mathbf{g} + \boldsymbol{\tau}' \tag{70b}$$

predict that the stresses will become infinite as the product of the stretch rate and Θ_f approaches a critical value. Marshall and Metzner (*120*) observe that if the stretch rate in flow through porous medium is approximated by $2V_0/\epsilon D_P$ the pressure drop needed to pump a viscoelastic fluid through a porous medium should rise to infinity as N_{DEB} approaches a value of about 0.25. That is, as solid-like behavior develops, one should observe a pressure drop significantly larger than that predicted on the basis of observed viscometric behavior, using Equation 32 or 42. Pearson (*169*) has also suggested that if a porous medium consists of long tubes, perhaps randomly oriented, then flow behavior should resemble that in a single tube—*i.e.*, giving rise to a decrease of the ratio of pressure drop to flow rate with increasing flow rate in the instance of a shear thinning fluid. By contrast, if the porous medium consists of alternate contractions and expansions—*i.e.*, rapidly converging and diverging channels (as in a sand bed), then flow behavior is more analogous to that in pure shear involving the elongational viscosity. For these conditions, the ratio of pressure drop to flow rate can increase catastrophically with increasing flow rate, even in the case of a fluid exhibiting shear thinning behavior in a viscometric flow.

A number of flow phenomena which have been related to viscoelastic effects are documented throughout the literature. Apparently Sadowski and Bird (*90, 121*) were the first to attempt to relate, in a fundamental study of a non-Newtonian flow through porous media, a departure from their modified Darcy's law to viscoelastic effects. Their data on moderately concentrated solutions of a medium molecular weight grade of hydroxyethylcellulose showed a consistent deviation from Equation 42 of this paper. The authors commented that the Θ_f values of these solutions were significantly larger than those of other solutions which correlated

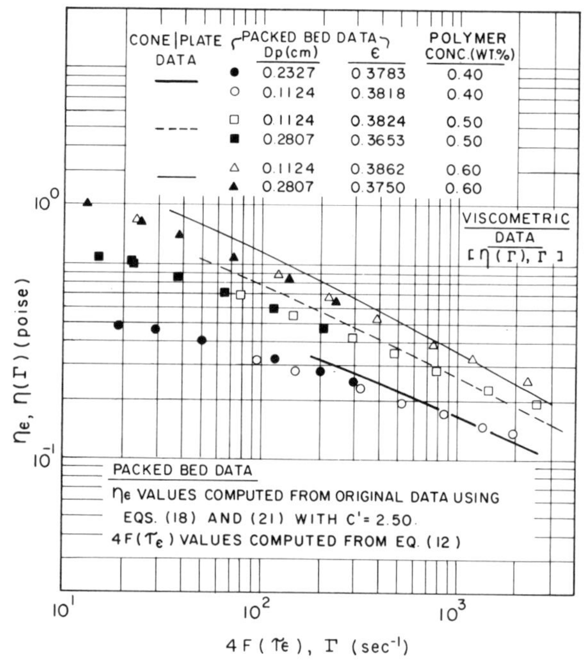

Figure 13. Anomalous flow behavior of hydroxyethylcellulose solutions in bead packs—comparison with behavior in cone/plate experiments (121)

with Equation 42—*e.g.*, compare their Figures 5 and 6. Sadowski (*121*) discovered that the deviation could be eliminated by introducing the Ellis number defined by Equation 68. Specifically, on introducing the term (−5El) inside the square brackets of Equation 44, excellent agreement was attained for all solutions. Marshall and Metzner (*120*) have commented strongly on what Sadowski and Bird (*90*) believe to be significant viscoelastic effects. As a matter of interest, the original data cited (*90*) for these particular solutions have been recast in the apparent viscosity-nominal shear rate coordinates of Figure 13. That is, the packed bed data have been presented in a form suggested by the generalized scale-up method. Sadowski's viscometric data show these solutions to be characterized by shear thinning behavior. In view of the other examples to be discussed, it is also important to note that the packed bed data also exhibit a shear thinning response. For the particular shift factor used ($C' = 2.5$), it appears the observed Darcy viscosities (Equation 15), are consistently lower than what would be expected from the viscometric behavior. However, this may be only an artifact, because a shift factor of $C' = 25/12$ applied in the same way to obtain η_ϵ values would lead one to conclude that the viscous resistance in the porous medium is still shear thinning, but consistently higher than that anticipated from viscometric experiments. As noted by Gaitonde and Middleman (*126*), it is difficult to assess the behavior Sadowski and Bird (*90*) indicate represents viscoelastic effects in non-Newtonian flow through porous media.

More dramatic examples of viscoelastic effects are illustrated in the experimental data of Dauben and Menzie (*127*), a portion of which have been recast in the forms illustrated in Figures 14 and 15. Their study pertains to flow behavior in porous media of solutions of different molecular weight grades of poly(ethylene oxide). Dauben's viscometric experiments showed the more dilute solutions of the highest molecular weight grade and the moderately concentrated solutions of the lower molecular weight grades to be characterized by constant viscosity behavior. The other solutions studied were found to exhibit a shear thinning behavior in a viscometric flow. Rheogoniometric experiments showed some solutions exhibited viscoelastic behavior. Unconsolidated porous media in the form of bead packs, characterized by porosities in the 34.7% to 39.2% range and permeabilities in the 2.43 darcies to 18.0 darcies range were used in these experiments. We observe that the data, when plotted on the $f_e - N'_{Re}$ coordinates of Figure 14, generally depart from the laminar curve defined by Equation 40. The friction factors for the solution of the highest molecular weight polymer are not particularly sensitive to Reynolds number. A series of branching curves, generally of low slope and seemingly departing from the laminar curve at a series of well defined, but very small, Reynolds numbers are obtained. Also observe the ratio of experimental f_e to that predicted by Equation 40; ratios as great as 10 are frequently obtained. The abnormal $f_e - N'_{Re}$ relationship observed in these experiments is further illustrated, but in a different way, in Figure 15. Here the experimentally derived Darcy viscosity-bed shear rate response—*i.e.*, Equations 21 and 19, respectively, for a moderately concentrated solution of the intermediate-molecular-weight grade, are compared with the behavior predicted from viscometric experiments. By contrast with the previously discussed results of Sadowski and Bird (Figure 13) the results from these packed bed experiments suggest a shear-thickening response. Additionally, the viscometric experiments show the purely viscous behavior for the solution of the higher molecular weight polymer seems to be coupled with the bed geometry, as evidenced by the distinct curve for each permeability. According to Dauben and Menzie (*127*), these effects result in part from interactions between the porous medium and the polymer molecules.

In a contemporary study, Marshall and Metzner (*120*) describe similar abnormal increases in flow resistance through porous media in viscoelastic solutions—*e.g.*, $0.95 \leq \Theta_f \leq 6.8$, of a partially hydrolyzed (polyacrylamide)—*e.g.*, see their Figures 2 and 3 and Figure 16 of this paper. Their experiments were performed with a consolidated porous medium in the form of a sintered porous bronze disk ($D_P = 0.012266$ cm, $\epsilon = 48.6\%$, $k = 46.4$ darcies). Analyses of the high flow rate data also suggest a shear-thickening response occurs in the porous medium, although the flow behavior was determined to be of the shear-thinning type in a viscometric experiment. The authors commented that the behavior at high flow rates suggests a transition from viscous flow

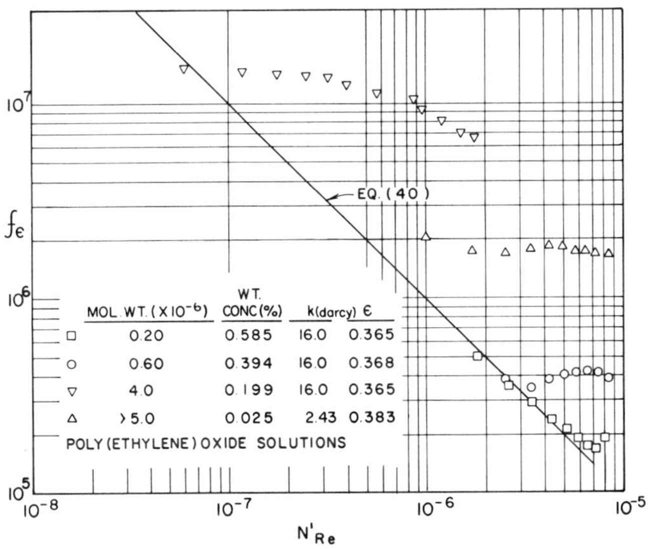

Figure 14. Reynolds number vs. friction factor for non-Newtonian flow through bead packs (119)

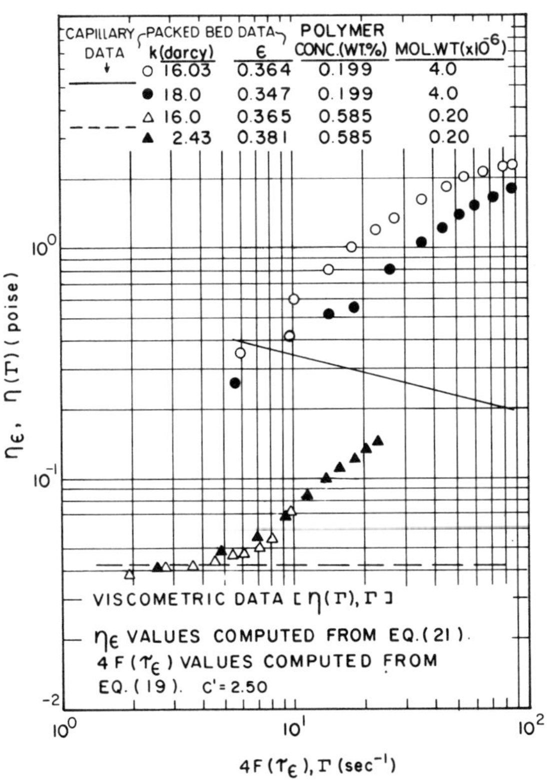

Figure 15. Flow behavior of polyethylene oxide solutions in bead packs—comparison with capillary flow experiments (119)

to conditions determined by the previously discussed stretch rate limitations. Interestingly, the point at which the abnormal effects observed by Marshall and Metzner (*120*) first occur corresponds to N_{DEB} values of 0.05 to 0.06. This is well within the range of values anticipated from their analysis of viscoelastic effects as outlined above. Very recently, Wissler (*170*) developed a perturbation analysis of flow of a Spriggs-type viscoelastic fluid, characterized by a constant viscosity and a single relaxation time, through a converging-diverging channel. His analysis indicates that the total mean pressure differences on two surfaces $r = r_1$ and r_2 $(r_2 > r_1)$ is given by the relation

$$\bar{p}(r_2) - \bar{p}(r_1) = \frac{-\eta \bar{V}_1}{r_1}\left[1 + 7.125\left(\frac{\pi}{4}\right)^2\left(\frac{\lambda \bar{V}_1}{r_1}\right)^2\right] \quad (71)$$

where the mean velocity in the throat of the restriction is given by $\bar{V}_1 = 4Q/\pi r_1$ and Q is defined here as the volumetric flow rate per unit width. Wissler suggests that Equation 71 might be applied to a porous medium by replacing $\bar{V}_1$ and r_1 in terms of superficial velocity, porosity, and particle diameter. Hence, for a porous medium there results

$$(\Delta P)_{VE} = (\Delta P)_{PV}\left[1 + A\left(\frac{\lambda V_0}{\epsilon D_P}\right)^2\right] \quad (72)$$

or equivalently, introducing Equation 66

$$f\mathrm{Re} = [1 + A(N_{\mathrm{DEB}})^2] \quad (73)$$

According to Equation 72, the purely viscous pressure loss must be multiplied by a factor $[1 + A(N_{\mathrm{DEB}})^2]$ to obtain the pressure loss in the presence of elastic effects. As shown by the correlation presented in Figure 16, Equation 73, with $A = 10$, is in remarkable agreement with the Marshall-Metzner results. However, not all published results agree with the rate of increase in pressure loss predicted by this model. Wissler suggests that a tentative explanation for the apparent lack of agreement in such cases is the effective value of Θ_f decreases with increasing shear rate. By contrast, the characteristic times reported (*120*) were evaluated for the fluids at appropriate shear rates so that one would anticipate that the Marshall-Metzner data would correlate with Equation 73.

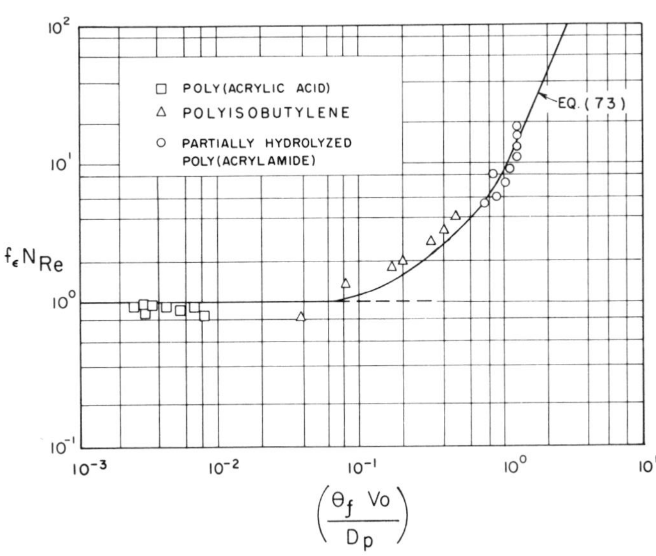

Figure 16. Dependence of viscoelastic effects ($f_e N'_{\mathrm{Re}} > 1.00$) on the Deborah number of the flow process. All data from Marshall and Metzner (120) in region $N_{\mathrm{Re}} > 0.05$

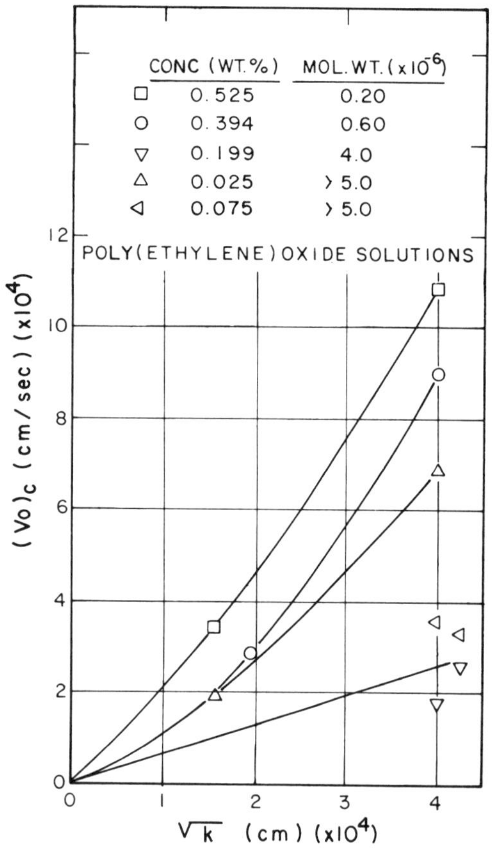

Figure 17. Dependence of viscoelastic effects in porous media on Deborah number and molecular weight of poly(ethylene oxide) solutions (119)

The Marshall-Metzner model of flow of viscoelastic fluids through porous media can be illustrated as follows. Inspection of Equation 66 indicates that, for a given fluid, the critical velocity $(V_0)_c$ at which these effects occur should be a linear function of ϵD_P or equivalently, of $\sqrt{k}$. Figure 17 tests the model with some of Dauben's data (*119*). The critical velocities indicated are only approximate values because they were indirectly obtained from the intersection of each experimentally determined porous media flow curve with the corresponding viscometric curve for the fluid of interest. Obviously the model is too crude to permit the quantitative prediction to be realized. However, the key features of the model seem to be obtained. It is worth noting from Equation 66 that, for a given fluid, the inverse slope of the correlation is proportional to the relaxation time Θ_f. Keeping this in mind, we observe the curves are approximately ordered according to molecular weight. Considering that these materials represent a homologous series of poly(ethylene oxides), it is anticipated from molecular theories (Equation 64), that the relaxation time should be proportional to the molecular weight of a particular species. Admittedly, this is a gross oversimplification because, as noted earlier, Θ_f is not expected to be constant except over limited ranges of conditions.

Both References *120* and *127* indicated that these results could not have resulted from spurious effects, such as progressive plugging or adsorption. Therefore, these effects are significantly different from the majority of the findings reported (*90*, *91*, *118*, *121*, *122*, *124–126*, *130*). Similar observations and conclusions, relating to flow of polymeric solutions through a variety of porous media, are scattered throughout the literature (*93*, *135*, *136*, *142*, *144*).

Harvey (*144*) has also observed major departures from the behavior of purely viscous fluids which are similar to those reported above. He studied the flow behavior of solutions of a polysaccharide, and solutions of high-molecular-weight grades of nonionic and partially carboxylated poly(acrylamides), through bead packs ($0.362 \leq \epsilon \leq 0.398$, $0.00635 \leq D_P \leq 0.0505$). All of the solutions studied exhibited shear-thinning behavior in viscometric experiments. A preliminary analysis indicates that the polysaccharide solutions exhibited a shear-thinning response in the packed bed experiments. By contrast, combinations of shear-thinning and shear-thickening behavior characterize the packed bed flow behavior of the other solutions studied. In an attempt to correlate the packed bed data, he modified the method of Christopher and Middleman (*122*) by replacing the particle diameter D_P with an effective particle diameter D_P^*. With the exception of the polysaccharide solutions, where D_P^*/D_P was in the range from 0.8 to 1.0, the ratios for the other solutions studied were usually in the range from 0.9 to 0.1. Harvey observed these effects are apparently related to some type of interaction between polymer molecules and the surface of the porous medium, and discussed possible mechanisms. A more thorough understanding of the observed behavior could probably result if the same flow behavior could be studied over a range of particle diameters—*e.g.*, in the manner of Dauben and Menzie (*127*). In experiments involving dilute solutions of certain polymeric materials, Gogarty (*93*), Pye (*135*), Sandiford (*136*), and Smith (*142*), among others, have also observed a flow behavior in porous media which is significantly different from that observed in viscometric experiments. The flow behavior is similar to that reported in References *120*, *127*. Significantly, Pye (*135*) comments that the effect is not related to plugging, and further that the effect appears to occur only in flow through tortuous passages. Smith (*142*) has observed significant abnormal increases in flow resistance on flowing saline solutions containing small concentrations of partially hydrolyzed poly(acrylamide) in sandstone cores. He also notes that both the onset and magnitude of the "shear-thickening" effect is coupled with molecular weight. As the molecular weight of the species is increased, the onset is shifted in the direction of lower velocities, and there is a significant increase in the magnitude of the effect. In these studies, involving consolidated porous media characterized by permeabilities on the order of 0.40 darcy, the abnormal increase in flow resistance occurred at velocities of about

10 ft/day. Experiments by Pearson and Paratello (*171*) on flow of viscoelastic and purely viscous polymeric solutions in porous media also suggest there need be no correlation between pressure drop/flow rate curves in porous media and similar flow curves derived from capillary viscometry experiments. In the instance of highly elastic shear-thinning solutions of poly(ethylene) oxide, similar to those investigated by Dauben (*119*), these authors also observed an increase in the pressure drop/flow rate ratio with increasing flow rate in porous media.

The general type of viscoelastic behavior observed in Figures 14 to 16 has been simulated in a primitive model of the flow passages in a bead pack (*172*). The model consists of a column of 0.3175-cm diameter spheres packed in single file inside a 0.3810 cm diameter smooth bore cylindrical tube. The slight difference in diameters produces a "staggered" arrangement of the column. The spheres are supported at the inlet and outlet to the tube with single layers of 100-mesh screen. Typical results are presented in Figure 18. The calibration runs with Newtonian fluids (denoted by the shaded circles) span a range in kinematic viscosities between approximately 0.01 cm²/sec and 0.07 cm²/sec. The Newtonian fluids serve to calibrate the flow cell, as well as to indicate the general trend in flow behavior observed with this general type of rheological behavior. Initially, the data points for the Newtonian fluids fall on the "laminar" curve described by Equation 40. At Reynolds numbers above 10^{-1}, they generally fall below the Ergun correlation (*157*), namely:

$$f_\epsilon = \frac{1}{N'_{\mathrm{Re}}} + 1.75 \qquad (74)$$

The other data pertain to different concentrations of aqueous solutions of a homopolymer of acrylamide (mol wt 6.5 to 10.0 million). The kinematic viscosities of these solutions were in the range from approximately 0.01 cm²/sec to 0.03 cm²/sec. The 0.01% solution exhibited a purely viscous behavior almost indistinguishable from that of the solvent. In a turbulent flow field this solution is spectacularly non-Newtonian in that significant drag reduction effects occur. The 0.075% solution was not appreciably more viscous but was markedly elastic in character. In contrast with the behavior of the Newtonian fluids, significant shear-thickening-like behavior and flow destabilization phenomena characterize the flow behavior of these non-Newtonian fluids. Furthermore, the deviations are seen to shift in the direction of smaller Reynolds numbers as the polymer concentration increases. As in the case of the Marshall-Metzner results (*90*), under these conditions the ratio J/Q is proportional to Q^m, where the exponent is between 2 and 3. Further, the results from the 0.01% solution agree quite well with Wissler's model (Equation 73).

Jones and Maddock (*128*, *129*) have associated visco-

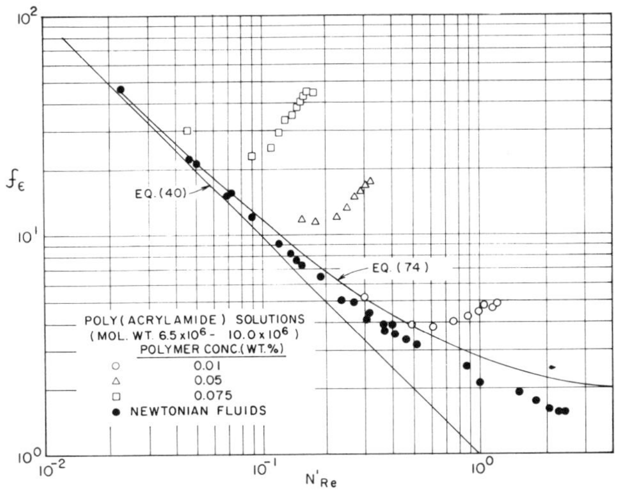

Figure 18. Reynolds number vs. friction factor for Newtonian and viscoelastic flow through a primitive model of a porous medium (172)

elastic effects with the flow destabilization-like phenomena illustrated in Figure 18. They observed similar effects on flowing dilute solutions of a high-molecular-weight grade of poly(acrylamide) through bead packs (see their Figures 5 to 8 in Reference *128*). In the case of the polymeric solutions, they discovered a departure from "laminar" flow at a value of Re about 32% lower than the Re where deviations associated with inertia effects first appeared with the solvent. The authors note that elasticity-induced destabilization can be predicted for various flow conditions, depending on the constitutive equation which is adopted in the mathematical analyses of the flow problem (*173*, *174*).

Pore blockage, adsorption, and related phenomena. There is considerable petroleum engineering literature related to adsorption or retention by a porous media of polymeric materials, and to the resultant permeability reduction effects which follow. It is not the intention of this review to consider these investigations in detail. Instead, we will focus attention on some of the frequently cited examples, noting that considerable controversy exists as to how these phenomena relate to flow behavior of non-Newtonian fluids through porous media. As noted in the preceding discussion of viscoelastic effects, both Marshall and Metzner (*120*), and Dauben and Menzie (*127*) considered the role of progressive plugging and adsorption of polymer aggregates in the flow phenomena they observed. They concluded these aberrations were absent from their experiments. By contrast, Sadowski (*121*) experienced difficulties with constant pressure experiments, but had no trouble with constant flow rate experiments. He concluded that polymer adsorption and gel formation was aggravated during constant pressure experiments because, under these conditions, any tendency towards gel formation would lower the permeability and thus automatically lower the flow rate. He reasoned that gel formation would be aggravated because, at the lower flow rates, there would be less tendency to remove polymer molecules from the particle surface. By contrast, in the constant flow rate experiment, adsorption might still occur, but the constant fluid motion would prevent a gel network from forming between particles. Thus the concept of a mechanical entrapment or pore blockage effect, developing under certain flow conditions, emerges from Sadowski's investigations. Burcik (*72*) is of the opinion that the shear-thickening behavior observed in flow of dilute polymeric solutions through porous media is related to polymer molecules retained within the pore structure. He suggests these bound molecules are uncoiled under the imposed (high) velocity gradient, thereby increasing the resistance to flow. Presumably polymer retention may occur either by adsorption or mecahnical entrapment. Burcik has pursued this thesis in a series of notes (*72, 137, 140, 141*). These relate to flow experiments in both consolidated and unconsolidated sandstone cores with dilute solutions of a presumably high molecular weight grade of partially hydrolyzed poly(acrylamide). This polymer retention effect is illustrated in Figures 1 and 2 of Reference *137*. Recently, Burcik and Walrond (*175*) have considered the role of microgel particles in the behavior of polymeric solutions in porous media. Presumably solutions of partially hydrolyzed polyacrylamides contain a filterable cross-linked material. Some of the conclusions have been questioned by Smith (*142*). Studies relating adsorption and permeability reduction effects in a variety of porous media with different polymeric materials are cited (*71, 91–93, 123, 135, 136, 139, 142–144*). The mechanism of permeability reduction has also been considered by Gogarty (*93*), who argues that in a constant flow rate experiment there is an initial phase of "stabilization." Presumably, polymer retention occurs by two mechanisms; namely, mechanical entrapment and adsorption, and at sites of small openings located between pores. After stabilization occurs—*i.e.*, constant pressure conditions attained, the polymer solution passes unchanged through the porous media, because flow takes place between pores through the larger openings. According to Gogarty, the overall permeability is reduced because of the diminution of polymer flow through the smaller openings between pores.

Kozicki *et al.* (*71*) have described complications relating to non-Newtonian flow through porous media which relate to polymer adsorption and gel formation effects. The approach is similar to that described earlier in which the generalized scale-up method was coupled with the capillary-hydraulic radius model of a porous medium. However, their analysis is different from that outlined earlier in three major respects. They allow for the previously discussed possibility of a coupling between rheological behavior and boundary or wall effects (Equation 3). The shift or correction factor is introduced in the manner suggested by Bird (Equation 16). Additionally, their generalized flow equation involves an aspect factor (ξ) which characterizes the shape of an arbitrarily defined flow cross section. For the circular capillary, where $\xi = 3$, it appears their scale-up equation, designed to also account for gel formation or adsorption effects in non-Newtonian flow through porous media, can be expressed in the form:

$$\frac{V'_0 - V_s}{\epsilon R_H} = \frac{2}{C'\tau^3_{RH}} \int_0^{\tau_{RH}} \tau^2 \Gamma(\tau) d\tau \tag{75}$$

(Compare Equations 12 and 16.) The authors observe that $\xi = 3$ is substantiated by Sadowski's packed bed data (*91*) and their results from packed bed experiments. Previously we have indicated the concept of the capillary-hydraulic radius model of a porous medium—*i.e.*, $\xi = 3$ is indeed approximately substantiated by the data of Christopher, Gaitonde, and Middleman (*122, 126*) and Sadowski and Bird (*90*), among others. As indicated earlier, the authors applied Equation 75, using a method well known in viscometry (*2*), to the analysis of their data on flow of moderately concentrated solutions of a cellulose derivative and a poly(acrylic acid). Some of their results, illustrating anomalous surface effects in capillaries and porous media, are illustrated in the previously cited Figure 4. Unconsolidated porous media in the form of bead packs ($\epsilon = 0.38$, $D_P = 0.3175$ cm, 0.3967 cm) and capillaries in the diameter range from 0.1478 cm to 0.2456 cm were used in the experiments. The thickness of the anomalous layer in the bead pack with $D_P = 0.3175$ cm was estimated to be in the range from 2.0×10^{-3} cm to 3.4×10^{-3} cm for the solutions studied. According to the authors, Sadowski's data on moderately concentrated solutions of a cellulose derivative (Figure 13 of this paper) showed evidence of a finite but considerably thinner anomalous surface layer.

In contrast to the negative effective velocity effects described by Kozicki *et al.* (*71*), are the experimental results of Williams and Taylor (*147*), shown in Figures 19 and 20. The data pertain to flow behavior in porous media of a *sec*-butylamine oleate system (*82*). Consolidated porous media in the form of alundum cores, characterized by porosities in the 26% to 35% range, and permeabilities in the 4.1-darcies to 0.115-darcy range were used in these experiments. We observe that when plotted on the $4F(\tau_\epsilon)$, τ_ϵ coordinates suggested by the generalized scale-up method, a distinct flow curve is generated for each permeability over most of the range of flow conditions studied. The characteristic spread suggests a wall effect behavior similar to that discussed above. However, in contrast to the results given in Figure 4, the ordering of the flow curves with respect to k suggests the presence of a positive wall effect. This is substantiated in the trends shown in Figure 20 where the $\mathfrak{L}^2K - \overline{V}/\mathfrak{L}$ coordinates are used. We observe that at a given "characteristic" shear rate $\overline{V}/\mathfrak{L}$, the apparent darcy viscosity ($\mathfrak{L}^2K$) decreases as k decreases. However, the flow behavior is even more complex. Combinations of shear thickening and shear thinning behavior are generated. Fluid systems exhibiting such

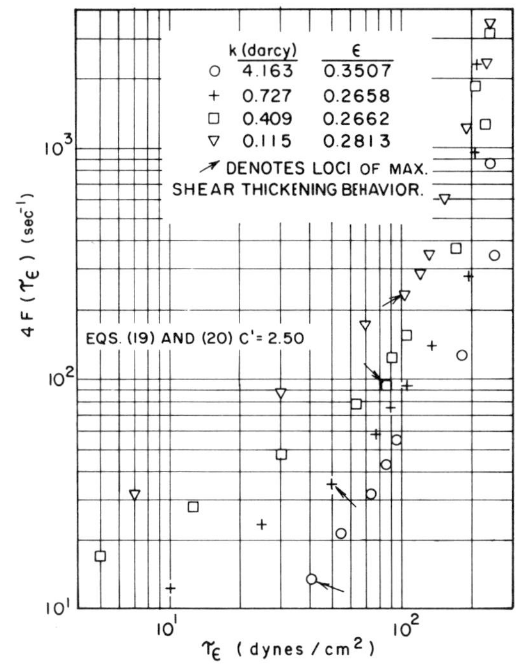

Figure 19. Anomalous flow behavior in consolidated porous media of sec-butylamine oleate system (147)

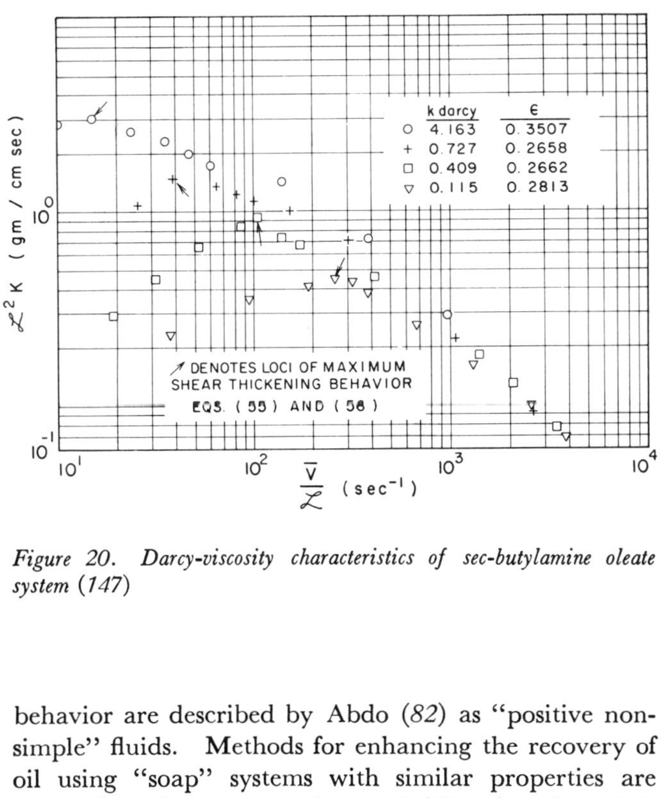

Figure 20. Darcy-viscosity characteristics of sec-butylamine oleate system (147)

behavior are described by Abdo (*82*) as "positive non-simple" fluids. Methods for enhancing the recovery of oil using "soap" systems with similar properties are outlined in the patents referenced (*73–84*). Superficially, it would appear that these data could be correlated by allowing for the existence of an anomalous layer, where instead of using Equation 12 for finite V_s, the following equation is applied

$$\frac{V'_0}{2^{3/2}\sqrt{C'k\epsilon}} = \frac{V_{0s}}{2^{3/2}\sqrt{C'k\epsilon}} + \frac{1}{\tau_\epsilon^{3}}\int_0^{\tau_\epsilon} \tau^2\Gamma(\tau)d\tau \qquad (76)$$

(Compare Equations 19 and 22.) However, analysis indicates the assumed functional relationship between V_s and τ_ϵ is not obtained. Rather, it appears the effective slip velocity is coupled in a complex way to both τ_ϵ and permeability. It is interesting that micellar systems also exhibit "diameter effects" under drag reducing conditions in turbulent pipe flow (*176*).

Applications

We now briefly consider some applications to illustrate the relevance of non-Newtonian flow through porous media to certain technology areas. In the instance of applications related to the petroleum industry, there exists considerable literature and patent art relating to the use in oil recovery processes of a variety of polymeric and chemical additives. Some of these materials are known to impart non-Newtonian flow properties to aqueous and hydrocarbon systems. However, as in the preceding discussion, the main emphasis of this review necessarily restricts us to a small portion of that literature. Additionally, however, we will comment on the patent literature available to us which discloses processes for exploiting non-Newtonian flow in the recovery of oil.

A number of studies relate to the use of additives, including polymeric materials, in waterflood processes directed at improving oil recovery. Typical are the frequently quoted studies of Pye (*135*) and Sandiford (*136*), and the investigations of Mungan *et al.* (*92*) and Gogarty (*93*). The significance of the information developed from studies of this kind has been discussed by Stahl (*133*), Dauben (*119*), Dauben and Menzie (*127*), Sherborne *et al.* (*131*), Lee and Claridge (*145*), and more recently by Slater and Farouq Ali (*146*), among others. Generally, studies of this kind have involved laboratory and/or field comparisons of how the presence of small concentrations (0.05% or less) of high-molecular-weight polymers in the waterflood affects such parameters as oil displacement efficiency, area sweep efficiency, and permeability distribution. As noted in Tables IV and V, a variety of porous media and fluid systems have been used in such studies. Generally, non-Newtonian flow behavior is but one factor in a set of other important considerations which include adsorption characteristics at rock and fluid interfaces, displacement mechanisms, lithology of the formation of interest, etc. Mungan *et al.* (*92*) comment that improved oil recovery is achieved by increasing the reservoir volume swept and increasing the oil recovery efficiency of the swept volume. According to these authors, improvements in sweep efficiency seem possible in polymer flood processes. The significance of non-Newtonian flow behavior in waterflood processes and oil recovery is briefly discussed by Dauben and Menzie (*127*). These authors note that an understanding of the flow behavior of complex fluids is an important factor in predicting their performance in stratified reservoirs. Recently, Slater and Farouq Ali (*146*) reviewed prior investigations related to polymer flooding. They commented on several features of polymer flooding, including the observation that the shear-thickening-like behavior exhibited by certain polymeric solutions seem to improve the displacement of oil by retarding the flow in the more permeable zones and in zones of high flow rates. Lee and Claridge (*145*) have compared the areal sweep efficiency of oil displacement by Newtonian and shear-thinning non-Newtonian fluids. Their experiments, in Hele-Shaw flow models, involved solutions of a high-molecular-weight polymeric material. These authors concluded that areal sweep at breakthrough was less for the non-Newtonian fluids than for Newtonian fluids of comparable viscosities. However, the authors claim that compared to the displacement process without water, areal sweep after one pore volume was improved with the non-Newtonian fluids. In a recent study of polymer floods, Mungan (*139*) concludes that radial polymer flooding in a stratified porous medium composed of layers of different permeabilities is more efficient than radial waterflooding. By contrast, it has been concluded that linear polymer flooding in a homogeneous porous medium does not recover more oil, compared to water flooding, if the oil viscosity is less than 60 cp (*136*, *92*). Again, these experiments involved experiments with shear-thinning, possibly elastic, solutions of high molecular weight polymers. Mungan also describes displacement experiments in Hele-Shaw models, with similar non-Newtonian solutions, designed to compare sweep efficiency and viscous fingering in Newtonian and non-Newtonian fluids. He claims these model studies reveal that the increased recovery results from improved sweep efficiency, and observes that these findings have been confirmed by theoretical studies (*146*). Some field studies have apparently shown that polymer flooding processes recover more oil than waterflooding processes. However the mechanism by which this occurs is controversial, probably because the phenomena which characterize the displacement process are poorly understood. Along somewhat different lines, van Poolen and Jargon (*177*) describe empirical relationships for use in predicting the steady and transient flow of purely viscous power law fluids in linear and radial geometries. These predictions are used to develop correlations of "productivity index" *vs.* rate. Several investigators have also considered other aspects of polymer flooding which relate to adsorption and polymer retention effects, sweep efficiency, effect of polymer molecular weight and composition, effects of electrolyte environment, and other variables on polymer solution properties, etc. References *72*, *92*, *93*, *123*, *131*, *135–137*, *139–141*, *143*, *144*, *175* should be consulted for details of this literature.

As noted earlier, the previously discussed *sec*-butylamine oleate system exhibits unusual combinations of flow behavior. As observed in Figures 19 and 20, the gradient-dependent viscosity exhibits a shear-thickening response at low shear rates and a shear-thinning response at high shear rates. Obviously the flow behavior is also coupled with permeability in a complicated manner. Additionally, it is not unusual for such systems to exhibit markedly elastic behavior. The patents cited under References *73* through *84* disclose processes which seek to exploit these non-Newtonian flow characteristics to improve the recovery of oil by (i) improving macroscopic sweep efficiency, enabling the recovery of a greater fraction of the oil before breakthrough occurs, and by (ii) improving microscopic flooding efficiency. For example, Abdo (*84*) notes an objective is to achieve more nearly uniform injection and flow profiles. Thus the shear-thickening effect enables the fluid to develop a higher viscosity when exposed to higher shear rate conditions in a formation. For example, at constant pressure drop, a higher shear rate is induced when the liquid flows in a more permeable stratum than when it flows in a less permeable stratum within a formation. The rates of flow in the two strata become more nearly "equivalent" because the fluid becomes selectively more viscous in the more permeable stratum. A useful result is that premature breakthrough effects are inhibited. However, the coupling between rheological properties and permeability is another useful result because permeability controls whether such a fluid becomes more viscous for the situation where equal shear rate conditions exist (*84*). Thus, both the shear-thickening property and the coupling between rheological behavior and permeability operate to achieve more nearly uniform injection and flow profiles. These rheological effects on permeability contrast can be readily illustrated by considering the situation where a constant-pressure gradient is applied across two porous beds of equal cross sectional areas. These media do not communicate with each other and are characterized by permeabilities k_H (high) and k_L (low), and corresponding porosities ϵ_H and ϵ_L. For these conditions we can use Equation 30, for example, to obtain the group

$$\left.\frac{k_L Q_H}{k_H Q_L}\right|_J = Q(R) \tag{77}$$

where

$$Q(R) = \left(\frac{k_H}{k_L}\right)^{(1-n)/2n} \left(\frac{\epsilon_L}{\epsilon_H}\right)^{(1-n)/2n} \tag{78}$$

Suppressing the group ϵ_L/ϵ_H for purposes of illustration, we observe that for a Newtonian fluid ($n = 1$), the group Q_R is unity. Similarly, for a purely viscous shear thinning fluid ($n < 1$), we find $Q_R > 1$, implying that the permeability contrast is aggravated for this case of fluid behavior. By contrast, $Q_R < 1$ implies that the permeability contrast is minimized. Some of Dauben's packed bed flow data pertaining to a poly(ethylene) oxide solution (*119*), and Williams-Taylor data (*147*) pertaining to the flow behavior of a *sec*-butylamine oleate system in alundum cores, have been recast in a form suggested by Equation 77. The results are shown in Figures 21 and 22. We observe that both sets of results generate Q_R values less than unity. Thus a favorable reduction of permeability contrast is achieved by fluid systems exhibiting a shear-thickening-like response in a porous medium and by combinations of such behavior and permeability-dependent flow behavior. The patent cited as Ahearn (*148*) also relates to processes which seek to exploit unusual rheological behavior in applications involving the recovery of oil from subterranean formations. Disclosed are systems containing mixtures of water soluble petroleum sulfonates and ethylene oxide polymers which exhibit shear-thickening and enhanced viscoelastic behavior when a threshold shear rate is exceeded.

Along somewhat different lines is the process described by Kirk *et al.* (*178*) for recovering oil utilizing non-Newtonian fluids. The disclosure relates to a technique for increasing fluid productivity at a production well by intermittent reduction and increase in pumping pressure in the injection and driving of a slug of shear-thinning fluid through an oil-bearing stratum. Finally, although there has been no opportunity to consider their work within the framework of the present review, we note the existence of some contemporary Russian investigations related to flow of complex fluids in porous media. References *179* to *183* encompass those currently known to us.

Other technology areas pertinent to non-Newtonian flow through porous media would seem to include naval architecture, polymer processing, lubrication, and waste disposal applications. The bulk of the applications involving naval architecture involve classified projects. However, in a review of hydrodynamic friction reduction techniques, Hoyt (*184*) suggests a problem with direct relevance to the studies reviewed here. Thus, Figure 14 of that reference shows a device designed to eject a polymeric solution into the boundary layer around a hydrodynamic vehicle. The properties of the dilute polymeric solution are such that changes in acceleration and velocity occur when polymer ejection occurs. Along similar lines, Wells (*185*) has recently considered skin friction reduction and mass flux requirements relating to uniform injection of a solution of a high-molecular-weight polymer into the turbulent boundary layer on a slender body of revolution. His analysis predicts that porous wall injection requires significantly less polymer solution than does slot injection to maintain an equivalent level of drag reduction activity. Depending upon certain design parameters, including the ratio of ejection surface area to total vehicle area, the morphology of the ejection surface, ejection mass flow rate, and polymer solution characteristics, flow conditions ranging from the purely viscous behavior observed, for example, by Christopher and Middleman

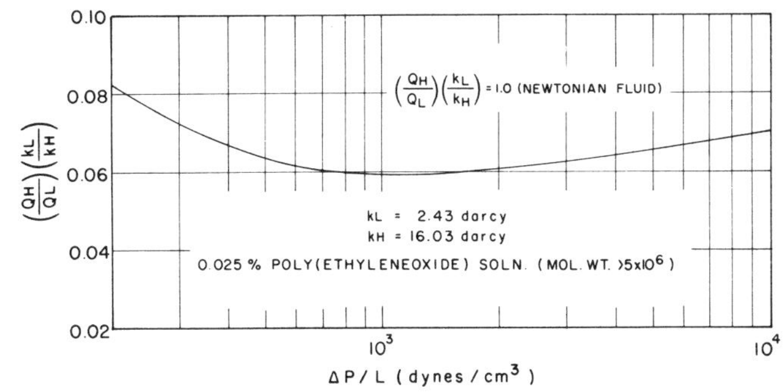

Figure 21. Flow of rheologically complex polymeric solution through bead packs—effect on permeability contrast (*119*)

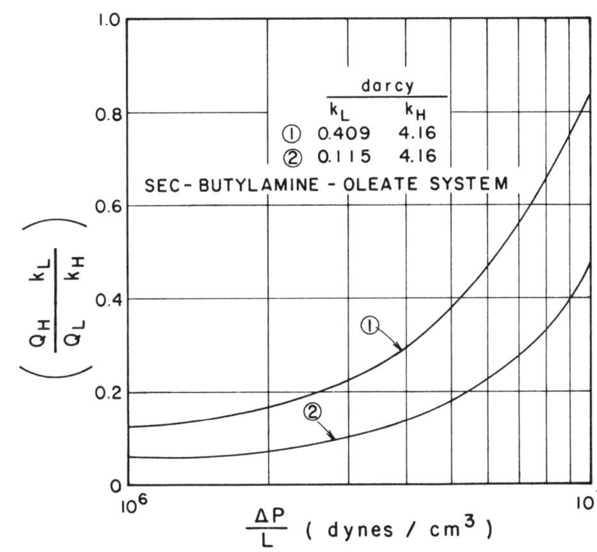

Figure 22. Flow of rheologically complex micellar system through consolidated porous media—effect on permeability contrast (*147*)

(*122*) to the extremal viscoelastic effects observed by Marshall and Metzner (*120*) could be encountered in this application area. The shear rate range of interest could be on the order of several thousand reciprocal seconds, and large Deborah number effects would seem to be characteristic of such processes. It also appears that flow of complex fluids through porous media is of interest in fiber-forming processes. Particulate matter, in the form of sand packs, is used at different stages of processing. The pore size and thickness of the porous medium, and the characteristics of the process fluid, usually non-Newtonian, are considered in design criteria related to the pressure drop across the bed and the back pressure on the extruder. According to Pearson (*169*), the characteristics of porous media composed of layers of particulate material of different sizes and shapes is indeed significant in the processing of non-Newtonian polymeric materials. White (*132*) also notes another relevant example is flow of molten polymer in an artificial fiber spinneret. Here the filter medium is composed of layers of sand of different sizes.

Nomenclature

A = constant in Equation 72
A_1, A_2 = parameters in Equation 27
B = parameter in Equation 26
C' = shift factor in Equation 18
c = concentration
$\mathbf{d}$ = deformation rate tensor
D_p = particle diameter
$\frac{D}{Dt}$ = material derivative
f_e = friction factor defined by Equation 39
$f(k)$ = function defined by Equation 27
$F(\ )$ = functions defined by Equations 5, 12, 16, 22, 25
$\mathbf{g}$ = metric tensor
$\mathbf{I}$ = identity transformation
J = pressure drop per unit length, defined as $\Delta P/L$
k = permeability
k_r = reference permeability
K = parameter in power law model
K' = power law parameter defined by Equation 31
K = resistance transformation
L = capillary or bed length
$\mathcal{L}$ = characteristic length of porous medium
M = molecular weight
n = parameter in power law model
N_{DEB} = Deborah number defined by Equation 66
N'_{Re} = Reynolds number defined by Equation 41
N_{WS} = Weissenburg number
P, p = pressure
p_0 = reference pressure
$\overline{\mathcal{P}}$ = modified pressure
$|\nabla\overline{\mathcal{P}}|$ = magnitude of the volume averaged modified pressure gradient
R = tube radius, gas constant
R_H = hydraulic radius
R_{eq} = equivalent pore size
R_e = Reynolds number defined by Equation 43
S = wetted surface per unit volume of porous material
S_0 = total particle surface per volume of particles
T = tortuosity
$\mathbf{t}$ = stress tensor in Equation 45
t_0 = characteristic time of incompressible Noll simple fluid
t = time
v_1, v_2, v_3 = velocity components
V_0 = superficial velocity, defined by Equation 7
$\langle V\rangle$ = pore velocity
V_s = effective velocity of slip
V' = average velocity in presence of slip
V = average velocity
$\mathbf{v}$ = velocity vector
$\overline{V}$ = magnitude of the volume-averaged velocity vector
$\overline{V}_1$ = mean velocity in throat of converging-diverging channel
x_1, x_2, x_3 = space coordinates
y = parameter in Equation 26

Greek Symbols

α = parameter in Ellis model
α_0 = parameter in Equation 24
β = parameter in second-order fluid model
γ = parameter in second-order fluid model
Γ = shear rate
$\Gamma(\ )$ = shear rate function
δ = normal distance from wall
ϵ = porosity
Δ = denotes a difference of the quantity following
η = viscosity coefficient
η_c = parameter in BKZ model
η_0 = limiting viscosity at zero shear rate
η_∞ = limiting viscosity at infinite shear rate
$\eta(\ )$ = viscosity function
$\eta_\alpha(\ )$ = function of BKZ model
$\eta_\beta(\ )$ = function in BKZ model
η_{cap} = apparent viscosity in Poiseuille flow
η_{eff} = effective viscosity defined by Equation 44
Θ = relaxation time
Θ_f = characteristic time of fluid
Θ_p = characteristic time of process
λ, λ_1, λ_2 = time constants
μ_0 = characteristic viscosity of incompressible Noll simple fluid
ρ = fluid density
σ = used in Equation 45
τ_{12}, τ = shear stress
$\tau_{1/2}$ = parameter in Ellis model
τ_W, τ_ϵ, τ_{RH} = shear stress defined by Equation 4, 13, 23, respectively
τ_{11}, τ_{22}, τ_{33} = normal stress components
$\boldsymbol{\tau}$, $\boldsymbol{\tau}'$ = stress tensors; total stresses and stresses relative to isotropic pressure
ϕ = yield stress
Φ_{cm}, Φ_{BSL} = functions defined by Equations 37, 38, respectively
ψ = functions defined by Equations 33, 34, 35, respectively

REFERENCES

(1) Metzner, A. B., *Rheo. Acta* **1**, (2/3), 205 (1958).
(2) Savins, J. G., "Fluid Rheological Measurements," in *Encyclopedia of Industrial Chemical Analysis*, F. D. Snell and C. L. Hilton, Eds., 111, 408–63, Wiley, New York, 1966.
(3) Savins, J. G., *Rheo. Acta*, **7**, (1) 87 (1968).
(4) Graessley, W. W., and Prentice, J. S., *J. Poly. Sci.*, *A2*, **6**, 1887 (1968).
(5) Middleman, S., "The Flow of High Polymers," Interscience, New York, 1968.
(6) Peterlin, A., Preprint No. **149**, Research Triangle Inst., Camille Dreyfus Lab., Research Triangle Park, N.C., January, 1967.
(7) Subirana, J. A., *J. Chem. Phys.*, 41, (12), 3852 (1964).
(8) Peterlin, A., Preprint No. **16**, Research Triangle Inst., Camille Dreyfus Lab., Research Triangle Park, N. C., June 1963.
(9) Van Oene, H., *J. Appl. Poly. Sci.*, **9**, 2607 (1965).
(10) Maruyama, T., and Yamomoto, M., *Jap. J. Appl. Phys.*, **7**, (7), 695 (1968).
(11) Graessley, W. W., *J. Chem. Phys.*, **47**, (6), 1942 (1967).
(12) Lodge, A. S., "Elastic Liquids," Academic Press, New York, 1964.
(13) Williams, M. C., *A.I.Ch.E. J.*, **12**, (6), 1064 (1966).
(14) Reiner, M., "Deformation and Flow," H. K. Lewis Co., London, 1949.
(15) Meter, D., and Bird, R. B., *A.I.Ch.E. J.*, **10**, 878 (1964).
(16) Oldroyd, J. G., *Proc. Roy. Soc. (London)*, **A245**, 278 (1958).
(17) Williams, M. C., and Bird, R. B., *Phys. Fluids*, **5**, 1126 (1962).
(18) Zapas, L. J., *J. Res. Nat. Bur. Stds.*, *70A*, (6), 525 (1966).
(19) Bogue, D. C., Ind. Eng. Chem. Fundam., **5**, 253 (1966).
(20) Pao, Y.-H., *J. Poly Sci.*, **61**, 413 (1962).
(21) White, J. L., and Metzner, A. B., *J. Appl. Poly. Sci.*, **1**, 1867 (1963).
(22) Spriggs, T. W., Ph.D. thesis, University of Wisconsin, Madison, Wis., 1966.
(23) Spriggs, T. W., Huppler, J. D., and Bird, R. B., *Trans. Soc. Rheol.*, **10** (1), 191 (1966).
(24) Metzner, A. B., "Flow of Non-Newtonian Fluids," in *Handbook of Fluid Dynamics*, V. L. Streeter, Ed., Section 7, McGraw-Hill, 1961.
(25) Matsuhisa, S., and Bird, R. B., *A.I.Ch.E. J.*, **11**, (4), 588 (1965).
(26) Skelland, A. H. P., "Non-Newtonian Flow and Heat Transfer," Wiley, New York, 1967.
(27) Spriggs, T. W., and Bird, R. B., Ind. Eng. Chem. Fundam., **4**, 182 (1965).
(28) Bogue, D. C., and Doughty, J. O., *ibid.*, **5**, 243 (1966).
(29) Metzner, A. B., White, J. L., and Denn, M. M., *Chem. Eng. Progr.*, **62**, (12), 81 (1966).
(30) Peterlin, A., *J. Lub. Tech.*, *Trans. A.S.M.E.*, Series F, **90**, 571 (1968).
(31) Bungenberg de Jong, J. G., and Van den Berg, H. J., *Proc. Kon. Ned. Akad. Wetensch.*, Amsterdam, **52**, 363 (1949).
(32) Eliassaf, J., Silberberg, A., and Katchalsky, A., *Nature*, **176**, 1119 (1955).
(33) Zubov, P. I., Lipatov, Yu S., and Kanevskaya, E. A., *Doklady Akad. Nauk. SSSR*, **141**, 387 (1961).
(34) Kanevskaya, E. A., Lipatov, Yu. S., and Zubov, P. I., *Poly. Sci., USSR*, **4**, 1264 (1963).
(35) Eliassaf, J., *Polym. Letters*, **3**, 767 (1965).
(36) Katz, D., and Steg, I., *Israel J. Chem.*, **2**, 282 (1964).
(37) Steg, I., and Katz, D., *J. Appl. Poly. Sci.*, **9**, 3177 (1965).
(38) Bourgoin, D., *J. Chemie Phys.*, **59**, 923 (1963).
(39) Peterlin, A., and Turner, D. T., *Nature*, **197**, 488 (1963).
(40) Peterlin, A., and Turner, D. T., *J. Chem. Phys.*, **38**, 2315 (1963).
(41) Komuro, K., Todani, Y., and Nagata, N., *Poly. Letters*, **2**, 643 (1964).
(42) Ballman, R. L., *Nature*, **202**, 288 (1964).
(43) Quadrat, O., and Bohdanecky, M., *Poly. Letters*, **4**, 833 (1966).
(44) Peterlin, A., Quan, C., and Turner, D. T., *ibid.*, **3**, 521 (1965).
(45) Peterlin, A., Turner, D. T., and Philippoff, W., *Kolloid Z. u Z. Polymere*, **204**, 21 (1965).
(46) Burow, S. P., Peterlin, A., and Turner, D. T., *Polymer*, **6**, 35 (1965).
(47) Nakayasu, H., *Zairyo Jnl. Soc. Mater. Sci.*, **14**, 300 (1965).
(48) Patterson, G. K., Zakin, J. L., and Rodriquez, J. M., Ind. Eng. Chem., **61**, (1), 22 (1969).
(49) Ginn, R. F., and Metzner, A. B., unpublished manuscript, 1969.
(50) Green, A. E., and Rivlin, R. S. Arch. Rat. Mech., **1**, 1 (1957).
(51) Coleman, B. D., Markovitz, H., and Noll, W., "Viscometric Flows of Non-Newtonian Fluids," Springer-Verlag, New York, 1966.
(52) Coleman, B. D. and Noll, W., *Ann. N. Y. Acad. Sci.*, **89**, 672 (1961).

(53) Oldroyd, J. G., *Proc. Roy. Soc. (London)*, **A200**, 523 (1950).
(54) *Ibid.*, **A245**, 278 (1958).
(55) Bernstein, B., Kearsley, E. A., and Zapas, L. J., *J. Res. Nat. Bur. Stds.*, **68B**, (4), 103 (1964).
(56) Walters, K., Mobil Res. and Develop. Corp., Dallas, Tex. unpublished materials, 1964.
(57) Markovitz, H. and Brown, D. R., in "Second-Order Effects in Elasticity, Plasticity, and Fluid Dynamics," M. Reiner and D. Abir, Eds., p. 585, Pergamon Press, New York, 1964.
(58) Ginn, R. F. and Metzner, A. B., in "Proc. 4th Int. Congress on Rheology," E. H. Lee, Ed., Part 2, p. 583, Interscience, New York, 1965.
(59) Middleman, S., *J. Appl. Poly. Sci.*, **11**, 417 (1967).
(60) Giesekus, H., in "Second-Order Effects in Elasticity, Plasticity, and Fluid Dynamics," M. Reiner and D. Abir, Eds., p. 553, Pergamon Press, New York, 1964.
(61) Ericksen, J. L., *Trans. Soc. Rheol.*, **6**, 275 (1962).
(62) Eringen, A. C., *J. Math. and Mech.*, **16**, 1 (1966).
(63) Leslie, F. M., *Arch. Ratl. Mech. Anal.*, **28**, (4), 265 (1968).
(64) Stokes, V. J., *Phys. Fluids*, **9**, 1709 (1966).
(65) Kotaka, T., and Osaki, K., *J. Poly. Sci.*, **C**, (15), 453 (1966).
(66) Ericksen, J. L., *Quart. J. Mech. Appl. Math.*, **19**, 455 (1966).
(67) Scott-Blair, G. W., *J. Phys. Chem.*, **34**, 1505 (1930).
(68) Mooney, M., *J. Rheol.*, **2**, 210 (1931).
(69) Oldroyd, J. G., *J. Colloid Sci.*, **4**, 333 (1949).
(70) Scott-Blair, G. W., *Rheo. Acta*, **1**, 123 (1958).
(71) Kozicki, W., Hsu, C. J., and Tiu, C., *Chem. Eng. Sci.*, **22**, 487 (1967).
(72) Burcik, E., *Earth Mineral Sci.*, **37**, (7), 57 (1968).
(73) Savins, J. G., U. S. Patent No. 3,292,698, Dec. 20, 1966.
(74) Abdo, M. K., and Dunlap, P. M., U.S. Patent 3,315,743, April 25, 1967.
(75) Dunlap, P. M., U.S. Patent No. 3,315,744, April 25, 1967.
(76) Townsend, H. L., McMillen, J. M., and Williams, S. A., U.S. Patent No. 3,333,634, Aug. 1, 1967.
(77) Williams, S. A., U.S. Patent 3,303,879, Feb. 14, 1967.
(78) Savins, J. G., Canada Patent No. 760,768, June 13, 1967.
(79) Abdo, M. K. and Blatz, P. E., Canadian Patent No. 760,906, June 13, 1967.
(80) Foster, W. R., U.S. Patent No. 3,362,473, Jan. 9, 1968.
(81) Abdo, M. K., U.S. Patent No. 3,391,736, July 9, 1968.
(82) Abdo, M. K., U.S. Patent No. 3,398,792, Aug. 27, 1968.
(83) Mobil Oil Corp., Great Britain Patent No. 1,100,051, Jan. 24, 1968.
(84) Abdo, M. K., U.S. Patent No. 3,410,343, Nov. 12, 1968.
(85) Toms, B. A., *J. Coll. Sci.*, **4**, 511 (1949).
(86) Winding, C. C., Bauman, G. P., and Kranick, W. L., *Chem. Eng. Prog.*, **43** (11), 613 (1947).
(87) Chu, J. C., Burridge, K. G., and Brown, F., *Chem. Eng. Sci.*, **3**, (6), 229 (1954).
(88) Hershey, D., and Cho, S. J., Preprint No. **7A**, 58th *A.I.Ch.E.* Meeting, Philadelphia, Pa., Dec. 1965.
(89) Savins, J. G., *Rheo. Acta*, **6**, 323 (1967).
(90) Sadowski, T. J., and Bird, R. B., *Trans. Soc. Rheol.*, **9**, (2), 243 (1965).
(91) McKinley, R. M., Jahns, H. O., Harris, W. W., and Greenkorn, R. A., *A.I.Ch.E. J.*, **12**, (1), 17 (1966).
(92) Mungan, N., Thompson, J. L., and Smith, F. W., *Trans. Soc. Petrol. Engr.*, **237** (Part I), 1143 (1966).
(93) Gogarty, W. B., *ibid.*, **240** (Part I), 151 (1967).
(94) Rowland, F. W., and Eirich, F. R., *J. Poly. Sci., Part A1*, **4**, 2033 (1966).
(95) Little, R. C., Naval Res. Lab. Report No. **6542**, Washington, D. C., May 1967.
(96) Shizgal, B., Goldsmith, H. L., and Mason, S. G., *Canadian J. Chem. Eng.*, **43**, 97 (1965).
(97) Karnes, A., Goldsmith, H. L., and Mason, S. G., *Nature*, **200**, 159 (1963).
(98) Starkey, T. V., *British J. Appl. Phys.*, **6**, 34 (1955).
(99) Starkey, T. V., Hewlett, V. A., Roberts, J. H. A., and James, R. E., *ibid.* **12**, 545 (1961).
(100) Starkey, T. V., *ibid.*, **13**, 182 (1962).
(101) Jeffrey, G. B., *Proc. Roy. Soc. (London)*, **A102**, 161 (1922).
(102) Vand, V., *J. Phys. Colloid Chem.*, **52**, 300 (1948).
(103) Higginbotham, G. H., Oliver, D. R., and Ward, S. G., *Brit. J. Appl. Phys.*, **9**, 372 (1958).
(104) Maude, A. D., and Whitmore, R. L., *ibid.*, **1**, 98 (1956).
(105) Segre, G., and Silberberg, A., *Nature*, **189**, 209 (1961).
(106) Whitmore, R. L., *Brit. J. Appl. Phys.*, **13**, 182 (1962).
(107) Oldroyd, J. G., "Non-Newtonian Flow of Liquids and Solids," *Rheology*, in F. R. Eirich, Ed., **1**, p. 653, Academic Press, New York, 1956.
(108) Goldsmith, H. L., and Mason, S. G., *Nature*, **190**, 1095 (1961).
(109) Goldsmith, H. L., and Mason, S. G., *J. Colloid Sci.*, **17**, 448 (1962).
(110) Chaffey, C. E., Brenner, H., and Mason, S. G., *Rheo. Acta*, **4**, 64 (1965).
(111) Rubinow, S. I., and Keller, J. B., *J. Fluid Mech.*, **11**, 447 (1961).
(112) Rubinow, S. I., *Biorheol.*, **2**, 117 (1964).
(113) Brenner, H., "Hydrodynamic Resistance of Particles at Small Reynolds Numbers," in *Advances in Chemical Engineering*, T. B. Drew, *et al.*, Eds., **6**, pp., 287–438, Academic Press, New York (1966).
(114) Cox, R. G., and Brenner, H., *Chem. Eng. Sci.*, **23**, 147 (1968).
(115) Ericksen, J. L., *Appl. Mech. Rev.*, **20** (11), 1029 (1967).
(116) Ericksen, J. L., Johns Hopkins Univ., private communication, 1968.
(117) Ariman, T., and Cakmak, A. S., *Rheo. Acta*, **1**, (3), 236 (1968).
(118) Christopher, R. H., Jr., M. S. thesis, University of Rochester, Rochester, N. Y., 1965.
(119) Dauben, D. L., Ph.D. thesis, University of Oklahoma, Norman, Okla., 1966.
(120) Marshall, R. J. and Metzner, A. B., IND. ENG. CHEM. FUNDAM., **6**, 393 (1967).
(121) Sadowski, T. J., Ph.D. thesis, University of Wisconsin, Madison, Wis. 1963.
(122) Christopher, R. H., Jr. and Middleman, S., IND. ENG., CHEM. FUNDAM., **4**, 422 (1965).
(123) Spitzl, J. M., M.S. thesis, Pennsylvania State University, University Park, Pa., 1965.
(124) Hassell, H. L., and Bondi, A., *A.I.Ch.E. J.*, **11**, (2), 217 (1965).
(125) Gaitonde, N. Y., M. S. thesis, University of Rochester, Rochester, N. Y., 1966.
(126) Gaitonde, N. Y., and Middleman, S., IND. ENG. CHEM. FUNDAM., **6**, (1),
(127) Dauben, D. L., and Menzie, D. E., *Trans. Soc. Petrol. Eng.*, **240**, (Part 1), 1065 (1967).
(128) Jones, W. M., and Maddock, J. L., *Nature*, **212**, 388 (1966).
(129) Jones, W. M., and Maddock, J. L., Paper presented at 5th Int. Rheology Congress, Kyoto, Japan, Oct. 1968.
(130) Gregory, D. R., and Griskey, R. G., *A.I.Ch.E. J.*, **13**, (1), 122 (1967).
(131) Sherborne, J. E., Sarem, A. M., and Sandiford, B. B., *7th World Petrol. Congress Proc.*, **3**, 509 (1967).
(132) White, D. A., *Chem. Eng. Sci.*, **23**, 243 (1968).
(133) Stahl, C. D., *Prod. Monthly*, **30**, 15 (1966).
(134) Hsu, C. J., M.S. Thesis, University of Ottawa, Ottawa, Canada, 1965.
(135) Pye, D. J., *Trans. Soc. Petrol. Eng.*, **231** (Part I), 911 (1964).
(136) Sandiford, B. B., *ibid.*, 917.
(137) Burcik, E. J., and Ferrer, J., *Prod. Monthly*, **32**, (3), 7 (1968).
(138) Jargon, J. R. and van Poollen, H. K., SPE Preprint No. **1567** (1966).
(139) Mungan, N., Third Symposium of the Assoc. de Recherche sur les Techniques de Forage et de Production, Pau, France, Sept. 1968.
(140) Burcik, E. J., *Prod. Monthly*, **29**, (6), 14 (1965).
(141) Burcik, E. J., **31**, (3), 27 (1967).
(142) Smith, F. W., SPE Preprint No. **2422**, Improved Oil Recovery Symp., Tulsa, Okla., April, 1969.
(143) Chain, S., M.S. thesis, Pennsylvania State University, University Park, Pa., 1965.
(144) Harvey, A. H., Ph.D. thesis, University of Oklahoma, Norman, Okla., 1967.
(145) Lee, K. S. and Claridge, E. L., *Trans. Soc. Petrol. Eng.*, **243**, (Part II), 52 (1968).
(146) Slater, G. E., and Farouq Ali, S. M., *Prod. Monthly*, **32**, (10), 22 (1968).
(147) Williams, S. A., and Taylor, B. F., (Mobil Res. and Develop. Corp., Dallas, Tex.), unpublished data, 1964.
(148) Ahearn, G. P., French Patent 1,539,568 (Aug. 5, 1968).
(149) Slattery, J. C., *A.I.Ch.E. J.*, **13**, (6), 1066 (1967).
(150) Benis, A. M., *Internatl. J. Nonlinear Mech.*, **3**, 31 (1968).
(151) White, D. A., *Chem. Eng. Sci.*, **22**, 669 (1967).
(152) Bird, R. B., Stewart, W. E., and Lightfoot, E. N., "Transport Phenomena," Wiley, New York, 1964.
(153) Eisenschitz, R., Rabinowitsch, B., and Weissenberg, K., *Mitt. deut. Materialprufungsanstalt*, **9**, 91 (1929).
(154) Scheidegger, A. E., "The Physics of Flow Through Porous Media," MacMillan, New York, 1957.
(155) Wyllie, M. R. J., and Gardner, G. H. F., *World Oil*, **147**, 121 (1958).
(156) Happel, J., and Brenner, H., "Low Reynolds Number Hydrodynamics," Prentice-Hall, Englewood Cliffs, N. J., 1965.
(157) Ergun, S., *Chem. Eng. Prog.*, **48**, 89 (1952).
(158) Brand, L., *Arch. Ratl. Mech. Anal.*, **1**, 34 (1957).
(159) Astarita, G., and Metzner, A. B., *Rend. Accad. Lincii*, **40**, 606 (1966).
(160) Truesdell, C., *Phys. Fluids*, **7**, 1134 (1964).
(161) Walters, K., and Kemp, R. A., in "Deformation and Flow of High Polymer Systems," R. W. Whorlow, Ed., pp. 238–50, London, 1967.
(162) Coleman, B. D., and Markovitz, H., *J. Appl. Phys.*, **35**, 1 (1964).
(163) Truesdell, C., "Second-Order Effects in Elasticity, Plasticity, and Fluid Dynamics," M. Reiner and D. Abir, eds., p. 1, Pergamon Press, New York, 1964.
(164) Bueche, F., "Physical Properties of Polymers," Interscience, New York, 1962.
(165) Bird, R. B., *Canadian J. Chem. Eng.*, **43**, (4), 161 (1965).
(166) Metzner, A. B., White, J. L., and Denn, M. M., *A.I.Ch.E. J.*, **12**, 863 (1966).
(167) Astarita, G., IND. ENG. CHEM. FUNDAM., **6**, 257 (1967).
(168) Marrucci, G., and Astarita, G., *A.I.Ch.E. J.*, **13**, 931 (1967).
(169) Pearson J. R. A., Cambridge Univ., Cambridge, England, private communication, 1969.
(170) Wissler, E., University of Texas, Austin, Tex., unpublished data (1969).
(171) Pearson, J. R. A., and Paratello, L., Cambridge, Univ. Cambridge, England, unpublished data.
(172) Savins, J. G., Dallas, Tex., unpublished data (1968).
(173) Chan Man Fong, C. F., *Rheo. Acta*, **7**, (4), 324 (1968).
(174) Rubin, H., Elata, C., and Poreh, M., ibid., p. 340.
(175) Burcik, E. J. and Walrond, K. W., *Prod. Monthly*, **32**, (9), 12 (1968).
(176) Savins, J. G., in "*Viscous Drag Reduction*," C. S. Wells, Jr., Ed., pp. 183–212, Plenum Press, New York, 1969.
(177) Van Poolen, H. K., and Jargon, Jr. R., *Trans. Soc. Petr. Eng.*, **246**, part II, 80 (1969).
(178) Kirk, W. B., Jr., Kinney, W. L., and Gogarty, W. B., U.S. Patent No. 3,279,537, (Oct. 18, 1966).
(179) Gheorghitza, St. I., *J. Fluid Mech.*, **20**, (2), 273 (1964).
(180) Mustaphaev, S. D., *Neft i Gaz.* **8**, (7), 71 (1965).
(181) Fomenko, I. E., *Neft. Khoz.*, **46**, (8), 28 (1968).
(182) Entov, V. M., Ibid., **46**, (10), 47 (1968).
(183) Mirzadzhanzade, A. Kh., Kremnev, V. I., Kovalev, A. G., Zaitsev, Yu. V., and Ismailov, R. D., *ibid.*, p. 45.
(184) Hoyt, J. W., Preprint No. **67-431**, AiAA 3rd Propulsion Joint Specialist Conf., Washington, D.C., July 1967.
(185) Wells, C. S., Jr., in "Viscous Drag Reduction," C. S. Wells, Jr., Ed., p. 361–82, Plenum Press, New York, 1969.

6

Physics and Thermodynamics of Capillary Action in Porous Media

Physics and Thermodynamics of Capillary

A thermodynamic theory and a detailed mechanism of immiscible displacement in porous media are developed in terms of a quantized model

A wide range of natural and technological processes involves capillary action associated with immiscible fluid displacement within the interstices of porous solids. Examples are the movement of air and water in soils and porous rocks, the recovery of hydrocarbons from subterranean reservoirs, the drying, centrifuging and dewatering of porous solids, the behavior of porous fuel cell electrodes, and the capillary properties of textile and paper fibers.

Although principles of capillary action are well understood, their application to practical problems is often limited by the complicated geometry of porous solids. Difficulties are compounded when there is uncertainty as to the surface energetics of the system which determine the property generally described as the wettability of a porous solid.

The structure and surface properties of numerous types of porous materials are commonly investigated by measurement of capillary displacement pressures as a function of displaced volume. Often it has been assumed that immiscible displacement can be treated as a smoothly reversible process. Thermodynamic relationships based on this assumption have been used to relate the work of displacement to changes in the area of the surfaces which bound the fluid phases contained by the solid. Although so-called capillary pressure equilibrium points appear to be part of a continuous relationship between capillary pressure and displaced volume, these relationships are not reversible. Furthermore, when examined in fine detail, they are not continuous. Much of the confusion surrounding the thermodynamic interpretation of capillary pressure data can be avoided by giving consideration to the frequently referenced general treatment of capillarity by J. W. Gibbs (*1*) and the investigations of displacement in porous solids by W. B. Haines (*2*).

This paper has three main sections. In the first section, the terminology used in discussing surface energetics is defined, and thermodynamic relationships for reversible displacements are derived from fundamental thermodynamic equations. In the second section, the mechanism and thermodynamics of displacement in porous media are discussed. To obtain infinitesimally slow displacement in porous media, special conditions are required which lead to a quantized model of displacement. Changes in volumetric and surface quantities can be classified within the framework of the quantized model to give a consistent account of the displacement mechanism. The third section describes an experimental investigation of displacement in random packings of equal-sized spheres, commonly known as the ideal soil. Changes in surface area of phases within the packings are measured directly to obtain the changes of free energy due to surface. Inherent thermodynamic efficiencies of displacement are then obtained by comparing changes in surface free energy with the work of displacement given by capillary pressure data.

SURFACE ENERGETICS

There is no generally accepted terminology by which surface energetics can be discussed without risk of ambiguity to the reader. The following terms have appeared in published literature (*1–10*): surface tension, superficial tension, interfacial tension, adhesion tension,

AUTHOR *Norman R. Morrow is Senior Research Scientist with the Petroleum Recovery Research Institute, The University of Calgary, Calgary, Alta., Canada.* *The author expresses his appreciation to J. O. Szabo for valuable assistance in preparation of thin sections and to Esso Production Research Co. for permission to publish.*

Figure 1. Gibbs model of surface between two fluids

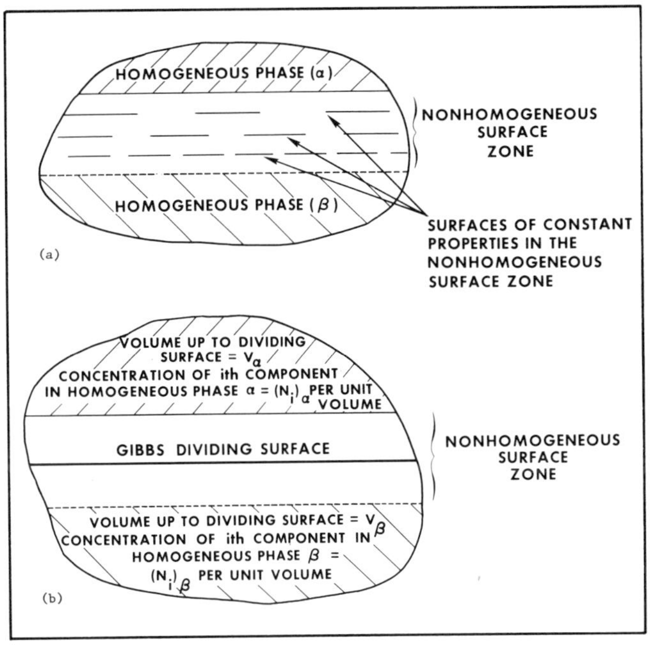

surface stress, stress tension, surface free energy, superficial densities of energy, specific surface energy, free surface energy, surface energy, surface energy density, Helmholtz free energy, surface Helmholtz free energy, specific surface free energy, pure surface energy, real surface energy, actual surface energy, theoretical free surface energy, free energy of formation, and relative surface free energy.

The terms surface tension, surface free energy, and superficial surface free energy will first be defined as used in this paper. The last term is defined to preclude any possible misunderstanding of the term surface free energy. We will begin by defining and comparing the above three terms as they relate to a system of two fluid phases, α and β, separated by a plane interface, S, of area A. In subsequent discussions, several pairs of fluids will be used as wetting and nonwetting phases with respect to a solid. The nomenclature will be defined as each pair is introduced, but for convenience in future references, they are listed in Table I.

Surfaces between Fluids

Superficial surface free energy. The physical surface of discontinuity is viewed as a nonhomogeneous film separating two homogeneous phases. The properties of the film vary along normals to the surface but are constant over any geometrical surface which is parallel to the physical surface (Figure 1a). Gibbs (*11*) introduced a mathematical model whereby thermodynamic quantities, which he termed superficial, could be assigned to the regions α and β and a geometrical surface which divides these regions (Figure 1b). The geometrical dividing surface lies in and is sensibly coincident with the physical surface of discontinuity (Figure 1b). All points in the mathematical surface are similarly situated with respect to the physical surface. The precise position of the Gibbs dividing surface is arbitrary, but, of course, all possible placements lie parallel to each other.

The geometrical surface precisely divides the volume, V, of the system into regions α and β with V_α, the volume

Table I. Nomenclature for Three-Phase Systems (Solid-wetting phase–nonwetting phase)

System fluids	Notation: Solid area, A_s	Notation: Wetting phase	Notation: Nonwetting phase	Pore space saturations: Wetting phase	Pore space saturations: Nonwetting phase	Surface areas between phases: Wetting-nonwetting	Surface areas between phases: Wetting solid	Surface areas between phases: Nonwetting solid
General	s	α	β	S_α	S_β	$A_{\alpha\beta}$	$A_{s\alpha}$	$A_{s\beta}$
Water–oil	s	w	o	S_w	S_o	A_{wo}	A_{sw}	A_{so}
Water–air	s	w	a	S_w	S_a	A_{wa}	A_{sw}	A_{sa}
Water–vapor	s	w	v	S_w	S_v	A_{wv}	A_{sw}	A_{sv}
Liquid–air	s	l	a	S_l	S_a	A_{la}	A_{sl}	A_{sa}
Liquid–vapor	s	l	v	S_l	S_v	A_{lv}	A_{sl}	A_{sv}

A = area, s = solid, w = water, o = oil, a = air, v = vapor, l = liquid.

associated with the α phase, and V_β, the volume associated with the β phase (see Figure 1b)

$$V = V_\alpha + V_\beta \tag{1}$$

The essential feature of the mathematical model is that bulk properties of each phase are maintained right up to the dividing surface. Let the number of moles per unit volume of the ith component be $(N_i)_\alpha$ and $(N_i)_\beta$ in the homogeneous parts of the α and β phases, respectively. Molar quantities for the regions α and β up to the dividing surface of the Gibbs model are given by

$$(n_i)_\alpha = (N_i)_\alpha V_\alpha \tag{2}$$

$$(n_i)_\beta = (N_i)_\beta V_\beta \tag{3}$$

If the total number of moles of component i in the system is n_i, a material balance defines a surface excess (or deficiency) of, say, the ith component which will be denoted by $(n_i)_S$

$$n_i = (n_i)_\alpha + (n_i)_\beta + (n_i)_S \tag{4}$$

Similarly, if F is the total Helmholtz free energy of the system, F_α is the Helmholtz free energy of a volume V_α of the α phase in bulk, and F_β is the Helmholtz free energy of a volume V_β of the β phase in bulk, a superficial surface free energy, F_S, of the surface is defined by

$$F = F_\alpha + F_\beta + F_S \tag{5}$$

The fundamental Gibbs equation for change in Helmholtz free energy of a system with j components is

$$dF = -S'dT + \sum_{i=1}^{j} \mu_i dn_i - P_\alpha dV_\alpha - P_\beta dV_\beta + \sigma dA_{\alpha\beta} \tag{6}$$

The intensive quantities given in the right-hand side of Equation 6 are temperature, chemical potential, pressure, and surface tension, respectively.

When changes are isothermal, for the two regions α and β, we may write

$$dF_\alpha = \sum_{i=1}^{j} \mu_i d(n_i)_\alpha - P_\alpha dV_\alpha \tag{7}$$

$$dF_\beta = \sum_{i=1}^{j} \mu_i d(n_i)_\beta - P_\beta dV_\beta \tag{8}$$

Change in superficial surface free energy F_S is given by

$$dF_S = dF - dF_\alpha - dF_\beta \tag{9}$$

From Equations 6–9,

$$dF_S = \sum_{i=1}^{j} \mu_i dn_i - \sum_{i=1}^{j} \mu_i d(n_i)_\alpha - \sum_{i=1}^{j} \mu_i d(n_i)_\beta + \sigma dA_{\alpha\beta} \tag{10}$$

From Equation 4

$$d(n_i)_S = dn_i - d(n_i)_\alpha - d(n_i)_\beta \tag{11}$$

Hence, writing A for $A_{\alpha\beta}$,

$$dF_S = \sum_{i=1}^{j} \mu_i d(n_i)_S + \sigma dA \tag{12}$$

Let the total area of the surface, S, be A and the surface excess per unit area of component i, as determined by the model, be Γ_i, thus

$$(n_i)_S = \Gamma_i A \tag{13}$$

$$d(n_i)_S = \Gamma_i dA + A d\Gamma_i \tag{14}$$

From Equation 12,

$$dF_S = \sum_{i=1}^{j} \mu_i \Gamma_i dA + \sum_{i=1}^{j} \mu_i A d\Gamma_i + \sigma dA \tag{15}$$

$$\left(\frac{\partial F_S}{\partial A}\right)_{T,V,n} = \sum_{i=1}^{j} \mu_i \Gamma_i + \sum_{i=1}^{j} \mu_i A \left(\frac{\partial \Gamma_i}{\partial A}\right)_{T,V,n} + \sigma \tag{16}$$

where the subscript n implies that the number of moles of each component, $i = 1$ to j, is constant.

Variation of Γ_i with surface area will occur if, say, one component resides only in the surface and not in the contiguous bulk masses. In theory, the surface could be supplied with this component so that the properties of the surface per unit area remain constant.

When the properties of the surface per unit area are independent of the total area,

$$\left(\frac{\partial \Gamma_i}{\partial A}\right)_{T,V,n} = 0 \tag{17}$$

Integration of Equation 16 then gives

$$\frac{F_S}{A} = \sum_{i=1}^{j} \mu_i \Gamma_i + \sigma \tag{18}$$

Let the superficial surface free energy per unit area be f_S defined by

$$f_S = \sum_{i=1}^{j} \mu_i \Gamma_i + \sigma \tag{19}$$

Recall that the location of the dividing surface which determines the magnitude of Γ_i and f_S is arbitrary. The dividing surface can be placed so that $\Gamma_i = 0$ for one component (12). For the single component system given by a pure liquid and its vapor, the dividing surface can be located so that $\Gamma = 0$ and then

$$f_S = \sigma \tag{20}$$

However, in general, the surface tension, which is numerically equal to the reversible work of extension per unit area of surface, is not equal to the superficial surface free energy.

Surface free energy and surface tension. From Equation 6, surface tension is defined by

$$\left(\frac{\partial F}{\partial A}\right)_{T,V,n} = \sigma \tag{21}$$

If the surface tension is constant,

$$(\Delta F)_{T,V,n} = \sigma \Delta A \tag{22}$$

Let the increase in Helmholtz free energy of the system due to a unit of area extension be f, then

$$\left(\frac{\Delta F}{\Delta A}\right)_{T,V,n} = f \tag{23}$$

Hence

$$\sigma = f \tag{24}$$

and from Equation 19, f and f_S are related by

$$f_S = \sum_{i=1}^{j} \mu_i \Gamma_i + f \tag{25}$$

The change in free energy of the system, $(\Delta F)_{T,V,n}$, is equal to the reversible work of extension of the surface. Gibbs described the product $\sigma\Delta A$, given by Equation 22 above, as the available energy due to the surface in the system (*13*). In many discussions of capillarity, the product, $\sigma\Delta A$, has been called surface free energy, and we shall follow that practice here. We do this somewhat reluctantly because Defay and Prigogine have called the quantity f_S the surface free energy. They describe confusion between surface tension and what they call surface free energy as particularly illogical (*14*). Johnson (7) points out that the term surface free energy is often used without further definition. In the majority of these instances, we believe the term surface free energy is intended to mean f, the free energy of the system due to the presence of the surface; this is particularly so if the Gibbs mathematical dividing surface has not been introduced.

In systems where the surface tension is dependent upon surface area, the surface free energy, as we have defined it, will not be equal to the product of surface tension and area. The change in surface free energy due to a change in area from A_1 to A_2 is now given by

$$(\Delta F)_{T,V,n} = \int_{A_1}^{A_2} \sigma(A)dA \tag{26}$$

Also, the surface free energy due to a unit area of surface, f, will depend upon the magnitude of the area and is not equal to the surface tension. Systems containing insoluble surfactants provide examples where the surface tension may vary significantly with area because surfactant concentration and the related spreading pressure of the surfactant depend upon absolute area. Such variations are commonly investigated experimentally with some form of surface tension trough (*3*).

Surface curvature and location of dividing surface. A completely general discussion of capillarity would require consideration of the effect of curvature on both surface tension and the location of the Gibbs reference surface which lies within the physical surface. However, these effects do not become significant until radii of curvature of a surface approach molecular dimensions (*15–17*). In porous media which are reasonably permeable to fluid flow, the curvatures of surfaces between fluids given by displacement pressures are relatively low. Thus the previous discussion of the energetics of plane surfaces extends to fluid interfaces developed during immiscible displacement.

Surface Energetics of Systems Containing Solid Surfaces

The conceptual difficulties involved in treating the surface energetics of solids are generally admitted.

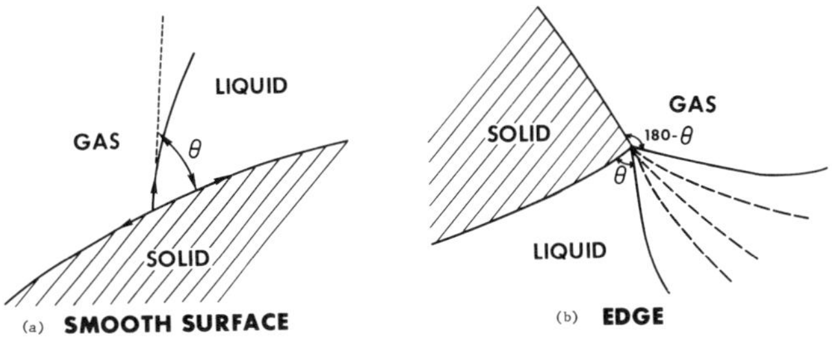

Figure 2. Equilibrium at a three-phase line of contact

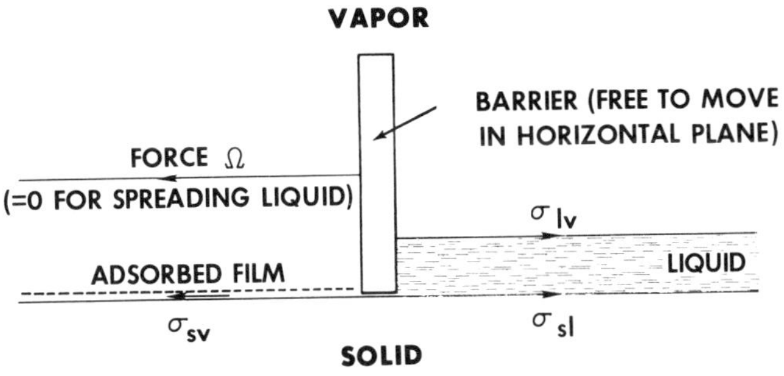

Figure 3. Hypothetical model illustrating stability for plane liquid and adsorbed film surfaces

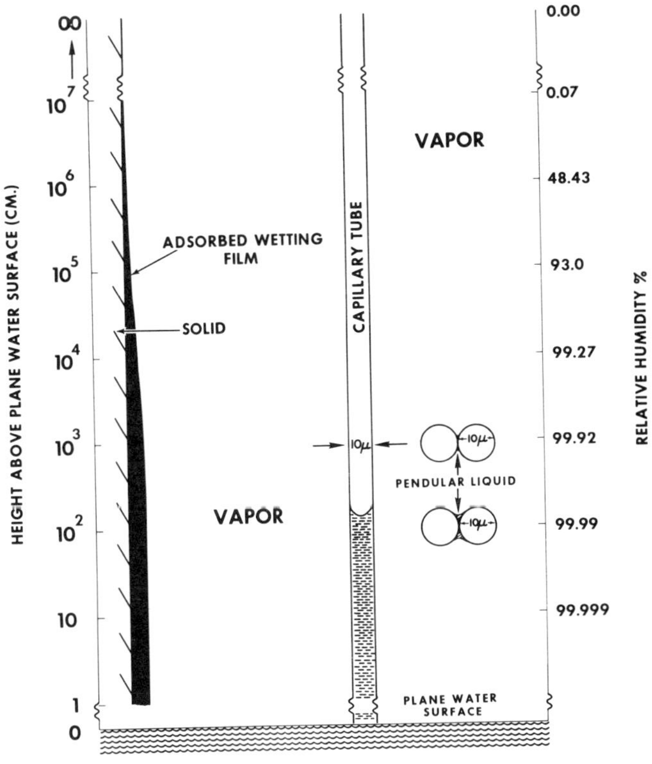

Figure 4(a). Equilibrium of adsorbed surface film and capillary-held liquid

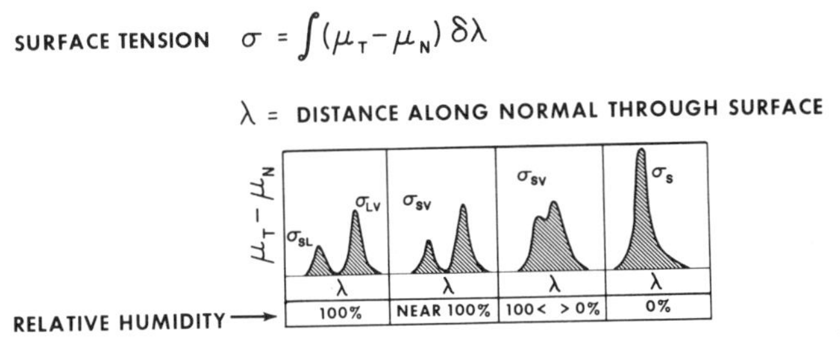

Figure 4(b). Sketches of variation in stress anisotropy with height for adsorbed film depicted in Figure 4a

Treatments of multiphase systems containing solids are found in the original work of Gibbs (*1*) and the more recent work of Defay and Prigogine (*18*). These general treatments will be tailored to the needs of the present development. Three-phase systems in which one of the phases is a solid will be considered first.

A complication of dealing with solids is that the state of strain of the solid in the region of the surface may be important. With fluids, surface tension acts uniformly in the surface, independent of direction, and can be measured directly as a force. In a crystalline solid with anisotropic elastic properties, surface tension cannot be measured directly, but the force required to stretch the surface infinitesimally may vary with direction in the surface.

Gibbs (*19*) defined a quantity, σ, as the work required to form a unit area of the surface, with the assumption that the properties of the surface are directly proportional to the area of the surface. Gibbs defined σ in terms of *integral* excess properties of the surface.

$$\sigma = f_S - \sum_{i=2}^{j} \mu_i \Gamma_i \tag{27}$$

The dividing surface was chosen to make the surface excess of the solid (component 1) equal to zero. In this way, the problems arising from variations in chemical potential with state of strain of anisotropic solids are circumvented.

We can also define σ as the contribution to the total Helmholtz free energy of the system due to the presence of a unit area of the surface in question. In other words, σ is the reversible work of formation of the surface

$$\sigma = \left(\frac{\Delta F}{\Delta A}\right)_{T,V,n} \tag{28}$$

If the state of stress is isotropic, σ is also equal to the surface tension. In accordance with the proposed terminology for fluid interfaces, σ will be called the surface free energy per unit area. (With regard to the problem of terminology, Defay and Prigogine call σ of Equation 27 the surface tension by analogy with the definition of σ of Equation 19 for fluid interfaces. They name the quantity f_S, as given by Equation 27, the relative surface free energy, where we suggest the name superficial surface free energy by analogy with Equation 19.)

Defay and Prigogine point out that one of the disadvantages of working in terms of σ as the free energy of formation is the practical difficulty of measuring the external work done in forming a solid-fluid surface by an isothermal and reversible process. This problem is absent for systems of interest here, since we consider only changes in surface free energy which accompany displacement of one overlying fluid by another from a solid surface assumed to maintain constant surface area. The reversible work of displacement of a three-phase line of contact across the solid surface will be given by the product of the changes in area and the difference in the two σ quantities of the solid surface.

Equilibrium at three-phase line of contact—Young's equation. It is frequently observed that the interface between two fluid phases meets an apparently smooth surface of a solid phase, s, at a finite angle, say θ, measured through the wetting phase as shown in Figure 2a.

In the region of the three-phase line of contact given by such a system, the model which enabled us to define a dividing surface within the physical surface breaks down. However, the selected dividing surfaces can be extrapolated into the three-phase confluent zone. The dividing surfaces will define exact surface areas between phases, and a mathematical line to which line excess quantities can be assigned (*20*). For example, the excess of the ith component with respect to the three-phase line of contact, $(n_i)_L$ is given by

$$n_i = (n_i)_\alpha + (n_i)_\beta + (n_i)_{S(\alpha\beta)} + (n_i)_{S(s\alpha)} + (n_i)_{S(s\beta)} + (n_i)_L \tag{29}$$

For the systems of interest here, the region occupied by the three-phase line of contact contains few molecules compared with the other surface regions. The contribution to the Helmholtz free energy of the system due to variation in the length of the three-phase line of contact, which may result from variation in surface areas, can therefore be neglected.

If we assume that the solid is smooth and rigid and that θ is single valued, the condition for thermodynamic equilibrium along the three-phase line of contact is given by Gibbs (*21*) as

$$\sum_{k=1}^{3} (\sigma_k \delta\tau)dx \geq 0 \tag{30}$$

where σ_k = surface free energy per unit area of the kth surface

$\delta\tau$ = element taken along the three-phase line of contact

dx = infinitesimal displacement over the solid surface normal to the element, $\delta\tau$

Let the surface area between the phases be A, with appropriate subscripts used to define a particular surface.

$$dA_{s\alpha} = \delta\tau dx \tag{31}$$

$$dA_{s\beta} = -\delta\tau dx \tag{32}$$

$$dA_{\alpha\beta} = \delta\tau dx \cos\theta \tag{33}$$

Hence, from Equation 30,

$$\sigma_{\alpha\beta} \cos\theta + \sigma_{s\alpha} - \sigma_{s\beta} = 0 \tag{34}$$

A more general derivation of this equation with gravitational potential and surface free energy minimized has been given by Collins and Cooke (*22*).

Equation 34 relates the difference in two quantities, $\sigma_{s\beta}$, $\sigma_{s\alpha}$ which cannot be measured individually, to the quantities $\sigma_{\alpha\beta}$ and θ which can be measured and is the equivalent of Young's equation (*23*). This equation is often presented on the basis of a simple force balance. For equilibrium, the fluid surface tension resolved in the plane of the solid surface must be balanced by some type of force acting in the surface. If the forces acting in the plane of the solid surface and normal to the three phase line of contact are denoted by γ, we obtain

$$\sigma_{\alpha\beta} \cos \theta = \gamma_{\beta s} - \gamma_{\alpha s} \tag{35}$$

Severe conceptual difficulties concerning the nature of the γ quantities offset the apparent simplicity of this relationship. These difficulties have led to several disagreements as to the interpretation and validity of Young's equation. A consistent interpretation of Young's equation, which includes the special case of zero contact angle, is given by Gibbs. Before discussing Gibbs treatment, two examples where Young's equation may not be appropriate in its above form will be mentioned.

Deformation of solid surfaces. The vertical component of tension between the fluids, $\sigma_{\alpha\beta} \sin \theta$, is balanced by forces which cause the solid to be strained. For example, deformation of solid surfaces along the three-phase line of contact is described for mercury on mica (*24*). When the modulus of elasticity is low, Lester calculated that the solid will deform to give a ridge along the three-phase line of contact. In such cases, the Neumann surface tension triangle for three fluids should be used to express the balance of forces at the three-phase line of contact, rather than Young's equation (*25*).

Spreading of solid substrates. If the solid is partially soluble in either of the fluid phases, adsorption may cause a significant reduction in the surface tension, $\sigma_{\alpha\beta}$, observed for the two fluids alone. Thus, measurements of $\sigma_{\alpha\beta}$ may not always be applicable to the equilibrium expressed by Equation 34. Fowkes (*26*) estimated the spreading pressure, π_L, of solids or possible atmospheric contaminants on a mercury-air interface by comparing measured contact angles with those calculated for a mercury-air surface tension of 484 dynes/cm with $\pi_L = 0$. The contact angles were calculated from interaction energies which had in turn been estimated from surface tension measurements on standard systems. Using liquid advancing and receding contact angle data, estimates of the spreading pressure of solids on mercury ranged from 45.5 dynes/cm for paraffin wax to 100.5 dynes/cm for polyethylene.

Gibbs formulation of Young's equation—the unchangeable solid. Gibbs (*27*) introduced the unchangeable solid as a model. The unchangeable solid is insoluble in the fluid phases and its state of strain when bounded by a vacuum is unaltered by the presence of an overlying fluid. Properties of the unchangeable solid will be assumed in all subsequent discussion of solids including porous media.

The superficial tension of the fluid, η, is given by the tension in the solid-fluid interface, $\sigma_{s\alpha}$, less the tension in the solid surface when it is bounded by a vacuum, σ_s.

$$\eta_{s\alpha} = \sigma_{s\alpha} - \sigma_s \tag{36}$$

In effect, the model divides the surface tension at the solid-fluid interface $\sigma_{s\alpha}$, into the two components, $\eta_{s\alpha}$ and σ_s. Under the stated conditions, the fluid has no influence on the disposition of the solid; but the presence of the solid does influence the contractile forces within the fluid to give rise to $\eta_{s\alpha}$, which Gibbs named the superficial tension of the fluid in contact with the solid. When the system is in equilibrium, a balance of forces along a tangent to the solid surface at, and normal to, the three-phase line of contact gives

$$\eta_{s\beta} + \sigma_s = \eta_{s\alpha} + \sigma_s + \sigma_{\alpha\beta} \cos \theta \tag{37}$$

Hence,

$$\eta_{s\beta} - \eta_{s\alpha} = \sigma_{\alpha\beta} \cos \theta \tag{38}$$

Melrose (*28*) points out that the quantity, η, with opposite sign, is equal to the more commonly encountered film pressure. For example, when a drop of liquid on a smooth solid surface is in equilibrium with its vapor, the tension in the vapor solid surface is given by the tension of the solid surface in vacuum less the film pressure of the adsorbed vapor on the solid surface.

Stability in systems containing solid edges. As in the treatment of surfaces between two fluids, the preceding discussion, which includes solids, applies strictly to plane surfaces. However, as with curved liquid surfaces, the effects of curvature on surface free energy can probably be safely neglected for smooth solid surfaces.

Many natural solids, especially if freshly broken, have regions of extremely high curvature. As a limiting case, without considering the question of smoothness at a molecular level, the solid can be regarded as having smooth surfaces which may be wholly or partially bounded by edges. Gibbs (*21*) considered the conditions for equilibrium at a three-phase (two fluids and a solid) line of contact for the two cases of smooth surfaces and edges. The simpler case of the smooth solid is given by Equation 34. In contrast, for equilibrium at an edge, the angles at which the interface between the fluids can possibly meet one of the edges forming the surface, cover a range which is determined by the angle subtended at the edge by the two solid surfaces. The range of angles through which a three-phase line of contact can be observed at an edge, provided other requirements of stability are met, is depicted in Figure 2b. Clearly, the contact angle condition for stability of an interface between two fluids in the presence of a solid has been relaxed. As a consequence, in systems containing angular solids, most of the three-phase line of contact will probably lie along particle edges.

Displacement pressures in porous solids are often assumed proportional to the cosine of the contact angle. When an attempt is made to deduce contact angles from capillary displacement pressures, special attention must be given to the possible effect of edges (*29*). Even if the solid is smooth, capillary displacement pressure is proportional to the cosine of the contact angle only for extremely simple geometries such as uniform capillary tubes.

Although the solid edges may be important in determining configurational stability of the interface separating the two fluids, their contribution to surface free energy can be neglected; as with three-phase lines of contact at smooth solid surfaces, there are relatively few molecules in the three-phase confluent zone compared with the total in all of the surface zones.

Contact angle hysteresis. Values of contact angles measured in real systems usually depend upon whether the three-phase line of contact is advancing or receding over the solid surface. Contact angle hysteresis can arise because of surface roughness, heterogeneous distribution of adsorbed impurities on the solid surface or the mechanism by which liquid molecules adsorb and desorb when the interface is displaced (*30*). If sufficient care is taken in preparation of a clean, smooth, solid surface, contact angle hysteresis can sometimes be reduced to less than one or two degrees (*31*).

Contact angles are clearly important in determining the capillary properties of porous media. However, in view of the difficulties of measuring reproducible contact angles on carefully prepared surfaces and the relaxed stability conditions at solid edges, it seems highly unlikely that all three-phase lines of contact within porous media will lie on smooth solid surfaces with a single-valued, uniform contact angle.

The present study is concerned with a quantitative treatment of displacement thermodynamics; hence, contact angle must be eliminated as an unknown variable since surface free energy is dependent on contact angle. Absence of contact angle hysteresis in experiments with porous media can only be confidently assumed when the liquid spreads over the solid. The contact angle is then zero if the three-phase line of contact lies on a smooth solid surface.

Surface energetics of spreading. In addition to disagreement on the general validity of Young's equation, there is also controversy as to its interpretation when the contact angle is zero (*7, 32*). Perhaps this is because Young's equation is commonly introduced by considering the balance of forces at the three-phase line of contact of a drop sitting on a plane solid surface. When the contact angle is zero, the drop will spread indefinitely. For the purpose of illustrating stability at plane surfaces, the above physical example is clearly inadequate.

A hypothetical model designed for general discussion of stability at plane interfaces was recently introduced by Melrose (*33*). The two fluids were a liquid, L and its vapor V. In this model, a plane liquid-vapor surface can coexist with a plane solid surface over which the liquid will spread. Spreading is prevented by a frictionless and inert mechanical barrier which can only move parallel to the solid surface. An essential requirement of the model is that the gravitational potential of the liquid be negligible. Only the main features of the model are depicted in Figure 3.

When the liquid is nonspreading, a force, Ω per unit length of the barrier, must be exerted to prevent contraction of the liquid surface. The force is given by

$$\Omega = \sigma_{lv} + \sigma_{sl} - \sigma_{sv} \tag{39}$$

From Young's equation 34 for finite contact angle, Ω is given by

$$\Omega = \sigma_{lv}(1 - \cos\theta) \tag{40}$$

On the other hand, for the spreading condition, adsorption of vapor will continue until the force, Ω, required to retain the barrier, vanishes. At this point, the system is asserted to be in equilibrium. For the nonspreading case, work is required to extend the liquid surface by moving the barrier; in the spreading system at equilibrium, however, the position of the barrier does not alter the free energy of the system since the barrier can be moved without expending work. In other words, if the adsorbed film is replaced by a thin liquid layer, there is no change in the surface free energy of the system.

Stability of the adsorbed film involves the limit of film thickness achieved at the expense of the liquid (*33*). Equilibrium requires that the outermost molecules of the film have the same chemical potential as the vapor and the surface molecules of the liquid. To attain this condition, the adsorbed film must thicken until its outermost molecules are unaffected by the attractive forces of the solid. Thus, the tension of the outermost part of the adsorbed film must be equal to that of the liquid-vapor surface. Similarly, the tension of the region of the film in contact with the solid is equal to the solid-liquid tension. Films of this nature were described by Harkins as duplex (*34*).

Spreading against Gravity. Figure 4a depicts spreading up a vertical infinite flat plate in a gravitational field with water and its vapor at equilibrium. If the vapor pressure at the level of the free water surface is p_o, the vapor pressure, p, at any height, h, above the plane water surface is given by

$$p = p_o \exp\left(-\frac{\Delta\rho g h M}{RT}\right) \tag{41}$$

where $\Delta\rho$ = density difference of liquid water and its vapor

g = acceleration due to gravity

M = molecular weight

R = universal gas constant

Vapor pressure is expressed in Figure 4a as percent relative humidity, H_R, where

$$H_R = \frac{p}{p_o} 100 \tag{42}$$

At an infinite height, the vapor pressure is zero and the tension is that of the solid against a vacuum denoted by σ_s. At finite heights, adsorption of vapor at the solid surface causes the thickness of the film to increase slowly with decreasing height as shown qualitatively in Figure 4a. The actual thickness of the adsorbed film and thickness variation with height are not known with any certainty; unpublished estimates for high relative humidities range from 10 to 30 Å (4 to 12 molecular layers of water) with 100 Å as an upper limit. There is mixed evidence that the solid can extend its range of influence on the structure of water to even greater distances but such considerations are beyond the scope of this paper (*35–37*).

Melrose (*33*) employs the relationship, due to Buff (*38*), between tension and stress distribution in an interface to illustrate qualitatively the formation of a duplex film when the vapor pressure over a plane solid is raised to its saturation value. Tension, σ, in the surface is given by

$$\sigma = \int (\mu_T - \mu_N)\, d\lambda \qquad (43)$$

where μ_T and μ_N are the transverse and normal components of the stress tensor in an interfacial region with integration over a normal, λ, to the interface.

Sketches of variation in stress anisotropy with vapor pressure given by Melrose (*33*) are included in Figure 4b to provide a rough, qualitative indication of how stress anisotropy in the surface varies with height. Although the presence of adsorbed material reduces the tension of the solid surface, the adsorbed water contributes to the total tension. As the film thickens, two maxima in stress anisotropy are conceived; at higher vapor pressures, the two contributions to tension become increasingly distinct (Figure 4b). If the plate touches the water surface, capillary rise against the plate to a maximum height of $\sqrt{2\sigma_{wv}/\Delta\rho g}$, precludes the existence of adsorbed vapor films below this height. At the water surface, the relative humidity is 100%, and a stable adsorbed film at this level would correspond to the duplex film given by the hypothetical model shown in Figure 3.

In this limiting case of film stability, the central minimum of the stress anisotropy diagram just touches the zero axis with separate surface-film and film-vapor regions contributing to the film's total tension. Further condensation of vapor would lead to the formation of a bulk liquid region separating liquid-solid and vapor-solid surfaces. The stress distribution in each of these surfaces is precisely that of the individual components of the duplex film; the stress distribution diagram for the system will now feature a central minimum of zero with finite length corresponding to the bulk water separating the solid-liquid and liquid-vapor surfaces.

In an adsorbed film, the distribution of mass in excess of the mass of displaced vapor is directly related to the variation of superficial tension with height. Gibbs (*39*) discusses the distribution of adsorbed material on a solid surface in a gravitational field. Young's equation is shown to follow from the special case of a local force balance at the three-phase line of contact that is equally valid for finite and zero contact angle. Gibbs discusses the variation in excess mass of the adsorbed material with height in terms of superficial tensions. At a particular height, h, the excess mass, m, of the adsorbed film is related to the superficial tension, η, by

$$d\eta = (mg)dh \qquad (44)$$

Thus, the excess mass, m', of adsorbed material between any two given heights (h_1 and h_2, $h_1 < h_2$), is equal to the difference in superficial tensions at these heights.

$$m'g = (\eta)_{h2} - (\eta)_{h1} \qquad (45)$$

A more physically appealing statement of the above force balance is given directly by the increase in area under the stress anisotropy curves for the two given heights. The superficial tension is related to the stress anisotropy distribution by

$$\eta_h = [\textstyle\int (\mu_T - \mu_N)\, d\lambda]_h - [\int (\mu_T - \mu_N)\, d\lambda]_{h=\infty} \qquad (46)$$

From Equations 45 and 46,

$$m'g = [\textstyle\int (\mu_T - \mu_N)\, d\lambda]_{h2} - [\int (\mu_T - \mu_N)\, d\lambda]_{h1} \qquad (47)$$

Capillary Equilibrium in Gravitational Field. Bulk water can also exist at equilibrium above the free water surface if suitable solid boundaries are provided. As a simple illustration, capillary rise for zero contact angle in a 10-μ diameter tube is included in Figure 4a. The vapor above the curved water surface is in equilibrium with the adsorbed film at this height on the flat plate. The equilibrium capillary pressure of any vapor-water surface is fixed by its height above the free water surface.

$$P_c = \Delta\rho g h \qquad (48)$$

The capillary pressure is related to surface curvature by the Laplace equation (*40*)

$$P_c = \sigma_{wv}\left(\frac{1}{r_1} + \frac{1}{r_2}\right) \qquad (49)$$

where σ_{wv} is the water-vapor surface tension and r_1 and r_2 are the principal radii of curvature at any point in the surface, usually defined as positive with respect to the nonwetting phase. If the surface between water and its vapor is stable, Equations 48 and 49 apply for any shape of solid surface and any wettability condition.

Also depicted in Figure 4a are two sets of touching spheres at heights of 100 and 1000 cm above the free water surface. Water will condense around the points of contact between the spheres until the pressure drop across the surface of the pendular ring so formed corresponds to capillary pressure given by Equation 48. The properties of pendular rings are well documented (*41–44*). Profiles for pendular rings at equilibrium between 10-μ diameter hydrophilic spheres at the two heights are included in Figure 4a.

Properties of adsorbed films on curved solid surfaces, such as the tube and sphere surfaces shown in Figure 4a, will vary slightly from those on the plane surface at the same height. The thickness of adsorbed film on a curved solid surface will be about equivalent to that given by the plane surface at a height corresponding to the equilibrium curvature of the vapor-water surface increased by the curvature of the solid (taking convex as positive). Thus, at any height, adsorbed films are slightly thinner on convex solid surfaces and slightly thicker on concave solid surfaces than the equilibrium film thickness on the corresponding plane surface.

When spreading occurs, the surface free energy per unit area of adsorbed film will be equal to that of the vapor-liquid surface at the same height. For the range of capillary pressures and fluid curvatures of interest,

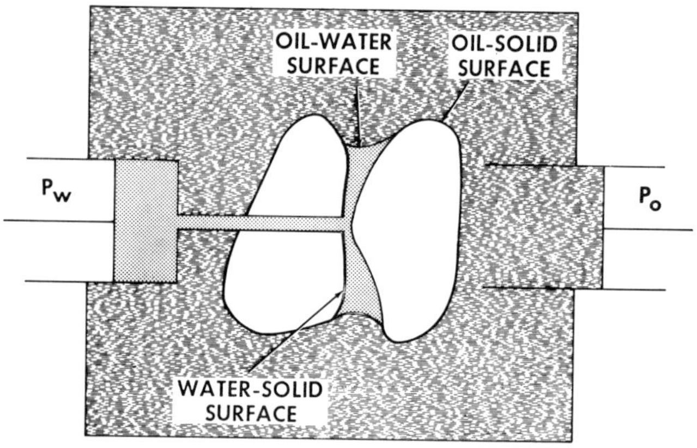

Figure 5. Idealized system for reversible immiscible displacement

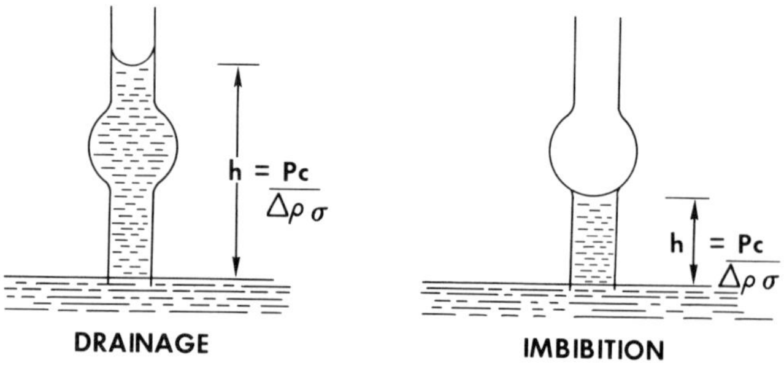

Figure 6. Bistability of an interface in a nonuniform tube

variation of vapor-liquid surface tensions with curvature are not measurable and can be neglected until the radii of curvature approach molecular dimensions (*15–17*). Also, for all practical purposes, the area of a surface is adequately defined by its physical position. Thus, for the spreading situation, the increase in free energy due to surface over the reference state given by solid overlain by bulk water is

$$\Delta F = \sigma_{wv}(A_{wv} + A_{sv}) \tag{50}$$

This result is often stated without formal derivation. The above discussion serves to demonstrate a fundamentally consistent treatment of surface energetics for spreading systems and clarifies the assumptions leading to Equation 50.

Thermodynamics of Reversible Displacement

The thermodynamics of reversible fluid displacement are now considered for an idealized model; changes of surface area within the model system are related to external work done on or by the system.

By the first law of thermodynamics, a change in the total energy of the system, dE, is given by

$$dE = dq + dW + \sum_{i=1}^{j} \mu_i dn_i \tag{51}$$

where dq is the heat flowing into the system. For an isothermal change

$$dq = TdS' \tag{52}$$

dW = the reversible work done on the system
μ_i = the chemical potential, and n_i is the number of moles of the ith component
j = the number of components

In the following analysis, gravitational, magnetic, and electric force fields are assumed to be absent or of negligible effect.

The work dW done on the system is given by

$$dW = -\sum_{p=1}^{3} P_p dV'_p + \sum_{k=1}^{3} \sigma_k dA_k \tag{53}$$

where p = the pth phase

and σ_k = interfacial tension of the kth surface
A_k = area of the kth surface
V' = volume

(We may note that two phases correspond to only one interface. In the above system, there are three phases and three interfaces. With four phases, six interfaces are possible, and so on.) If the system is closed,

$$dn_i = 0 \qquad i = 1 \ldots, j \tag{54}$$

and

$$dE = TdS' - \sum_{p=1}^{3} P_p dV'_p + \sum_{k=1}^{3} \sigma_k A_k \tag{55}$$

The Helmholtz free energy is defined by,

$$F = E - TS' \tag{56}$$

$$dF = dE - TdS' - S'dT \tag{57}$$

Thus from Equations 55 and 57,

$$dF = -S'dT - \sum_{p=1}^{3} P_p dV'_p + \sum_{k=1}^{3} \sigma_k dA_k \tag{58}$$

If displacement is isothermal,

$$dF = -\sum_{p=1}^{3} P_p dV'_p + \sum_{k=1}^{3} \sigma_k dA_k \tag{59}$$

To facilitate discussion, we will assume a model system consisting of a solid (s), wetted by water (w) with a unique contact angle θ, in the presence of oil (o). Figure 5 depicts a pendular ring of water held between two solid particles. The water can enter or leave the ring *via* a passage through one of the particles. The geometry of the particles allows the ring to vary smoothly and reversibly in volume and surface area. The walls of the piston are assumed to make no contribution to the surface free energy.

If the pistons were free to move, water would tend to be imbibed and the pendular ring would grow. If the piston in contact with the oil is free to move, an outward force proportional to $(P_o - P_w)$ must be exerted on the piston in contact with water in order to maintain equilibrium. The external work, W_{ext}, done on the enclosed system in moving the piston in contact with the water outward to give a displacement, $-dV_w$, where V_w is the volume of water in the pendular ring, is given by:

$$dW_{ext} = -(P_o - P_w)dV_w \tag{60}$$

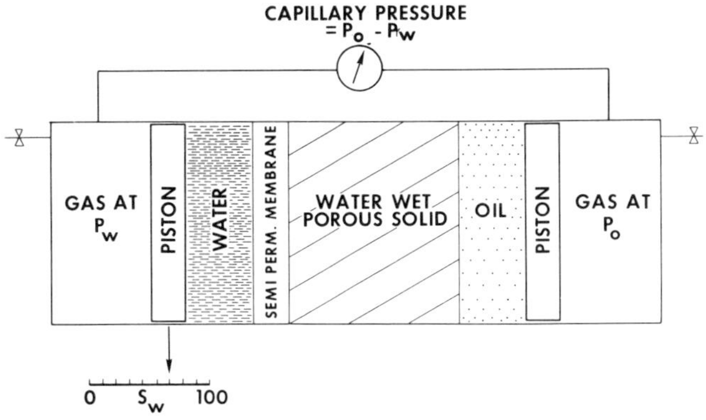

Figure 7. Idealized capillary pressure apparatus

The decrease in Helmholtz free energy of the surroundings of the closed system is equal but opposite in sign to the work done on the closed system. From Equation 59 the change in Helmholtz free energy, dF, of the closed system can be written

$$dF = -P_s dV'_s - P_w dV'_w - P_o dV'_o + \sigma_{wo} dA_{wo} + \sigma_{so} dA_{so} + \sigma_{sw} dA_{sw} \quad (61)$$

For a virtual displacement, the total change in the Helmholtz free energy of the system and its surroundings is zero.

$$dF_{\text{tot}} = 0 = (P_o - P_w)dV_w - P_s dV'_s - P_w dV'_w - P_o dV'_o + \sigma_{wo} dA_{wo} + \sigma_{so} dA_{so} + \sigma_{sw} dA_{sw} \quad (62)$$

If the work of compression on all three phases is neglected,

$$dV'_s = dV'_w = dV'_o = 0 \quad (63)$$

Also, the total solid surface area is constant so that,

$$dA_{so} = -dA_{sw} \quad (64)$$

and from Young's equation (see Equation 34)

$$\sigma_{so} - \sigma_{sw} = \sigma_{wo} \cos \theta \quad (65)$$

The pressure difference $(P_o - P_w)$ is the capillary pressure P_c across the curved water-oil interface.

Equation 62 can therefore be reduced to

$$- P_c dV_w = \sigma_{wo} d(A_{wo} + A_{so} \cos \theta) \quad (66)$$

If the water spreads over the solid, $\cos \theta = 1$, and

$$dF = - P_c dV_w = \sigma_{wo} d(A_{wo} + A_{so}) \quad (67)$$

For finite displacements which can be conducted reversibly Equations 66 and 67 can be integrated to give, respectively

$$\Delta F = - \int_{v_1}^{v_2} P_c dV_w = \sigma_{wo}(\Delta A_{wo} + \Delta A_{so} \cos \theta) \quad (68)$$

and

$$\Delta F = -\int_{v_1}^{v_2} P_c dV_w = \sigma_{wo}(\Delta A_{wo} + \Delta A_{so}) \quad (69)$$

The above thermodynamic analysis of reversible displacements will next be extended to a treatment of displacement in porous media.

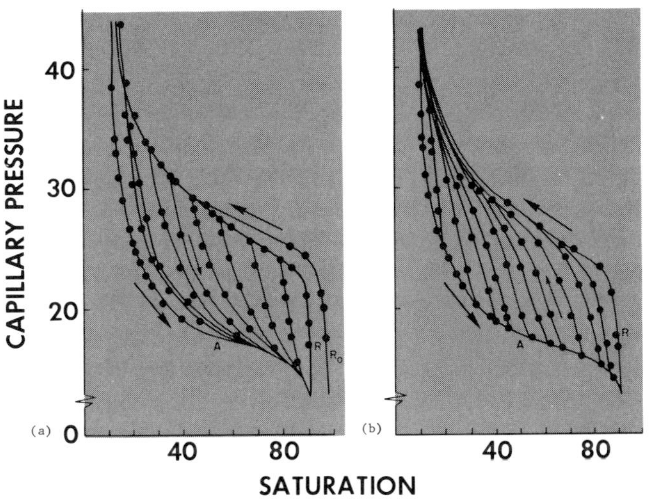

Figure 8. Capillary pressure hysteresis data for 48–60 mesh microbeads (64)

THERMODYNAMICS OF IMMISCIBLE DISPLACEMENT IN POROUS MEDIA

Relationships between capillary pressure and displacement volume can be obtained by a number of routine procedures which are adequately described elsewhere (*45–48*). Equations 68 and 69 have often been applied directly to determine changes in surface area (*47–54*); this necessarily assumes that displacement can be conducted smoothly and reversibly. The assumption of reversible displacement is at variance with commonly observed hysteresis in capillary pressure. Irreversibility at the microscopic scale is evident from observations made by Haines (*2*); during invasion of pore spaces, the development of unstable interface configurations caused small oscillations in pressure. Striking demonstrations of the discontinuities in capillary pressure *vs.* displaced volume were reported recently by Crawford and Hoover (*55*).

The bistability of an interface in a nonuniform capillary of the type shown in Figure 6 is commonly used to explain capillary pressure hysteresis. Liquid level depends upon its initial condition. Furthermore, when the filled tube is slowly raised above the free water surface, the interface passes into the bulbous region spontaneously unless restrained by carefully controlling the pressure in the tube above the interface. Similarly, when the drained tube is slowly lowered, the liquid will imbibe spontaneously into the bulb, but at a pressure distinctly different from the drainage pressure.

It follows that when given volumes of fluid are comingled in a complicated network of nonuniform pores typical of a porous media, many stable fluid configurations are possible. Unlike many thermodynamic systems which are stable at a single minimum free energy, there will be many local minima of free energy, and the global minimum for the system has no special significance.

Measurement of Capillary Pressures in Porous Solids

An idealized apparatus for measurement of capillary pressures in porous solids is shown in Figure 7. Normally, the porous solid is initially saturated with the wetting phase (water) and placed in close contact with a fine-pored membrane which remains saturated with water at all times. The pressure on the water phase (neglecting hydrostatic head) is the measured gas pressure acting on a frictionless piston. The nonwetting phase (oil) is in direct contact with the porous solid, and its pressure is given by the gas pressure P_o acting on a second piston. Since capillary pressure is defined as the pressure difference between oil and water, absolute pressures P_o and P_w are not required. Normally, one phase or the other is at atmospheric pressure throughout the measurements.

Displacement in the porous solid is usually expressed as per cent change in saturation, S_w, the volumetric fraction of pore space in the solid occupied by the wetting phase. Displacement of a volume of water ΔV_w is related to a saturation change ΔS_w by

$$\Delta V_w = V_B \phi \Delta S_w \qquad (70)$$

where V_B is the bulk volume of the porous solid and ϕ is the porosity of the solid. Changes in saturation result from changes in pressure. Each capillary pressure data point is obtained by holding the external pressure difference at a constant value until fluid flow to or from the porous solid ceases. In the apparatus of Figure 7, volumetric displacement and the time at which flow ceases is indicated by the position of the piston acting on the water phase.

Data points are usually referred to as capillary pressure equilibria; but this description requires qualification because not all of the stability requirements for equilibrium are met. Before discussing the nature of the stability represented by capillary pressure data points, the general form of capillary pressure *vs.* saturation curves will be reviewed.

General form of capillary pressure data. Saturation changes which result from variations in capillary pressure do not follow a unique functional relationship. Plots of capillary pressure data *vs.* saturation are shown in Figure 8a and b. To aid discussion of capillary pressure curves, we define the following terms.

- Irreducible saturation, S_{wi}: the volume of wetting phase retained at high capillary pressures where the wetting phase saturation appears to be independent of further increases in the externally measured capillary pressure.
- Residual saturation, S_{or}: the volume of nonwetting phase which is entrapped when the externally measured capillary pressure is reduced from a high value to zero (see Figure 8a).
- Initial drainage curve, R_o: the relationship characteristic of displacement of wetting phase from 100% saturation to the irreducible wetting phase saturation, S_{wi}.
- Imbibition, *A:* increase in wetting phase saturation from the irreducible wetting phase saturation to the residual nonwetting phase saturation, S_{or}.
- Secondary drainage, *R:* drainage from the residual nonwetting phase saturation, S_{ro}, to the irreducible wetting phase saturation, S_{wi}.

Most experimental evidence indicates that the irreducible saturation obtained by initial drainage is the same as that obtained by secondary drainage. When the two conditions are the same, imbibition after secondary drainage will follow exactly the imbibition curve obtained after initial drainage. Thus, the secondary drainage curve and the imbibition curve constitute a closed and reproducible hysteresis loop, *RA*.

Scanning curves within the main hysteresis loop, *RA*, are obtained by reversing the direction of pressure change at some intermediate point along the secondary drainage or imbibition curves. In the former case, they will be called primary imbibition scanning curves, and in the latter, primary drainage scanning curves.

Hysteresis. Many natural systems exhibit hysteresis in the relationship of one variable to another. Sometimes hysteresis, as judged by the magnitude of the hysteresis loop, can be reduced merely by increasing the time for traversal of a loop. If the loop degenerates into a line, then hysteresis is clearly time-dependent. Capillary pressure hysteresis in porous media is essentially independent of time in that the loop cannot be reduced by further increasing the experimental time for each data point. Such behavior has been named permanent hysteresis (*57*).

Everett and coworkers (*56–59*) developed an independent domain theory of permanent hysteresis which has recently been investigated as a model for capillary pressure hysteresis (*60–62*). Everett's theorems can be stated for experimentally observed capillary pressure behavior as follows:

1. The secondary drainage curve, *R*, and the imbibition curve, *A*, form a closed loop, *RA*, with all subsequent capillary pressure data points lying on or within this loop (Figures 8a and b).
2. The primary drainage scanning curves, which begin at the imbibition curve, *A*, either meet at the upper intersection of the *RA* curve (Figure 8b), or

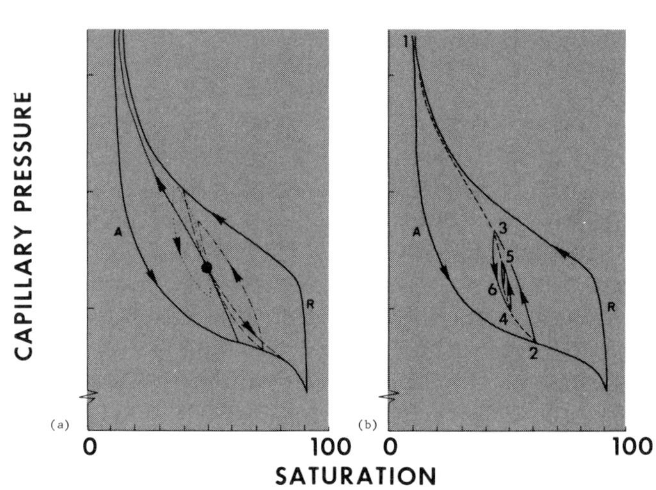

Figure 9. General behavior of capillary pressure scanning curves

Table II. Comparative Reductions in Relative Humidity of Water Due to Surface Curvature, Temperature Reduction, and Salinity

Capillary pressure, h, cm H_2O ($P_c/\Delta \rho g$)	Surface curvature, cm^{-1} (P_c/σ_{wv})	Cylindrical capillary radius, ($\theta = 0$), cm ($2\sigma_{wv}/P_c$)	Percent relative humidity, 20°C, (p/p_o) 100%	Equivalent temperature reduction ΔT below 20°C	Equivalent salinity, (gmol NaCl per liter)
0	0	∞	100	0	0
10	1.36×10^2	1.47×10^{-2}	99.999	1.6×10^{-4}	3×10^{-4}
10^2	1.36×10^3	1.47×10^{-3}	99.99	1.6×10^{-3}	3×10^{-3}
10^3	1.36×10^4	1.47×10^{-4}	99.92	1.3×10^{-2}	2.4×10^{-2}
10^4	1.36×10^5	1.47×10^{-5}	99.27	1.2×10^{-1}	0.22
10^5	1.36×10^6	1.47×10^{-6}	93.0	1.16	2.1

converge tangentially onto the secondary drainage curve, R, in a region close to the intersection. Behavior of imbibition scanning curves, which begin at the drainage curve, R, is analogous (Figure 8a).

3. Any point within a hysteresis loop can be reached by many paths. Complete specification of the system at a given pressure-saturation coordinate must include the path by which the point was attained, since the path determines the microscopic distribution of the fluids and future behavior of the system. (Four possible paths to the same pressure coordinate are shown in Figure 9a.)

4. If the system were taken through a series of pressure oscillations of decreasing amplitude, after the nth pressure reversal, the system moves toward the point at which the $n - 1$ reversal occurred, and if the system is carried through this point, it moves toward the $n - 3$ reversal point, and so on. (This behavior is shown in Figure 9b.)

The above theorems describing the general behavior of capillary pressure curves are consistent with the expected displacement mechanism. Closed scanning loops can be explained as follows. Consider a region which has been drained to a saturation S_{w1} by a pressure P_1. If the pressure were then lowered to P_2, the wetting phase would imbibe to S_{w2}. When the pressure is raised again to P_1, precisely that region which became occupied by imbibition (when the pressure was lowered from P_1 to P_2), can be expected to drain. The liquid in this region must have continuity because it imbibed; the previous drainage demonstrated that reapplication of pressure P_1 will be sufficient to remove only that liquid from the region where liquid was freshly imbibed. Equally straightforward, but not entirely rigorous, explanations can be given for the other general statements made above.

For a porous medium to satisfy the conditions of Everett's independent domain model, each pore must have a characteristic drainage pressure p_1 and a filling pressure p_2. The model does not require that p_1 and p_2 be functionally related. If the conditions of the independent domain theory hold for porous media, then imbibition scanning curves could be used to predict drainage scanning curves and *vice versa*. Satisfactory predictions were obtained by Poulovassilis (*60*) in tests on beds of sintered glass beads; however, later work by Topp and Miller (*62*) showed that the theory is not generally applicable.

In general, assignment of draining and filling pressures to a given region may be unrealistic, since pressures for displacement are also determined by accessibility and phase continuity. When the behavior of a domain is influenced by other domains, the system is said to exhibit cooperative behavior. In other words, the behavior of any given pore depends on its neighbors. A model of hysteresis which includes cooperative behavior has been investigated by Enderby (*63*).

Another fundamental weakness in applying the independent domain theory to displacements in porous media is that pore space cannot be divided precisely into volumetric zones which exhibit one to one correspondence with respect to drainage and imbibition behavior. Although capillary behavior is commonly discussed in terms of pores, no definition of the bounds of a single pore exists for real porous media. If a definition were attempted in terms of capillary behavior, a first requirement would be that a distinction should be made between imbibition and drainage conditions.

Stability of capillary pressure points. When two homogeneous fluid phases are held within a porous solid, a necessary condition for equilibrium, neglecting gravity, is that capillary pressure across the interface separating the two fluids be the same everywhere. The form of capillary pressure curves indicates that this condition is not satisfied at each point. For example, when water is displaced by oil, some water becomes trapped in small pores and around points of contact between particles. Once hydraulic communication *via* bulk water paths is lost, volumes of water elements cannot adjust quickly to attain equilibrium with the externally measured pressure. As drainage progresses, increasing amounts of wetting phase are isolated and an irreducible wetting phase saturation is reached (*2, 47*). At this stage only continuous oil and hydraulically discontinuous water are retained. Hence, the externally measured capillary pressure bears no relationship to the capillary pressures of oil-water interfaces within the porous medium. Similarly, during imbibition, the oil phase is not completely displaced. When the external capillary pressure is reduced to zero, globules of residual oil are retained at pressures higher than the water which surrounds them. Both trapped oil and trapped water will tend to diffuse within the porous solid; however, the time required to reach equilibrium by diffusion is long and experimentally prohibitive.

Although discontinuous elements are not at equilibrium, capillary pressure data points can be regarded as stable if we assume that changes in their volume can be neglected for the duration of an experiment. Experimental data points can then be assumed to correspond to an idealized model, with no mass transfer across phase boundaries or *via* films retained at the surfaces of drained solids (*64*). Under these conditions, only those interfaces which can adjust curvature by hydraulic flow of both fluids maintain capillary equilibrium with the externally measured capillary pressure; elements of fluid at differing pressures can coexist indefinitely until the region they occupy is encroached upon by an invading continuous region of the same phase.

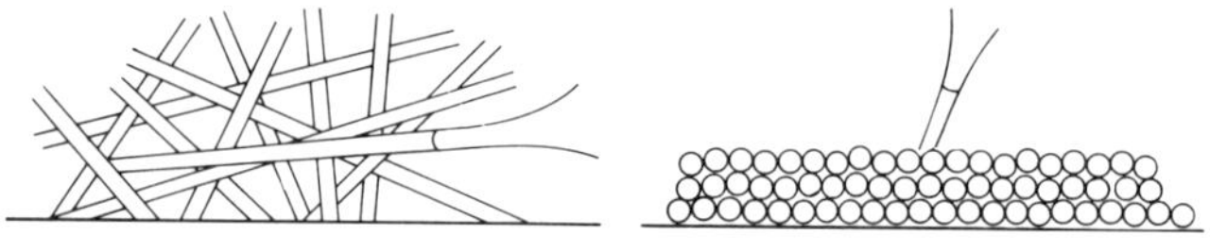

Figure 10. Experiments illustrating fluctuations in pressure during drying

The above model appears to hold well for so-called clean solids. When the solid surface of a porous system is rough and channeled, or has a coating of colloidal particles, hydraulic conductivity may never be lost completely. The almost isolated wetting liquid may approach equilibrium at much faster rates than in a comparable situation with smooth surfaces; this could even permit stability to be achieved through attainment of a true capillary equilibrium. (There is no violation of the model in these cases because the almost discontinuous liquid is still strictly classified as continuous.) In general, the applicability of the above capillary behavior model must be judged by examining the stability of experimental capillary pressure data points. Another indication of applicability is the development of an irreducible saturation which is distinctly independent of both time and further increases of the external capillary pressure.

Stability can also be affected by temperature. Sometimes it is claimed that a true capillary equilibrium would be obtained if capillary pressure experiments were conducted over sufficiently long times. For example, if a column of particles were drained of water with air as the nonwetting phase, the observed and apparently stable distribution of water would slowly change by diffusion (caused by differences in vapor pressure due to curvature) until a true capillary equilibrium is reached (*65*). Aside from the long time required for diffusion, the need for adequate temperature control would make such an experiment difficult to execute. The temperature reduction necessary to lower the vapor pressure by the same amount as that due to surface curvature alone is given in Table II (*66*). Vapor pressure is extremely sensitive to temperature. Thus, changes due to diffusion that are observed over long time periods are probably caused by ambient temperature variations. It follows that accurate temperature control is needed for capillary pressure experiments that take several days or even months to complete even though the measurement of true capillary pressure equilibria is not the objective.

Vapor pressure of water is also sensitive to salt concentration. Salt concentration (gmol of sodium chloride/l.), necessary to reduce the vapor pressure by an amount equivalent to effects of surface curvature alone, is also listed in Table II. Vapor pressure above a curved surface of a salt solution depends on both curvature and concentration; thus liquid volumes at different capillary pressures and salinities could exert equal vapor pressures and be in stable equilibrium with each other. For isolated volumes of water, each of which contains a fixed weight of salt, any variation in volume will result in a change in salt concentration which will tend to stabilize the volume variations. Thus, if saline water were used in capillary pressure experiments, applicability of the idealized model of capillary behavior would be enhanced.

Haines Jumps

The basis for treating experimental capillary pressure points as stable has now been established. We will now consider the behavior of the system during displacement. A basic feature of fluid interface motion during displacement is the Haines jump resulting from unstable fluid configurations (*2*, *67*, *68*). These jumps are illustrated by the following simple experiments.

In the first experiment, melting point tubes were heated and pulled into long fine capillaries which were broken into short lengths. These tube segments were then piled in a heap, saturated with a volatile liquid, and the system was observed through a microscope as the wetting-phase saturation decreased by evaporation (Figure 10a). The tubes were slightly tapered and the air-liquid menisci moved slowly toward the narrower ends of the tubes as the saturation decreased and the capillary pressure increased. Occasionally, a meniscus would suddenly recede or jump back toward the wider ends, indicating a sudden pressure change. The meniscus would then continue to displace smoothly from its new position until it suddenly kicked back again, and so on. In effect, each tapered tube served as a pressure measuring device which showed that capillary pressure does not rise smoothly and continuously with decrease in saturation.

In a second experiment on packings of 3 mm spheres, sudden flexures in the interface as described by Haines again indicated that pressure was not increasing mono-

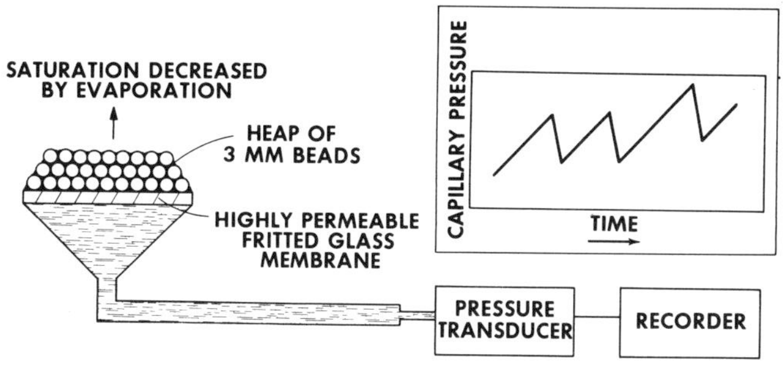

Figure 11. Apparatus for recording pressure fluctuations

tonically with saturation. When the tapered end of a melting point capillary was inserted into the pile of spheres (Figure 10b), the liquid in the tube behaved erratically as it followed pressure changes in the bead pack.

The same spheres were heaped on a highly permeable fritted glass membrane connected with a sensitive pressure transducer as depicted in Figure 11. The bead pile was saturated with volatile liquid. As the liquid evaporated, the recorded variation of pressure with time was of the form shown in Figure 11. The pressure rose relatively slowly, and then suddenly fell, then slowly rose again, and so on.

The above experiments clearly demonstrate that displacement cannot be conducted smoothly and reversibly because of spontaneous changes in fluid configuration. In the next section a new treatment of displacement as a distinctly quantized process is developed from application of thermodynamic fundamentals.

Infinitesimally Slow Displacement

First we consider infinitesimally slow withdrawal of liquid from the capillary model shown in Figure 12. Liquid can be removed reversibly from both tubes of the model until the meniscus in the tube on the right meets the enlargement in tube cross section at *A*. Boundary conditions at *A* cause the interface to become unstable, and the liquid descends rapidly through the upper bulb to a stable position in the vicinity of *B*. If the hole through which fluid is withdrawn were infinitesimally small, we would have the limiting case of no liquid flow during the spontaneous motion of the interface. Alternatively, we could envision use of an idealized valve which would only permit flow if the pressure drop across it were infinitesimally small.

If a finite pressure developed across the valve because of an unstable fluid configuration, the valve would close instantly; the valve would reopen only after the fluid had attained a stable configuration, and the fluid pressure on the outlet side of the valve was adjusted to equal the system pressure. From this point, further removal of fluid could be effected under reversible conditions until the system again becomes unstable as the lower bulb drains spontaneously.

Again, a spontaneous redistribution of fluid results in a reduction of capillary pressure. If capillary pressure were plotted against volume of liquid in the model, we would see that displacement under these idealized circumstances consists of smooth, reversible changes linked by spontaneous changes in pressure at constant saturations. The sequence of alternate reversible and irreversible changes during this quantized displacement would be determined by the detailed structure of the model.

The above argument can be extended in principle to the complicated structures of porous media. Instability can be regarded as localized to a single pore or a group of pores. The capacity of the rest of the system to receive or give up a small volume of fluid, represented by the tube on the left in Figure 12, determines the size of the finite pressure change caused by the instability.

Rheons and Isons

The form of the pressure-volume relationship characteristic of the above model is similar to that obtained experimentally for the heap of 3-mm glass beads. In both cases, the pressure slowly increases as drainage progresses until an unstable fluid configuration precipitates a rapid pressure change. Reducing the drainage rate causes the sudden pressure changes to occur at almost constant saturation as in the model of Figure 12.

The sudden motions of unstable interfaces reflected by small changes in pressure have been described as Haines jumps or jerks (*67, 68*). Heller (*69*) mentions that the term rheon has been suggested by J. C. Melrose as an alternative name for the Haines jump. In this treatment, we adopt the term *rheon* for a Haines jump which takes place at constant saturation. Under this condition, no external work can be done on or by the system during a Haines jump. The segments of the pressure–volume displacement curve between rheons correspond to smoothly reversible displacements; they will be called *isons*.

When capillary pressures are measured in fine-pored media, the discontinuities in pressure are generally too small to observe, and experimental capillary pressure curves, although not reversible, appear to be smooth and continuous. In theory, displacements from porous media could be conducted as described above for the tube model to give capillary pressure curves consisting

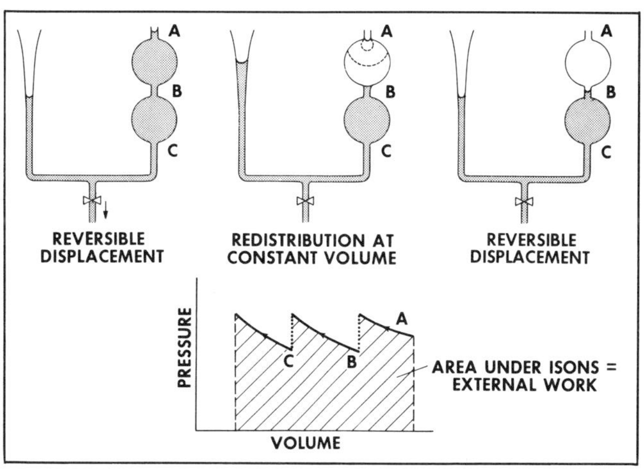

Figure 12. Infinitesimally slow drainage of a model capillary system

of series of isons and rheons. Since capillary pressure hysteresis is permanent with respect to time, we postulate that the experimental data give the same path, apart from loss of detail, as that which would be given by the isons and rheons. Thermodynamic relationships which apply to individual isons and rheons provide the basis for thermodynamic relationships which apply to experimental capillary pressure data.

Rheons. A rheon is defined as a spontaneous redistribution of fluid within a porous medium. Instabilities arise from fluid interfaces that are unable to change curvature smoothly with variations in pressure. The extent to which an interface can vary its curvature without spontaneous motion depends upon the boundary conditions at the surface of the porous solid. A general discussion of the stability of interfaces is given by Gibbs (*70*).

Once an unstable configuration is reached, the interface moves rapidly to a new equilibrium position. The volume of liquid spontaneously displaced from a region may vary from a fraction of a single pore space to an assembly of pores. For example, with rough surfaces, displacement from regions of a single pore could be likened to a ratchet mechanism. On the other hand, cooperative effects could cause displacement in large groups of pores during a single rheon. In general, the number of rheons contained by a capillary pressure *vs.* saturation curve can be expected to increase directly with sample size and be inversely proportional to the cube of the average pore diameter.

The magnitude of the finite pressure change accompanying a rheon depends on both the size of the region spontaneously vacated by a given fluid and the capacity of the system to advance its phase boundaries elsewhere. Generally in fine-pored solids the system capacity is large compared to the net volume of fluid redistributed by a rheon; hence, the pressure change is very small. The sign of the small changes in capillary pressure will be negative for rheons on drainage curves and positive for rheons on imbibition curves.

From the second law of thermodynamics the free energy of the system must be diminished as a result of the spontaneous change. Since the change occurs at constant saturation, no external work is done on or by the system and loss in free energy, ΔF, is given by

$$\Delta F = \sigma_{ow}(\Delta A_{ow} + \Delta A_{so} \cos \theta) \qquad (71)$$

where the quantity $(\Delta A_{ow} + \Delta A_{so} \cos \theta)$ must be negative. This loss of free energy is the most important characteristic of the rheon. As the pressure change associated with the rheon approaches zero, the decrease in surface free energy has a definite value.

Total surface energy including both surface free energy and surface heat of formation will finally appear as heat. For water against air, surface free energy at 15°C is approximately 74 ergs/cm² and heat of formation about 43 ergs/cm², giving a total energy of 117 ergs/cm². In general, the heat capacity of the system will be so large that the temperature changes accompany-

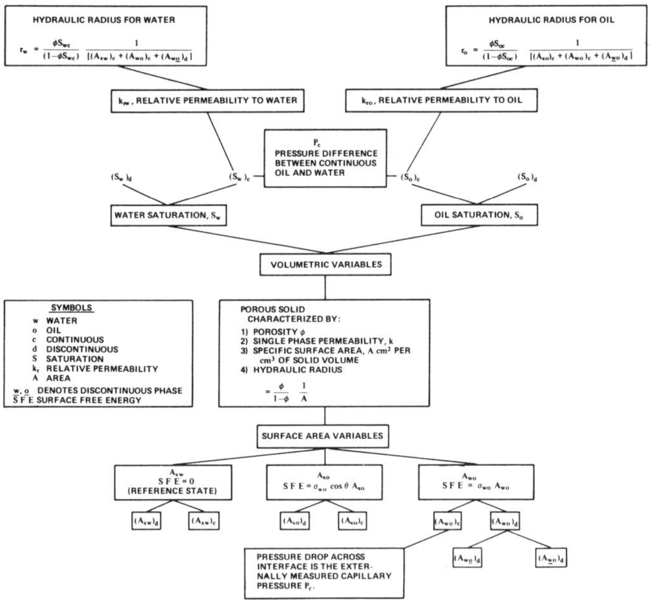

Figure 13. Relationships between volumetric and surface variables of two fluid phases in a porous solid

ing loss of surface energy will be miniscule. Recovery of work because of the temperature difference of the system and its surroundings is not a serious possibility; we assume that the heat escapes until thermal equilibrium is established.

Isons. Isons are defined as displacements of one phase by another from porous media under balanced, stable conditions. These displacements are smooth and reversible with continuous variation of pressure with saturation. An ison's domain of stability with respect to saturation change extends in both directions from the saturation of the previous rheon; but the defined saturation range of an ison begins with its initial saturation and is terminated by the next rheon. Since isons and rheons occur alternately, the number of isons in a particular capillary pressure-saturation curve is governed by the same factors that determine the number of rheons.

For each individual ison the thermodynamic relationships developed for reversible displacements hold. Thus, under the conditions for which Equation 68 was derived, we have with Equation 70

$$\phi V_B \int_{S_{w1}}^{S_{w2}} P_c dS_w = \int \sigma_{wo} d(A_{wo} + A_{so} \cos \theta) \qquad (72)$$

where S_{w1} and S_{w2} lie within the domain of a single ison. For drainage, work is done on the porous system and the surface free energy of the system is increased; for imbibition, the system does work on the surroundings and surface free energy decreases. Changes are conducted isothermally with heat of surface formation continuously given up to or adsorbed from the surroundings.

Displacement Mechanism in Porous Media

A consistent account of the mechanism of displacement in porous media can be formulated within the framework of the quantized model. Interrelated changes

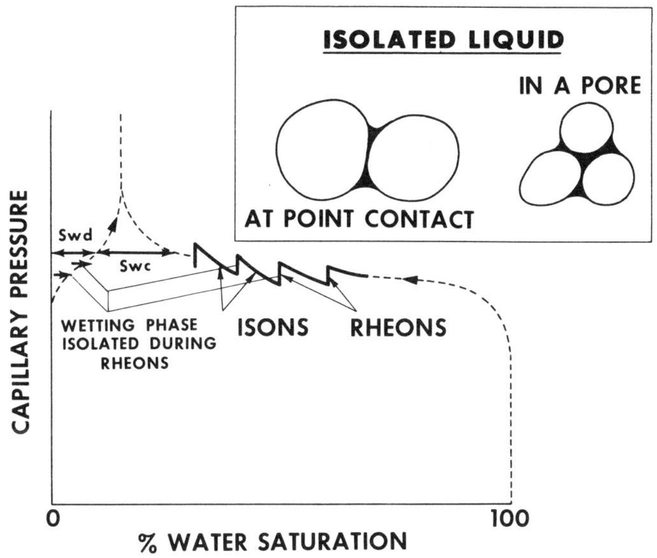

Figure 14. Detailed mechanism of drainage, showing isons and rheons of drainage curve and isolation of wetting phase

in the volumetric and surface area variables that occur during displacement are shown as dimensionless quantities in Figure 13. Volumetric variables are wetting phase (say water) saturation S_w, and nonwetting phase (say oil) saturation S_o. Portions of the volumes making up these saturations may have hydraulic continuity with the surroundings; other portions may not. The continuous and discontinuous portions are denoted by $S_{w,c}$ and $S_{w,d}$ respectively for water, and $S_{o,c}$ and $S_{o,d}$ respectively for oil. Surface area variables are the water-solid, oil-solid, and water-oil interfaces which bound the above volumes; these variables are expressed as fractions of the total solid surface area, A, and denoted by A_{sw}, A_{so} and A_{wo}, respectively. Each of these three surface areas can be subdivided into contributions from the continuous and discontinuous parts of each phase, denoted by suffixes c and d, to give six surface area variables in all: $(A_{sw})_c$, $(A_{sw})_d$, $(A_{so})_c$, $(A_{so})_d$, $(A_{wo})_c$, and $(A_{wo})_d$. The discontinuous oil-water interface area, $(A_{wo})_d$ can be further subdivided into $(A_{\underline{w}o})_d$ and $(A_{w\underline{o}})_d$ where the discontinuous phases are $\underline{w}$ater and $\underline{o}$il respectively (Figure 13). (Other quantities included in Figure 13 relating to relative permeability will be discussed in a later section of this paper.)

Variations of the above volumetric and surface quantities with respect to rheons and isons will first be considered for the initial drainage capillary pressure curve. For this curve, the porous solid is initially saturated with water and quantities $S_{o,d}$, $(A_{w\underline{o}})_d$, and $(A_{so})_d$ are always zero.

As drainage from 100% saturation progresses, some water is cut off by the invading oil and is retained around particle contacts and in the smaller pores as sketched in Figure 14. The manner in which the components S_{wc} and S_{wd} of the water saturation vary as drainage progresses is indicated by the wetting phase isolation curve of Figure 14. As the water saturation S_w decreases, the fraction of water which is discontinuous, S_{wd}, increases from zero until at the irreducible saturation S_{wi}, all of the retained wetting phase is held as disconnected volumes. With the disappearance of the continuous interface, $(A_{wo})_c$, the externally measured capillary pressure loses significance with respect to conditions within the porous solid.

The capillary pressure of an individual element of water will be approximately equal to the capillary pressure, P_c, across the continuous interface, $(A_{wo})_c$, at the time the element became isolated (*64, 71*). This correspondence cannot be exact, however, because an element of wetting phase loses continuity by an irreversible rupturing process. Since irreversible changes are restricted to rheons, formation of a discontinuous element of liquid can only occur during a rheon. A rheon also occurs at constant saturation, S_w; thus, any increase in the discontinuous water saturation, ΔS_{wd}, must be exactly equal to the decrease in continuous water saturation, ΔS_{wc}. These exchanges must be finite

Table III. Relationships between Continuous and Discontinuous Fluid Saturations, Surface Free Energy and Work of Displacement for Isons and Rheons

	Drainage	Imbibition
Ison, (smoothly reversible displacement)	$S_{wd}, S_{od}, (A_{wo})_d, (A_{so})_d, (A_{sw})_d$ are constant; $\Delta W = V_B\phi \int P_c dS_w$	
	$dS_o = [dS_w]$ $\Delta F = V_B\phi \int P_c[dS_w]$ $= \int \sigma_{wo} d\{(A_{wo})_c + (A_{so})_c \cos\theta\}$	$dS_w = [dS_o]$ $[\Delta F] = V_B\phi \int P_c dS_w$ $= \int \sigma_{wo}[d\{(A_{wo})_c + (A_{so})_c \cos\theta\}]$
Rheon, (spontaneous change at constant saturation)	S_o, S_w are constant; $\Delta W = 0 \quad \Delta F < 0$; $[\Delta F] = [\Delta A_{wo} + \Delta A_{so} \cos\theta]$	
	$\Delta S_{wd} = [\Delta S_{wc}]$ $\Delta S_{oc} = [\Delta S_{od}]$	$\Delta S_{wc} = [\Delta S_{wd}]$ $\Delta S_{od} = [\Delta S_{oc}]$

Quantities in square brackets [ ] are negative unless equal to zero.

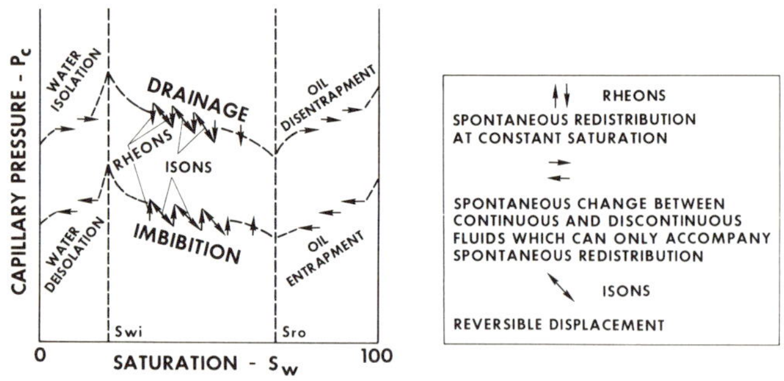

Figure 15. Detailed mechanism of displacement showing relationships between continuous and discontinuous parts of each phase during a drainage-imbibition cycle

and any number of discrete volumes of water could be isolated during a single rheon.

When water is imbibed, the fraction of discontinuous water decreases and we can postulate the existence of a water de-isolation curve which complements the measured capillary pressure imbibition curve. As with isolation of water during drainage, any union of the invading water with a discontinuous element of water will be irreversible and must be associated with a rheon. Some imbibition rheons may also include entrapment of oil as globules. Entrapment of oil tends to occur at relatively high water saturation (*72*). A capillary imbibition curve is depicted as a series of isons and rheons in Figure 15, together with water de-isolation and oil entrapment curves. Changes in the discontinuous saturations are shown as discrete steps. For simplicity we assume that no water remains trapped within regions occupied by the discontinuous oil.

Secondary drainage from the residual oil saturation S_{or} is similar to the initial drainage curve except that rheons may include both disentrapment of oil globules and isolation of water. The secondary drainage curve together with water isolation and oil disentrapment curves are also depicted in Figure 15. In all, Figure 15 includes three hysteresis loops. The central loop is the capillary pressure hysteresis loop which can be measured experimentally. In fine detail, these curves consist of isons and rheons as shown. The curves illustrating variations of discontinuous saturations (horizontal arrows) represent finite saturation changes with capillary pressure coordinates defined by the capillary pressure P_c measured at the time of formation or elimination of discontinuous elements.

The relationships between the surface area and volume quantities shown in Figure 13 for drainage and imbibition isons and rheons are summarized in Table III.

Free Energy Changes of Displacement

The area under a series of isons is exactly equal to the external work done on or by the system during displacement. Neglecting loss of fine detail in experimental capillary pressure data, the area under the isons also corresponds exactly to the area under a capillary pressure curve. This observation provides a consistent thermodynamic basis for the common assumption that the area under a capillary pressure curve corresponds to reversible external work.

The quantum approach is also useful in setting definite limits on surface free energy changes that accompany displacement. If a closed capillary pressure hysteresis loop is measured from wetting phase volume V_1 to a smaller volume V_2, the difference in surface free energy, ΔF, of the states corresponding to V_1 and V_2, lies between the work done on the system for drainage and the work recovered from the system for imbibition.

$$\int_{V_1}^{V_2} P_c dV > \Delta F > \int_{V_2}^{V_1} P_c dV \qquad (73)$$

(Pearce and Donald (*73*) attempt to define free energy changes more closely by comparing different work paths to the same pressure-saturation coordinate on the main hysteresis loop. Primary scanning curves are shown as closed loops which span intermediate points on a main drainage-imbibition loop. However, such behavior is not in accord with the general form of experimental capillary pressure hysteresis data.)

The area of the hysteresis loop is the minimum possible amount of work lost in describing the hysteresis loop. The area of the loop also gives the maximum error which arises from the often made assumption that work given by the area under a capillary pressure curve equals the change in surface free energy.

Any displacement represented by a capillary pressure curve will have a characteristic efficiency which is determined by the geometry of the solid. For example, the efficiency of initial drainage is given by the ratio of the increase in surface free energy of the system to the work of drainage.

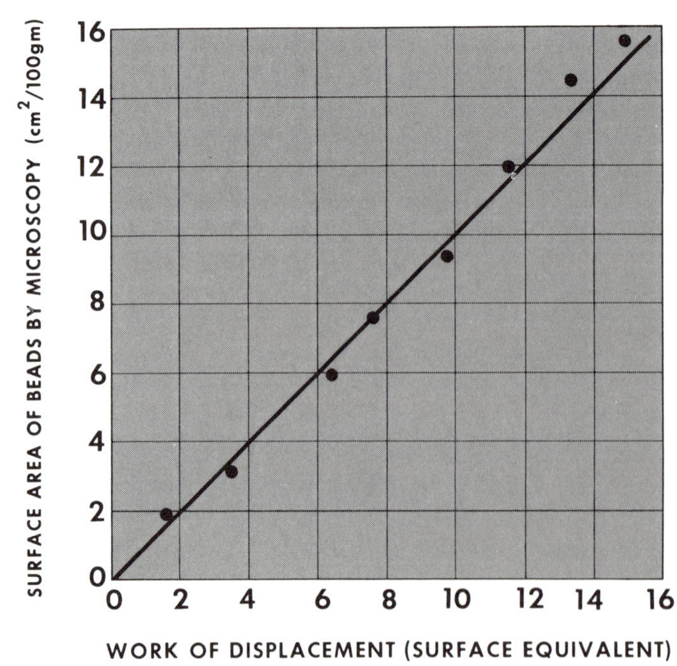

Figure 16. Solid surface area versus work of immiscible displacement for random packings of glass beads

Table IV. Surface Areas at Irreducible Wetting-Phase Saturation as Percent of Total Solid Surface Area from Analysis of Thin Sections

Thin section no.	A_{sa}	A_{sl}	A_{la}	Porosity	Irreducible saturation
1	70.8(243)	29.2(100)	8.45(29)	40.5(153:62)	16.1(10)
	67.8(250)	32.2(119)	11.1(41)	45.5(154:70)	11.4(8)
2	79.8(253)	20.2(64)	3.5(11)	37.5(141:53)	3.8(6)
	78.3(252)	21.7(70)	6.5(21)	42.5(153:65)	4.6(3)
3	68.8(237)	31.2(107)	7.55(26)	36.6(156:57)	5.3(3)
	70.3(253)	29.7(107)	6.4(23)	29.4(143:42)	4.8(2)
4	71.8(267)	28.2(105)	7.25(27)	35.1(148:52)	7.7(4)
	74.6(294)	25.4(100)	4.55(18)	41.8(146:61)	4.9(3)
5	72.4(270)	27.6(103)	8.85(33)	43.6(149:65)	7.7(5)
	72.5(275)	27.5(104)	9.25(35)	36.5(145:53)	0.0(0)
6	67.4(269)	32.1(130)	7.25(29)	33.6(152:51)	7.8(4)
	66.7(263)	33.3(131)	7.6(30)	34.0(147:50)	2.0(1)
Av. values	71.6(3126)	28.4(1240)	7.4(323)	38.1(1787:681)	7.2(49)

EFFICIENCY OF DISPLACEMENT IN RANDOM PACKINGS OF EQUAL SPHERES

Data which relate solid surface area to work of displacement have been reported by Payne (*50*). Drainage curves for water displaced by air were measured for several random packings of equal spheres with sphere size varied from one packing to the next. Sphere surface area was measured directly by microscopic analysis. The work required to drain the spherical beads was almost equal to the product of the interfacial tension and the solid surface area of the beads. Average values of Payne's data are shown in Figure 16 as bead surface area versus work of displacement. These results appeared to indicate that, for a packing formed from spheres of a given size, work of displacement equals the increase in surface free energy of the system. The method has since received some acceptance as a general technique for determining the surface areas of coarse solids (*51*, *74*, *75*). However, the one-to-one relationship between surface area and work observed for random sphere packings is mainly coincidental. A critical assumption in evaluating the surface free energy of the drained bead pack was that essentially all of the solid surface had been exposed to the invading phase. In fact, the beads cannot be drained beyond their irreducible saturation; thus, some of the solid area must remain covered by wetting phase. Also, an interface exists between the discontinuous water and air which contributes to the total surface free energy of the drained bead pack. The following section describes experiments in which surface free energy changes were determined directly.

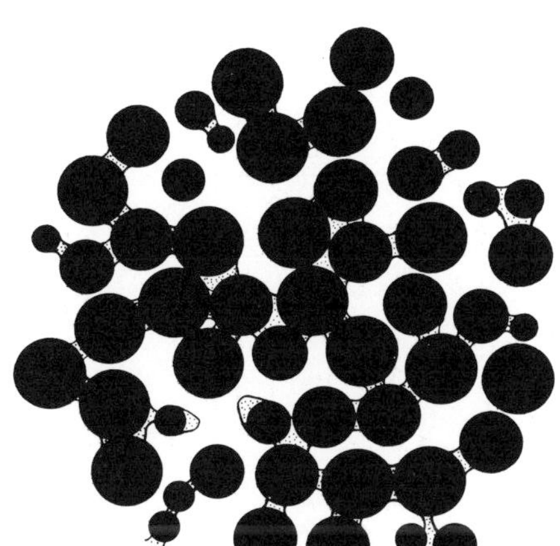

Figure 17. Distribution of fluid phases in a packing of 3 mm beads at the irreducible wetting phase saturation

Determination of Surface Areas at Irreducible Minimum Saturations

A random packing of 3-mm beads was held in a 1-1/2 in. diameter lucite tube with a supporting wire screen at one end. The column was saturated with a liquid epoxy resin containing a blue organic dye. The epoxy was allowed to drain by gravity. In the upper parts of the column liquid epoxy was retained as an irreducible wetting phase saturation. The blue epoxy hardened with very little shrinkage and occupied essentially the same position as when in the liquid state. The air space in the core was evacuated and filled with an orange dyed epoxy resin primarily to strengthen the sample.

After the orange epoxy had hardened, six thin sections were prepared. They showed clearly the regions originally occupied by the spheres (transparent), the irreducible liquid epoxy saturation (blue), and the invading air (orange). Photographic enlargements were made directly from the thin sections. A drawing made from an enlargement is shown in Figure 17. On the enlargements, the region within about two particle diameters from the lucite wall was covered by a circular mask. A square grid was cast at random over the central region of the thin section photograph. Points on the grid were used to determine the relative fractions of space occupied by the solid spheres and the two colors of epoxy. The point count was used to determine porosity and saturation. Ratios of solid-liquid, solid-air, and liquid-air surface areas were determined by counting the number of times the lines on the grid intersected the various interfaces. Results are presented in Table IV. Surface areas are expressed as a percentage of the solid surface and are recorded as A_{sl}, A_{sa}, A_{la} where the subscripts, s, l, and a correspond to solid, liquid and air, respectively.

Bracketed figures in Table IV under columns labeled A are the number of line intersections for the indicated interface. The two bracketed numbers in the porosity column are the total number of points counted and the number lying within the pore space of the packing. Percentage porosity is calculated as the ratio of the second to the first. The bracketed figure in the irreducible saturation column is the number of points that fell in the region occupied by wetting phase. Although counts for individual thin sections resulted in fairly wide variations in porosity and irreducible saturation, the average values of 38.1% and 7.2%, respectively, are in close agreement with direct measurements of these quantities. Random packings of equal spheres usually have porosities in the range of 36–40% and irreducible saturations in the range of 6–8%. The average percentage of solid surface drained at the irreducible saturation was 71.6%, leaving 28.4% still covered by the liquid phase. The area of the air-liquid interface at the irreducible saturation was 7.4%.

An independent check of the fraction of solid surface which drains was made using a packing of 6-mm diameter ball bearings. The packing was saturated with a dilute acid and drained to its irreducible saturation. The retained acid etched the undrained region of the bearing surfaces. The bearings were dried, laid on a plane, and the uppermost point of each bearing used in point counting to obtain the ratio of drained to undrained surface. The result of 500 point counts indicated that 69.2% of the solid surface was drained; this value agrees substantially with the average value of 71.6% obtained from thin sections of the bead packs.

Efficiency of Initial Drainage

The liquid epoxy resin spread on glass; hence, the increase in surface free energy due to drainage should be

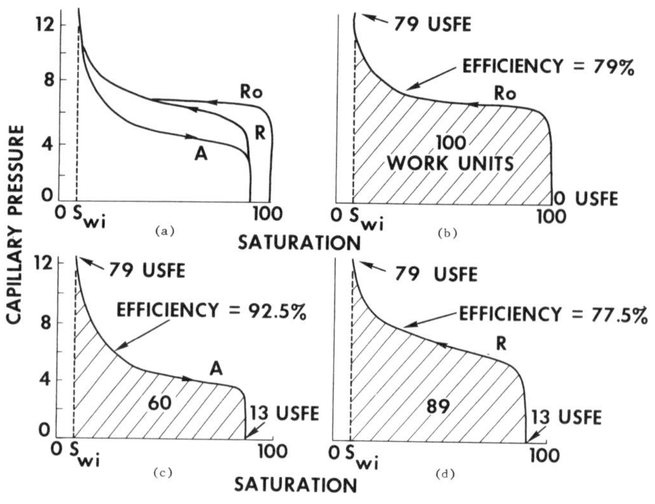

Figure 18. Relationships between work of displacement and changes in surface area for initial drainage, imbibition and secondary drainage (USFE = units of surface free energy)

directly proportional to the sum of the air-liquid and air-solid surfaces (cf. Equation 50). Expressing these areas as percentage units of the total solid surface, $\Delta \bar{F}$ units of surface free energy,

$$\Delta \bar{F}_{(100 \rightarrow S_{wi})} = \Delta A_{sa}\,(71.6\%) + \Delta A_{la}\,(7.4\%) = 79\ USFE \quad (74)$$

The inherent efficiency, E_D of conversion of work to creation of surface is given by the increase in free energy divided by the work of displacement

$$E_D = \frac{\sigma_{la}(\Delta A_{sa} + \Delta A_{la})}{V_B\,\phi \int_{100}^{S_{wi}} P_c dS_w} \quad (75)$$

But from Payne's results shown in Figure 16, the work of displacement is equivalent in energy to the surface free energy of totally drained spheres. For purposes of comparing inherent displacement efficiencies, we can take the area under the capillary pressure initial drainage curve as 100 units of work and the surface energy of the totally drained solid (zero wetting phase saturation) as 100 units of surface free energy, equivalent to 100 units of work. (In general, the work given by the area under a capillary pressure drainage curve is not likely to be equivalent to the surface energy of any porous solid at zero wetting phase saturation.)

If the completely saturated solid is taken as a reference state for zero surface energy, then a drained bead pack will have 79 units of surface free energy at irreducible saturation. Since 100 units of work were required for drainage, the inherent thermodynamic efficiency of drainage from 100% saturation to S_{wi} is 79%. (See Figure 18b.)

Surface Area of Entrapped Nonwetting Phase

Determination of the inherent thermodynamic efficiency of imbibition requires measurement of residual nonwetting phase surface area. Gravitational effects in 3-mm beads are such that very little air is entrapped by imbibition of a wetting liquid. Entrapment of air by liquid epoxy was therefore studied on two packings of microbeads. (The compressibility of air has little effect on the measured volume and surface area.) In one packing, bead diameters were in the range of 297 to 350 μ and in the other 125 to 149 μ.

A given packing was first saturated with liquid epoxy resin and then drained. The liquid epoxy was then imbibed into the packing and allowed to solidify. Attempts to prepare thin sections from packings of microspheres at residual nonwetting phase saturation were unsuccessful. Even in preparing a polished surface (which could only be examined by reflected light), portions of beads become dislodged during polishing; the resulting void spaces were not clearly distinguishable from entrapped air pockets.

The method finally chosen for preparation of a surface cross section was as follows. The packing was cut with a diamond saw, and the exposed surface was given a thick coating of red-dyed epoxy resin. After the resin hardened, it was ground away until the upper surface of the exposed packing had been lowered by a depth of about 2/3 of a bead diameter; this procedure removed the red-dyed epoxy which had filled spaces left by dislodged sphere segments. Examination of the exposed surface revealed the following distinguishable regions: (1) sphere segments or spherical bowls where sphere segments had broken away, (2) the blue epoxy of the originally imbibed wetting phase, (3) the red epoxy which had invaded pockets of air exposed when the sample surface was coated with red epoxy, and (4) air pockets (containing no red epoxy) which had been freshly exposed by grinding. The freshly exposed air pockets were relatively rare and generally distinct from spaces left by dislodged sphere segments. Finally, the surface was briefly exposed to hydrofluoric acid vapor which frosted the visible surface of bead segments to make them distinct when observed under reflected light.

To determine surface areas, the prepared cross section was examined with a binocular microscope using reflected light. The eye piece contained a 10 × 10 arithmetic grid. The objective was chosen so that one square on the grid was a little larger than one particle. For each line on the grid, the number of intersections with the solid surface and the number of intersections between the wetting phase and the nonwetting phase were counted. If grid lines cut positions where a nonwetting phase was touching the solid, it was counted both as a solid-liquid interface and as a liquid-air interface.

Results are shown in Table V. Six sets of measurements were made on the 297 to 350 μ bead packs. Only two measurements were made for the packing of 125 to 149 μ beads because of difficulty in obtaining sufficient illumination with reflected light. For the smaller bead packings, the surface area of the entrapped nonwetting phase was 13.8% of the solid surface and for the larger beads, 13.1%, to give an average overall count of 13.2%.

Work of Imbibition and Secondary Drainage

From surface areas determined thus far, efficiencies of imbibition and secondary drainage were obtained from respective areas under these curves. Results for three sets of capillary pressure tests on glass beads giving curves of the form shown in Figure 18a are presented in Table VI. The average area expressed as a percentage of the area under the initial drainage curve was 60% for imbibition and 89% for secondary drainage. These percentages are shown in Figure 18 as units of work.

Table V. Surface Area of Residual Nonwetting Phase as Percent of Total Solid Surface from Analysis of Thin Sections

Microbead diameter	Thin section no.	Counts ($A_{sa} + A_{sl}$), total solid surface	Counts ($A_{sa} + A_{la}$), residual nonwetting-phase surface	($A_{sa} + A_{la}$) as % of total solid area
297–350 μ	1	395	57	14.4
	2	390	55	14.1
	3	397	46	11.6
	4	393	58	14.7
	5	388	53	13.7
	6	401	40	10.0
All regions (1–6)		2364	309	13.1
125–149 μ	1	356	44	12.4
	2	342	52	15.2
Regions (1–2)		698	96	13.8
	Averaged values	3062	405	13.2

A = area, s = solid, l = liquid, a = air.

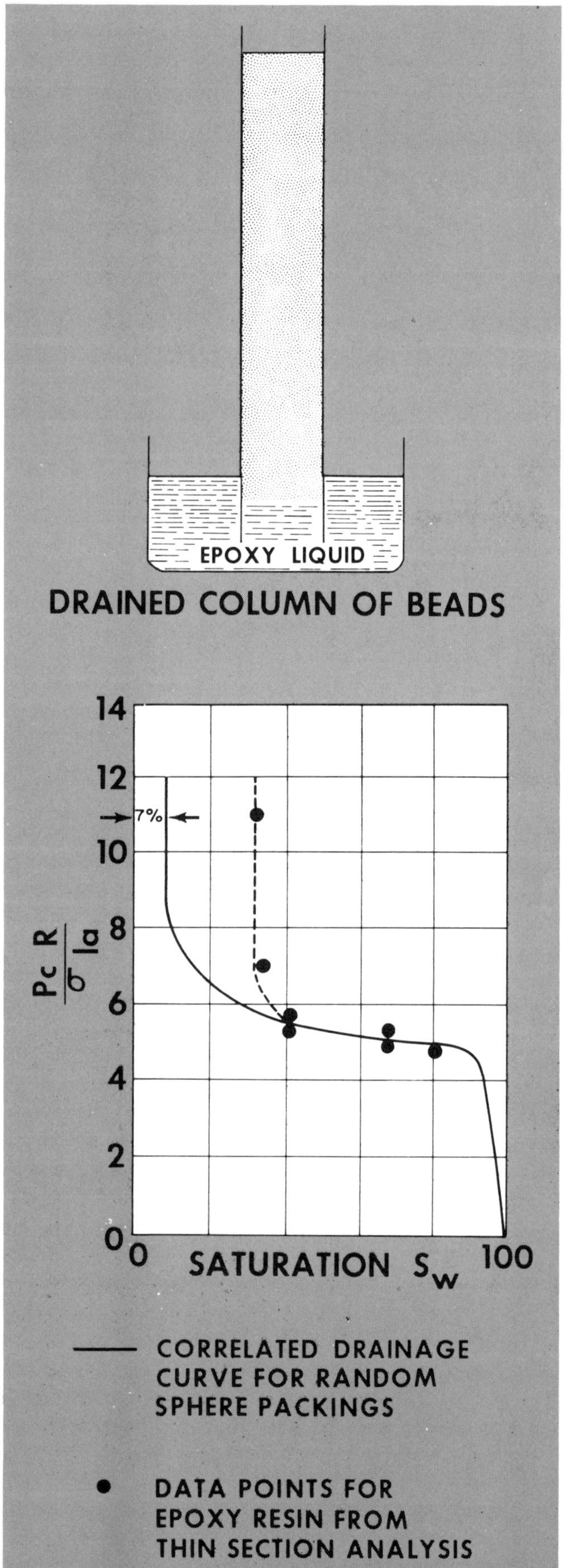

Figure 19. Distribution of solidified epoxy resin in a column of glass beads

Table VI. Percent Area under Capillary Pressure Curves Relative to Initial Drainage Curve

Microbead diameters, μ	Imbibition, A	Secondary drainage, R
250–297	62	92
177–210	63	85
177–210	56	90
Averaged values	60	89

Efficiency of Imbibition

Imbibition corresponds to the conversion of surface free energy of the porous system into external work. The area under the imbibition curve is about 60% of the area under the drainage curve (see Figure 18c). However, after imbibition is complete, the system still has surface free energy due to entrapped nonwetting phase equal to 13% of the solid surface. Adding the 60% for recovered work to the 13% for the stored surface energy, the imbibition efficiency is calculated as 73/79 or 92.5%; this value is considerably higher than the efficiency of initial drainage.

Efficiency of Secondary Drainage

The work required for secondary drainage is about 89% of that required for initial drainage. The initial surface free energy of the system was 13 units. Efficiency of secondary drainage can be defined as the ratio of the final surface free energy (79 units) to the sum of the initial free energy (13 units) and the work done on the system (89 units) to give a value of 79/102 or 77.5%. The presence of 13 units of residual nonwetting phase surface reduces the work of drainage by 11 units.

Displacement Efficiencies at Intermediate Saturation

Inherent thermodynamic efficiencies of displacement so far determined apply to maximum changes in saturation. To determine the efficiency of displacement at intermediate saturations, a column of 30–35 mesh microbeads was saturated with liquid epoxy resin and allowed to drain under gravity, as shown in Figure 19. Beads of this size give a transition zone of about 2 cm depth through which saturation should vary from nearly 100% to the irreducible saturation. After the epoxy had set, the space in the upper part of the column was evacuated and an orange epoxy injected under pressure and allowed to harden. Thin sections were prepared at one-half cm intervals along the length of the column and their saturation and surface areas determined by counting point and line intersections. Figure 20 is an illustration of part of a thin section in which both phases are mainly continuous. Analysis of the thin sections showed that epoxy had set in the upper region of the column before drainage to the irreducible saturation could occur. However, using the surface tension (38 dynes/cm) and density (1.1 g/cm^3) of the liquid epoxy and the average bead radius, R, a plot of $\Delta\rho ghR/\sigma_{la}$ ($= P_cR/\sigma_{la}$) *vs.* saturation approached a plot of P_cR/σ *vs.* saturation given by conventional capillary pressure drainage tests on sphere packings. Figure 19 shows the close agreement obtained for saturations above 35%.

The saturations and surface areas obtained by thin section analysis are summarized in Figure 21. In Figure 21, work of displacement at any saturation is expressed as percentage of the total work of displacement to irreducible saturation.

$$\text{work of displacement} = \frac{\int_{100}^{S_w} P_c dS_w}{\int_{100}^{S_{wi}} P_c dS_w} \cdot 100 \qquad (76)$$

The surface free energy is expressed as percentage of the total solid surface, A_s.

$$\text{surface free energy} = \frac{A_{la} + A_{sa}}{A_s} \cdot 100 \qquad (77)$$

Data points corresponding to irreducible saturation ($S_{wi} = 7\%$) are those obtained for the 3-mm diameter beads. The surface area between the fluid phases, shown in Figure 21, never exceeds about 10% of the solid surface. Since the surface area of residual nonwetting phase was 13%, the interfacial area between the fluids is probably about 10% at all stages of displacement.

Net efficiency, the ratio of increase in free energy to the work done, declines as the saturation decreases, Figure 21. A plot of net efficiency *vs.* saturation is shown in Figure 22.

Thermodynamic efficiency for a small displacement at a saturation S_w is given by

$$(E_D)_{S_w} = \frac{\sigma_{la} d(A_{la} + A_{sa})}{V_B \phi P_c dS_w} \cdot 100 \qquad (78)$$

The variation of $(E_D)_{S_w}$ with saturation is shown in Figure 22 as the stagewise efficiency curve. The inherent efficiency of conversion of work to surface free energy is highly nonlinear, declining from nearly 100% at high saturations to less than 60% near the irreducible saturation.

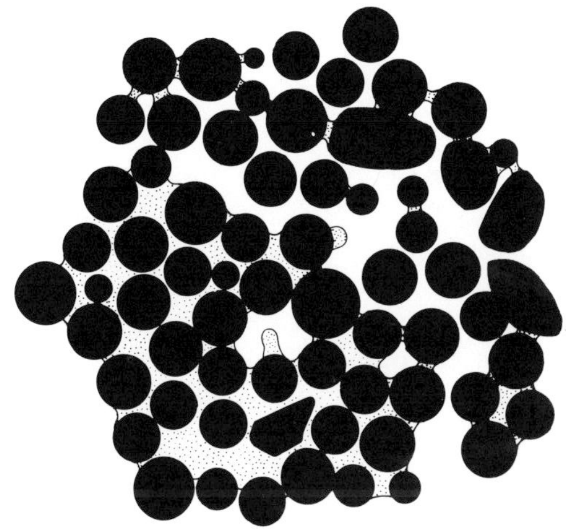

●SOLID ○WETTING PHASE ○NONWETTING PHASE

Figure 20. Distribution of fluid phases in a packing of 30–35 mesh microbeads when both phases are mainly continuous

DISCUSSION

Techniques used in the present study could be extended to many other systems such as porous sedimentary rocks. Details of a solidified liquid technique for preparing thin sections illustrating fluid distribution in porous rock are reported by Le Fournier (*76*). Studies on other porous systems could determine whether the inherent thermodynamic efficiencies of drainage and imbibition found for random sphere packs are general or vary significantly from one rock sample to the next. Determination of changes in surface area of individual phases may find application in studies of both wettability and two-phase flow.

Wettability

Wettability of a porous medium relates to the surface energetics of displacement but is not a well-defined term. Capillary displacement pressures are often used as a measure of wettability; recently an empirical method of defining wettability according to ratios of areas under capillary pressure drainage and imbibition curves was presented by Donaldson, Thomas, and Lorenz (*77*). The present study also suggests that these areas provide a method of determining wettability if the inherent thermodynamic efficiencies of displacement and the surface areas of residual wetting and nonwetting phases are known.

If it could be assumed that the inherent thermodynamic efficiency of imbibition is always relatively high, then the work done by the system during imbibition might give an acceptable estimate of the surface free energy of the system at its irreducible saturation. A theoretical upper limit to the area under an imbibition curve would be largely determined by the product of the fluid surface tension and the contact angle integrated over the total exposed solid surface area. However, calculations of displacement pressures in sphere packings suggest that the inherent efficiency of imbibition will be considerably reduced when the contact angle is greater than zero (*67*). On the other hand, experimental studies of the effect of contact angle change with cubic shaped galena particles indicate that efficiency of imbibition may not change greatly with wettability (*78*).

The above observations point to fundamental difficulties that must be overcome in obtaining relationships among wettability, pore geometry, displacement pressures, work of displacement, and changes in surface free energy.

Fluid Flow—Hydraulic Radii of Individual Phases

The presented detailed mechanism of displacement is based on the assumption that capillary pressure data points are time independent and that capillary pressure curves correspond to infinitely slow displacement. Capillary pressure data are commonly used in solving equations of two-phase flow (*79, 80*). Instances of distinct change in displacement mechanism with flow rate have been reported (*81*). For the present, we will assume that variations in the volume and surface properties shown in Figure 13 remain essentially unchanged.

Solutions to equations of flow also require relative permeability data for the two phases as a function of saturation. A variety of direct methods of measuring relative permeability have been described, mostly in soil science and oil production literature. Understanding of the mechanism of two-phase flow might be enhanced by comparing measured relative permeability to the volume and surface areas of individual flowing phases. Single-phase permeability of clean sands correlates well with hydraulic radius, r_h, given by the ratio of the volume of the pore space to the wetted surface area (*82*, *83*). If ϕ is the porosity and A_s is the surface area of the solid per unit solid volume then,

$$r_h = \frac{\phi}{1 - \phi} \frac{1}{A_s} \tag{79}$$

Darcy's equation for single-phase flow is:

$$\frac{dV}{dt} = \frac{kA'}{\mu} \frac{dP}{dL} \tag{80}$$

where dV/dt is the volumetric flow rate through a cross section, A', under pressure gradient dP/dL, μ is viscosity, and k is permeability.

The Kozeny-Carman equation incorporating hydraulic radius into the Darcy equation is usually written,

$$\frac{dV}{dt} = \frac{A'\phi r_h^2}{K\mu} \frac{dP}{dL} \tag{81}$$

The dimensionless parameter K, known as Kozeny's constant, usually has a value of about 5 for clean sands. If the surface area and porosity of a particle packing are known, permeability can be predicted from

$$k = \frac{\phi r_h^2}{5} \tag{82}$$

The concept of hydraulic radius has been extended to the flowing portions of the individual phases in two-phase flow (*52*).

Expressions for the hydraulic radius of individual phases are included in Figure 13. Only the continuous portion of each phase provides a flow path. The continuous portion of a flowing phase, say oil, is bounded mainly by the solid surface, $(A_{so})_c$. The remaining boundary of the continuous oil is given by the continuous oil-water surface and the oil-water surface of the discontinuous water. Thus, the hydraulic radius for oil, r_o, is given by

$$r_o = \frac{\phi S_{oc}}{(1 - \phi S_{oc})} \frac{1}{[(A_{so})_c + (A_{wo})_c + (A_{\underline{wo}})_d]} \tag{83}$$

Data relating changes in surface area to saturation (Figure 21) permit calculation of r_o as a function of saturation for initial drainage from 100% saturation. Since all of the oil is continuous, S_{oc} is given by the oil saturation, S_o. The sum of the surface areas $(A_{so})_c$, $(A_{wo})_c$, $(A_{\underline{wo}})_d$ relative to the surface area of the solid can be taken directly from Figure 21.

The equation

$$\frac{dV_o}{dt} = \frac{A'S_{oc}\phi r_o^2}{K_o\mu} \frac{dP}{dL} \tag{84}$$

defines a parameter K_o analogous to the Kozeny constant K of Equation 81 for single-phase flow. The value of K is believed to reflect the tortuosity of flow paths within a porous medium. For displacements corresponding to a given capillary pressure curve, K_o probably varies with

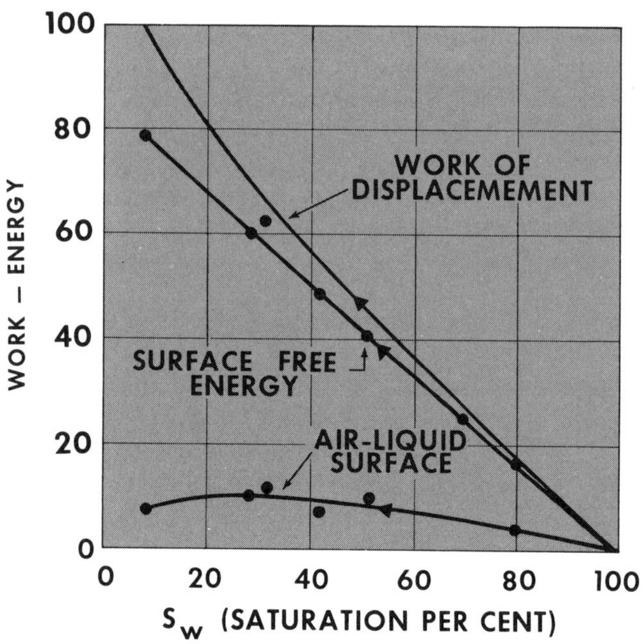

Figure 21. Relationships between work of displacement, surface free energy, and saturation for initial drainage

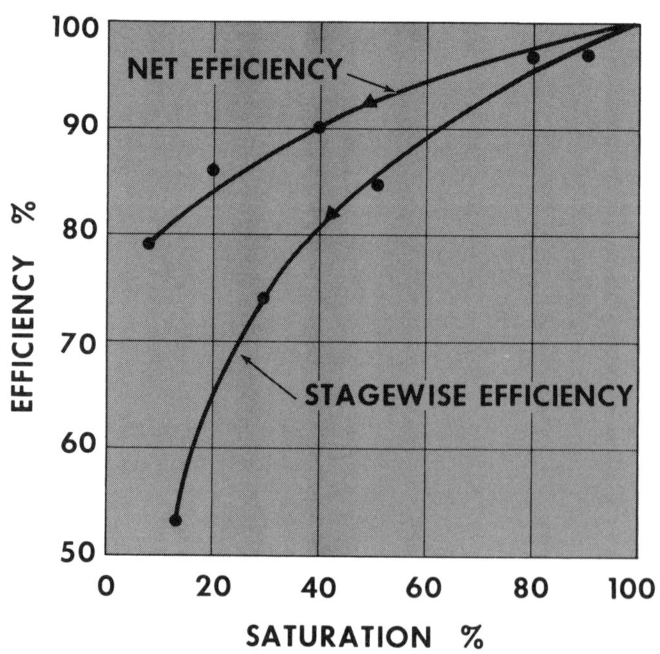

Figure 22. Net efficiency and stagewise efficiency versus saturation for initial drainage

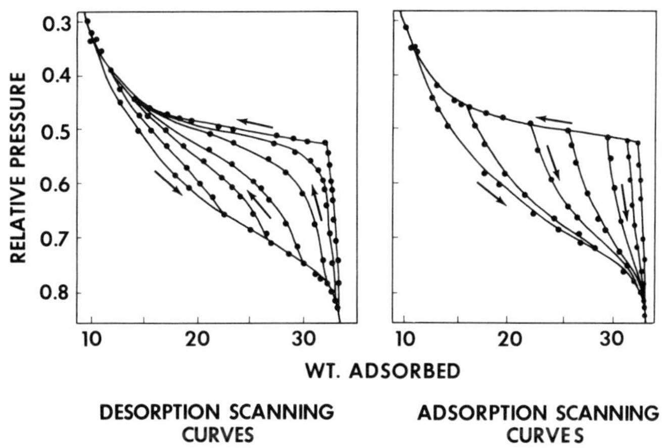

Figure 23. Hysteresis in sorption isotherms for xenon on porous glass at 151°K. (57)

saturation but this variation may similarly reflect changes in the tortuosity of flow paths available for the oil phase.

The value of K_o can be determined by comparing measured values of relative permeability k_{ro} with the hydraulic radius at a given saturation from the relationship

$$K_o = \frac{S_{oc}\phi r_o^2}{k_{ro}} \tag{85}$$

Relative permeability to oil, k_{ro}, is defined by

$$\frac{dV_o}{dt} = \frac{A'k_{ro}}{\mu}\frac{dP}{dL} \tag{86}$$

Equations 84 and 86 are equations of flow for individual phases and are analogous to Equations 81 and 80 for single-phase flow.

If both variations of K_o and displacement efficiencies are reasonably consistent, relative permeability could be predicted from capillary pressure data through Equation 85.

Hysteresis in Adsorption by Capillary Condensation

A number of general theories of hysteresis have been proposed. [See references cited by Everett (*56–59*).] A review of these theories and their relevance to the specific problem of immiscible displacement in porous media is beyond the scope of this paper. At present, no existing theory leads to accurate prediction of capillary pressure data. However, several observations of the present investigation bear importantly on the theory of hysteresis in adsorption by capillary condensation.

Hysteresis in adsorption isotherms, a topic recently reviewed by Everett (*57*), has been ascribed to several causes. Where the pore diameter of a solid adsorbent is large compared with the molecular diameter of the adsorbate, hysteresis has often been explained by capillary models. (When the solid has a high surface area, multilayer surface adsorption may contribute significantly to the total weight of adsorbed material (*84*); present considerations will be restricted to capillary condensate.) Striking similarities exist between the form of adsorption and capillary pressure hysteresis. (Compare Figures 8 and 23.)

Since adsorption isotherms represent a vapor in equilibrium with its liquid, all vapor-liquid curvatures are given by the externally measured vapor pressure. Whereas mass transfer across a phase boundary could be neglected for immiscible displacement, it dominates the mechanism of saturation change in adsorption. Thus, the distinction made between discontinuous and continuous fluid, defining stability conditions for capillary pressure data points, is unnecessary for adsorption. For example, a pendular ring of liquid held between two particles, after immiscible displacement, was assumed constant in volume, but the surface curvature and volume of capillary condensed liquid varies with vapor pressure.

Experimental observations of adsorption by capillary condensation are made in the range of capillary pressures where relative humidity is less than about 95% and vapor pressure varies significantly with surface curvature. Packings of spheres with diameters ranging from 100 to 1000 Å would provide suitable boundaries for the development of high surface curvatures. As a model, assume a sphere packing in which the absolute size of the spheres can be varied while maintaining similarity. If liquid is desorbed from this packing, essentially the same unstable configurations arise as in immiscible displacement, resulting in spontaneous movements of liquid. When vapor is adsorbed, the capillary mechanism differs somewhat from imbibition because isolated regions of fluid grow in volume as vapor pressure is increased, but imbibition and adsorption will still have comparable features. As with immiscible displacement, adsorption isotherms cannot be smoothly continuous on a microscopic scale.

A model proposed by Everett in discussion of macroscopic changes during adsorption (*57*) can also be used to discuss adsorption from a microscopic point of view as a quantized process (Figure 24). The adsorbent and adsorbate are contained in a cylinder, and vapor pressure can be changed by moving a frictionless piston. As in immiscible displacement, movement of the piston is only permitted when the vapor pressure is exactly balanced by the force on the piston. When an unstable configuration arises, the piston does not move until equilibrium is re-established. The force on the piston is then altered to balance the pressure of the vapor, and the piston continues displacement smoothly and reversibly until the next instability occurs. The area under the segmented pressure-volume curve so obtained is the reversible work done on the system. The increase in free energy of the system equals this work less the free energy lost during spontaneous transition. The size of the pressure change accompanying a transition is determined essentially by the factors that determined the pressure change associated with the rheons of immiscible displacement.

The hypothetical nature of this discussion is emphasized by the fact that these pressure changes for adsorption may be smaller than statistical fluctuations. However, as with rheons, the important feature of each

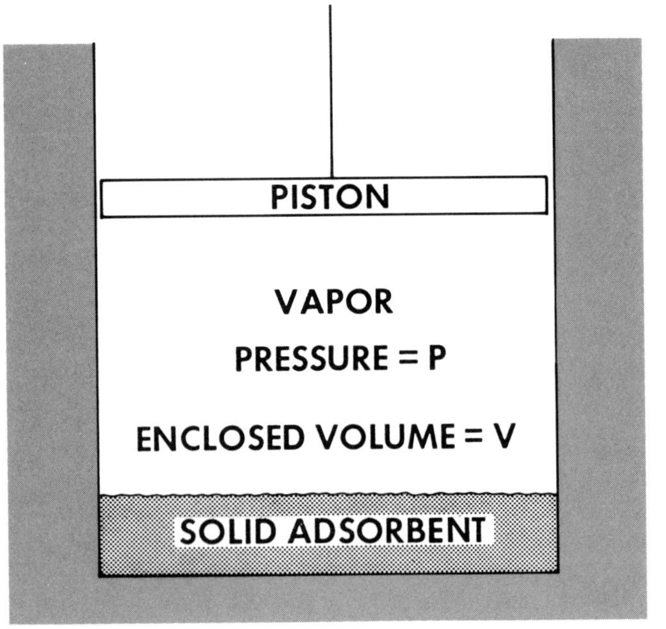

Figure 24. Idealized adsorption apparatus

instability is not the size of the pressure change, but the characteristic loss of surface free energy. This loss has a definite value as the pressure change approaches zero. As with capillary pressure hysteresis, the origin of hysteresis in adsorption by capillary condensation can be ascribed to the losses of surface free energy which accompany spontaneous changes of interface configuration. The simultaneous measurements of adsorption isotherms and surface areas reported by Wade and coworkers should permit calculation of inherent thermodynamic efficiencies of adsorption (*85, 86*).

In seeking the origin of adsorption hysteresis, possible differences in the degree of irreversibility of adsorption and desorption isotherms have been investigated. Kington and Smith (*87*) made a calorimetric study of sorption of argon on porous glass. They concluded that adsorption occurred reversibly whereas desorption was accompanied by a spontaneous process. Although it does not seem likely that any one side of a hysteresis loop can actually correspond to reversibility, the above observation is consistent with the results of the present investigation in which imbibition was characterized by a considerably higher inherent thermodynamic efficiency than drainage.

REFERENCES

(1) Gibbs, J. W., "Scientific Papers," Vol. 1, pp 219–331, Dover, reprint, New York, N. Y., 1961.
(2) Haines, W. B., *J. Agr. Sci.*, **20**, 97 (1930).
(3) Adam, N. K., "The Physics and Chemistry of Surfaces," 3rd ed., Oxford University Press, London, 1941.
(4) Herring, C., "Structure and Properties of Solid Surfaces," (Eds., R. Gomer, and C. S. Smith), University of Chicago Press, 1952.
(5) Shuttleworth, R., *Proc. Phys. Soc., London*, **A, 63**, 444 (1950).
(6) Pethica, B. A., and Pethica, T. J. P., Second International Congress on Surface Activity, Vol. III, Academic Press, New York, N. Y., p 151, 1957.
(7) Johnson, R. E., *J. Phys. Chem.*, **63**, 1655 (1959).
(8) Gray, V. R., Discussion, Second International Congress on Surface Activity, Vol. III, p 192, Academic Press, New York, N. Y., 1957.
(9) Gray, V. R., *S.C.I. Monograph*, No. 25, p 123, London, 1967.
(10) Defay, R., and Prigogine, I., with collaboration of Bellemans, A., translated by D. H. Everett, "Surface Tension and Adsorption," Wiley, New York, N. Y., 1966.
(11) Gibbs, J. W., ref 1, p 219.
(12) Gibbs, J. W., *ibid.*, p 235.
(13) Gibbs, J. W., *ibid.*, p 237.
(14) Defay, R., and Prigogine, I., ref 10, p 61.
(15) Gibbs, J. W., ref 1, p 232.
(16) Tolman, R. C., *J. Chem. Phys.*, **17**, 333 (1969).
(17) Melrose, J. C., Ind. Eng. Chem., **60** (3), 63 (1968).
(18) Defay, R., and Prigogine, I., ref 10, chap 17.
(19) Gibbs, J. W., ref 1, pp 314–15.
(20) Buff, F. P., and Saltsburg, H., *J. Chem. Phys.*, **26**, 23 (1957).
(21) Gibbs, J. W., ref 1, p 326.
(22) Collins, R. E., and Cooke, C. E., *Trans. Faraday Soc.*, **55**, 1602 (1959).
(23) Young, T., *Phil. Trans.*, p 65 (1805).
(24) Lester, G. R., *S.C.I. Monograph*, No. 25, p 123, London, 1967.
(25) Bailey, A. I., Discussion, Second International Congress on Surface Activity, Vol. III, p 189, Academic Press, New York, N. Y., 1957.
(26) Fowkes, F. M., "Surfaces and Interfaces I—Chemical and Physical Characteristics," Syracuse University Press, New York, N. Y., 1967.
(27) Gibbs, J. W., ref 1, p 328.
(28) Melrose, J. C., *J. Colloid Sci.*, **20**, 801 (1965).
(29) Morrow, N. R., Discussion, Symposium on Wetting, *S.C.I. Monograph*, No. 25, p 210, 1967.
(30) Gibbs, J. W., ref 1, p 327.
(31) Fox, H. W., and Zisman, W. A., *J. Colloid Sci.*, **5**, 514 (1950).
(32) Bikerman, J. J., Discussion, Second International Congress on Surface Activity, Vol. III, Academic Press, New York, N. Y., 1957.
(33) Melrose, J. C., *S.C.I. Monograph*, No. 25, pp 123–43, London, 1967.
(34) Harkins, W. D., "The Physical Chemistry of Surface Films," Reinhold, New York, N. Y., 1952.
(35) Henniker, J. C., *Rev. Mod. Phys.*, **21**, 322 (1949).
(36) Low, P. F., *Advan. Agron.*, **13**, 269 (1961).
(37) Deryagin, B. V. Ed., "Research in Surface Forces," Consultants Bureau, New York, N. Y., 1963.
(38) Buff, F. P., *J. Chem. Phys.*, **23**, 419 (1955).
(39) Gibbs, J. W., ref 1, p 330.
(40) Laplace, "Mechanique Celeste," supplement to 10th book, 1806.
(41) Fisher, R. A., *J. Agr. Sci.*, **16**, 492 (1926).
(42) Melrose, J. C., and Wallick, G. C., *J. Phys. Chem.*, **71**, 3676 (1967).
(43) Melrose, J. C., *A.I.Ch.E. J.*, **12**, 986 (1966).
(44) Erle, M. A., Dyson, D. C., and Morrow, N. R., Preprint 26A, 61st Ann. Meeting, A.I.Ch.E., Los Angeles (1968); *A.I.Ch.E. J.* (to be published).
(45) Croney, D., Coleman, J. D., and Bridge, Pamela M., *DSIR Road Research Laboratory Tech. Paper*, **24**, HMSO, London (1952).
(46) Scheidegger, A. E., "The Physics of Flow Through Porous Media," Macmillan Co., New York, N. Y., 1960.
(47) Leverett, M. C., *Trans. AIME*, **142**, 152 (1941).
(48) Calhoun, J. C., Lewis, M., and Newman, R. C., *Trans. AIME*, **186**, 189 (1949).
(49) Brooks, C. S., and Purcell, W. R., *Trans. AIME*, **195**, 289 (1952).
(50) Payne, D., *Nature*, **172**, 261 (1953).
(51) Ringqvist, G., *Proc. Swedish Cement Res. Inst.*, Stockholm, **28** (1955).
(52) Rapoport, L. A., and Leas, W. J., *Trans. AIME*, **192**, 83–98 (1951).
(53) Ksenzhek, O. S., Kalinovskii, E. A., and Tsyganok, O. P., *Russian J. Phys. Chem.*, **38**, No. 11 (1964).
(54) Rootare, H. M., and Prenzlow, C. F., *J. Phys. Chem.*, **71**, 2733 (1967).
(55) Crawford, F. W., and Hoover, G. M., *J. Geophys. Res.*, **71** (12), 2911 (1966).
(56) Everett, D. H., *Trans. Faraday Soc.*, **50**, 1077 (1954); **51**, 1551 (1955).
(57) Everett, D. H., "Adsorption Hysteresis in the Solid-Gas Interface," Marcel Dekker, New York, N. Y., 1967.
(58) Everett, D. H., and Smith, F. W., *Trans. Faraday Soc.*, **50**, 187 (1954).
(59) Everett, D. H., and Whitton, W. I., *ibid.*, **48**, 749 (1952).
(60) Poulovassilis, A., *Soil Sci.*, **93**, 405 (1962).
(61) Philip, J. R., *J. Geophys. Res.*, **69** (8), (1964).
(62) Topp, G. C., and Miller, E. E., *Soil Sci. Soc. Amer. Proc.*, **30**, 156 (1966).
(63) Enderby, J. A., *Trans. Faraday Soc.*, **51**, 835 (1955); **52**, 106 (1956).
(64) Morrow, N. R., and Harris, C. C., *Soc. Petrol. Eng. J.*, **5**, 15 (1965).
(65) Smith, W. O., *Physics*, **4**, 425 (1933).
(66) Weast, R. C., Ed., "Handbook of Chemistry and Physics," The Chemical Rubber Co., Cleveland, 1964.
(67) Melrose, J. C., *Soc. Petrol. Eng. J.*, **5**, 259 (1965).
(68) Miller, E. E., and Miller, R. D., *J. Appl. Phys.*, **27**, 324 (1956).
(69) Heller, J. P., *Soil Sci. Soc. Amer. Proc.*, **32**, 778 (1968).
(70) Gibbs, J. W., ref 1, pp 242–52.
(71) Harris, C. C., and Morrow, N. R., *Nature*, **203**, 706 (1964).
(72) Raimondi, P., and Torcaso, M. A., *Soc. Petrol. Eng. J.*, **4**, 49 (1964).
(73) Pearce, K. W., and Donald, M. B., *Chem. Eng. Sci.*, **10**, 212 (1959).
(74) Orr, Clyde, Jr., and Dallavalle, J. M., "Fine Particle Measurement," Macmillan Co., New York, 1959.
(75) Eisenklam, Paul, "Chemical Engineering Practice," Vol. 2, Solid State, Butterworths, London, 1956.
(76) Le Fournier, M. J., *Bull. Ass. Fr. Tech. Petrole*, No. 175, New Ser., **17** (1966).
(77) Donaldson, E. C., Thomas, R. D., and Lorenz, P. B., *Soc. Petrol. Eng. J.*, **9**, 13 (1969).
(78) Harris, C. C., Jowett, A., and Morrow, N. R., *Trans. I.M.M.*, **73**, 335 (1964).
(79) Muskat, Morris, "Physical Principles of Oil Production," McGraw-Hill, New York, N. Y., 1949.
(80) Collins, R. E., "Flow of Fluid Through Porous Materials," Reinhold, New York, N. Y., 1961.
(81) Handy, L. L., and Datta, P., *Soc. Petrol. Eng. J.*, **5**, 261 (1965).
(82) Carman, P. C., *Trans. Inst. Chem. Eng., (London)*, **23**, 150 (1937).
(83) Coulson, J. M., and Richardson, J. F., "Chemical Engineering," Vol. 2, Pergamon, London, 1955.
(84) Venable, R., and Wade, W. H., *J. Phys. Chem.*, **69**, 1395 (1965).
(85) Wade, W. H., *ibid.*, **68**, 1029 (1964).
(86) Ferguson, C. B., and Wade, W. H., *J. Colloid Sci.*, **24**, 366 (1967).
(87) Kington, G. L., and Smith, F. S., *Trans. Faraday Soc.*, **60**, 705 (1964).

7

Flow of Suspensions Through Porous Media—Application to Deep Filtration

Flow of Suspensions through Porous Media—

New differential equation for clogged beds is derived

The flow of suspensions through porous media is a very complex phenomenon owing to the diversity of the mechanisms involved. However, the study of these mechanisms shows that it is possible to define two types of deep filtration—a mechanical filtration for large particles (diameter $\geqslant 30\ \mu$), and a physicochemical filtration for small particles (diameter about 1 μ)—for mean particles ($3\ \mu <$ diameter $< 30\ \mu$), both mechanical and physicochemical phenomena intervene. The clogging (colmatage) of a porous medium is described mathematically by the mass balance equation and the kinetic equations describing the rate of clogging and the rate of decolmatage. Generally, the mass balance equation simplifies since diffusion of particles is not important nor are suspended particles with respect to retained ones. Usually, the kinetic equation is assumed to be of first order and decolmatage of previously retained particles is considered to be negligible as is usually found experimentally. With these assumptions, a new differential equation for clogged beds is derived. It is valid for all relationships between the retention probabilities and the retention. From this equation, general macroscopic properties are deduced.

However, it is difficult accurately to connect the experimentally determined macroscopic parameters with the elementary mechanisms of clogging—retention sites, retention forces, and rate of elementary capture processes. Likewise, it is difficult to express the variation in the pressure drop of the fluid through the clogged bed as a function of the retention—A number of improvements of the Kozeny model have been tried, with limited success.

PROBLEM STATEMENT

Liquid-solid separation by filtration can be accomplished either by solid accumulation in front of a filter medium, or by retention inside a deep porous bed. In the first case, a filter cake is built-up and it stops the suspended particles. In the second case, the suspension flows through the porous medium in which the particles are retained. The flow of suspensions through porous media is quite a natural phenomenon since it often occurs when water seeps into geological masses. The use of deep filtration in the treatment of polluted water is not new, but this phenomenon has only been studied recently.

When flowing through porous medium, the particles are brought into contact with the possible retention sites; they stop there or are carried away by the stream. The deep filtration is therefore the result of several mechanisms:

the contacting of particles with the retention sites
the fixing of particles on sites

and eventually,

the breaking away of previously retained particles

Thus, the problem to be solved consists in relating the clogging rate—*i.e.*, the number of retained particles per unit of time and per unit of volume of porous medium—to the various factors which define the system, namely

the carrier fluid (flowrate, viscosity, density)
the suspension of particles (concentration, size, and shape of particles)

J. P. HERZIG
D. M. LECLERC,
P. LE GOFF

Application to Deep Filtration

the porous medium filter (porosity, diameter of pores, size and shape of grains, retention—*i.e.*, volume fraction already clogged)

Furthermore, the important problem of the pressure drop of the fluid through the bed has to be studied as well as its change with the clogging. To approach this problem rationally, the authors will first establish the mass balance equations for a medium in the process of clogging. With reasonable assumptions, these equations are simplified to allow convenient comparison with experiments.

In a second part, the elementary mechanisms which cause the retention of particles in a porous medium are identified. From the basic laws of particles mechanics and the physical chemistry of interfaces, the authors derive the general form of the equations governing the clogging "declogging" phenomena.

The third part combines these equations and the mass balance equations to describe the general properties of these systems. The conditions in which the complete integration of these differential equations is possible are also discussed.

The fourth part examines the pressure drop and its variation with clogging.

MASS BALANCE EQUATIONS

Exact Equations

Mass balance equations for the particles and the carrier fluid.

Particles balance. Consider a porous medium element of depth Δz and area Ω (Figure 1), which contains:

a volume $\Omega\sigma\Delta z$ of retained particles
a volume $\Omega\epsilon y\Delta z$ of moving particles entrained by the liquid

where

σ is the retention—*i.e.*, the volume of deposited particles per unit of filter volume
y is the volume fraction of particles in the suspension—*i.e.*, the volume of particles per unit of suspension volume
ϵ is the porosity of the bed—*i.e.*, the volume occupied by the flowing suspension per unit of filter volume

But the retained particles imprison between themselves some liquid called "dead liquid" and form a deposit of a real volume $\Omega\beta\sigma\Delta z$. β is the inverse of the

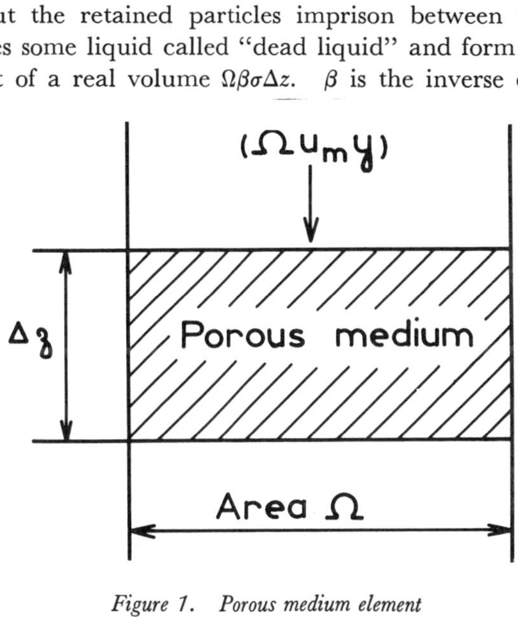

Figure 1. Porous medium element

compaction factor of the retained matter. The porosity of the clogged bed can be expressed by the relation

$$\epsilon = \epsilon_o - \beta\sigma \qquad (1)$$

where ϵ_o is the porosity of the clean bed.

The general mass balance equation is

$$\text{accumulation rate} = (\text{flow in}) - (\text{flow out})$$

or

$$\frac{\partial}{\partial t}[\Omega(\sigma + \epsilon y)] + \frac{\partial}{\partial z}\left[\Omega u_m y - \Omega D \frac{\partial y}{\partial z}\right] = 0$$

where

u_m = approach velocity of the suspension

$u_m y$ = barycentric flow of particles entrained by the fluid

$-D\frac{\partial y}{\partial z}$ = diffusional flux of particles

Usually the suspension flow rate is kept constant in deep filtration and the exact form of the mass balance for the particles is

$$\frac{\partial}{\partial t}(\sigma + \epsilon y) + u_m \frac{\partial y}{\partial z} - D\frac{\partial^2 y}{\partial z^2} = 0 \qquad (2)$$

Liquid balance equation. The same porous medium element contains a volume $\Omega(\beta - 1)\sigma\Delta z$ of dead liquid and a volume $\Omega\epsilon(1 - y)\Delta z$ of flowing liquid.

Thus, the liquid balance equation is

$$\frac{\partial}{\partial t}[\Omega\epsilon(1 - y) + \Omega(\beta - 1)\sigma] + \frac{\partial}{\partial z}\left[\Omega u_m(1 - y) + \Omega D\frac{\partial y}{\partial z}\right] = 0$$

The term $u_m(1 - y)$ describes the barycentric flow of liquid, and $+D(\partial y/\partial z)$ represents the diffusional flux opposed to that of the particles, since the porous bed is always filled with suspension.

If the flow rate is kept constant

$$\frac{\partial}{\partial t}\epsilon(1 - y) + \frac{\partial}{\partial t}(\beta - 1)\sigma + u_m\frac{\partial}{\partial z}(1 - y) + D\frac{\partial^2 y}{\partial z^2} = 0 \qquad (3)$$

which is the exact balance equation for the liquid.

With Equation 1 between ϵ and ϵ_o taken into account, Equation 3 becomes

$$\frac{\partial \epsilon_o}{\partial t} - \frac{\partial}{\partial t}(\sigma + \epsilon y) - u_m\frac{\partial y}{\partial z} + D\frac{\partial^2 y}{\partial z^2} = 0$$

or since ϵ_o is a constant

$$\frac{\partial}{\partial t}(\sigma + \epsilon y) + u_m\frac{\partial y}{\partial z} - D\frac{\partial^2 y}{\partial z^2} = 0$$

which is identical with particles balance Equation 2. This is not surprising because the porous medium is fully filled with suspension, and every time a particle stops, it drives away a volume of carrier liquid equal to its own volume, and consequently, it is normal that the balance equations are the same for the particles and the liquid.

The mass balance in the porous bed is characterized completely by Equation 2.

Approximate Equations

Three simplification modes for exact Equation 2 are considered.

Third approximation. Particles diffusion is negligible when their size is larger than 1 μ; even for ultrafine particles, the barycentric flow is always the most significant; thus, the diffusional term may be neglected. This case is denoted "third approximation" which is characterized by Equation 4:

$$\frac{\partial(\sigma + \epsilon y)}{\partial t} + u_m\frac{\partial y}{\partial z} = 0 \qquad (4)$$

This assumption is usually made; only a few investigations take into consideration particles diffusion (*38, 45, 44, 48*). Even with this simplification, the balance equation remains complicated and is little used under this form (*22, 63*) because it is usually further simplified by additional assumptions.

Second approximation. The term $\beta\sigma$ is often neglected with respect to ϵ_o in the balance equation (*3, 48, 71, 72*) and with $\epsilon_o \simeq \epsilon$, the balance equation becomes

$$\frac{\partial \sigma}{\partial t} + \epsilon_o\frac{\partial y}{\partial t} + u_m\frac{\partial y}{\partial z} = 0 \qquad (5)$$

This simplification, called the "second approximation," consists of neglecting $\beta\sigma y$ with respect to $\sigma + \epsilon_o y$. It is justified only when clogging starts, when σ is small and ϵ close to ϵ_o, but is always valid for deep filtration since concentrations are low. The error is

$$\frac{\beta\sigma y}{\sigma + \epsilon_o y} = \frac{\beta y}{1 + \epsilon_o y/\sigma}$$

and always remains lower than βy for all values of σ. Generally, concentrations are about 0.1% and β may be estimated at 2 or 3; thus, the maximum error is 0.3% and therefore negligible.

First approximation. If, as it is often assumed, all moving particles are neglected with respect to the retained particles (*22, 24, 28, 35, 54, 57, 70*), the balance equation is

$$\frac{\partial \sigma}{\partial t} + u_m\frac{\partial y}{\partial z} = 0 \qquad (6)$$

This simplification, called the "first approximation," is only reasonable if ϵy is small compared with σ. Since

---

AUTHORS *P. Le Goff is the Director of the Centre de Cinétique Physique et Chimique, D. M. Leclerc, and J. P. Herzig, are researchers at the Département de Génie Chimique at the University of Nancy in France.* *J. P. Herzig acknowledges the generosity of the French Commissariat ă l'Energie Atomique which has provided him a fellowship for study in Nancy.*

ϵy is always less than $\epsilon_o y_i$ (y_i is the concentration of the suspension applied to the filter) the error is at most $\epsilon_o y_i/\sigma$; thus the simplification is best for large values of σ.

For example, if $\epsilon_o = 0.40$ and $y_i = 0.1\%$ the error is 8% at most for $\sigma = 0.50\%$ and only 1% for $\sigma = 4.0\%$.

Comparison of first and second approximations. The second approximation is justified for the duration of the filtration, whereas the first is only valid for σ sufficiently large and is invalid when clogging starts. Consequently, it seems logical to use the first as soon as the required conditions are satisfied (*i.e.*, when σ is large enough) and the second for the beginning of clogging. These two approximations are complementary, and it is even possible to group them under the same equation.

Retention age. Generally, before the suspension enters the porous bed, the latter is filled with clear liquid which is driven away by the suspension. If the bed is deep, the suspension needs a substantial time to pass through it; accordingly, the instant when the suspension reaches a bed, element is not the same all along the porous medium. Thus, it is necessary to compare the bed elements at the same "retention age"—*i.e.*, at different instants t such as $t - \epsilon_o z/u_m$ = constant.

The term $\epsilon_o z/u_m$ is the required time so the suspension arrives at the bed depth, z.

The "retention age," τ, is defined by

$$\tau = t - \frac{\epsilon_o z}{u_m} \tag{7}$$

The term $\epsilon_o z/u_m$ is substantial when clogging starts—*e.g.*, if $u_m = 0.1$ cm/sec and $\epsilon_o = 0.40$, the suspension needs 200 sec to reach the depth of 50 cm; but as deep filtration operations normally last several hours this difference becomes negligible fairly quickly and for most of the filtration, t and τ may be identified.

To follow a suspension element when it passes through the clean bed, it is necessary to keep τ = constant. This is not exact for a clogged bed because the porosity and therefore, the suspension velocity are changed. At constant flow rate, it is necessary to choose instants t such as

$$t - \int_0^z \frac{\epsilon z}{u_m}\, dz = \text{constant}$$

or

$$\tau + \int_0^z \frac{\beta\sigma}{u_m}\, dz = \text{constant}$$

Now the term

$$\int_0^z \frac{\beta\sigma}{u_m}\, dz$$

is a correction term that occurs only when σ is large enough—*i.e.*, when the filtration has lasted for some time. The difference between t and τ is now negligible; consequently, this correction is unnecessary and even for a clogged bed, a suspension element passes through the bed at τ = constant.

Transformation of balance Equation 5. Substituting $t = \tau + \epsilon_o z/u_m$ for t in balance Equation 5 of the second approximation yields Equation 5′

$$\frac{\partial\sigma}{\partial\tau} + u_m \frac{\partial y}{\partial z} = 0 \tag{5'}$$

Equations 5′ and 6 have an identical mathematical form but in the former, the variables are τ and z and in Equation 6, t and z. This may be explained as follows: writing the balance equation with the variable τ consists of following a suspension element that passes through the bed; for this element, the decrease in concentration is equal to the increase in bed retention (Equation 5′). On the other hand, writing the balance equation with the variable t consists in taking a fixed-bed element; the suspension which passes through it may have a concentration changing with the time (term $\partial\epsilon y/\partial t$ in Equation 2). In the first approximation, these changes in concentration with the time are neglected, and the filtration is treated as if the decrease in concentration was only due to clogging. Thus, it is expected that Equations 5′ and 6 have the same mathematical form.

CLOGGING MECHANISMS—THE KINETIC EQUATION

Although the mass balance equation is independent of the clogging mechanisms, the kinetic equation, which describes the rate of the transfer of particles to the porous medium, is a function of the elementary processes of this transfer. Therefore, it cannot be obtained in a strictly theoretical way as the mass balance equation and must be inferred from experiments.

Retention probabilities definition. Consider again the porous bed element of area Ω and depth Δz (Figure 1); during the small time interval Δt, the retention $\Omega\sigma\Delta z$ increases to $\Omega(\sigma + (\partial\sigma/\partial t)\,\Delta t)\Delta z$ and a volume $\Omega u_m y \Delta t$ of suspended particles enters. The retention probability q of a particle in this element will be defined by

$$q = \frac{\Omega\Delta z \dfrac{\partial\sigma}{\partial t}\Delta t}{\Omega u_m y\, \Delta t} = \frac{\Delta z}{u_m y}\frac{\partial\sigma}{\partial t}$$

or by dividing by Δz the retention probability per unit of depth, k, is obtained:

$$k = \frac{1}{u_m y}\frac{\partial\sigma}{\partial t} \tag{8}$$

It is also possible to define a retention probability, k', of a particle per unit time: if a particle has a probability k to be retained in the unit bed depth, then it has the probability k to be retained during the time interval required to pass through a layer of unit depth; the average time interval for passage is ϵ/u_m. Thus

$$k' = k\frac{u_m}{\epsilon}$$

$$k' = \frac{1}{\epsilon y}\frac{\partial\sigma}{\partial t} \tag{9}$$

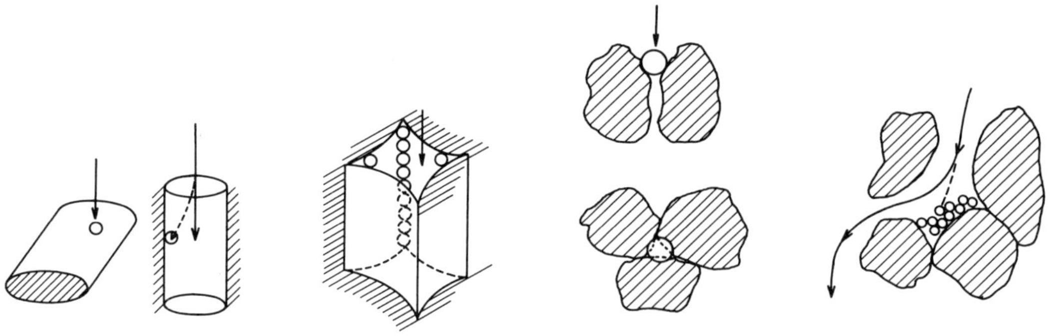

Figure 2. Retention sites

Such a definition of retention probabilities does not distinguish the two directions of the particles exchange:

suspension → porous medium (colmatage or clogging or capture)

porous medium → suspension (decolmatage or scouring)

Definition of the colmatage (capture) and decolmatage (de-clogging) probabilities. If in the porous bed element colmatage and decolmatage are simultaneous, the variation in the retention per unit time $\Omega\Delta z(\partial\sigma/\partial t)$ can be considered as the sum of the increase due to the capture of particles, indicated by $\Omega\Delta z(\partial\sigma/\partial t)_c$ and of the decrease owing to the decolmatage, indicated by $\Omega\Delta z(\partial\sigma/\partial t)_r$.

Thus, the probability of capture of a particle per unit time will be defined by

$$k_c' = \frac{1}{\epsilon y}\left(\frac{\partial\sigma}{\partial t}\right)_c \tag{10}$$

and the probability of decolmatage of a particle per unit time will be

$$k_r' = -\frac{\Omega\Delta z\left(\frac{\partial\sigma}{\partial t}\right)_r}{\Omega\sigma\Delta z}$$

or

$$k_r' = -\frac{1}{\sigma}\left(\frac{\partial\sigma}{\partial t}\right)_r \tag{11}$$

Under these conditions, the retention probability per unit time k' will be written

$$k' = k_c' - \frac{k_r'\sigma}{\epsilon y} \tag{12}$$

Description of Elementary Mechanisms

Consider the inventory of all parameters which describe the elementary processes of clogging and decolmatage—*i.e.*, the retention sites, the retention forces exerted on the particles retained in these sites, the capture mechanisms which bring the particles into contact with the sites, as well as the decolmatage processes of retained particles.

Retention sites. It is possible to distinguish several retention sites (Figure 2):

Surface sites. The particle stops and is retained on the surface of a porous bed grain.

Crevice sites. The particle becomes wedged between the two convex surfaces of two grains.

Constriction sites. The particle cannot penetrate into a pore of a smaller size than its own.

Cavern sites. The particle is retained in a sheltered area, a small pocket formed by several grains.

Retention forces. They include:

Axial pressure of the fluid. The fluid pressure may hold an immobilized particle against the opening at a constriction.

Friction forces. A particle wedged in a crevice may have been slightly deformed when stopped, and may remain in place by friction.

Surface forces. These include the Van der Waals forces, which are always attractive, and the electrical forces (electrostatic or electrokinetic) which are either attractive or repulsive according to the physicochemical conditions of the suspension.

Chemical forces. In the case of colloidal particles or in other particular cases, actual chemical bonding may occur.

Capture processes. These are:

Sedimentation. If the particles have a density different from that of the liquid, they are subjected to gravity and their velocity no longer is that of the fluid; thus, by sedimentation they can meet the filter medium.

Inertia. Still owing to their apparent weight, the particles cannot follow the same trajectories as the fluid, they deviate from the streamlines (when the directions of the trajectories change suddenly) and can be brought into contact with the bed grains.

Hydrodynamic effects. Owing to the nonuniform shear field and the nonsphericity of particles, hydrodynamic effects may occur; these effects cause a lateral migration of suspended particles which may be brought into contact in this way with retention sites.

Direct interception. Even with exactly the same density as the fluid, the particles would not be able, owing to their size, to follow the smallest tortuosities of the streamlines of the carrier fluid and they will thus collide with the walls of the convergent areas of the pores.

Diffusion by Brownian motion. The particles diffuse and can reach areas which are not normally irrigated by the suspension, and they are retained there.

Decolmatage processes. It is necessary to distinguish between the spontaneous decolmatage due to the normal flow of suspension through the clogged bed, and the decolmatage caused by the operator who suddenly changes the flow conditions.

Table I. Ratio of Inertial to Gravity Forces ($\epsilon = 0.40$)

u_m:d_g Cm/sec	100 μ	500 μ	1000 μ	1500 μ
0.001	1.3×10^{-6}	2.6×10^{-7}	1.3×10^{-7}	0.85×10^{-7}
0.010	1.3×10^{-4}	2.6×10^{-5}	1.3×10^{-5}	0.85×10^{-5}
0.100	1.3×10^{-2}	2.6×10^{-3}	1.3×10^{-3}	0.85×10^{-3}
1.0	1.3	0.26	0.13	0.085

Spontaneous decolmatage may occur if local variations in pressure or flow rate change the flow in the neighborhood of retained particles or if a moving particle collides with a retained particle.

Provoked decolmatage results from impulses—*i.e.*, from sudden variations in pressure or flow rate in the whole bed caused by the operator or by reversal of the flow direction.

The processes of spontaneous or provoked decolmatage are similar but the extent of the first is local, whereas the second occurs everywhere in the whole bed.

Factors acting upon clogging kinetics. Consider the experimental parameters which are easily varied and which have an effect upon clogging kinetics:

The carrier fluid is characterized by its dynamic viscosity η_l, its specific mass ρ_l, and its flow rate.

The suspension is defined by the concentration y of particles which are assumed spherical (of mean diameter d), their specific mass is ρ_S.

The filter porous medium is characterized at the initial instant (designed by the index $_o$) by its permeability B_o (Darcy's Law) and its porosity ϵ_o. The bed is assumed to be constructed with spherical grains characterized by a mean diameter, d_g. As fast as the bed clogs, the parameters B, ϵ, and k vary with the retention σ.

The nature of the fluid motion will be characterized by the carrier fluid Reynolds number Re; for flow-through granular porous media the Reynolds number is often written

$$\mathrm{Re} = \frac{u_m d_g \rho_l}{6(1 - \epsilon_o)\eta_l}$$

The physicochemical properties are characterized by Φ. In the general case, it can therefore be written

$$k = \text{function}\ (\eta_l, \rho_l, u_m, y, \rho_S, d, d_g, B_o, \epsilon_o, \Phi, \sigma, \mathrm{Re})$$

This general function simplifies to different forms depending on the deep filtration types which are considered later.

General kinetic equations of deep filtration. *Effect of the suspension concentration.* It is usually assumed that the retention probabilities, k and k', or other related parameters (see Equation 24′), such as the filtration coefficient λ, are independent of the concentration y of the particles in suspension. This hypothesis, confirmed by experiments, assumes that the particles do not interact. Thus, the process is a kinetic one analogous to a first-order reaction in chemical kinetics. This hypothesis seems the more reasonable as the filtered suspensions are always dilute (y is typically of the order of 0.1%).

Relative significance of gravity and inertial forces. In a bed of grains of diameter d_g, the radius of trajectories is of the order of $d_g/2$ and the centrifugal inertial force which moves the particles away from the streamlines is of the order

$$\pi \frac{d^3}{3} (\rho_S - \rho_l) \frac{u^2}{d_g}$$

where u is the mean interstitial fluid velocity and is equal to u_m/ϵ.

But particles are also subjected to the gravity force, $\pi d^3(\rho_S - \rho_l)g/6$, where g is the acceleration of gravity; therefore, the ratio of the inertial force to the gravity force is

$$\frac{2}{gd_g} \left[\frac{u_m}{\epsilon}\right]^2$$

In Table I, this ratio is calculated for several values of d_g and u_m with $\epsilon = 0.40$.

Generally, deep filtration is practiced with d_g near 500 or 1000 μ and with a laminar flow (u_m of the order of 0.1 cm/sec), therefore, according to Table I, inertial forces are perfectly negligible with regard to gravity forces, as is often assumed (*14, 20, 22, 28, 53, 60, 69*).

Relative significance of sedimentation with respect to bulk flow. The particles subjected to gravity are moving with respect to the carrier fluid with the sedimentation velocity v_S, this velocity can be calculated by setting the gravity force equal to the fluid drag (Stokes Law).

$$v_S = \frac{d^2}{18\eta_l} (\rho_S - \rho_l)g \qquad (13)$$

To compare the sedimentation velocity with the mean interstitial velocity of the fluid, the minimum flow rate u_m such that $v_S/u \leqslant 0.01$ is calculated for water (η_l = 0.01 P) for several values of d, and ($\rho_S - \rho_l$), again assuming $\epsilon = 0.40$ (Table II).

The values in Table II show that for typical deep filtration conditions, sedimentation is negligible only for small particles (approx. 1 μ) regardless of their density. For larger particles ($d < 25$ μ), sedimentation is negligible only if ($\rho_S - \rho_l$) is small and for particles larger

Table II. Minimum Flow Rate u_m (in Cm/Sec) Such That $u/v_S \geqslant 100$ ($\epsilon = 0.40$, $\eta_l = 0.01$ P)

$\rho_S - \rho_l$: d (G/cm³)	1 μ	10 μ	25 μ	50 μ	75 μ	100 μ
0.5	0.0011	0.11	0.68	2.72	6.15	10.9
1	0.0022	0.22	1.36	5.44	12.3	21.8
2	0.0044	0.44	2.72	10.9	24.5	43.6
4	0.0087	0.87	5.44	21.8	49.0	87.2
6	0.013	1.31	8.16	32.6	73.6	130.8

Table III. Ratio (in Per Cent) of Diffusional Mean Path of Particles to Mean Path Due to Bulk Flow

(d_g = 1000 μ, ϵ = 0.40)

u_m: d (Cm/s)	0.01 μ	0.10 μ	1.0 μ
0.001	6.0%	1.9%	0.60%
0.010	1.9%	0.60%	0.19%
0.100	0.60%	0.19%	0.06%
1.00	0.19%	0.06%	0.02%

than 25 μ, sedimentation is always appreciable. Hall (*20*) has suggested that the effect of sedimentation upon λ_o can be written by

$$\lambda_o \propto \frac{v_S}{d_g u_m}$$

Relative significance of Brownian motion with respect to bulk flow. In Brownian motion, characterized by a diffusivity D_B, the quadratic mean $\bar{x}^2$ of the particle path in a time interval Δt is given by the classical law

$$\bar{x}^2 = 2D_B \Delta t \tag{14}$$

If it is assumed that this time interval is equal to the residence time of a particle in an elementary layer of bed depth equal to the grains diameter, the particle entrained by the bulk flow will have moved forward of a length $d_g = u\Delta t$.

The ratio, $\bar{x}/d_g$, which characterizes the diffusion significance with respect to bulk flow is

$$\frac{\bar{x}}{d_g} = \frac{\sqrt{2D_B\Delta t}}{u\Delta t} = \sqrt{\frac{2D_B}{ud_g}} \tag{15}$$

Note that D_B/ud_g is the inverse of a Peclet number.

D_B can be estimated from the Stokes-Einstein relation (*17, 28*).

$$D_B = \frac{Ⓗ h_B}{3\pi\eta_l d} \tag{16}$$

where

h_B = Boltzmann constant
Ⓗ = absolute temperature

By substituting Equation 16 into Equation 15 we get

$$\frac{\bar{x}}{d_g} = \sqrt{\frac{2Ⓗ\epsilon h_B}{3\pi\eta_l d u_m d_g}} \tag{17}$$

In Table III, some values of this ratio are given for water at 20°C for d_g = 1000 μ and ϵ = 0.40.

This table shows clearly that Brownian motion always is insignificant with respect to bulk flow and is negligible in normal deep filtration conditions. By comparison, in aerosol filtration the efficiency is proportional to $[D_B/(u_m d_g)]^{2/3}$ when diffusion is the main mechanism (*17, 73*); with deep filtration notations, this result would be written

$$\lambda_o \propto \frac{1}{d_g^{5/3}(u_m \eta_l d)^{2/3}}$$

In the three preceding paragraphs, it has been assumed that the fluid velocity was the same everywhere in the porous medium and equal to the mean value $u = u_m/\epsilon$. However, in such a complex medium as a granular bed, there are areas where the fluid velocity is larger than the mean velocity and others where the velocity is smaller. Besides in the close proximity of grains, the fluid velocities diminish rapidly due to the laminar flow velocities distribution. Therefore, it is probable that the preceding conclusions are not valid everywhere in the porous medium; but as more accurate information is not available, the comparisons may only be made with mean values.

Significance of direct interception. Even if particles have the same density as fluid, they meet the filter medium when the streamlines they follow become nearer than $d/2$ from the grains surface; this process occurs in constrictions and in flow-past obstacles (Figure 3). The particles brought into contact with the filter medium will stop if there are some available retention sites.

Stein (*70*) has shown that the fraction of suspended particles which contact the wall is proportional to $(d/r_c)^2$ (here r_c is the constriction radius), if a parabolic velocity distribution is assumed. As r_c depends on d_g, it may be shown that $\lambda_o \propto (1/d_g)(d/d_g)^2$. Stein has demonstrated the same result for a flow around a sphere of diameter d_g.

In aerosol filtration, the direct interception of particles by a cylinder fiber of diameter d_g may be expressed by an analogous relation (*17, 73*)

$$\lambda_o \propto \frac{1}{d_g}\left(\frac{d}{d_g}\right)^2$$

Stein's assumption (parabolic velocity distribution) is probably not exact since bed pores are short with respect to their diameter. If a flat velocity profile is assumed, this relation becomes

$$\lambda_o \propto \frac{1}{d_g}\left(\frac{d}{d_g}\right)$$

Straining in Constriction. Consider a triangular constriction formed by three tangent spherical grains (Figure 4), a diameter limit exists above which no particle can penetrate the constriction. This limit is such that $d/d_g = 0.154$. It is also possible to imagine straining by several particles, for three particles the limit is $d/d_g = 0.10$ and for four particles $d/d_g = 0.082$.

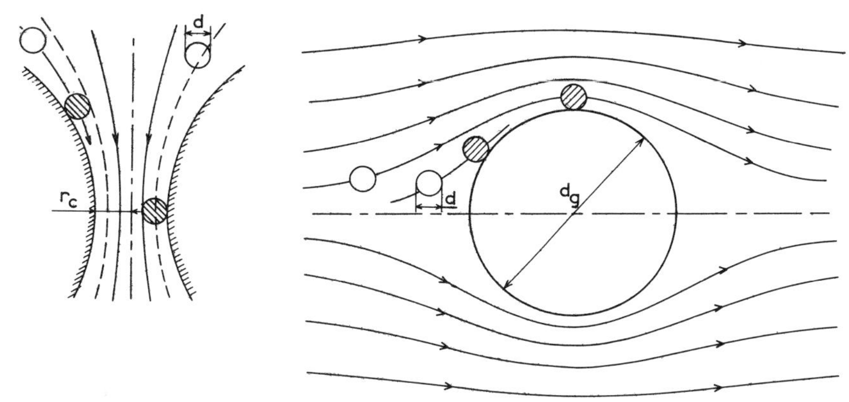

Figure 3. Direct interception

Straining in such a constriction is not necessarily the result of the simultaneous arrival of particles but may occur by successive stoppages of particles in each crevice. As the constriction is blocked, all particles entrained by the fluid are retained and fill up the upper cavern.

For the value of $d/d_g < 0.082$, it seems unlikely that constrictions are blocked, but particles may always become wedged in crevices.

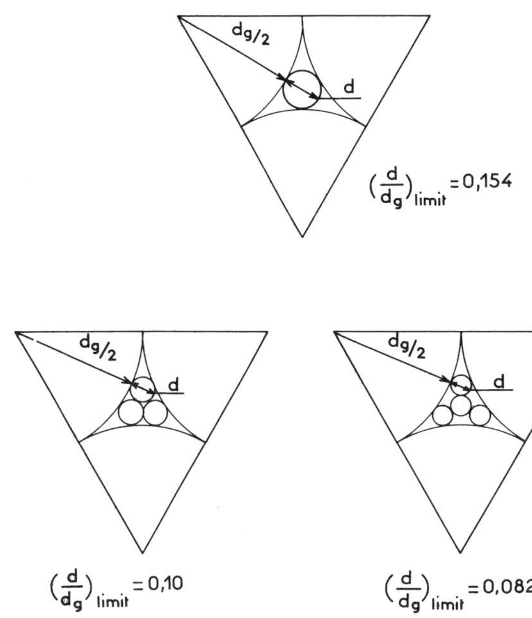

Figure 4. Straining in a triangular constriction

Wedging in Crevices. Consider the void between two tangent grains (Figure 5). Theoretically, a particle of any diameter may wedge in a crevice site; this interpretation is no longer valid if d/d_g is small because it can be assumed that the particle lies on a surface site.

For a particle to remain wedged in such a site, the fluid must bring it to the site with a sufficient kinetic energy.

To estimate the significance of this effect, the retention has been calculated in case all available sites are occupied on the first row.

Thus,

$$\sigma = \frac{1 - \epsilon_o}{2} \pi C \left(\frac{d}{d_g}\right)^2 \sqrt{\left(1 + \frac{d}{d_g}\right)^2 - 1} \qquad (18)$$

C is the coordination.

For $\epsilon_o = 0.40$ and $C = 7.0$, the σ values are given in Table IV. These values show that the retention in crevice sites is important if $d/d_g \geqslant 0.05$.

To express the filtration coefficient λ_o when particles are wedged in crevices, Hall (*20*) proposed the relation

$$\lambda_o \propto \frac{1}{d_g} \left[\frac{d}{d_g}\right]^{3/2}$$

which is similar to relations discussed above for direct interception.

Surface forces significance. Van der Waals Forces. In the case of a particle and a porous bed grain, the interaction is modeled by a sphere and a flat plate, thus

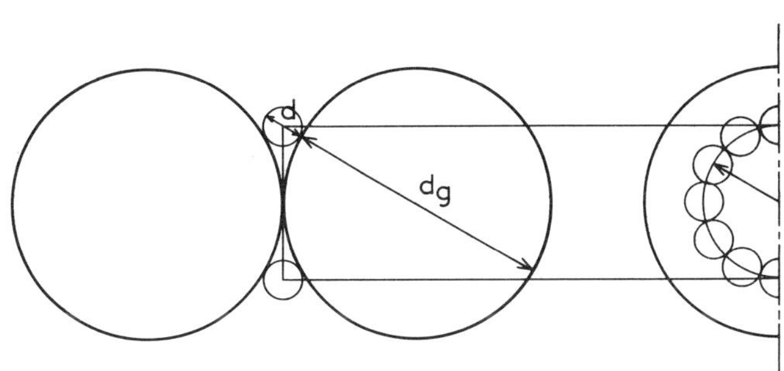

Figure 5. Retention in a crevice site

Table IV. Retention Values When All Crevice Sites Are Occupied

(ϵ_0 = 0.40, C = 7.0)

d/d_g	0.02	0.05	0.08	0.10
σ	0.053%	0.53%	1.72%	3.02%

the attractive Van der Waals forces can be calculated by Equation 19 (*32, 40*)

$$F_W = \frac{h_H d}{12r^2} \qquad (19)$$

where h_H is the Hamaker constant and r is the separation between the particle and the grain.

Unfortunately, h_H determination is generally unreliable and only an order of magnitude is known (*40*). The range of these forces is smaller than 0.05 μ, typically the order of 0.01 μ; moreover Equation 19 is not correct if r is larger than 0.05 μ (*32, 40, 49*). These attractive forces decrease as $1/r^2$ and thus rather rapidly. To compare Van der Waals forces to gravity force, the separation has been calculated so that Van der Waals forces are equal to gravity force (Table Va). For particles suspended in water, $h_H = 10^{-13}$ erg is a reasonable order of magnitude; this separation is given by

$$r^2 = \frac{h_H}{2\pi g d^2(\rho_S - \rho_l)} \qquad (20)$$

In spite of the fact that r is larger than the validity limit of the calculation, the results in Table Va indicate that for small particles ($d < 1\ \mu$), molecular forces dominate gravity force when the particles are close to the grains. By contrast, for larger particles ($d > 10\ \mu$) Van der Waals forces can be neglected, since gravity force varies as d^3 and molecular forces as d.

It is also interesting to compare Van der Waals forces with thermal energy forces. The suspended particles will be subjected to molecular forces if the Van der Waals forces energy is larger than the thermal energy of Brownian motion. As in the above discussion of Brownian motion with respect to bulk flow, the analogy with the gas kinetic theory may be continued because on the one hand, only an order of magnitude is desired and on the other hand, Van der Waals forces themselves are incompletely known. Therefore, it may be assumed that the mean kinetic energy of particles is $3h_B\text{Ⓗ}/2$, whereas the attraction energy is $h_H d/12r$. Consequently, the ratio of molecular attraction energy to kinetic energy of Brownian motion is $h_H d/(18\ r h_B \text{Ⓗ})$.

Several values of this ratio are given in Table Vb for $r = 0.01\ \mu$ which is the order of magnitude of Van der Waals forces range. These values show that for particles of diameter larger than 0.1 μ the attraction energy overcomes the random-movement energy, but for smaller particles it is the contrary.

Although these forces act over a small range (of the order of $10^{-2}\ \mu$) some investigators (*22, 49, 50*) have built filtration models by assuming that particles are attracted towards grains by Van der Waals forces and move towards them at the limit velocity given by Stokes law. A suspension which is constantly agitated owing to the flow sinuosities would always bring particles into the range of these forces.

for Heertjes-Lerk (*22*)

$$\lambda_o \propto \frac{(1 - \epsilon_o)\epsilon_o h_H}{\eta_l u_m d_g^3}$$

for Mackrle (*49, 50*)

$$\lambda_o \propto \frac{a_c}{\epsilon_o \, \mathrm{Re}^{\xi}}$$

where Re = Reynolds number
a_c = specific surface of the bed
ξ = a constant

According to experimental values

$$\lambda_o \propto \frac{\eta_l^{1/2}}{d_g^2 u_m}$$

ELECTROKINETIC FORCES—IONIC DOUBLE LAYER. Any solid immersed in an electrolyte solution adsorbs some ions and by compensation, a layer, richer in ions of opposite charge is formed around the solid. This layer, known as the "ionic double layer," is bound to the solid and moves with it (*32, 40, 66, 68*). The potential at the boundary between solution and double layer is the electrokinetic potential ζ and characterizes the electrokinetic interactions of the solid. The double layer thickness and the potential ζ depend on the ions present, their concentration, and on their valency. Thus, electrokinetic effects between particles and filter bed on the one hand, and between the particles themselves on the other hand, depend on pH, ionic strength of the solution, and the nature of the particles and ions (*32, 40*).

The electrokinetic interaction force between a sphere and a flat plate is proportional to the sphere diameter and decreases exponentially with their separation (*40*). The range of electrokinetic forces is generally greater than that of Van der Waals forces but is not considered to exceed 0.1 μ (*32, 40*). Electrokinetic forces are most often repulsive because most solids immersed in electrolyte solutions exhibit negative potentials ζ; but in some cases (*i.e.*, pH, ionic strength), for some solids, they may become attractive. Their order of magnitude is near that of molecular forces and depending upon the solids, they will be smaller or larger than Van der Waals

Table Va. Grain-Particle Separation (in μ) when Van der Waals Forces Are Equal to Gravity Force ($h_H = 10^{-13}$ erg)

$\rho_S - \rho_l$:d G/Cm³	0.10 μ	1.0 μ	10 μ	100 μ
0.5	5.7	0.57	0.06	0.006
1.0	4.0	0.40	0.04	0.004
5.0	1.8	0.18	0.02	0.002

Table Vb. Ratio of Molecular Attraction Energy to Kinetic Energy of Brownian Motion ($h_H = 10^{-13}$ erg, $r = 0.01\ \mu$, Ⓗ $= 293$°K)

d	0.01 μ	0.1 μ	1.0 μ	10 μ
Ratio	0.14	1.4	14	140

Table VI. Deep Filtration Types

Filtration type	*Particle size*	*Retention sites*	*Retention forces*	*Capture mechanisms*	*Spontaneous decolmatage*	*Provoked decolmatage*
Mechanical	$\geqslant 30\ \mu$	Constrictions, crevices, caverns	Friction, fluid pressure	Sedimentation, direct interception	Improbable	Flow direction reversal
Physicochemical	$\sim 1\ \mu$ ($< 3\ \mu$)	Surface sites	Van der Waals forces, electrokinetic forces	Direct interception	Possible	Increase in flowrate
Colloidal	$< 0.1\ \mu$	Surface sites	Van der Waals forces, electrokinetic forces, chemical bonding	Direct interception, diffusion	Possible	Increase in flowrate

forces. For each case, it is necessary to combine both surface effects to calculate the total effect (*32, 40*). Physicochemical factors are important when small particles are filtered, because they can greatly influence the particle-bed interactions and the particle interactions (*via* flocculation).

The surface forces, molecular or electric, have small ranges compared to grain diameter; thus, they do not attract particles from the bulk of suspension towards the grains wall, but act only when particles are in close proximity of the surface.

Classification of Deep Filtration Types

Consideration of the mechanisms, which have been described in the preceding paragraphs, allows a classification according to filtered particle size (Table VI).

Large particles ($d \geqslant 30\mu$)—volume phenomena prevail over surface phenomena

Small particles (diameter of the order of 1 μ)—surface phenomena prevail

Colloidal particles ($d < 0.1\ \mu$)—the laws of the physical chemistry of colloids are most important

For particles with $3\ \mu < d < 30\ \mu$, volume phenomena and surface phenomena often have the same order of magnitude.

Analysis of Previous Work

Classification. The most important previous work is grouped according to the preceding classification by referring to the mean diameter of the particles used (Table VII).

Investigations with colloidal particles not first subjected to flocculation are not considered here. For each study it is also indicated whether flocculant or coagulant have been added because these additions may conceal some processes.

Flocculation. As previously discussed electrokinetic effects exist simultaneously between the filter bed and the particles themselves. A change in these effects may thus influence both particle-bed and particle-particle interactions. For instance, Smith (*67, 68*) finds that only particle-bed interactions take place when particles are filtered without any coagulant. On the contrary, he notices the interactions between particles (flocculation) play a great part when coagulants or flocculants are added. In this case the particle-bed interactions are found to have no effect upon bed efficiency and the particle diameter is observed to increase. Ghosh (*18, 19*), Borchardt-O'Melia (*4*) believe that adding certain ions causes flocculation and formation of larger particles which can be subjected to sedimentation or to straining between grains, whereas single particles are not. Stanley (*69*) and Stein (*70*), however, claim that flocculation does not occur inside the bed owing to the small particle size, low residence time, and small suspension concentrations. However, Camp (*6*), analyzing the results of Eliassen (*13*), considers that flocculation is probably an important mechanism. Finally O'Melia-Crapps (*60*) found their results to be better explained by bed-particle interactions than by particle interactions between themselves.

Elementary mechanisms. *Large particles.* RETENTION SITES. According to Delachambre (*10*) and Sakthivadivel (*63*), particles follow the streamlines but are stopped in the passageways too narrow for passage (crevices and constrictions). The resulting deposits continuously reduce the free passage and eventually cause blocking and filling up of the caverns above these passageways. Maroudas (*53, 55*) assumes that particles settle by gravity on the grain tops or on the horizontal pore surfaces and thus narrow, or block the channels. They found the important parameter to be the ratio d/d_g; for $d/d_g > 0.15$, the porous medium is irreversibly blocked and a filter cake is formed; for $d/d_g < 0.065$, the retention remains always low and for the intermediate values a partial blocking of the porous bed may occur (this depends on particle shape and bed porosity). The bed becomes, in all cases, saturated—*i.e.*, it retains no more particles.

EVOLUTION WITH CLOGGING. Sakthivadivel (*63*) shows that the retained matter improves the bed efficiency which is assumed to increase linearly with σ; but this effect is small and decreases further as d/d_g decreases from 0.11 to 0.07. Maroudas (*52, 54, 56*) believes there is a limit u_M for the interstitial velocity beyond which no particle may be retained because the deposit reduces the porosity, the interstitial velocity thus increases with σ, and, as a result, the efficiency decreases. Her proposed relation, $\lambda \propto 1/u - 1/u_M$, may also be written under a linear form $\lambda = \lambda_o (1 - b\sigma)$ where b is a constant.

Delachambre (*10*) assumes also that the clogging rate decreases linearly with the retention since it is assumed to

Table VII. Experimental Conditions of Some Investigations

Author	Particles Type	Particles Size (μ)	Porous Medium Type	Porous Medium Size (μ)	Liquid	Filtration rate (cm/s)	Flow	Flocculant or ions
			Large Particles					
Sakthivadivel	Styron	900–1400	Spherical plastic balls	12,500	Mineral oil	0.1–0.4	Laminar	None
Maroudas	Spherical polystyrene Angular polystyrene Angular quartz	125–390 65–1100 <50	Model filter Sintered disk Packed bed (glass spheres)	Channels > 3000 Channels > 200 2000	Water	0.06–30	Laminar and transitional	None
Delachambre	Spherical polystyrene	60–350	Glass rods Glass spheres	3300 2280–3600	Organic mixture		Turbulent and transitional	None
Leclerc	Pollen	32	Glass spheres	500–1000	Organic mixture	0.05–2.0	Laminar	None
Herzig	Pollen	31	Glass spheres	500	Organic mixture	0.02–0.1	Laminar	None
			Small Particles					
Edwards-Monke	Clay	1	Sand	350	Water	0.007	Laminar	Ions
Heertjes-Lerk	Iron hydroxide	0.1–10	Glass spheres	540, 650, 780	Water	0.14–0.3	Laminar	Flocculant
Ives	Polyvinylchloride	1.3	Glass spheres, anthracite sand	425–1100	Water	0.06–0.200	Laminar	Ions
Jorden	Clay	1	Gravel	5500	Water	0.007	Laminar	Ions
O'Melia Stumm	Ferric precipitate		Glass spheres	4000	Water	0.18	Laminar	Ions
Trzaska	Clay	2	Glass spheres	800, 1300	Water	0.007–0.025	Laminar	
			Mean Particles					
Borchardt, O'Melia	Algae	15–60	Sand	320, 400, 525	Water	0.01–0.14	Laminar	Without or with flocculant
Cleasby-Baumann	Hydrous ferric oxide, calcium carbonate	5 1–10	Sand	500	Water	0.07–0.7	Laminar	Flocculant
Eliassen	Ferric hydroxide	6–20	Sand	460	Water	0.14	Laminar	Flocculant
Fox-Cleasby	Ferric hydroxide	4–25	Sand	700	Water	0.2–0.4	Laminar	Flocculant
Ghosh	Diatomite	20	Sand, glass spheres	460, 650, 770	Water	0.07–0.2	Laminar	None
Herzig	Alumina, silica	4	Glass spheres	500	Water	0.05–0.3	Laminar	None
Hsiung	Ferric hydroxide		Sand	386, 458, 545, 649, 771	Water	0.4–0.8	Laminar	Flocculant
Ives	Algae Kaolinite Sepiolite Diatomite Quartz	4 × 10, 5 2.5–10 1–30 2–60 2–22	Sand, glass spheres, anthracite	250–1300	Water	0.05–0.25	Laminar	None
Iwasaki	Algae, clay, bacteria	1–40	Sand	100–800	Water	0.0035–0.012	Laminar	None
Krone	Bacteria	1 × 4	Sand	63, 190, 400, 1100	Water	0.01–0.20	Laminar	None
Ling	Diatomite	10	Sand	350–550	Water	0.07–0.40	Laminar	Flocculant
Mackrle	Ferric hydroxide, aluminum hydroxide	5–20	Sand, anthracite, calcium carbonate, glass spheres	900–1750	Water	0.10–0.400	Laminar	Flocculant
Mintz	Clay, humus, ferric hydroxide		Sand	1000–2150	Water	0.15–0.25	Laminar	Without or with flocculant
O'Melia-Crapps	Ferric hydroxide	20	Sand	700	Water	0.14	Laminar	Flocculant, ions
Ornatskii	Clay		Sand					
Stanford-Gates	Bacteria, alum floc	>10	Sand	500	Water	0.14	Laminar	Flocculant, surfactant
Shekhtman	Crayon paste	<10	Sand	800–1200				
Smith	Clay	5	Sand	600	Water	0.14	Laminar	Flocculant, ions
Stanley	Ferric hydroxide		Sand	250–750	Water	0.07–0.4	Laminar	Flocculant
Stein	Ferric hydroxide	10–20	Lucite rods Sand	600 450	Water	0.07–1.0 0.13–1.0	Laminar	Flocculant

be proportional to the unclogged volume fraction. Leclerc's (*41*) and Herzig's (*23*) work have also shown that the retention probabilities decrease as soon as clogging starts. Thus, it appears that the presence of retained particles in the bed decreases its efficiency.

Small particles (about 1μ). RETENTION SITES. Most investigators (*12, 22, 32, 36*) cite the importance of the specific surface of porous bed and agree that retained particles occupy surface sites.

RETENTION FORCES. Typically, only surface forces are considered. Heertjes and Lerk (*22*) consider Van der Waals forces but neglect electrokinetic forces, whereas in other conditions Jorden (*36*) assumes the contrary; for Ives (*29, 32*), Edwards and Monke (*12*), the attractive Van der Waals forces and the electrokinetic forces are taken into account.

CAPTURE MECHANISMS. Generally, sedimentation is neglected. Diffusion effects are also neglected for the particles of diameter $> 1\mu$. Ives (*34*), O'Melia, and Stumm (*61*) believe that particles are brought into contact with the filter medium by direct interception. Ives (*34*) takes λ_o to be proportional to specific surface, therefore to $1/d_g$, and inversely proportional to u_m and η_l^2. Edwards and Monke (*12*) and Heertjes and Lerk (*22*) assume that Van der Waals forces cause the contact between particles and filter bed, in spite of their small range.

EFFICIENCY EVOLUTION WITH RETENTION. Ives (*34*) found the bed efficiency increases with the retention when clogging begins, perhaps due to an increase in specific surface, then to decrease because the deposit finally deteriorates it. For others (*22, 71, 72*), this initial period is not observed. Ives (*34*) describes this evolution by

$$\lambda = \lambda_o\left(1 + b_1\sigma - \frac{b_1'\sigma^2}{\epsilon_o - \sigma}\right)$$

(b_1 and b_1' are constants). Heertjes and Lerk (*22*) believe that the decrease in λ is proportional to the increase in the interstitial velocity

$$\lambda = \lambda_o\left(1 - \frac{\beta\sigma}{\epsilon_o}\right)$$

while Trzaska (*71, 72*) assumes a linear decrease in k with respect to the retention.

Mean particles (3-30 μ). RETENTION SITES. Generally retained particles are considered to occupy only surface sites but some investigators suggest that the particles can be wedged between the grains in crevices or constrictions (*2, 6, 15, 20, 39*). Ghosh (*19*) and Ling (*43*) assume that retention occurs on grain surfaces or in crevices and sheltered areas where interstitial velocity is low.

RETENTION FORCES. When deposition on grain surfaces is assumed, only surface forces are usually considered. Mintz (*58*) and Mackrle (*49*) account for only Van der Waals forces, while others consider that these forces must be combined with electrokinetic forces (*29, 60, 69*). Sanford and Gates (*64*) assume that surface effects may be neglected when $d > 10\ \mu$.

CAPTURE MECHANISMS. Sedimentation is neglected by some (*16, 49, 60, 67, 68, 70*) but is considered a probable process by others (*6, 14, 15, 20, 28, 31, 39, 64*). Krone (*39*) assumes that sedimentation should not act upon the single particles but instead should act upon the agglomerates torn from the deposit; Diaper (*11*) and Ison (*27*) found it to be important.

Direct interception is considered by some authors (*14, 20, 27, 28, 49, 70*). Mackrle (*49, 50*) assumes that Van der Waals forces, in spite of their low range, are sufficient to explain the contacting of particles with bed grains. In addition to the expressions already mentioned for λ_o as a function of the experimental parameters (see the sections "General kinetic equations of deep filtration"), other relations are found in previous work; Mintz (*58*) finds the clogging rate to be proportional to $u_m^{0.3}/d_g^{1.7}$ at the initial instant. Ives (*57*) takes λ_o to depend on specific surface, interstitial velocity, fluid viscosity, and pores tortuosity

$$\lambda_o \propto \frac{1}{\eta_l u_m d_g^{1-3}}$$

The d_g exponent varies with experimental conditions.

Ison and Ives (*27*) find for a fixed gravitational parameter, v_s/u_m, and a given Reynolds number the bed efficiency decreases when the particle diameter increases according to $\lambda_o \propto (d/d_g)^{-2.3}$; this phenomenon is assumed to be due to hydrodynamic effects. Experimentally they observed

$$\lambda_o \propto \frac{1}{d_g}\left(\frac{d}{d_g}\right)^{-2.3}\left(\frac{v_s}{u_m}\right)^{1.3} \mathrm{Re}^{-2.7}$$

or

$$\lambda_o \propto \frac{\eta_l^{1.4} d^{0.3}}{d_g^{1.4} u_m^4}$$

According to Ison and Ives (*27*), hydrodynamic effects play a great part in the capture of particles.

Yao (*74*) studies theoretically the transport of suspended particles to porous medium grains. According to him interception and sedimentation play the most significant part in the capture of relatively large particles ($>5\ \mu$); for these particles the transport efficiency is proportional to d^2. On the contrary, for the submicroscopic particles ($<1\ \mu$), only diffusion intervenes and the transport efficiency is proportional to $d^{-2/3}$. For intermediate particles (1 or 2 μ) the transport efficiency reaches a minimum. Yao (*74*) really observes by experiment the existence of such a minimum. Iwasaki (*35*) found $1/(d_g u_m^{0.3})$ but Ives (*34*), taking again Iwasaki's experimental results, finds $\lambda_0 \propto 1/(d_g u_m)$.

EFFICIENCY EVOLUTION WITH RETENTION. The existence of an initial period during which the filtrate concentration decreases is often observed. This period is sometimes assumed (*2, 19, 35, 39*) to result from a narrowing of pores which favors the straining or the wedging of other particles; while at other times (*28, 31, 43, 70*), it is attributed to an increase in specific surface.

Generally, this initial effect is written under the form of a linear expression (*28, 31, 35, 62, 70*)

$$\lambda = \lambda_o(1 + b'\sigma)$$

Iwasaki (*35*) does not give a limit to this period, but the others consider that the interstitial velocity increases with the deposit and hinders particle retention; thus, the filtrate concentration curve *vs.* time presents a minimum.

Stein (*70*) proposes an expression for the final clogging stage

$$\lambda = \lambda_c \left(\frac{1 - b_2\sigma}{1 - b_2\sigma_c}\right)^4$$

where λ_c is the maximum value of λ which occurs when $\sigma = \sigma_c$ and b_2 is a constant.

Ives (*31*) postulates that the interstitial velocity increases and the tortuosity as well as the specific surface decrease at the same time. With these assumptions, he derives

$$\lambda = \lambda_o \left(1 + b_1\sigma - \frac{b_1'\sigma^2}{\epsilon_o - \sigma}\right)$$

He has later (*28*) generalized the relation by writing

$$\lambda = \lambda_o \left(1 + \frac{b''\sigma}{\epsilon_o}\right)^{\xi_1} \left(1 - \frac{\beta\sigma}{\epsilon_o}\right)^{\xi_2} \left(1 - \frac{\sigma}{\sigma_M}\right)$$

where b'', ξ_1, and ξ_2 are constants; σ_M is the maximum value of σ.

Ornastkii (*62*) found that taking this initial period into account does not change the results significantly and Ives (*30*) has also shown that it can be often neglected. Ornatskii (*62*) attributes the variations in k with σ to an increase in interstitial velocity and these variations are fairly well described by

$$k = k_o \frac{\epsilon}{\epsilon_o} \quad \text{or} \quad k = k_o \left(\frac{\epsilon}{\epsilon_o}\right)^2$$

Ives (*30*) takes the relation between λ and σ to be

$$\lambda = \lambda_o(1 - b_3{}^2\sigma^2)$$

Mackrle (*49*) first believed that the initial increase in λ did not exist because when clogging started, the decrease in porosity and in specific surface caused the retention probabilities to diminish and he proposed

$$\lambda = \lambda_o \left(1 - \frac{\beta\sigma}{\epsilon_o}\right)^{1.2}$$

Recently (*51*) he takes into account this initial period and suggests the relation

$$\lambda = \lambda_o \left(1 + 7.5\frac{\beta\sigma}{\epsilon_o}\right)\left(1 - \frac{\beta\sigma}{\epsilon_o}\right)^2$$

also based on variations in porosity and specific surface.

Mintz (*57, 58, 59*) states that the retained matter has an unequally strong structure. When the retained matter accumulates the interstitial velocity increases as well as the hydrodynamic forces. These forces become large enough to cause the breakaway of the retained particles which are less strongly attached to the others. However, the capture of suspended particles continues, and colmatage and decolmatage are simultaneous. According to Mintz the decrease in bed efficiency is due to a certain spontaneous decolmatage; the bed efficiency is maximum when colmatage starts and afterward decreases continuously.

EMPIRICAL RELATIONS. Historically, experimental results have been grouped under empirical formulas. Unfortunately, these formulas often have been established with poorly defined parameters. Stanley (*69*) looked for a relation between d_g, u_m and the penetration—*i.e.*, the depth to which particles are retained—which was found by him to be proportional to $d_g^{2.46}u_m^{1.56}$.

Fair (*14*) noticed that the minimum depth required to produce a satisfactory filtrate in the usual conditions varies as

$$\frac{60}{\theta_F + 10} d_g^{5/3}$$

where θ_F is the Fahrenheit temperature.

Baylis (*2*) and Ling (*43*) proposed analogous formulas to calculate the time t_H at the end of which the pressure drop through the bed reaches a certain value. For Baylis

$$t_H \propto d_g^{2.15}u_m^{-1.50}$$

and for Ling

$$t_H \propto d_g^{2.32}u_m^{-1.48}$$

Hudson (*26*) considered that the length of filter run depends on $d_g^2\epsilon_o^3u_m^{-1.5}$.

Hsiung and Cleasby (*24*) recently proposed a new approach to the problem when analysing the experimental results of Eliassen (*13*) and Ling (*43*). They noticed the curves $(1/y_i)(\partial y/\partial z)$ against z have the same shape as the χ^2 distribution curves provided that the time (in hours) and the degree of freedom in χ^2 distribution are identified. Using this analogy they succeeded in grouping their experimental results in two curves. They found for a given suspension the grain diameter and the flow rate are connected by the relation $u_md_g^{2.1} = c^t$ if the effluent quality and the bed depth are fixed. They also noticed the difference between the fluid pressure drop at the beginning and at the end of the operation is proportional to $u_m^{1.5}y_i^{1.4}z^{0.5}d_g^{-1.9}t$; this result is quite similar to those above mentioned.

Decolmatage. *Large particles.* It has been observed that the spontaneous decolmatage of large particles (diameter $\geqslant$ 30 μ) is improbable in usual deep filtration conditions (*23, 52, 63*). Even when the fluid flow rate is increased by as much as a factor 3 or 4, decolmatage is low and sometimes nonexistent (*23, 63*). "Impulses"—*i.e.*, sudden variations of pressure or flow rate—are sometimes not observed to give any effect. When the same impulse is repeated several times and decolmatage occurs, it occurs only with the first impulse (*23*). On the contrary, flow direction reversal causes an appreciable decolmatage (*23, 41, 63*). Delachambre (*10*) has shown that it was possible to reach an equilibrium between colmatage and decolmatage in transitional and turbulent flow, but his work is hardly comparable to deep filtration processes.

Mean and small particles. With particles of diameter $<$ 30 μ, a consensus does not exist in the literature, with some experiments showing spontaneous decolmatage and others showing its impossibility. Mintz (*57, 59*) reports the following experiments: if deep filtration is performed with a constant flow rate (u_m about 0.23 cm/sec) and with sand grains (d_g = 1500 μ), fine flakes are observed to emerge from the bed. After 2 hr, deep filtration is stopped. At this instant, the pressure drop of the fluid through the bed is found to be almost 25 times larger than at the filtration beginning. Then clear liquid is passed through the clogged bed at the same rate as the suspension; no flake is carried out by the fluid. However, if the flow rate is increased (about 30%) a certain decolmatage of previously retained particles is observed and many flakes emerge from the bed. If the flow rate is increased six or seven times further by progressive stages, a decolmatage occurs at each stage. Mintz deduces from these experiments the retained matter has an unequally strong structure and any increase in interstitial velocity and therefore, in pressure drop causes a certain decolmatage due to an increase in hydrodynamic forces. In particular, the capture of suspended particles causes a decrease in porosity and an increase in pressure drop. Consequently, colmatage and decolmatage are simultaneous. In Mintz's opinion, spontaneous decolmatage occurs during deep filtration. However, Stanley (*69*) found that radioactive particles of the same kind (hydroxide floc) used by Mintz, which were retained when filtration began, remained at their place during the whole operation. Mackrle (*49*) noticed that clear liquid flow at the same flow rate did not cause decolmatage in a clogged bed. With a microscope, Stein (*70*) observed the decolmatage of the least-linked flocs when interstitial velocity increased. Others (*2, 15, 39, 43*) believe that decolmatage can occur when the interstitial velocity exceeds a certain limit. Ives (*57*) claims that it is unreasonable that particles, retained in the bed in spite of the hydrodynamic forces, are torn off by the same forces. Heertjes and Lerk (*57*) are also of this opinion. The authors (*23*) have also studied the behavior of small and mean particles retained in the bed. They are not broken away as the bed is fed with clear liquid at the same flowrate as the suspension, but in case of impulses, a great part of them is torn off.

These experimental results show the flow of clear liquid through a clogged bed at the same rate as the suspension does not cause any decolmatage. However, notable increases in interstitial velocity are observed to cause a certain decolmatage. Spontaneous decolmatage will occur during a deep filtration operation if such variations actually take place in the bed. It seems improbable that such increases really occur when colmatage starts because the flow is hardly disturbed by the retained matter. Thus, spontaneous decolmatage is normally improbable at the beginning of filtration, but when the bed is nearly saturated and when the interstitial velocity is considerably increased, decolmatage may occur. Typically, deep filtration is halted far before the bed is saturated (otherwise, the loss of pressure head would be too high and the filtrate much too turbid). Thus, it seems reasonable to assume that spontaneous decolmatage does not occur in normal deep filtration.

Importance of decolmatage study. It is obvious that knowledge of decolmatage conditions of a clogged bed is important because if they are accurately determined, they show how the particles are retained in bed. The behavioral difference between the large and small particles in the case of an impulse has just been seen; it shows that small particles are less strongly fixed in bed than large ones. Thus, the colmatage mechanisms are different in these two cases: large particles appear to be irreversibly fixed in the porous medium, whereas small particles seem to be fixed in a provisional way.

Decolmatage probability. Generally it is assumed for spontaneous decolmatage that k_r' is independent of σ and y and that it remains constant during the whole experiment (*3, 44, 58, 65*). For example, Mintz (*58*) observes that k_r' is proportional to u_m/d_g but Delachambre (*10*) finds that the decolmatage rate is proportional to $(\sigma - \sigma_n)$ and not to σ. $(\sigma - \sigma_n)$ represents the particles which can be broken away and σ_n designates the particles that are irreversibly fixed.

Comparison of some experimental values of λ_o. When values of λ_o, calculated from various filtration experiments with particles of a diameter between 1 and 1000 μ with grains of a diameter between 200 and 12500 μ, are compared (Table VIII), it is seen that they are spaced between 2 and 0.002 cm^{-1} and principally between 0.5 and 0.005 cm^{-1}—that it is to say in an interval of two orders of magnitude. This shows that the porous medium efficiency does not vary much with the experimental parameters in spite of the diversity of the possible mechanisms.

However, it is necessary to note that this efficiency is sensitive to the nature of the bodies present, especially when the surface effects play a great part. For example, under the same conditions of grain diameter, of particles, and of flowrate, Ives (*32*) found that λ_o varies in the ratio of 1:3 according to the dissolved ions.

Kinetic Equation

If it is assumed that spontaneous decolmatage ("de-clogging") is improbable and that the retention probabilities, k and k', are independent of the suspension concentration (as assumed by most authors and verified by experiments), Equations 8 and 9 can be written

$$\frac{\partial \sigma}{\partial t} = k u_m y \qquad (8')$$

and

$$\frac{\partial \sigma}{\partial t} = k' \epsilon y \qquad (9')$$

where k and k' depend on the parameters which characterize the system (see above) and particularly on the retention σ. Differential Equations 8′ and 9′ are analogous with classical equations of chemical kinetics and

Table VIII. Some Experimental Values of λ_o

Author	Particles Type	Particles Size (μ)	Porous Medium Type	Porous Medium Size (μ)	Filtration rate (cm/sec)	λ_o (cm^{-1})
		Large Particles				
Sakthivadivel	Styron	1060	Spherical, plastic balls	12500	0.0985	0.015
		900		12500	0.0985	0.010
Maroudas	Angular polystyrene	65	Glass spheres	2000	0.63	0.026
		65		2000	2.50	0.007
		65		2000	8.85	0.002
		90		2000	2.50	0.012
		90		2000	10.0	0.004
		125		2000	2.28	0.021
		125		2000	8.35	0.013
		Small Particles				
Ives	Polyvinylchloride (PVC)	1.3	Glass spheres	460	0.060	0.024
				460	0.060	0.033
				460	0.060	0.070
				460	0.140	0.020
				460	0.200	0.140
	(PVC)	1.3	Anthracite	460	0.060	0.067
				460	0.140	0.30
				460	0.200	0.20
	(PVC)	1.3	Sand	425	0.060	0.110
				600	0.140	0.044
Iwasaki	Colon bacillus	1	Sand	227	0.0035	0.100
				227	0.0058	0.095
				227	0.0081	0.080
				104	0.0035	0.512
				104	0.0058	0.480
				104	0.0081	0.450
		Mean Particles				
Eliassen	Ferric hydroxide	6–20	Sand	460	0.140	0.200
Fox-Cleasby	Ferric hydroxide	4–25	Sand	700	0.270	0.10
				700	0.140	0.0060
Ives	Algae	5	Sand	540	0.140	0.14
				700	0.140	0.147
	Quartz	2–22	Anthracite	770	0.140	0.064
				1020	0.140	0.074
Iwasaki	Algae	15	Sand	829	0.0035	0.55
				829	0.0081	0.52
				227	0.0035	1.99
				227	0.0081	1.40
Mackrle	Ferric hydroxide	10	Calcium carbonate	1330	0.140	0.10
				1330	0.280	0.044
Maroudas	Angular quartz	<20	Glass spheres	2000	0.25	0.017
				2000	0.36	0.010
				2000	0.80	0.004

for this reason are called the kinetic equations of clogging. In fact, they form one kinetic equation only:

$$\frac{\partial \sigma}{\partial t} = Ky \tag{21}$$

Here K is called the clogging rate constant; the dimension of K is T^{-1} and K is related to k and k' by Equation 22

$$K = ku_m = k'\epsilon \tag{22}$$

Thus, K depends on the same parameters as k and k', especially on the retention. To take into account this fact, K can be written

$$K = K_oF(\sigma) \tag{23}$$

Such a kinetic equation has been proposed previously by some authors (*38, 45, 48, 62, 72*) and has also been derived from statistical considerations (*24, 46*).

When writing Equation 21 with the retention age τ, Equation 21′ of unchanged form is obtained

$$\frac{\partial \sigma}{\partial \tau} = Ky \tag{21'}$$

Combining Equation 21′ with mass balance Equation 5′ yields

$$\left(\frac{\partial y}{\partial z}\right)_\tau = \frac{-K}{u_m}y \quad \text{or} \quad \left(\frac{\partial y}{\partial z}\right)_\tau = -ky \tag{24}$$

Kinetic equations analogous to Equation 24 are often proposed (*20, 22, 27, 28, 35, 49, 54, 70*) but written with the variables, z and t, in the form:

$$\left(\frac{\partial y}{\partial z}\right)_t = -\lambda y \tag{24'}$$

where λ is the filtration coefficient introduced for the first time by Iwasaki (*35*), λ has the dimension L^{-1} and it is assumed to be independent of y but a function of retention σ and other parameters which characterize the system.

In all cases, Equation 24′ cannot be a kinetic equation because by eliminating $(\partial y/\partial z)_t$, between it and exact Equation 4 of mass balance, one obtains

$$\frac{\partial(\sigma + \epsilon y)}{\partial t} = \lambda u_m y$$

which is a kinetic equation between the whole quantity of particles (moving and retained) and the suspension concentration, whereas only retained particles must be taken into account for clogging.

However, if the first approximation is justified (*i.e.*, if ϵy is negligible with respect to σ, or if t and τ are very close), Equation 24′ becomes the kinetic equation of clogging since it can be written:

$$\frac{\partial\sigma}{\partial t} = \lambda u_m y$$

which is identical with Equation 21 and shows that λ becomes identical with k, the retention probability per unit of bed depth.

Recently, another type of kinetic equation has been proposed (*63*)

$$\left(\frac{\partial y}{\partial z}\right)_t = -\lambda_o[1 + b_4(\sigma + \epsilon y)]y \qquad (25)$$

where b_4 is a constant.

This equation contains the same error as the equations of type 24′. Moreover, λ depends on the total volume of particles (moving and retained) whereas in reality only retained particles can change the retention conditions.

However, if ϵy is small with respect to σ, the kinetic equations of the types 24′ and 25 are identical with type 21; all these equations have nearly the same form and are grounded on the hypothesis that spontaneous decolmatage (declogging) is improbable and retention probabilities are independent of the concentration.

If colmatage (clogging) and decolmatage (declogging) are simultaneous and if it is assumed as usual that k_c' is independent of y and k_r' is independent of σ and y, the kinetic equation may be written

$$\frac{\partial\sigma}{\partial t} = k_c'\epsilon y - k_r'\sigma \qquad (26)$$

Here $k_c'\epsilon y$ is the clogging rate, $k_r'\sigma$ the decolmatage rate.

Such a kinetic equation has previously been proposed (*3, 10, 44, 47, 65*). In addition to the assumptions discussed so far regarding $k_c'\epsilon$ and k_r' as functions of the experimental parameters, it is interesting to note that some authors (*3, 10, 44, 47*) have also assumed that k_c' is independent of σ although Shekhtman (*65*) assumes that $k_c'\epsilon$ does not depend on σ.

It is also noted that the kinetic equation proposed by Mintz (*57, 58*) under the form

$$\left(\frac{\partial y}{\partial z}\right)_t = -b_5 y + b_5'\sigma \qquad (27)$$

(where b_5 and b_5' are constants) contains the same error as the equations of type 24′.

In spite of their apparent simplicity, these kinetic equations of colmatage-decolmatage phenomena are very difficult to integrate and their application to real problems is not straightforward. Thus, as the categorical proof of the existence of spontaneous decolmatage has not yet been demonstrated (see "Decolmatage"), it does not appear advantageous to continue their study.

INTEGRATION OF THE DIFFERENTIAL SYSTEM

It has been seen that the most frequently used kinetic equation is of the first order and may be written

$$\frac{\partial\sigma}{\partial t} = Ky \qquad (21)$$

or

$$\frac{\partial\sigma}{\partial\tau} = Ky \qquad (21')$$

with

$$K = K_oF(\sigma) \qquad (23)$$

From this equation another differential equation is now derived.

Differential Equation in σ

Consider a porous bed, of depth Z, at the instant t_1. At $t = 0$, suspension begins to penetrate the bed and thence at $t = 0$, $\sigma = 0$ for all z.

The total mass balance for the time interval $[0, t_1]$ is given by Equation 28, (flow in) = (flow out) + (accumulation):

$$\int_o^{t_1} u_m y_i dt = \int_o^{t_1} u_m y(Z,t)dt + \int_o^Z [\sigma(z,t_1) + \epsilon(z,t_1)y(z,t_1)]dz \qquad (28)$$

At the depth Z, the kinetic equation can be written in the following form

$$y(Z,t) = \frac{1}{K_oF[\sigma(Z,t)]}\frac{\partial\sigma(Z,t)}{\partial t}$$

therefore

$$\int_o^{t_1} u_m y(Z,t)dt = \int_o^{t_1} \frac{u_m}{K_oF[\sigma(Z,t)]}\frac{\partial\sigma(Z,t)}{\partial t}dt = \int_o^{\sigma(Z,t_1)} \frac{u_m}{K_o}\frac{d\sigma(Z,t)}{F[\sigma(Z,t)]}$$

Then Equation 28 becomes

$$\int_o^{t_1} u_m y_i dt = \int_o^{\sigma(Z,t_1)} \frac{u_m}{K_o}\frac{d\sigma(Z,t)}{F[\sigma(Z,t)]} + \int_o^Z [\sigma(z,t_1) + \epsilon(z,t_1)y(z,t_1)]dz \qquad (29)$$

Equation 29 is established for a bed of fixed depth Z, but if it is derived with respect to Z (by replacing Z by z), it becomes

$$0 = \frac{u_m}{K_o}\frac{1}{F[\sigma(z,t_1)]}\frac{\partial\sigma(z,t_1)}{\partial z} + \sigma(z,t_1) + \epsilon(z,t_1)y(z,t_1)$$

since in steady operation u_m is the same in all bed sections and since in deep filtration u_m is kept constant during the experiment. This differential equation is valid for an arbitrary instant t_1, and therefore for every point (z,t)

$$\frac{\partial \sigma}{\partial z} = -\frac{K}{u_m}(\sigma + \epsilon y) = -\frac{K_o F(\sigma)}{u_m}(\sigma + \epsilon y) \quad (30)$$

This differential equation in σ is very interesting and presents the following characteristics:

(1) It is valid whatever the function $F(\sigma)$ is, so long as the function $F(\sigma)$ does not become zero. This condition is always realized since $F(\sigma) = 0$ means that no separation occurs. Thus, Equation 30 is exact for any expression of K provided that K is independent of y, as assumed when a kinetic equation of type 21 is used.

(2) It is also valid if y_i changes with the time as long as K_o does not vary with y_i, which will be true if K is independent of y.

(3) It is exact with respect to the mass balance because no assumption has been made to derive it, except the hypothesis that diffusion is negligible. The Equation 30 becomes simpler in the first and second approximations.

(4) In the second approximation, previously discussed (see section on Approximate Equations), ϵ_o replaces ϵ, thus Equation 30 becomes

$$\frac{\partial \sigma}{\partial z} = -\frac{K}{u_m}(\sigma + \epsilon_o y) \quad (31)$$

Furthermore with a change of variables (*i.e.*, if τ is substituted for t), Equation 31 becomes

$$\frac{\partial \sigma}{\partial z} = \frac{\epsilon_o}{u_m}\frac{\partial \sigma}{\partial \tau} - \frac{K}{u_m}(\sigma + \epsilon_o y)$$

and according to kinetic Equation 21′, it can be written

$$\left(\frac{\partial \sigma}{\partial z}\right)_\tau = -\frac{K}{u_m}\sigma \quad (32)$$

(5) In the first approximation, all moving particles are neglected, (*i.e.*, ϵy is neglected with respect to σ), thence, Equation 30 becomes

$$\left(\frac{\partial \sigma}{\partial z}\right)_t = -\frac{K}{u_m}\sigma \quad (32')$$

Differential Equations 32 and 32′ have exactly the same mathematical form but in Equation 32, the variables are z and τ and in 32′ z and t; one equation may be written for both cases

$$\frac{\partial \sigma}{\partial z} = -\frac{K}{u_m}\sigma \quad (35)$$

without specifying t or τ.

Differential System

Thus, clogging is described by three equations

$$\frac{\partial \sigma}{\partial \tau} = -u_m\frac{\partial y}{\partial z} \quad \text{local mass transfer} \quad (5')$$

$$\frac{\partial \sigma}{\partial \tau} = Ky = K_o F(\sigma) y \quad \text{kinetic equation} \quad (21')$$

$$\frac{\partial \sigma}{\partial z} = -\frac{K}{u_m}\sigma \quad (33)$$

The boundary conditions of the differential system are:

$$\sigma = 0 \quad \text{for} \quad (z,\tau) = (z,0)$$

$$y = y_i \quad \text{for} \quad (z,\tau) = (0,\tau)$$

The first condition means that there are no retained particles at z prior to the suspension arrival and the second that the suspension concentration in the inlet layer is that of the suspension applied to the filter.

Relation between σ, y, σ_i and y_i

Equations 24 and 33 have the same mathematical form:

$$\frac{1}{y}\frac{\partial y}{\partial z} = -\frac{K}{u_m} \quad (24)$$

$$\frac{1}{\sigma}\frac{\partial \sigma}{\partial z} = -\frac{K}{u_m} \quad (33)$$

Integrating, $\ln y = \ln \sigma +$ constant.

The constant may be determined from the conditions in the bed layer at any z (*i.e.*, $z = 0$ is the inlet layer conditions σ_i, y_i). Thus, constant $= \ln(y_i/\sigma_i)$ and

$$\frac{y}{y_i} = \frac{\sigma}{\sigma_i} \quad (34)$$

where σ_i is the retention in the inlet layer.

Equation 34 applies when the following conditions y and σ are taken at the same depth z, and y, y_i, σ, and σ_i are taken at the same retention age τ, therefore at different instants t such as

$$t - \frac{\epsilon_o z}{u_m} = \text{constant}$$

Equation 34 allows the determination of distribution of the concentrations y along the bed from that of the retentions σ, or the reverse, as long as y and σ are known in a reference layer (for example y_i and σ_i).

This equation has been previously derived (*22, 30*) but only in particular cases. Here it is general and independent of the function $F(\sigma)$.

Existence of a Constant Clogging Front

If σ is plotted against z, for various retention ages, the curves are parallel (Figure 6). Equation 33 states that the retention curves, σ against z, have an identical slope at fixed σ. These curves are consequently parallel and can be deduced from each other by translations parallel to the axis $0z$. Thus, clogging of porous media progresses as a constant form front, analogous to a fully developed chromatography front.

Front velocity. The front velocity is determined from mass balance Equation 5′ which can be written under the form

$$\left(\frac{\partial z}{\partial \tau}\right)_\sigma = u_m\frac{\partial y}{\partial \sigma}$$

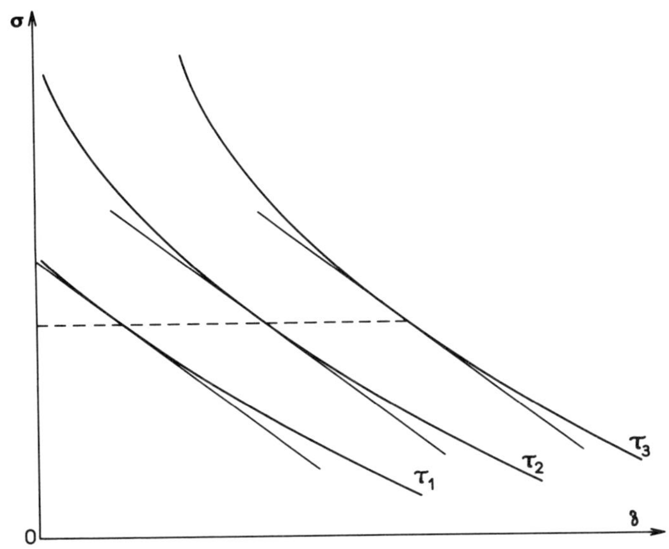

Figure 6. Retention curves

or

$$\left(\frac{\partial z}{\partial \tau}\right)_\sigma = u_m \frac{\partial y}{\partial z} \frac{\partial z}{\partial \sigma}$$

Using Equations 24 and 33 and noting that $(\partial z/\partial \tau)_\sigma$ is the front velocity V_F, we see that $V_F = u_m(y/\sigma)$ or from Equation 34

$$V_F = u_m \frac{y}{\sigma} = u_m \frac{y_i}{\sigma_i} \tag{35}$$

From Equations 21′ and 33, it can be seen that

$$V_F = -\frac{\partial \sigma/\partial \tau}{\partial \sigma/\partial z} \tag{36}$$

Equation 36 states that the partial derivatives of σ with respect to τ and z are proportional all along a curve of σ against z at a fixed retention age τ.

These properties are strictly applicable only to curves of σ against z at fixed τ. As the operating time increases, t and τ approach each other and the above properties can be applied to the retention curves at fixed t.

The existence of such a constant clogging front is only visible on the curves of σ against z. The plots of y against z do not present this feature since $\partial y/\partial z$ depends on both y and σ. However, if y_i and σ_i are constant (as happens when the inlet layer has reached saturation), the curve of y against z will have a constant form, since according to Equation 34, this curve can be deduced by proportionality to the retention curve of σ against z.

Differential System Integration

Integration with respect to z. Equation 33 $\partial \sigma/\partial z = -K\sigma/u_m$ is integrated to obtain

$$\frac{\partial \sigma}{\partial z} = -\frac{K_o}{u_m} F(\sigma)\sigma \tag{33}$$

or

$$\frac{d\sigma}{\sigma F(\sigma)} = -\frac{K_o}{u_m} dz$$

thence

$$\int \frac{d\sigma}{\sigma F(\sigma)} = -\frac{K_o}{u_m} z + c^t$$

The dependence of the constant upon τ is determined by substituting an expression of σ as a function of τ for a fixed-bed layer. This reference may be the inlet layer as taken for Equation 34 or another layer such as the outlet layer. The inlet layer is usually more convenient since the concentration of the suspension applied to the filter is well known.

Thus, an implicit function between σ and σ_i is obtained:

$$\varphi(\sigma, \sigma_i) = -\frac{K_o}{u_m} z$$

If it can be resolved, σ can be written under the form

$$\sigma = F_1\left(\sigma_i, \frac{K_o}{u_m} z\right) \tag{37}$$

Determination of σ_i. σ_i is calculated from kinetic Equation 21′

$$\frac{d\sigma_i}{d\tau} = K_o F(\sigma_i) y_i$$

$$\frac{d\sigma_i}{F(\sigma_i)} = K_o y_i d\tau$$

from which

$$\sigma_i = F_2(K_o y_i \tau)$$

where σ_i depends only on the dimensionless term $K_o y_i \tau$.

Expression of σ. By substituting the expression of σ_i into Equation 37, σ is obtained:

$$\sigma = F_1'\left(K_o y_i \tau, \frac{K_o}{u_m} z\right) \tag{38}$$

here σ depends only on two dimensionless terms, $K_o y_i \tau$ and $(K_o/u_m)z$. If y_i changes with the time, Equation 38 is still valid if $y_i \tau$ is replaced by $\int_0^\tau y_i dt$

Determination of y. According to Equation 34, $y = y_i(\sigma/\sigma_i)$. Thus, y/y_i, like σ, is only dependent on both dimensionless terms $K_o y_i \tau$ and $(K_o/u_m)z$.

Numerical integration. The total analytic resolution of the differential system can only be made for some single functions $F(\sigma)$ (see Appendix 1). This function is always inferred from experiments since it is impossible to obtain it theoretically. The inferred function is often complex and total analytic resolution is difficult. Even if the first stage of the calculation is easy (integration of $1/F(\sigma)$) the second is less easy with the integration of $1/(\sigma F(\sigma))$ and the resolution of the implicit function $\varphi(\sigma, \sigma_i) = -z(K_o/u_m)$. However, from the expression of σ_i as a function of τ and from the unresolved implicit function, the retention curve (σ against z) can be numerically determined. For any value of τ, σ_i can be calculated and for every value of σ_i, z can be determined with the implicit function for various $\sigma \leqslant \sigma_i$. In other words, the clogging front is determined at fixed τ.

Flow Rate Effect

It is assumed that the flow rate alters only K_o and that $F(\sigma)$ is independent of u_m, the clogging front curves for any flowrate $(u_m)_2$ can be deduced from that of a different flowrate $(u_m)_1$, since according to Equation 33 the ratio of their tangent slopes is constant at fixed σ and is equal to $(K_o/u_m)_1(u_m/K_o)_2$.

In the same way, curves of K against σ can also be deduced from each other.

Effect of Suspension Concentration

At fixed flow rate u_m. As σ and the ratio y/y_i depend only on two dimensionless terms, K_oz/u_m and K_oy_it (or $K_o\int_0^\tau y_idt$ if y_i varies with the time), all experimental points performed under identical conditions, except for various concentrations y_i, must lie on the same curves σ and y/y_i against $\int_0^\tau y_idt$ at fixed z. The colmatage is thus only dependent on the quantity of particles which have entered the bed for a fixed flowrate.

At variable flow rate. If experiments made under the same conditions except for different flowrates and concentrations are considered, they can be conveniently represented on the same diagram by plotting the curves of $(K_o/u_m)z$ against $K_o\int_0^\tau y_idt$ at fixed σ or at fixed y/y_i since σ and y/y_i depend only on these two terms. Should K_o be proportional to $u_m^{\xi'}$ (ξ' is a constant), it would be sufficient to plot $zu_m^{\xi'-1}$ against $u_m^{\xi'}\int_0^\tau y_idt$ at fixed σ or y/y_i.

These properties are only exact if K_o is independent of y_i, as usually observed, and if $F(\sigma)$ is independent of u_m.

Experimental points performed with various values of other parameters, such as d_g, ϵ_o, η_l, will also lie on the same curves $\sigma = c^t$ and $y/y_i = c^t$ when plotting $(K_o/u_m)z$ against K_oy_it provided that these parameters alter only K_o and do not intervene in $F(\sigma)$.

Conclusion

These properties which have been derived above are valid whatever the form of the function $F(\sigma)$. Therefore, they apply to all deep filtration cases, provided that the kinetic equation is similar to Equation 21′—*i.e.*, of first order—and that spontaneous decolmatage does not occur. It has been assumed that particle diffusion as well as a part of the suspended particles (second approximation) are negligible, but these assumptions are generally verified by experiment and consequently, these results apply to all common deep filtration cases. They characterize the kinetic equation of first order without decolmatage and diffusion.

PRESSURE DROP

Experimental Results—Empirical Formulas

In constant rate filtration, the pressure drop P across the filter increases with the clogging. Various investigators (*9, 22, 25, 28*) have proposed purely empirical formulas to represent the influence of the retention σ on the pressure drop through a filter layer Δz sufficiently thin such that the retention may be considered constant in it.

Most of these formulas may be written

$$\frac{\Delta P}{(\Delta P)_o} = \frac{1}{(1 - j\sigma)^m} \tag{39}$$

where ΔP is the pressure drop through the element of thickness Δz, and σ is the retention. $(\Delta P)_o$ is the pressure drop across a thickness Δz of clean porous bed, and j and m are constants.

Expanding Equation 39 when $j\sigma$ is small,

$$\frac{\Delta P}{(\Delta P)_o} = 1 + mj\sigma + \frac{m(m+1)}{2}j^2\sigma^2 + \ldots \tag{40}$$

Experience indicates that often only the first term of the expansion is significant and thus that ΔP increases linearly with σ.

Ives (*30*), Heertjes and Lerk (*22*) have shown under certain conditions that if

$$\frac{\Delta P}{(\Delta P)_o} = 1 + jm\sigma \tag{41}$$

the total pressure drop P across the entire filter is in certain conditions a linear function of time. At the beginning of filtration through a deep filter, the filtrate concentration, y_f, is very low. It may be assumed that all the particles are retained in the filter. If the initial velocity, u_m, and the initial concentration, y_i, remain constant in time, the total retention in the bed at time t is

$$u_my_it = \int_0^Z \sigma dz \tag{42}$$

Since $P = \int_0^Z (dP/dz)dz = P_o/Z\int_0^Z(1 + mj\sigma)dz$, the pressure drop may be written

$$P = P_o\left(1 + \frac{mju_my_it}{Z}\right) \tag{43}$$

This last expression is easier to verify experimentally than Equation 41, since the knowledge of σ is not required in Equation 43.

Cleasby and Baumann (*9*), Eliassen (*13*) and Ling (*43*) have observed a linear increase of the pressure drop through the whole bed with time. Experimentally Mintz (*58*) finds that the proportionality coefficient, here denoted by mj, varies as $d_g^{0.5}/u_m$, Heertjes and Lerk (*22*) observe it to be independent of d_g but proportional to $1/u_m$ whereas Hsiung and Cleasby (*24*) find it to be proportional to $(y_i^{0.4}Z^{0.5}d_g^{0.1})/u_m^{0.5}$. Stanley (*69*) observed that the pressure drop increased linearly with the quantity of particles which are retained but notes that the proportionality coefficient varied from one experi-

Table IX. Some Experimental Values of (mj)

Author	y_i	Particles, d	u_m, cm/mn	$\frac{P_o}{\rho_1 gz}$	d_g/mm	Grains Type	mj
Mackrle	0.3 10^{-3}	Iron hydroxide	8.2	0.14	1.33	Calcium carbonate	57.8
		10 μ	16.7	0.215			28.4
Ives	0.135 10^{-3}	Chlorella	8.2	0.65	0.54	Sand	61.5
	0.055 10^{-3}	5 μ	8.2	0.44	0.70		432
Robinson	0.38 10^{-3}	Quartz powder	8.2	0.57	0.77	Anthracite	41.2
	0.76 10^{-3}	2 to 22 μ	8.2	0.35	1.02		79.3

ment to another. He ascribes this variation to a variation of the entrance concentration of particles. Borchardt and O'Melia (*4*) obtained nearly linear variations when coagulants were not used in their suspensions. Heertjes and Lerk (*22*), Ives (*31*) find Equation 43 invalid except at the beginning of filtration, when all particles are retained. Ives (*30*) calculated values of $P_o jm/Z$ from his own work and compared his results with those of Mackrle (*49*) and Robinson. Table IX gives the corresponding values of (jm) calculated from Ives (*30*) and the experimental conditions.

We note that values of (jm) are widespread over a considerable range, possibly due to the diversity of experimental conditions. Because Equation 39 does not indicate any dependence on other conditions, a more realistic model is now considered—the Kozeny-Carman model of clogging beds.

Utilization of Kozeny-Carman Model

In the domain of laminar flow the pressure drop $(\Delta P)_o$ through a clean layer Δz of porous bed can be written with the fundamental expression of Kozeny-Carman (*8*):

$$\frac{(\Delta P)_o}{\Delta z} = (h_K)_o \eta \frac{u_m}{\epsilon_o} \frac{(a_c)_o^2}{\epsilon_o^2} \tag{44}$$

in which u_m is the suspension approach velocity, η the dynamic viscosity of the suspension, $(a_c)_o$ the bed specific surface (total grain surface per unit bed volume), ϵ_o porosity of clean porous bed, $(h_K)_o$ the Kozeny constant.

This relation is based on the use of simple model in which the void space is described by a bundle of identical, cylindrical, nonconnected capillaries. In this model,

$$(h_K)_o = 2\gamma T^2 \tag{45}$$

where γ is a circularity coefficient and T a tortuosity coefficient. If the bed specific surface is replaced by the grain specific surface $(a_g)_o$,

$$(a_g)_o = \frac{(a_c)_o}{1 - \epsilon_o}$$

the Kozeny-Carman relation becomes Equation 44′

$$\frac{(\Delta P)_o}{\Delta z} = (h_K)_o \eta u_m (a_g)_o^2 \frac{(1-\epsilon_o)^2}{\epsilon_o^3} \tag{44'}$$

By changing the rate through a clogged bed, Cleasby and Baumann (*9*) verified that the pressure drop remains proportional to the fluid velocity. The pressure drop may be thus written

$$\frac{\Delta P}{\Delta z} = h_K \eta \frac{u_m}{\epsilon} \frac{(a_c)^2}{\epsilon^2} \tag{46}$$

in which ϵ is the new porosity of the clogged bed, a_c the actual or fictitious bed specific surface, and h_K the new value of Kozeny constant.

From Equations 44 and 46,

$$\frac{\Delta P}{(\Delta P)_o} = \frac{h_K}{(h_K)_o} \left[\frac{a_c}{(a_c)_o}\right]^2 \left(\frac{\epsilon_o}{\epsilon}\right)^3 \tag{47}$$

To obtain a theoretical relation between ΔP and σ, it is necessary to establish first the relations between the retention σ and the Kozeny constant, the bed specific surface a_c, and the porosity. The decrease in porosity can be easily calculated by:

$$\frac{\epsilon}{\epsilon_o} = 1 - \frac{\beta\sigma}{\epsilon_o} \tag{48}$$

where β is the inverse of compaction factor of retained particles. The two other factors of Equation 47 are difficult to determine. If it is assumed that the Kozeny constant does not vary with retention, then

$$\frac{\Delta P}{(\Delta P)_o} = \left[\frac{a_c}{(a_c)_o}\right]^2 \left(\frac{\epsilon_o}{\epsilon}\right)^3 \tag{49}$$

It remains to derive the variation of bed specific surface with the retention.

The first hypothesis assumes that the Kozeny model is not simply a mathematical or fictitious model but that porous media are actually constituted of bundles of capillaries. It is further supposed that the particles diameter, d, is small compared to the pore diameter, d_p, and that the capillaries are uniformly coated internally by the deposit.

If there are N capillaries of length Z_p and diameter d_p in the unit volume, the initial porosity is

$$\epsilon_o = NZ_p \frac{\pi d_p^2}{4} \tag{50}$$

If deposited particles diminish the diameter of the capillaries to $(d_p - \Delta d_p)$, the porosity becomes

$$\epsilon = NZ_p \frac{\pi(d_p - \Delta d_p)^2}{4} \tag{51}$$

or

$$\epsilon = \epsilon_o \left(1 - \frac{\Delta d_p}{d_p}\right)^2 \tag{52}$$

Similarly, the initial bed specific surface is

$$(a_c)_o = NZ_p\pi d_p$$

and with deposition it becomes

$$a_c = NZ_p\pi(d_p - \Delta d_p) \quad (53)$$

or

$$a_c = (a_c)_o\left(1 - \frac{\Delta d_p}{d_p}\right) \quad (54)$$

combining Equations 51 and 54,

$$\frac{a_c}{(a_c)_o} = \left(\frac{\epsilon}{\epsilon_o}\right)^{1/2} \quad (55)$$

or

$$\frac{a_c}{(a_c)_o} = \left(1 - \frac{\beta\sigma}{\epsilon_o}\right)^{1/2} \quad (56)$$

With the aid of Equations 48, 49 and 56, the pressure drop as a function of retention, bed porosity, and particle property, β is given by Equation 57

$$\frac{\Delta P}{(\Delta P)_o} = \left(1 - \frac{\beta\sigma}{\epsilon_o}\right)^{-2} \quad (57)$$

Empirical Equation 39 is thus verified with

$$m = 2$$

$$j = \frac{\beta}{\epsilon_o}$$

and thus,

$$mj = \frac{2\beta}{\epsilon_o} \quad (58)$$

Unfortunately, such a result is incompatible with values of (mj) presented in Table IX. β is the coefficient by which it is necessary to multiply the retention σ to obtain the fraction of porous medium which does not participate to the flow of suspension. In the case of a regular and uniform coating of pores, the maximum value of β corresponds more or less to the value for a cubic packing arrangement of spheres, which is

$$\beta = \frac{6}{\pi} = 1.91$$

This value is too small, since for a porosity of 0.40, it should lie in the range $6 < \beta < 87$. Is it possible to improve the Kozeny model?

Generalization from Uniform Clogging of Surface Sites to Case of Arbitrary Texture

In the Kozeny-Carman model, the bed specific surface a_c decreases when the retention σ increases since the pores are concave volumes. However, if void space was included between nonjoined spheres, the void volumes would be convex and the specific surface would increase with σ.

Table X presents some expressions for $a_c/(a_c)_o$ and $\Delta P/(\Delta P)_o$ as functions of σ for the cases of different geometrical porous media which are regularly and uniformly coated. We have included in the table the value of $a_c/(a_c)_o$ for $\beta\sigma/\epsilon_o = 0.1$. In the case of spherical joined grains, the hypothesis of uniform coated retention does yield a decrease of bed specific surface when σ increases and the same value of $a_c/(a_c)_o$ is found (0.96 instead of 0.95) as in the Kozeny-Carman model. This coincidence does not imply similarity of the models.

Generalizing, the hypothesis of retention by uniform coating may be written:

$$\frac{a_c}{(a_c)_o} = \left[1 + \frac{\beta\sigma}{f(\epsilon_o)}\right]^{m_1} \quad (59)$$

where m_1 may vary between 1/2 and 2/3 and $f(\epsilon_o)$ between $-\epsilon_o$ and $1 - \epsilon_o$. Substituting Equation 59 in the Kozeny-Carman relation,

$$\frac{\Delta P}{(\Delta P)_o} = \left(1 - \frac{\beta\sigma}{\epsilon_o}\right)^{-3}\left[1 + \frac{\beta\sigma}{f(\epsilon_o)}\right]^{2m_1} \quad (60)$$

Equation 60 does not account for variation of tortuosity. The term which describes the variation of bed specific surface varies less rapidly than the term describing the variation of porosity. Investigators agree neither on which model should be chosen nor on the importance of the variation of specific surface. Some take the model of cylindrical pores of Kozeny, others use the model of nonjoined spheres, and still others assume that specific surface does not vary.

Kozeny's model. In his thesis, Mackrle (*49*) used a model of narrowed pores but proposed replacing the exponent 1/2 of Equation 56 by a constant m_2 such that

$$\frac{a_c}{(a_c)_o} = \left(1 - \frac{\beta\sigma}{\epsilon_o}\right)^{m_2} \quad (61)$$

and therefore,

$$\frac{\Delta P}{(\Delta P)_o} = \left(1 - \frac{\beta\sigma}{\epsilon_o}\right)^{2m_2-3} \quad (62)$$

Experimentally, Mackrle found $m_2 \simeq 1$.

More recently, Mackrle (*51*) proposed another relation between bed specific surface and retention:

$$\frac{a_c}{(a_c)_o} = \left(1 - \frac{\beta\sigma}{\epsilon_o}\right)^{1.5}\left(1 + 7.5\frac{\beta\sigma}{\epsilon_o}\right)^{0.5} \quad (63)$$

In this expression, the first factor describes the reduction of bed specific surface when pore diameter decreases and the second the increase of specific surface when the spherical grains are being coated. Values of the coefficients are determined by experiment.

Ives and Pienvichitr (*33*) utilized Mackrle's Equation 61 but their experimental results yield $m_2 \simeq -4$. It is noted that m_2 has a value far different from that found by Mackrle and the sign of m_2 means that the specific surface increases. This is at variance with the hypothesis of the pore model in which specific surface is assumed to decrease when retention increases.

Nonjoined spheres model. Stein (*70*) and Camp (*7*) proposed a model for spherical grains which are uniformly coated, obtaining:

$$\frac{a_g}{(a_g)_o} = \frac{1}{1 + \dfrac{\beta\sigma}{3(1 - \epsilon_o)}} \quad (64)$$

The denominator in Equation 64 is the first term in the expansion of

$$\left(1 + \frac{\beta\sigma}{1 - \epsilon_o}\right)^{1/3}$$

and therefore,

$$\frac{a_g}{(a_g)_o} = \left(1 + \frac{\beta\sigma}{1 - \epsilon_o}\right)^{-1/3} \tag{65}$$

where a_g is the grain specific surface, by definition

$$a_g = a_c \cdot \frac{1}{1 - \epsilon} \tag{66}$$

Combining Equations 65, 66, and 48,

$$a_c = (a_c)_o \left(1 + \frac{\beta\sigma}{1 - \epsilon_o}\right)^{2/3} \tag{67}$$

which is also given in Table X. Stein (*70*) showed that his model is in agreement with Eliassen's results (*13*). He also tried to take into account the proximity of the grains which forbids a uniform coating. His calculation yields equations which do not represent the experimental results well and he concludes that the equations of Kozeny are not particularly well suited to describe a clogged bed.

Maroudas (*52, 56*) worked with particles of relatively large dimensions (>25 μ) and also used the model of nonjoined spheres. She described the pressure drop by

$$\frac{\Delta P}{(\Delta P)_o} = \left(\frac{1 - \epsilon}{1 - \epsilon_o}\right)^{4/3} \left(\frac{\epsilon_o}{\epsilon}\right)^3 \tag{68}$$

or

$$\frac{\Delta P}{(\Delta P)_o} = \left(1 + \frac{\beta\sigma}{1 - \epsilon_o}\right)^{4/3} \left(1 - \frac{\beta\sigma}{\epsilon_o}\right)^{-3} \tag{69}$$

Other models. Sakthivadivel (*63*) studied the retention of large particles, and proposed two models in which the tortuosity remains constant. One is a model of uniform coating of parallel pores, but the expression that he gives:

$$\frac{\Delta P}{(\Delta P)_o} = \left(\frac{\epsilon_o}{\epsilon}\right)^3 \tag{70}$$

does not correspond to his hypothesis, since it does not take into account the variation of bed specific surface. Equation 70 is experimentally verified for very fine particles (*63*).

The other is an "hydraulic radius model." In this model, it is assumed that the grain diameter, d_g, remains constant, and thus that the grain specific surface is constant. Then

$$\frac{\Delta P}{(\Delta P)_o} = \left(\frac{1 - \epsilon}{1 - \epsilon_o}\right)^2 \left(\frac{\epsilon_o}{\epsilon}\right)^3 \tag{71}$$

or

$$\frac{\Delta P}{(\Delta P)_o} = \left(1 + \frac{\beta\sigma}{1 - \epsilon_o}\right)^2 \left(1 - \frac{\beta\sigma}{\epsilon_o}\right)^{-3} \tag{72}$$

This model gives a better representation of Sakthivadivel's experimental results than Equation 70.

Ghosh (*18, 19*) and Ling (*43*) have also used the "hydraulic model" since they assume that grain specific surface varies very slowly whereas $(1 - \epsilon)^2/\epsilon^3$ varies rapidly.

Mintz (*57*) assumed that the bed specific surface, a_c, remains constant and used Equation 70.

In conclusion, even when investigators have agreed to adapt the fundamental Kozeny-Carman relation to

Table X. Expressions of Bed Specific Surface and Pressure Drop as Function of Retention σ

Models	$\frac{a_c}{(a_c)_o}$ *vs.* σ	$\frac{a_c}{(a_c)_o} =$ *for* $\frac{\beta\sigma}{\epsilon_o} = 0.1$, $\epsilon_o = 0.4$	$\frac{\Delta P}{(\Delta P)_o} = f(\beta,\sigma,\epsilon_o)$	*Limit expansion of* $\frac{\Delta P}{\Delta P_o}$
Circular capillaries	$\frac{a_c}{(a_c)_o} = \left(1 - \frac{\beta\sigma}{\epsilon_o}\right)^{1/2}$	$\frac{a_c}{(a_c)_o} = 0.95$	$\frac{\Delta P}{(\Delta P)_o} = \left(1 - \frac{\beta\sigma}{\epsilon_o}\right)^{-2}$	$\frac{\Delta P}{(\Delta P)_o} = 1 + \frac{2\beta\sigma}{\epsilon_o} + \ldots$
Spherical caverns	$\frac{a_c}{(a_c)_o} = \left(1 - \frac{\beta\sigma}{\epsilon_o}\right)^{2/3}$	$\frac{a_c}{(a_c)_o} = 0.94$	$\frac{\Delta P}{(\Delta P)_o} = \left(1 - \frac{\beta\sigma}{\epsilon_o}\right)^{-5/3}$	$\frac{\Delta P}{(\Delta P)_o} = 1 + 1.67 \frac{\beta\sigma}{\epsilon_o} + \ldots$
Hyperbolic paraboloid	$\frac{a_c}{(a_c)_o} = 1$	$\frac{a_c}{(a_c)_o} = 1$	$\frac{\Delta P}{(\Delta P)_o} = \left(1 - \frac{\beta\sigma}{\epsilon_o}\right)^{-3}$	$\frac{\Delta P}{(\Delta P)_o} = 1 + 3 \frac{\beta\sigma}{\epsilon_o} + \ldots$
Void between cylindrical nonjoined rods	$\frac{a_c}{(a_c)_o} = \left(1 + \frac{\beta\sigma}{1 - \epsilon_o}\right)^{1/2}$	$\frac{a_c}{(a_c)_o} = 1.033$	$\frac{\Delta P}{(\Delta P)_o} = \left(1 - \frac{\beta\sigma}{\epsilon_o}\right)^{-3} \times \left(1 + \frac{\beta\sigma}{1 - \epsilon_o}\right)$	$\frac{\Delta P}{(\Delta P)_o} = 1 + 3.67 \frac{\beta\sigma}{\epsilon_o} + \ldots$
Void between nonjoined spherical grains	$\frac{a_c}{(a_c)_o} = \left(1 + \frac{\beta\sigma}{1 - \epsilon_o}\right)^{2/3}$	$\frac{a_c}{(a_c)_o} = 1.044$	$\frac{\Delta P}{(\Delta P)_o} = \left(1 - \frac{\beta\sigma}{\epsilon_o}\right)^{-3} \times \left(1 + \frac{\beta\sigma}{1 - \epsilon_o}\right)^{4/3}$	$\frac{\Delta P}{(\Delta P)_o} = 1 + 3.89 \frac{\beta\sigma}{\epsilon_o{}^a} + \ldots$
Void between spherical joined grains coordination C^b	$\frac{a_c}{(a_c)_o} = \left(1 + \frac{\beta\sigma}{1 - \epsilon_o}\right)^{2/3} \times \left(1 - \frac{C}{2} \frac{\beta\sigma}{3(1 - \epsilon_o) + \beta\sigma}\right)$	$\frac{a_c}{(a_c)_o} = 0.96$	See annexes	$\frac{\Delta P}{(\Delta P)_o} = 1 + 2.11 \frac{\beta\sigma}{\epsilon_o{}^a} + \ldots$

[a] $\epsilon_o = 0.40$.
[b] $C = 8$.

the case of a clogged bed, they have differed in the choice of the model to be used. The various solutions which they propose, do not give any more information than the purely empirical expression Equation 39. Their models are all based on the assumed regular and uniform coating of the interior surfaces of the porous medium. It seems unreasonable that clogging occurs only on surfaces sites, especially for the case of large particles in the suspensions.

Nonuniform Clogging

In the section on Generalization from the Uniform Clogging of Surface Sites to the Case of Arbitrary Texture, we tried eliminating a restriction to the Kozeny-Carman model. We replaced the uniformly concave surface of the capillaries by various types of concave or convex surfaces.

There still remains another restriction. In reality, the porous medium is not composed of pores of uniform size. It would be better represented by a group of interconnected pores which present widenings or constrictions, or better by a combination of caverns connected ones to the others by longitudinal and transverse constrictions (Figure 7).

In the longitudinal constrictions, fluid velocity will be higher. The pressure drop will depend essentially on the number and shape of these constrictions. In the transverse constrictions and in the dead zones of the caverns, the fluid velocity is less. Then depending on whether the kinetic energy more or less favors capture processes, clogging will occur preferentially in the first or second zone. In these conditions, a model of uniform coating of the cavern surfaces is hardly credible and must be replaced by two different models.

Blocking of constrictions. In this case, the entire flowrate through a long canal can be stopped by the presence of one large grain in a constriction of this canal. Fluid then will be deflected and pass through parallel canals which will undergo an increased head loss.

This mode of clogging also may be represented as the clogging of a pile of perforated plates, supposing that the holes of the plates have nearly the same diameter as the suspension particles.

Heertjes (*21*), studying the blocking of cloth with particles verified the relation:

$$\frac{\Delta P}{(\Delta P)_o} = \frac{1}{\left(1 - m' \frac{\sigma}{N_p}\right)^{m''}} \tag{73}$$

where N_p is the number of pores per unit section and m' and m'' are constants.

Sakthivadivel (*63*) and Maroudas (*52, 54, 56*) also have a model of blocking pores to describe filtration of large particles ($d/d_g > 0.08$). Each time a pore is blocked in constant flow rate filtration, the interstitial fluid velocity, u, through the other pores increases. At every instant, interstitial velocity is inversely proportional to the number of pores which remain free.

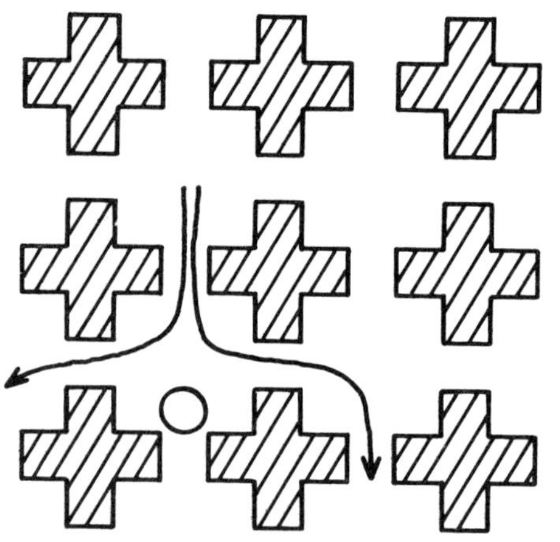

Figure 7. Model of caverns connected with constrictions

$$\frac{u}{u_o} = \frac{1}{1 - n_b} \tag{74}$$

where n_b is the fraction of pores blocked.

Assuming that the pressure drop is proportional to the interstitial fluid velocity (laminar flow),

$$\frac{\Delta P}{(\Delta P)_o} = \frac{1}{1 - n_b} \tag{75}$$

Stein (*70*) and Camp (*7*) have tried to apply this model of blocked pores to the retention of mean-sized (10 μ) particles. Stein was unsuccessful in applying it to Eliassen's (*13*) experiments, since he found that Equation 75 gave too small an increase of the pressure drop.

The problem lies in relating the fraction of blocked pores to the retention. The fraction of blocked pores corresponds to the decrease of the volume fraction which remains free to the fluid flow. By analogy with Equation 48, we write

$$\frac{\epsilon}{\epsilon_o} = 1 - n_b = 1 - \frac{\beta'\sigma}{\epsilon_o} \tag{76}$$

and

$$\frac{\Delta P}{(\Delta P)_o} = \frac{1}{1 - \frac{\beta'\sigma}{\epsilon_o}} \tag{77}$$

Here β' is the coefficient by which it is necessary to multiply σ to obtain the fraction of porous volume which no longer participates in the fluid flow. In the case of blocked pores which are entirely filled with particles, β' will be equal to β, the inverse of the compaction factor of the packing of particles in the pores. However, if only one particle is enough to block a canal, β' is then equal to the ratio of the canal volume to the particle volume. So β' will be difficult to determine since it will depend not only on the particle shape and size but also on the texture of the porous medium. Further, β' cannot be considered constant since the blocking of a pore by one or more particles does not exclude further deposition of particles in the dead zone created. These particles, which do not affect the pressure drop, will modify the value of β'.

Hudson (*25*) did not take β' into account, and he proposed

$$\frac{\Delta P}{(\Delta P)_o} = \frac{1}{1 - \dfrac{\sigma}{\epsilon_o}}$$

Le Goff and Delachambre (*42*) consider that the pressure drop in a bed of spheres depends only on the passage of fluid through triangular, square, and pentagonal constrictions formed by the spheres. Large particles are retained in the caverns upstream of these constrictions which are progressively clogged beginning with triangular constrictions first. They calculate the number of triangular and square constrictions as well as the values of retention corresponding to the clogging of the caverns upstream of these constrictions. By summing the contribution to the velocity increase from each type of blocked constrictions they relate the pressure drop ΔP to the retention σ. Finally, their experimental results verify the derived expression for ΔP.

Deposition in dead zones. Dead zones do not participate in the flow of suspension. The presence of particles in these zones will not modify the pressure drop but will complicate the estimation of a relation between pressure drop and retention.

APPENDIX I

Analytic Integration of the Differential System for Some Simple Cases

$$\frac{\partial \sigma}{\partial \tau} = -u_m \frac{\partial y}{\partial z}$$

$$\frac{\partial \sigma}{\partial \tau} = K_o F(\sigma) y$$

$F(\sigma) = 1$, then $K = K_o$

$$\sigma = K_o y_i \tau \exp\left(-\frac{K_o}{u_m} z\right)$$

$$y = y_i \exp\left(-\frac{K_o}{u_m} z\right)$$

$F(\sigma) = 1 - b\sigma$ (b is a constant)

$$\sigma = \frac{1}{b} \frac{1 - \exp(bK_o y_i \tau)}{1 - \exp\left(\dfrac{K_o}{u_m} z\right) - \exp(bK_o y_i \tau)}$$

$$y = y_i \frac{\exp(bK_o y_i \tau)}{-1 + \exp\left(\dfrac{K_o}{u_m} z\right) + \exp(bK_o y_i \tau)}$$

$F(\sigma) = 1 - b^2\sigma^2$

$$\sigma = \frac{1}{b}\left[1 + (G^2 - 1)\exp\left(\frac{2K_o}{u_m} z\right)\right]^{-1/2}$$

$$y = y_i G\left[1 + (G^2 - 1)\exp\left(\frac{2K_o}{u_m} z\right)\right]^{-1/2}$$

with

$$G = \frac{1 + \exp(2bK_o y_i \tau)}{\exp(2bK_o y_i \tau) - 1}$$

Partial Analytic Integration for Some Other Cases

The analytic expression for σ_i and the implicit relation between σ and σ_i are given.

$F(\sigma) = (1 - b\sigma)^{1/2}$ (b is a constant)

$$\sigma_i = K_o y_i \tau \left(1 - \frac{1}{4} bK_o y_i \tau\right)$$

$$\ln \frac{\sigma}{\sigma_i} + 2 \ln \frac{1 + \sqrt{1 - b\sigma_i}}{1 + \sqrt{1 - b\sigma}} = -\frac{K_o}{u_m} z$$

$F(\sigma) = (1 - b\sigma)^{3/2}$

$$\sigma_i = \frac{K_o y_i \tau (4 + bK_o y_i \tau)}{(2 + bK_o y_i \tau)^2}$$

$$\ln \frac{\sigma}{\sigma_i} + 2 \ln \frac{1 + \sqrt{1 - b\sigma_i}}{1 + \sqrt{1 - b\sigma}} + 2\left[\frac{1}{\sqrt{1 - b\sigma}} - \frac{1}{\sqrt{1 - b\sigma_i}}\right] = -\frac{K_o}{u_m} z$$

$F(\sigma) = (1 - b\sigma)^2$

$$\sigma_i = \frac{K_o y_i \tau}{1 + bK_o y_i \tau}$$

$$\ln \frac{\sigma}{\sigma_i} + \ln \frac{1 - b\sigma_i}{1 - b\sigma} + \frac{1}{1 - b\sigma} - \frac{1}{1 - b\sigma_i} = -\frac{K_o}{u_m} z$$

$F(\sigma) = (1 - \sigma/\sigma_m)(1 - \sigma/\sigma_M)$ (σ_m and σ_M are constant)

$$\sigma_i = \frac{\sigma_m \sigma_M [1 - \exp(G')]}{\sigma_M - \sigma_m \exp(G')}$$

with
$$G' = \frac{\sigma_m - \sigma_M}{\sigma_m \sigma_M} K_o y_i \tau$$

$$\ln \frac{\sigma}{\sigma_i} + \frac{\sigma_m}{\sigma_M - \sigma_m} \ln \frac{\sigma_M - \sigma}{\sigma_M - \sigma_i} - \frac{\sigma_M}{\sigma_M - \sigma_m} \ln \frac{\sigma_m - \sigma}{\sigma_m - \sigma_i} = -\frac{K_o}{u_m} z$$

APPENDIX 2

Calculation of Variation of Bed Specific Surface with Retention for Different Models

Model of spherical caverns

We suppose that the porous medium is composed of N spherical caverns of diameter d per unit of bulk volume. By regular deposition on the walls of the caverns, the diameter diminishes to $(d - \Delta d)$ and the porosity changes

from $$\epsilon_o = N\frac{\pi d^3}{6} \quad (1)$$

to $$\epsilon = N\pi\frac{(d - \Delta d)^3}{6} \quad (2)$$

or $$\epsilon = \epsilon_o\left(1 - \frac{\Delta d}{d}\right)^3 \quad (3)$$

Similarly, the specific surface varies

from $$(a_c)_o = N\pi d^2 \quad (4)$$

to $$a_c = N\pi(d - \Delta d)^2 \quad (5)$$

or $$a_c = (a_c)_o\left(1 - \frac{\Delta d}{d}\right)^2 \quad (6)$$

Combining Equations 3 and 6 yields 7:

$$\frac{a_c}{(a_c)_o} = \left(\frac{\epsilon}{\epsilon_o}\right)^{2/3} \quad (7)$$

Model of nonjoined cylindrical rods

We suppose that the porous medium is composed by N cylindrical rods of diameter d and length Z per unit of bulk volume. By regular coating of the rods, the diameter increases to $d + \Delta d$, solid fraction of porous medium changes

from $$1 - \epsilon_o = NZ\frac{\pi d^2}{4} \quad (8)$$

to $$1 - \epsilon = NZ\pi\frac{(d + \Delta d)^2}{4} \quad (9)$$

or $$1 - \epsilon = (1 - \epsilon_o)\left(1 + \frac{\Delta d}{d}\right)^2 \quad (10)$$

In the same way, the specific surface goes

from $$(a_c)_o = NZ\pi d \quad (11)$$

to $$a_c = NZ\pi(d + \Delta d) \quad (12)$$

or $$a_c = (a_c)_o\left(1 + \frac{\Delta d}{d}\right) \quad (13)$$

Using Equations 10 and 13, we get

$$\frac{a_c}{(a_c)_o} = \left(\frac{1 - \epsilon}{1 - \epsilon_o}\right)^{1/2} \quad (14)$$

Model of nonjoined spherical grains

We suppose that the porous medium is composed of N_S spherical grains of diameter d_g per unit volume of filter. By regular coating of the grains, the solid fraction of the porous medium varies

from $$1 - \epsilon_o = N_S\frac{\pi d_g^3}{6} \quad (15)$$

to $$1 - \epsilon = N_S\pi\frac{(d_g + \Delta d_g)^3}{6} \quad (16)$$

or $$1 - \epsilon = (1 - \epsilon_o)\left(1 + \frac{\Delta d_g}{d_g}\right)^3 \quad (17)$$

and the specific surface varies

from $$(a_c)_o = N_S\pi d_g^2 \quad (18)$$

to $$(a_c) = N_S\pi(d_g + \Delta d_g)^2 \quad (19)$$

or $$a_o = (a_c)_o\left(1 + \frac{\Delta d_g}{d_g}\right)^2 \quad (20)$$

Combining Equations 17 and 20, we get

$$\frac{a_c}{(a_c)_o} = \left(\frac{1 - \epsilon}{1 - \epsilon_o}\right)^{2/3} \quad (21)$$

Model of spherical-joined grains (coordination C)

Figure 8 shows a cutaway view of two joined spherical grains which are partially coated. The coated thickness is $\Delta d_g/2$. The new grain volume is

$$\frac{\pi(d_g + \Delta d_g)^3}{6} - v \quad (22)$$

If there are C points of junction (coordination C), the new volume of a grain will be

$$\frac{\pi(dg + \Delta dg)^3}{6} - Cv \quad (23)$$

with $$v = \frac{\pi(\Delta dg)^2}{4}\left(\frac{dg + \Delta dg}{2} - \frac{1}{3}\frac{\Delta dg}{2}\right) \quad (24)$$

Thus the new solid fraction of the porous medium is

$$1 - \epsilon = \frac{N_S\pi(dg + \Delta dg)^3}{6} - \frac{N_S C\pi(\Delta dg)^2}{8}\left(dg + \frac{2}{3}\Delta dg\right) \quad (25)$$

$$\text{or} \quad 1 - \epsilon = \frac{N_S\pi dg^3}{6}\left[\left(1 + \frac{\Delta dg}{dg}\right)^3 - \frac{3C}{4}\left(\frac{\Delta dg}{dg}\right)^2 - \frac{C}{2}\left(\frac{\Delta dg}{dg}\right)^3\right] \quad (26)$$

Since Δdg is very small in comparison to dg, the corrective terms which take into account the proximity of grains are unnecessary and thus using Equation 15,

$$(1 - \epsilon) \simeq (1 - \epsilon_o)\left(1 + \frac{\Delta dg}{dg}\right)^3 \quad (27)$$

which is identical to Equation 17,

$$\text{or} \quad \frac{1 - \epsilon}{1 - \epsilon_o} \simeq 1 + \frac{3\Delta dg}{dg} \quad (28)$$

$$\text{or} \quad \frac{3\Delta dg}{dg} \simeq \frac{\beta\sigma}{1 - \epsilon_o} \quad (29)$$

Similarly, the new surface of a grain is

$$\pi(dg + \Delta dg)^2 - w \quad (30)$$

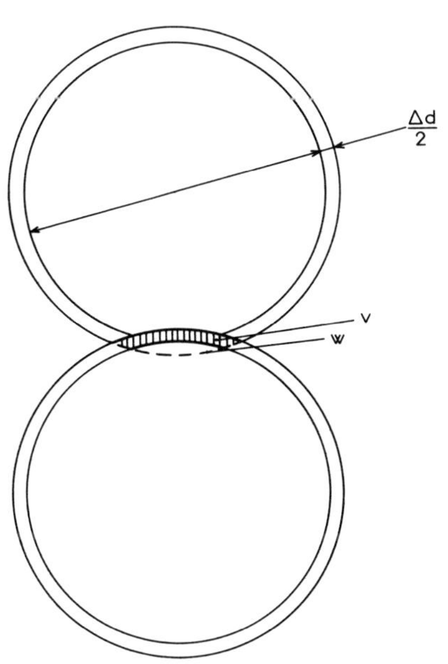

Figure 8. Cutaway view of two joined and coated grains

$$\text{with} \quad w = \frac{\pi(dg + \Delta dg)\,\Delta dg}{2} \quad (31)$$

For N_S grains with coordination C per unit volume of filter, the specific surface will be

$$a_c = N_S\pi(dg + \Delta dg)^2 - N_S C w \quad (32)$$

$$\text{or} \quad a_c = N_S\pi(dg + \Delta dg)^2\left[1 - \frac{C\Delta dg}{2(dg + \Delta dg)}\right] \quad (33)$$

Using Equation (18), we see that

$$a_c = (a_c)_o\left(1 + \frac{\Delta dg}{dg}\right)^2\left[1 - \frac{C\Delta dg}{2(dg + \Delta dg)}\right] \quad (34)$$

combining Equation 34 with Equations 29 and 27 yields Equation 35

$$a_c \simeq (a_c)_o\left(1 + \frac{\beta\sigma}{1 - \epsilon_o}\right)^{2/3} \times \left[1 - \frac{C}{2}\,\frac{\beta\sigma}{3(1 - \epsilon_o) + \beta\sigma}\right] \quad (35)$$

Since

$$\frac{\epsilon}{\epsilon_o} = 1 - \frac{\beta\sigma}{\epsilon_o} \quad (36)$$

and admitting that

$$\frac{\Delta P}{\Delta P_o} = \left[\frac{a_c}{(a_c)_o}\right]^2\left(\frac{\epsilon_o}{\epsilon}\right)^3 \quad (37)$$

and combining Equations 35, 36, and 37 yield Equation 38

$$\frac{\Delta P}{\Delta P_o} \simeq \left(1 - \frac{\beta\sigma}{\epsilon_o}\right)^{-3}\left(1 + \frac{\beta\sigma}{1 - \epsilon_o}\right)^{4/3} \times \left[1 - \frac{C}{2}\,\frac{\beta\sigma}{3(1 - \epsilon_o) + \beta\sigma}\right]^2 \quad (38)$$

Assuming that $\beta\sigma/\epsilon_o$ is very small in comparison to unity, we find that

$$\frac{\Delta P}{\Delta P_o} \simeq 1 + \frac{\beta\sigma}{\epsilon_o}\left[3 + \frac{4\epsilon_o}{3(1 - \epsilon_o)} - \frac{C\epsilon_o}{3(1 - \epsilon_o)}\right] \quad (39)$$

$$\text{or} \quad \frac{\Delta P}{\Delta P_o} \simeq 1 + \frac{\beta\sigma}{\epsilon_o}\left[3 + \frac{\epsilon_o(4 - C)}{3\,(1 - \epsilon_o)}\right] \quad (40)$$

Conclusion—Theory of Pressure Drop

A universal theory which deals with the problem of variation of pressure drop with retention does not exist. Two models may be used. The first states that there is an uniform and regular coating of internal surfaces of the porous filter. It may be adapted to filtration of small particles, but it always yields semiempirical results. The second model corresponds to the clogging of filters by large particles. Since a very good knowledge of the texture of the filter is needed, this latter model is more difficult to use.

Unfortunately, these models allow a description of the change of pressure drop with retention, for the experiment from which they are derived, but they do not allow an accurate prediction of how ΔP varies with σ in a new situation.

SUMMARY

The various studies about flow of suspensions through porous media allow the following conclusions:

1. The possible elementary mechanisms of deep filtration are now known, and they could be brought in evidence. However, it is always difficult to evaluate accurately their significance in any system. Therefore, experimental studies are needed before any filtration.

2. Nevertheless, two theoretical filtration types may be defined:

(*a*) a mechanical deep filtration for large particles (over 30 μ); for them, volume phenomena prevail and spontaneous decolmatage is improbable.

(*b*) a physicochemical deep filtration for small particles (approx. 1 μ); for them, surface effects prevail and spontaneous decolmatage may occur in case of sudden variations of flowrate or pressure.

Typical deep filtrations are performed for mean particles of an intermediary size; thus, volume phenomena and surface effects have the same order of magnitude.

3. However, in most cases, kinetic equation of colmatage has been verified by experiment to be of first order—*i.e.* there is no interaction between suspended particles.

4. Mass balance equation may be simplified in usual deep filtration. The third approximation (no diffusion) and the second approximation ($\epsilon \simeq \epsilon_o$) are valid for all common deep filtrations, whereas the first approximation is justified only if the retentions are large enough.

5. General macroscopic properties are deduced from kinetic equation of first order and mass balance equation; they are independent of the studied system and describe the evolution of colmatage in deep porous beds. They allow an easy representation of experimental results and a rapid calculation of industrial deep filter beds.

6. The fluid pressure drop through the porous medium increases with retention; various attempts have been made to describe this evolution:

(*a*) Empirical formulas are practical but do not explain what really occurs in porous media. Besides, the coefficients vary greatly from one investigator to another.

(*b*) Kozeny model accurately describes the flow of liquid through clean porous bed; however, it may not be applied to clogged beds because too small an increase in pressure drop is obtained when calculating the pressure drop from a model based on the uniform coating of cylindrical pores.

(*c*) The model based on the uniform coating of bed grains with retained particles is sometimes in agreement with experimental results in the case of small particles. However, a consensus does not exist between the investigators to explain the variations of specific surface.

(*d*) For large particles, a model based on the blocking of constrictions is preferable; however, it is difficult to apply this model because the volume occupied by a retained particle is unknown.

7. The study of colmatage and the study of pressure drop are connected as they are two aspects of the same phenomenon—*i.e.*, the retention of suspended particles smaller than the pores of the porous medium.

NOMENCLATURE

a_c	= bed specific surface
a_g	= grain specific surface
b, b', b_1 ... etc.	= constants
d	= particle mean diameter
d_g	= grain mean diameter
d_p	= pore diameter (Kozeny model)
g	= gravity acceleration
h_B	= Boltzmann constant
h_H	= Hamaker constant
h_K	= Kozeny constant
j	= constant
k	= retention probability par unit depth
k'	= retention probability per unit time
k_c'	= capture probability per unit time
k_r'	= decolmatage probability per unit time
$m, m', m'', m_1, m_2, \ldots$	= constants
n_b	= fraction of blocked pores
q	= retention probability in a porous medium element
r	= grain-particle separation
r_c	= constriction radius
t	= time (variable)
t_H	= time at the end of which the pressure drop reaches a certain value (Baylis, Ling)
t_1	= particular value of t
u	= mean interstitial velocity of suspension
u_m	= suspension approach velocity
$(u_m)_1, (u_m)_2$	= particular values of u_m
u_M	= critical interstitial velocity (Maroudas)
v_S	= sedimentation velocity
$\bar{x}$	= mean diffusional path
y	= volume fraction of particles in suspension
z	= depth (variable)
B	= permeability
C	= coordination
D	= diffusion coefficient
D_B	= diffusivity (Brownian motion)
$F(\sigma)$	= function of σ in $K = K_oF(\sigma)$
F_1, F_1', F_2	= functions
F_w	= Van der Waals forces
G, G'	= dimensionless parameters
K	= clogging rate constant
N	= number of pores per unit volume of filter
N_p	= number of pores per unit of filter section
N_S	= number of spheres per unit volume of filter

P = total pressure drop across filter depth
Re = Reynolds number
T = tortuosity coefficient
V_F = front velocity
Z = filter bed depth
Z_p = pore length

Greek Letters

β = inverse of compaction factor (1 − porosity) of retained matter
β' = particle property
γ = circularity coefficient
ϵ = bed porosity
ζ = electrokinetic potential
η = dynamic viscosity of suspension
η_l = dynamic viscosity of carrier fluid
θ_F = fahrenheit temperature
λ = filtration coefficient
λ_c = maximum value of λ (Stein)
ξ, ξ', ξ_1, ξ_2 = constants
ρ_l = specific mass of carrier fluid
ρ_S = specific mass of particles
σ = retention—*i.e.*, volume of deposited particles per unit filter volume
σ_m = constant (Appendix 1)
σ_n = volume of irreversibly fixed particles per unit filter volume (Delachambre)
σ_M = maximum value of σ
τ = retention age
τ_1, τ_2, τ_3 = particular values of τ
φ = implicit function
Ⓗ = absolute temperature (°K)
Φ = factor characterizing physico-chemical properties
Ω = bed area
Index f = outlet layer
Index i = inlet layer
Index o = clean filter bed

REFERENCES

(1) Abutaliev, F. B., Klenov, V. B., On the operation of individual layers of rapid sand filters, *Izv. Akad. Nauk. Uzb. SSR, Ser. Tekh. Nauk.*, **2**, 34 (1963).

(2) Baylis, J. R., Experiences in filtration, *J. Amer. Water Works Assoc.*, **29**, 7, 1010–1048 (1937).

(3) Bodziony, J., Kraj, W., Equation describing colmatage and suffosion phenomenon, *Bull. Acad. Pol. Sci., Ser. Sci. Tech.*, **14**, 7, 417–677 (1966).

(4) Borchardt, J. A., O'Melia, C. R., Sand filtration and algal suspensions, *J. Amer. Water Works Ass.*, 1493–1502 (Dec. 1961).

(5) Boreli, M., Jovasovic, D., Clogging of porous media, Int. Ass. for Hydraulic Res., 9th Convention, Dubrovnik, 1961.

(6) Camp, T. R., Discussion of the filter sand for water purification plants, *Amer. Soc. Civil Eng.*, 769–777, April 1937.

(7) Camp, T. R., Theory of water filtration, *J. Sanit. Eng. Div.* (*Proc. Amer. Soc. Civil. Eng.*), **SA4**, paper 3990, p 1 (August 1964).

(8) Carman, P. C., Fluid flow through granular beds, *Trans. Amer. Inst. Chem. Eng.*, **15**, 150–166 (1937).

(9) Cleasby, J. L., Baumann, E. R., Selection of sand filtration rates, *J. Amer. Water Works Ass.*, 579–602, May 1962.

(10) Delachambre, Y., Contribution à l'étude de l'écoulement d'une suspension à travers un milieu poreux et du mécanisme de la filtration Thése Fac. des Sci., Nancy (France), 1966.

(11) Diaper, E. W. J., Ives, K. J., Filtration through size graded media, *J. Sanit. Eng. Div., Proc. Amer. Soc. Civil Eng.*, **SA3**, paper 4378, 89–114, June 1965.

(12) Edwards, D. M., Monke, E. J., Electrokinetic studies of slow sand filtration process, *J. Amer. Water Works Ass.*, 1310–1319, Oct. 1967.

(13) Eliassen, R., Clogging of rapid sand filters, *J. Amer. Water Works Ass.*, **33**, 5, 926–941 (1941).

(14) Fair, G. M., The hydraulics of rapid sand filters, *J. Inst. of Water Eng.*, **5**, 171–213 (1951).

(15) Feben, D., Theory of flow in filter media, *J. Amer. Water Works Ass.*, 940–958, July 1960.

(16) Fox, D. M., Cleasby, J. L., Experimental evaluation of sand filtration theory, *J. Sanit. Eng. Div., Proc. Amer. Soc. Civil Eng.*, **SA5**, paper 4941, 61–82 (1966).

(17) Friedlander, S. K., Pasceri, R. E., Aerosol Filtration by fibrous filters in Biochemical and Biological Engineering Science, Vol 1, N. Blackbrough, Ed., Academic Press, New York, 1967.

(18) Ghosh, G., Mechanism of rapid sand filtration, *Water Water Eng.*, 147–153 (1958).

(19) Ghosh, G., Media characteristics in water filtration, *J. Sanit. Eng. Div., Proc. Am. Soc. Civil Eng.*, **SA1**, paper 1533, 1–25 (1958).

(20) Hall, W. A., An analysis of sand filtration, *J. Sanit. Eng. Div. Proc. Am. Soc. Civil Eng.*, **SA3**, paper 1276, 1–9 (1957).

(21) Heertjes, P. M., Studies in filtration; blocking filtration, *Chem. Eng. Sci.*, **6**, 190–203 (1957).

(22) Heertjes, P. M., Lerk, C. F., The functioning of deep bed filters, *Trans. Inst. Chem. Eng.*, **45**, T129–T145 (1967).

(23) Herzig, J. P., Thesis Fac. Sci., Nancy (France), to be published.

(24) Hsiung, K., Cleasby, J. L., Prediction of filter performance, *J. Sanit. Eng. Div., Proc. Am. Soc. Civil Eng.*, **94**, SA6, 1043–1069 (1968).

(25) Hudson, H. E., Jr., A theory of the functioning of filters, *J. Am. Water Works Ass.*, 868–872 (1948).

(26) Hudson, H. E., Jr., Functional design of rapid sand filters, *J. Sanit. Eng. Div., Proc. Am. Soc. Civil Eng.*, **SA1**, paper 3384, 17–28 (1963).

(27) Ison, C. R., Ives, K. J., Removal Mechanisms in deep bed filtration, *Chem. Eng. Sci.*, **24**, 717–729 (1969).

(28) Ives, K. J., Deep filters, Filtr. Separ., 125–135, (March/Apr., 1967).

(29) Ives, K. J., Research on deep filters, *Trans. Inst. Chem. Eng.*, **43**, 238–247 (1965).

(30) Ives, K. J., Simplified rational analysis of filter behaviour, *Proc. Inst. Civil Engrs.*, **25**, 345–364 (1963).

(31) Ives, K. J., A theory of the functioning of deep filters, Proc. of the Symposium on the Interaction between fluids and particles, *Inst. Chem. Eng.*, 260–268 (1962).

(32) Ives, K. J., Gregory, J., Surface forces in filtration, *Proc. Soc. Water Treat. Exam.*, **15** (2), 93–116 (1966).

(33) Ives, K. J., Pienvichitr, V., Kinetics of the filtration of dilute suspensions, *Chem. Eng. Sci.*, **20**, 965–973 (1965).

(34) Ives, K. J., Sholji, I., Research on variables affecting filtration, *J. Sanit. Eng. Div., Proc. Am. Soc. Civil Eng.*, **SA4**, Paper 4436, 1–18 (1965).

(35) Iwasaki, T., Some notes on sand filtration, *J. Am. Water Works Ass.*, **29** (10), 1591–1602 (1937).

(36) Jorden, R. H., Electrophoretic studies of filtration, *J. Am. Water Works Ass.*, 771–782, 1963.

(37) Klenov, V. B., Various questions on the calculation of filters, *Izv. Akad. Nauk Uzb. SSR, Ser. Tekh. Nauk*, **62** (5), 55 (1960).

(38) Kraj, W., A simplified model of colmatage accompanied by diffusion process, *Bull. Acad. Pol., Sci. Ser. Sci. Techn.*, **14** (8), 477–773 (1966).

(39) Krone, R. B., Orlob, G. T., Hodgkinson, C., Movement of coliform bacteria through porous media, *Sewage Ind. Wastes*, **30** (1), 1–13 (1958).

(40) Krupp, H., Particle adhesion, Theory and experiment, *Advan. Colloid Interface Sci.*, **1** (2) (1967).

(41) Leclerc, D., Thesis Fac. Sci., Nancy (France), to be published.

(42) Le Goff, P., Delachambre, Y., Etude sur modèle du colmatage d'un milieu filtrant. Ecoulement d'une suspension de microsphères à travers un empilement de macrosphères, *Rev. Fr. Corps Gras*, **1**, 3–11 (1965).

(43) Ling, J. T., A study of filtration through uniform sand filters, *Proc. Am. Soc. Civil Eng.*, **81**, Paper 751, 1–35 (1955).

(44) Litwiniszyn, J., Cauchy's problem in the process of colmatage accompanied by diffusion, *Bull. Acad. Pol., Sci. Ser. Sci. Techn.*, **14** (7), 427–687 (1966).

(45) Litwiniszyn, J., Colmatage accompanied by diffusion, *Bull. Acad. Pol., Sci. Ser. Sci. Techn.*, **14** (4), 295–381 (1966).

(46) Litwiniszyn, J., Colmatage considered as a certain stochastic process, *Bull. Acad. Pol. Sci., Ser. Sci. Techn.*, **11** (3), 81–117 (1963).

(47) Litwiniszyn, J., Colmatage—scouring kinetics in the light of stochastic Birth—Death process, *Bull. Acad. Pol. Sci., Ser. Sci. Techn.*, **14** (9), 561–907 (1966).

(48) Litwiniszyn, J., The phenomenon of colmatage, *Arch. Mechaniki Stosowanej*, **18** (4), 479–495 (1966).

(49) Mackrle, V., L'étude du phénomène d'adhérence—Colmatage dans le milieu poreux, Thesis Fac. Sci., Grenoble, France, 1960.

(50) Mackrle, V., Mackrle, S., Adhesion in filters, *J. Sanit. Eng., Div. Proc. Am. Soc. Civil Eng.*, **SA5**, paper 2940, 17–32 (1961).

(51) Mackrle, V., Dracka, O., Svec, J., Hydrodynamics of the disposal of low level liquid radioactive wastes in soil, *Int. At. Energy Ag., Contract Rep.* 98 (1965).

(52) Maroudas, A., Clarification of suspensions: a study of particle deposition in granular filter media, Ph. D. Thesis, Univ. of London, 1961.

(53) Maroudas, A., Eisenklam, P., Clarification of suspensions: a study of particle deposition in granular media. Pt. I. Some observations on particle deposition, *Chem. Eng. Sci.*, **20**, 867–873 (1965).

(54) Maroudas, A., Eisenklam, P., Clarification of suspensions: a study of particle deposition in granular media. Pt. II. A theory of clarification, *Chem. Eng. Sci.*, **20**, 875–888 (1965).

(55) Maroudas, A., Particle deposition in granular filter media, Pt. I, *Filtr. Separ.*, **2**(5), 369–372 (1965).

(56) Maroudas, A., Particle deposition in granular filter media, Pt. II, *Filtr. Separ.*, **3**(2), 115–121 (1966).

(57) Mintz, D. M., Modern theory of filtration, *Int. Water Supply Ass., Congr.*, London (1966).

(58) Mintz, D. M., Krishtul, V. P., Investigation of the process of filtration of a suspension in a granular bed, *J. Appl. Chem.* (English transl.), **33**, 303–314 (1960).

(59) Mintz, D. M., Paskutskaya, L. N., Chernova, Z. U., Mechanism of the filtration process in high speed water filters, *Zh. Prikl. Khim.*, **40**, 1695–1700 (1967).

(60) O'Melia, C. R., Crapps, D. K., Some chemical aspects of rapid sand filtration, *J. Am. Water Works Ass.*, **56**, 1326–1344 (1964).

(61) O'Melia, C. R., Stumm, W., Theory of water filtration, *J. Am. Water Works Ass.*, **59**, 1393–1412 (1967).

(62) Ornatskii, N. V., Sergeev, E. M., Shekhtman, Yu. M., Theory of clogging of a sand bed when a clay suspension is filtered through it, *Izv. Mosk. Univ.*, chap III, 49–80 (1955).

(63) Sakthivadivel, R., Theory and mechanisms of filtration of noncolloidal fines through a porous medium, Tech. Rep. HEL 15–5, Hydraulic Eng. Lab., Univ. of Calif., Berkeley, 1966.

(64) Sanford, L. H., Gates, C. D., Effects of synthetic detergents on rapid sand filter performance, *J. Amer. Water Works Assoc.*, 45–54 (1956).

(65) Shekhtman, Yu. M., The filtration of a liquid containing suspended solid particles, *Izv. Akad. Nauk SSSR, Otd. Tekh. Nauk, Mekh. Mashinostr.*, (2) 205–207 (1959).

(66) Smith, C. V., Determination of filter media zeta potential, *J. Sanit. Eng., Div. Proc. Am. Soc. Civil Eng.*, **93** (SA5), 91–107 (1967).

(67) Smith, C. V., Electrokinetic phenomena in particulate removal by rapid sand filtration, *J. N. Engl. Water Works Ass.*, **81**, 170–212 (1967).

(68) Smith, C. V., Electrokinetic phenomena in particulate removal from water by rapid sand filtration, Ph. D. Thesis, The Johns Hopkins Univ., 1966.

(69) Stanley, D. R., Sand filtration studied with radiotracers, *Proc. Am. Soc. Civil Eng.*, **81** (592), 1–23 (1955).

(70) Stein, P. C., A study of the theory of rapid filtration of water through sand, D.Sc. Thesis, M.I.T., 1940.

(71) Trzaska, A., Colmatage phenomena in an artificial porous medium, *Zesz. Probl. Gornictwa Pan Tom*, **4**, 257–293 (1966).

(72) Trzaska, A., Experimental research on the phenomenon of colmatage, *Bull. Acad. Pol. Sci., Ser. Sci. Tech.*, **13** (9), 451–775 (1965).

(73) White, P. A. F., Smith, S. E., High Efficiency Air Filtration, Butterworths, London, 1964.

(74) Yao, K. M., Influence of suspended particles size on the transport aspect of water filtration, Ph. D. Thesis, Univ. of North Carolina, 1968.

8

Dispersion in Heterogeneous Nonuniform Anisotropic Porous Media

R. A. GREENKORN
D. P. KESSLER

Dispersion in Heterogeneous Nonuniform Anisotropic Porous Media

This is a state-of-the-art review of dispersion in real porous media to examine present knowledge and to indicate research directions.

We limited the discussion to macroscopic mixing caused by uneven cocurrent laminar flow exemplified by flow of petroleum in reservoirs, movement of fresh water and waste fluids in aquifers, and flow of fluids in fixed beds. Countercurrent gas–liquid, liquid–liquid, and fluidized systems are not considered. Further, the discussion is directed at dispersion in nonideal media.

We do not pretend to be exhaustive in this review—rather, we attempt to cover selected papers which illustrate the major directions of research. The interested reader can find more material in the bibliography than can be given fair treatment within the scope of this paper.

We start with definitions of dispersion and of the three nonidealities, heterogeneity, nonuniformity, and anisotropy. After a brief general introduction, we consider statistical models of dispersion, continuum models of dispersion, and data which apply to systems containing nonidealities.

Definitions. In a sentence, dispersion, as considered here, is the macroscopic mixing caused by uneven cocurrent laminar flow in fixed beds of real media. The behavior of a traced single fluid in one-dimension is indicated schematically in Figure 1 with some boundary conditions. We imagine fluid flowing according to Darcy's law and further imagine we can somehow mark a line of particles (the *X*'s in the figure). As a result of velocity variation and path selection, this line spreads as the flow progresses so that the *X*'s are now contained in a region *a*. If the line had not spread and the *X*'s coincided with line *b* (having arrived there at the Darcy velocity), a plot of the concentration of marked particles, C (normalized by the input concentration, C_0) *vs.* throughput measured in multiples of total pore volume would give a step increase from $C/C_0 = 0$ to $C/C_0 = 1$ at $V/V_0 = 1$, where V is fluid throughput and V_0 is the pore volume of the medium. If only diffusion were operative, there would not be a sharp line at *b*, but rather the line would be fuzzy. The resulting C/C_0 *vs.* V/V_0 would give a breakout of marked particles before and a "tail" after $V/V_0 = 1$. With dispersion, the breakout curve takes on a distinct *S*-shape with breakout much before $V/V_0 = 1$ and with a long "tail" after $V/V_0 = 1$. We often apply the diffusion model to dispersion with velocities such that there is no significant diffusion effect, so:

$$\frac{\partial C}{\partial t} = D\frac{\partial^2 C}{\partial x^2} - v_x\frac{\partial C}{\partial x} \tag{1}$$

where D is the dispersion coefficient. The approximate solution of this equation with the boundary conditions of Figure 1*a* is

$$C/C_0 = \frac{1}{2}\left[1 - \text{erf}\,\frac{(L - v_x t)}{2\sqrt{Dt}}\right] \tag{2}$$

where L is the length of the media. At $V/V_0 = 1$, for this idealized situation:

$$D = \frac{1}{4\pi}\frac{Lv_x}{\left.\dfrac{\partial C/C_0}{\partial V/V_0}\right|_{V/V_0 = 1}} \tag{3}$$

In two-dimensional dispersion (during one-dimensional flow), we imagine a dot of marked particles will spread in the direction of flow (longitudinal dispersion) and perpendicular to the direction of flow (transverse dis-

This review examines dispersion and defines the nonidealities of heterogeneity, nonuniformity, and anisotropy. Statistical models are considered and continuum theories are reviewed

persion) as shown in Figure 1*c*. This situation is often modeled by:

$$\frac{\partial C}{\partial t} = D_L \frac{\partial^2 C}{\partial x^2} + D_T \frac{\partial^2 C}{\partial y^2} - v_x \frac{\partial C}{\partial x} \quad (4)$$

where D_L and D_T are called the longitudinal and transverse dispersion coefficients, respectively. The above is an oversimplified picture since Equations 1 and 4 may not be the correct equations, the boundary conditions may be improper, and the approximation of Equation 2 may not be valid (since there is an additional term). However, Equation 2 is most often used to curve-fit break-out curves like Figure 1*b*, and Equation 3 is used to calculate dispersion coefficients. This procedure does provide a picture of what we mean by dispersion and the dispersion coefficient.

Since we must discuss porous media that are nonideal (that is heterogeneous, nonuniform, and anisotropic), it is necessary to define our particular use of these terms since some ambiguity exists in their use. We will define the terms based on macroscopic properties of the media, since it is the macroscopic properties of the media that are usually of interest to us (except, perhaps, for some problems involving catalyst beds—such as local high temperatures). We will attempt to give some insight into what these definitions may mean from a microscopic standpoint but with considerable imprecision.

First, by macroscopic we imply averaging over elemental volumes of radius ϵ about a point in the media, where ϵ is large enough that Darcy's law can be applied for appropriate Reynolds numbers. In other words, we are at volumes large with respect to that of a single pore. Further, we intend ϵ to be the minimum radius that satisfies such a condition; otherwise, by making ϵ too

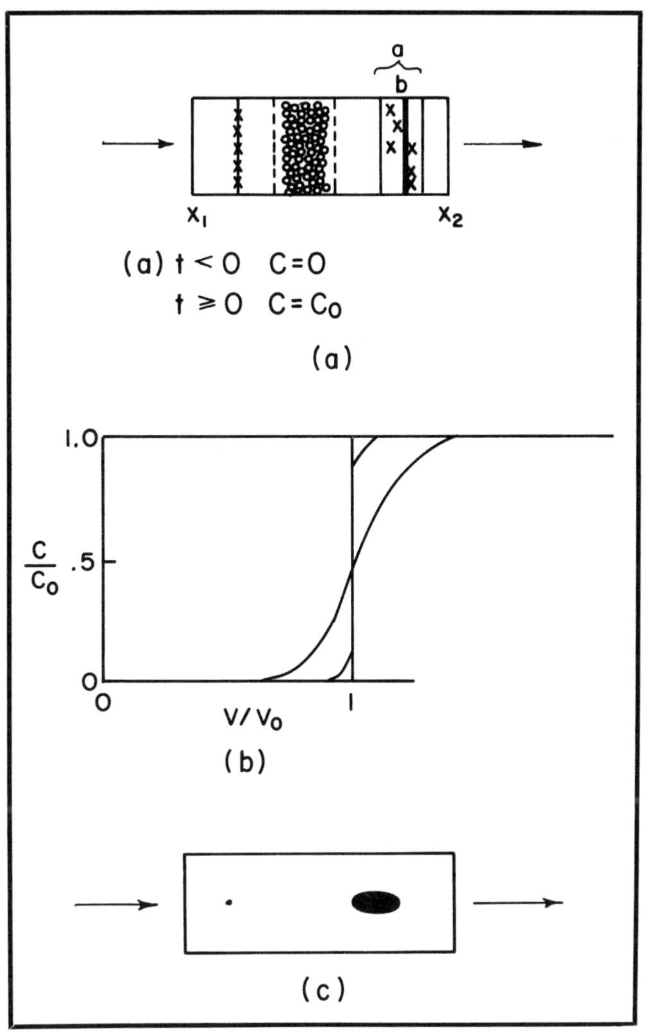

Figure 1. Dispersion and flow

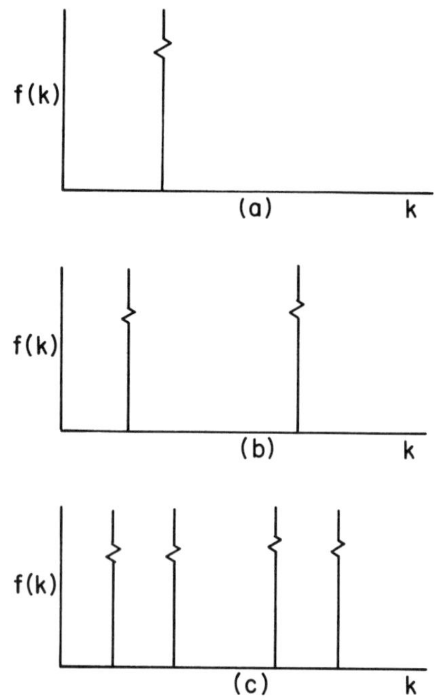

Figure 2. Observed permeability for uniform media

(*a*) *Homogeneous*
(*b, c*) *Heterogeneous*

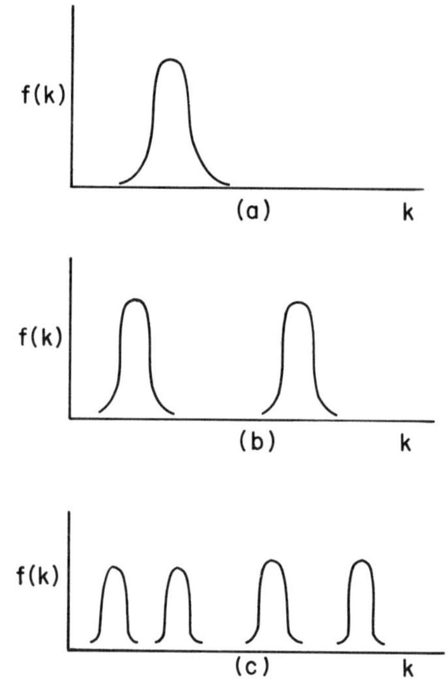

Figure 3. Observed permeability for nonuniform media

(*a*) *Homogeneous*
(*b, c*) *Heterogeneous*

large, we may obscure certain nonidealities by burying their effects far within the elemental volume. Obviously one can, for all practical purposes, remove certain effects by scale.

For example, consider a single ping-pong ball buried in a bed of 0.01-in. glass spheres. An elemental volume of radius of 1 in. will certainly not behave in a homogeneous, uniform, isotropic (ideal) fashion if it includes the ping-pong ball; however, in an elemental volume of radius 500-ft, one will probably never detect the effects of one lone ping-pong ball.

We base our definitions of heterogeneity, nonuniformity, and anisotropy on the probability density distribution of permeability of random macroscopic elemental volumes selected from the medium, the permeability being expressed by the one-dimensional form of Darcy's law

$$v_x = -\frac{k}{\mu}\frac{dp}{dx} \qquad (5)$$

By a uniform medium we mean one in which the probability density function for permeability is either a Dirac delta function or a linear combination of N functions that satisfy the relation:

$$f(k) = \sum_{i=1}^{N} \xi_i \delta, \qquad N \text{ finite} \qquad (6)$$

where

$$\sum_{i=1}^{N} \xi_i = 1$$

(The ξ_i are constants, and the side condition is necessary to ensure that the integral of the density function is equal to one.) Examples of this sort of behavior are shown in Figure 2. By nonuniform we mean a medium in which the probability density function cannot be constructed with a finite number of weighted delta functions (Figure 3).

By a heterogeneous medium we mean one in which the permeability distribution is at least bimodal (Figure 2*b*,*c*, and Figure 3*b*,*c*).

Figure 2*a* is homogeneous and uniform; Figure 3*a* is homogeneous and nonuniform.

We have drawn the distributions in Figure 3 as if all have the same variance (homoscedastic). We can also speak of a second-order nonideality where the variance of the permeability distribution depends on spatial location or orientation in the medium.

Thus far, we have said nothing about anisotropy. By anisotropy we mean the permeability varies with direction in an elemental volume in the medium.

We now can make a more mathematical statement of our definitions. The probability density function for permeability is, in general, a function of location and orientation. We can describe this function with five independent variables: the rectangular coordinates x_i ($i = 1, 2, 3$) for location, and the angular coordinates θ, ψ for orientation:

$$P(k_1 \leq k \leq k_2) = \int_{k_1}^{k_2} f(x_i,\theta,\psi)dx_i d\theta d\psi \qquad (7)$$

If the probability density distribution is independent of θ and ψ, the medium is isotropic; if the distribution is expressible by a finite linear combination of delta functions, the medium is uniform; if the distribution is monomodal, the medium is homogeneous; and if the variance of the distribution is not constant, the medium has second-order nonidealities. The second-order nonidealities may also be classified with respect to the probability density function for *variance* of the permeability distribution:

(1) Nonuniformity (variance density function not a finite series of delta functions)

(2) Heterogeneity (variance density function bimodal)

(3) Anisotropy (variance density function a function of orientation) where the variance distribution is:

$$P(\text{var}_1 \leq \text{var } k \leq \text{var}_2) = \int_{\text{var}_1}^{\text{var}_2} g(x_i,\theta,\psi)dx_i d\theta d\psi \qquad (8)$$

(To get the probability density function for the variance we obviously must consider elemental volumes much larger than for the parent permeability distribution.) Extension to higher order nonidealities is obvious but not useful at this point.

Mechanisms of dispersion. One might imagine a variety of mechanisms on a microscopic scale which lead to observable macroscopic dispersion. Some of them might be:

(1) MOLECULAR DIFFUSION: If time scales are sufficiently long, dispersion results from molecular diffusion.

(2) EDDIES: If the flow within the individual flow channels of the porous medium becomes turbulent, dispersion results from eddy migration.

(3) MIXING DUE TO OBSTRUCTIONS: The fact that the flow channels in a porous medium are tortuous means that fluid elements starting a given distance from each other and proceeding at the same velocity will not remain the same distance apart, as shown in Figure 4.

(4) PRESENCE OF AUTOCORRELATION IN FLOW PATHS: Dispersion can result from the fact that all pores in the porous medium are not accessible to a fluid element after it has entered a particular flow path. In other words, the connectivity of the medium is not complete (Figure 5).

(5) RECIRCULATION CAUSED BY LOCAL REGIONS OF REDUCED PRESSURE: Dispersion can be caused by a recirculation arising from flow restrictions. The conversion of pressure energy into kinetic energy gives a local region of low pressure, and if this region is accessible to fluid which has passed through the region previously, a recirculation is set up much as in a venturi-manometer combination which contains no manometer fluid (Figure 6).

(6) DEAD-END PORES: Dead-end volumes cause dispersion in unsteady flow (concentration profiles varying) because, as a solute-rich front passes the pore, diffusion into the pore occurs by molecular diffusion. After the front passes, this solute will diffuse back out, thus dispersing, for example, a step concentration input to the system. This pore volume also causes confusion in experimental interpretation because it is measured as porosity, but this porosity does not contribute to the available flow cross section.

(7) ADSORPTION: Dispersion by adsorption is again an unsteady-state phenomenon. Just as with dead-end pores, a concentration front will deposit or remove material and therefore tends to flatten concentration profiles in the interstitial fluid.

(8) HYDRODYNAMIC DISPERSION: Macroscopic dispersion is produced in a capillary even in the absence of molecular diffusion because of the velocity profile produced by the adhering of the fluid to the wall. This causes fluid particles at different radial positions to move relative to one another so a series of mixing-cup samples at the end of the capillary exhibits dispersion.

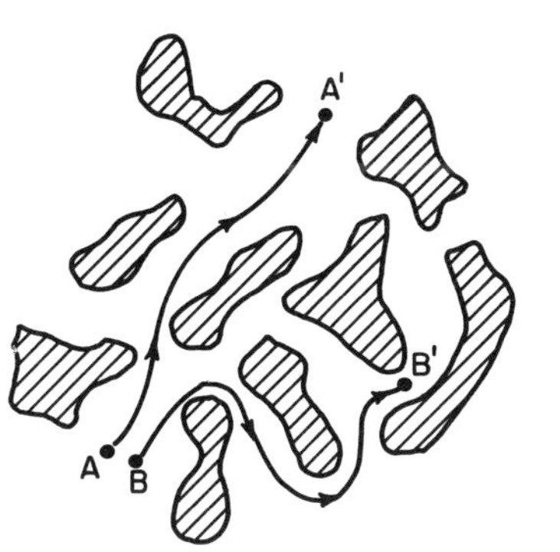

Figure 4. Mixing caused by obstructions

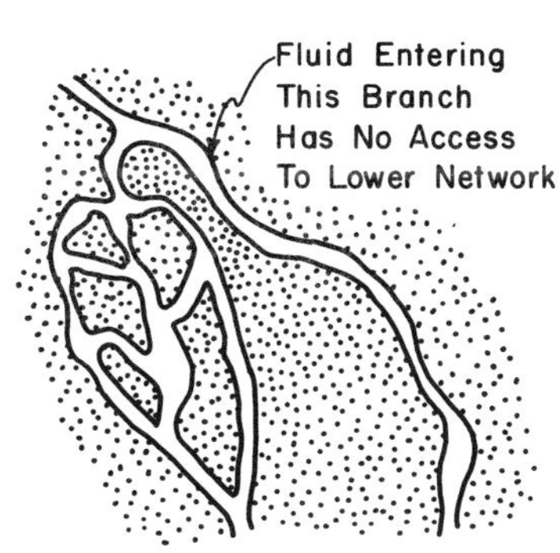

Figure 5. Incomplete connectivity of medium

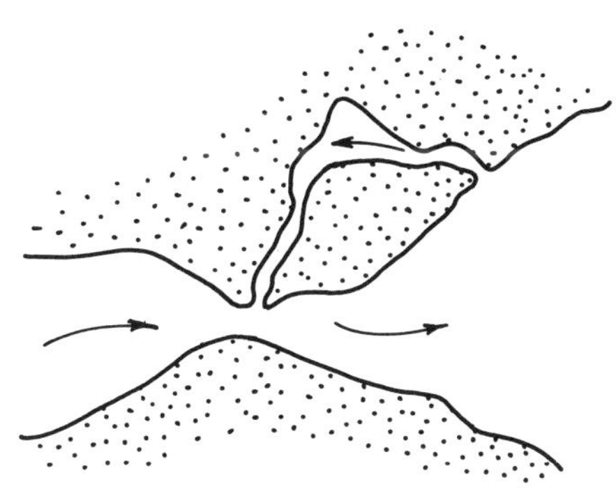

Figure 6. Recirculation caused by local regions of reduced pressure

(9) Macroscopic Dispersion: Caused by nonidealities which change gross streamlines.

Literature review. Two relatively recent papers have summarized some of the work in the field of dispersion: first, the comprehensive review by Perkins and Johnston in 1963 (*75*), and the more recent review of longitudinal dispersion by Chung and Wen in 1968 (*24*). Perkins and Johnston summarize some of the mathematical treatment and discuss the effects of fluid and media properties. The article of Chung and Wen contains an extensive tabulation of experimental work including such items as systems and ranges of variables studied.

We will not list data on longitudinal dispersion since many are available in the review articles cited above. Here we briefly mention recent data which include transverse measurements and/or nonideal media. (Later some of these data will be discussed in more detail.)

Hassinger and Von Rosenberg (*50*) have investigated transverse dispersion in a column packed with relatively uniform-size spherical beads. They present plots of the transverse dispersion coefficient *vs.* the usual Peclet number (*50*) and compare their results to those of Pozzi and Blackwell (*82*) and those of Perkins and Johnston (*76*). Qualitative agreement was achieved. They point out that it is not realistic to expect to characterize all porous media by a single parameter, such as effective particle size, in light of all the potential variations in pore configuration.

Similar data on longitudinal and transverse dispersion were obtained by Grane and Gardner (*42*), who studied dispersion in beds of uniform-size glass spheres and in a sandstone core. The limit at low Peclet numbers is reported as 0.6 to 0.7 by various authors (*105*). The dispersion coefficient is calculated by applying one of the mathematical models to the data and therefore contains within it the assumptions of the model used.

Several investigators have attempted to correlate the dispersion exhibited at large Peclet numbers with the properties of the media by using the single parameter, effective diameter. Most authors (*47, 80, 84, 91*) usually take as the effective diameter that of the 50% point in the number distribution of diameter. The correlation for homogeneous media is of the form:

$$\text{Dispersion coefficient} = \text{parameter} \times (\text{interstitial velocity})^n \quad (9)$$

where n ranges between one and two and is a function of effective diameter. Sufficient agreement does not exist to give an adequate functional description of n with this one-parameter model.

Niemann (*68*) has extended the data of Pleshek (*80*) to show that nonuniform heterogeneous media exhibit reciprocity (no difference on reversal of flow direction), but that simple averaging procedures applied to dispersion coefficients or Peclet numbers for homogeneous media to predict dispersion coefficients for heterogeneous media were not fruitful.

The effects of fluid properties are discussed by Perkins and Johnston (*76*). For multiphase flow, the viscosity ratio must be favorable (displacing fluid more viscous) or instabilities exist which can lead to fingering. The stability problem is an extremely complicated one and will not be discussed here. For the case of favorable viscosity ratio, the dispersion coefficient can be referred to the dispersion coefficient at unit mobility ratio as a function of mobility ratio.

Density ratio of fluids also can be unfavorable (denser fluid above), and this instability will not be covered. For favorable density ratios, Grane and Gardner (*42*) show data supporting a theoretical limit of 11 for Peclet number for turbulent completely mixed interstitial flow.

Blackwell *et al.* (*19*) give breakthrough curves for both homogeneous and heterogeneous systems but do not present a correlation of dispersion coefficients.

Shamir and Harleman (*97*) have studied longitudinal and lateral dispersion in layered (heterogeneous) media and suggest that consideration of such media for very thin layers may provide an approach to anisotropic media.

Statistical Approach

Even though single-continuum models for dispersion in porous media find utility, the single-continuum approach ignores the microscopic nature of the process—namely, that the dispersion occurs not in one continuous medium, but in a medium which exhibits abrupt changes in fundamental properties when one goes from fluid to the porous structure and *vice versa*. For fluid flowing in a porous medium under conditions where the fluid and the porous medium each behave individually as continua (in other words, not under conditions where random molecular motion is important), if sufficient macroscopic parameters (pressure drop, fluid-solid boundary location, etc.) are specified, the problem is deterministic rather than stochastic in nature; in principle, given complete information about the system, we could calculate the detailed path of each fluid particle.

Such an approach, although possible in principle, is, in practice, useless for several reasons: Determination of the precise solid-fluid boundary is, and will probably remain, impossible; the boundary (even if known) within which we would have to solve the equations of motion of the fluid would be so tortuous as to render the problem mathematically intractable; and even if solution of the paths of individual fluid particles could be obtained, such detail is of almost no practical value—only macroscopic or integral properties of the flow field are of interest.

For these reasons, one approach to description of flow in porous media is to attempt to construct a stochastic model which represents, in its average properties, the properties of a porous medium. Instead of solving differential equations representing behavior of a continuum to obtain individual path lines for fluid elements and then taking appropriate configuration-space/time averages to get information such as average flow rate, seepage velocity, etc., we rather feed a large number of fluid elements to the stochastic model and average over the probability density distribution of the model. If the ergodic assumption that the averages in configuration-space/time and probability space coincide, we have a useful tool to study gross flow behavior.

The first question is what features to incorporate in the stochastic model. One can choose to regard flow in porous media as either flow in channels or flow around submerged objects or some combination of the two. Almost all stochastic models focus on flow in channels.

Once one decides to consider flow in elemental channels, it is then necessary to decide how elaborate the description of an elemental region in probability space must be. A real porous medium is a hornets' nest of voids, interconnecting passages, and cul de sacs. (We will, incidentally, exclude from our discussion that large body of literature concerned with flow interior to catalyst pellets and focus our attention on the interstices. In principle, the pellet interior is simply a porous medium of different scale.)

Several approaches toward developing stochastic models for channel networks exist. First, it is possible to regard the passage of fluid through the medium as a series of steps in a random walk. Second, the process may be regarded as analogous to turbulent flow, and the same mathematical model may be applied (*91*). Third, by concentrating on the void spaces and modeling them as mixing cells, certain of the mathematics from statistical thermodynamics may be applied, including such concepts as the Hamilton partition function (*91*). Fourth, one can model the medium by networks of capillaries built up of common subelements—*e.g.*, capillaries connected so that they form the edges of certain regular polyhedra. Fifth, a model may be constructed by considering flow serially through a succession of capillaries drawn from some distribution of size, length, and orientation.

The fourth and fifth of these approaches seem to have enjoyed the greatest success and will be discussed in order. Although in principle the processes considered need not be Markoffian, for the sake of simplicity this assumption is usually made. In other words, we usually limit ourselves to processes in which the transition probabilities for a fluid element from one state to the next are constant—in particular, they are independent of time (*e.g.*, we do not treat swelling media). In fact, not only is the Markoffian assumption made, but also the complete absence of autocorrelation is usually assumed—*i.e.*, the probability of entrance of the fluid into a pore of given size and orientation is independent of the size and orientation of the pore it is leaving.

This is not true for media which have been deposited by some mechanism which is strongly nonrandom–extreme layering, for example.

Network models are exemplified by the work of Fatt (*36*, *37*). Network models can be applied either as stochastic models or not, depending on whether the pores which constitute the connections in the network are taken as having other than a point-probability distribution of sizes. In a sense, the randomly oriented capillary models are simply more general network models in which the connecting links in the network are permitted a distribution of orientations and lengths rather than being fixed as, for example, the edges of regular polyhedra.

Fatt used two-dimensional polygonal (not polyhedral) networks based on square elemental units, as well as isolated and overlapping hexagonal elemental units (one square, three hexagonal). He calculated capillary pressure curves using a radius distribution. This type of model has the advantage that the intruding fluid is not assumed to have access to all pores in the network at once—for example, a large pore isolated by connection only to smaller pores would not be invaded by fluid until the pressure exceeds the capillary pressure of the largest of the connecting smaller pores.

Fatt also calculates relative permeability but does not calculate dispersion. There is, however, no reason that dispersion cannot be calculated for this approach.

Where one wishes to vary either length or orientation of the elemental pores in the network, it becomes impossible to continue to use networks constituting edges of regular polyhedra. One gains flexibility in description of the elemental pore, but the question of how the pores are interconnected becomes agonizing. The usual tactic is to assume a serial path for each fluid element in which no account is taken of the multiplicity of pores connected at any given point (or alternatively, this is assumed to be accounted for by the probability density function).

This more general model of randomly oriented pores is exemplified by the work of Scheidegger, Saffman, De Josselin De Jong, and Haring and Greenkorn (*28*, *46*, *89*, *91*). Scheidegger (*91*) treated the statistical groundwork for the models of Saffman and De Josselin De Jong which were developed in parallel and independently. Saffman considered an ensemble of randomly mixed straight pores. The pressure gradient in the medium was taken as linear with distance with an imposed fluctuation described by a Gaussian and isotropic probability density function. A random walk was taken with this model, assuming that successive steps were statistically independent. The lateral and longitudinal dispersions were then calculated and comparison with data was made.

The model of Saffman and De Josselin De Jong has been extended by Haring and Greenkorn to the case of nonuniform media by use of the beta distribution, both for the radius and for the length distribution individually. Capillary pressure, permeability, and longitudinal and transverse dispersion are calculated in terms of the

---

AUTHORS *R. A. Greenkorn and D. P. Kessler are on the faculty at Purdue University, Lafayette, Ind. 47907. Dr. Greenkorn is Professor and Dr. Kessler is Associate Professor of Chemical Engineering.* *This work was supported in part by Federal Water Pollution Control Administration grant WPO1048-81.*

parameters of the beta distribution. In this model, they assume an elemental pore as shown in Figure 7. The length, radius, and orientation angles θ and ψ are assumed independent. The dimensionless length $l^* = l/L$ where L is the longest pore and dimensionless radius $r^* = r/R$, where R is the largest radius pore are distributed according to the beta function. (The model can be made anisotropic by distributing the orientation angles.) The choice of the beta function is arbitrary and was made since it is a distribution which gives a range of skew and symmetric shapes, depending on values of the parameters. It is also conveniently normalized.

The probability distribution functions for l^* and r^* are

$$f(l^*) = \frac{(a + b + 1)!}{a!b!}(l^*)^a(1 - l^*)^b \qquad (10)$$

$$g(r^*) = \frac{(\alpha + \beta + 1)!}{\alpha!\beta!}(r^*)^\alpha(1 - r^*)^\beta \qquad (11)$$

The capillary pressure, permeability-porosity ratio, and tortuosity of the model can be calculated. We are interested in seeing the effect of nonuniformity on the dispersion during flow through the model.

The probability of existence of a pore with properties in the range $l^* \rightarrow l^* + dl^*$, $r^* \rightarrow r^* + dr^*$, $\theta \rightarrow \theta + d\theta$, $\psi \rightarrow \psi + d\psi$ is given by

$$dE = \frac{1}{2\pi}[f(l^*)dl^*][g(r^*)dr^*] \sin \theta d\theta d\psi \qquad (12)$$

If we assume a marked particle takes a random walk through the model selecting pores proportional to volumetric flow rate, the probability of pore selection is

$$dP = \frac{q}{M} dE = \frac{v\pi r^2}{M} dE \qquad (13)$$

where M is the normalization constant. The dispersion is the variance of the average position of a marked particle; thus the longitudinal dispersion is

$$D_L = \frac{\overline{(Z - \bar{V}T)^2}}{2T} \qquad (14)$$

and the transverse dispersion is

$$D_T = \frac{\overline{X^2}}{2T} = \frac{\overline{Y^2}}{2T} \qquad (15)$$

The position of a marked particle at the end of a number of steps is

$$Z_n = \sum_{i=1}^{n} z_i \qquad (16)$$

and the average position is

$$\bar{Z}_n = nL \int_P l^* \cos \theta dP \qquad (17)$$

Likewise

$$\bar{X}_n = \bar{Y}_n = 0 \qquad (18)$$

and

$$\bar{T}_n = nL \int_P \left(\frac{l^*}{v}\right) dP \qquad (19)$$

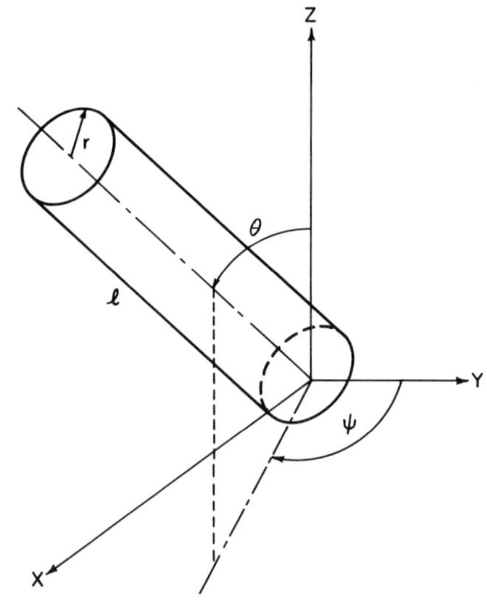

Figure 7. Elemental pore for statistical model

The variances and co-variance may also be found by

$$\overline{(Z_n - \bar{Z}_n)^2} = n \int_P (z - \bar{z})dP = nL^2\sigma_z^2 \qquad (20)$$

and from similar integration over P

$$\overline{X_n^2} = nL^2\sigma_x^2 \qquad (21)$$

$$\overline{(T_n - \bar{T}_n)^2} = n \frac{L^2}{V^2} \sigma_T^2 \qquad (22)$$

$$\overline{(Z_n - \bar{Z}_n)(T_n - \bar{T}_n)} = \sigma_{zT}^2 \qquad (23)$$

The expressions for the dispersion coefficients in terms of the parameters of the distribution function are

$$D_L = \frac{1}{12} \frac{(a + 2)(a + b + 2)}{(a + 1)(a + b + 3)\, J^2} \langle l \rangle V \times \ln\left[\frac{27}{2} \frac{(a + b + 2)^2}{(a + 1)^2 J^3} \frac{vT}{\langle l \rangle}\right] \qquad (24)$$

and

$$D_T = \frac{3}{16} \frac{(a + 2)(a + b + 2)}{(a + 1)(a + b + 3)J} \langle l \rangle V \qquad (25)$$

and

$$\frac{D_L}{D_T} = \frac{4}{9} J \ln\left[\frac{27}{2} \frac{(a + b + 2)^2}{(a + 1)^2 J^3} \frac{VT}{\langle l \rangle}\right] \qquad (26)$$

where

$$J = \frac{(\alpha + 1)(\alpha + 2)(\alpha + \beta + 4)(\alpha + \beta + 5)}{(\alpha + 3)(\alpha + 4)(\alpha + \beta + 2)(\alpha + \beta + 3)} \qquad (27)$$

The dispersion coefficients for the model are definitely functions of the nonuniformity as shown in Table I. Interestingly, the particle following the most probable path does not travel at the Darcy velocity but at a velocity

$$\tilde{V} = \frac{V}{J} \tag{28}$$

where V is the Darcy velocity. The model does predict a linear dependence of dispersion on velocity. (It inherently assumes radial diffusion in the pore is fast.) Finally, the tortuosity of the model is constant.

Continuum Approaches. Babbitt (*6*) investigated the differential equations for diffusion of gases through solids. Although the problem of dispersion is not discussed, he does raise this issue—at the molecular level the number of molecules transferred is proportional to a concentration gradient; but if we assume diffusion (and dispersion) to be represented by an equation where the dependent variable is concentration, then:

$$\frac{\partial C}{\partial t} = D\frac{\partial^2 C}{\partial x^2} \tag{29}$$

we have tacitly assumed concentration is directly proportional to pressure, since the fundamental equation of flow expresses the flux as proportional to the gradient of a potential function. The potential function for diffusion should be pressure. According to Darcy's equation, the flux is proportional to the velocity of flow. Thus, if we use a form of Equation 29 to describe dispersion the resulting coefficient must be dependent on the velocity.

Kramers and Alberda (*56*), in studying frequency response of continuous flow systems, rationalized that the equation for concentration in a packed tube with uniform flow is

$$\frac{\partial C}{\partial T} = D\frac{\partial^2 C}{\partial x^2} - v_x\frac{\partial C}{\partial x} \tag{30}$$

on the basis of a cell-mixing model. The solution for this problem yields the same frequency response diagram as a number n of perfect mixers, each with the same residence time. Thus, if a packed bed is divided into n parts of equal length, L/N, the average residence time is L/nv_x and the dispersion time constant is $L^2/2Dn^2$.

Aris and Amundson (*3*) show if one adopts a mixing cell model for dispersion, the resulting concentration distribution is given by Poisson's distribution. They show for a Peclet number of 2 the mixing cell model and diffusion model give the same results. They also point out that the usual boundary conditions used with Equation 30 (Figure 1*a*) are improper.

Coats and Smith (*25*) considered the effect of dead-end pores and investigated the three-parameter capacitance model proposed by Deans (*30*), as well as solution of Equation 30 with the three sets of boundary conditions:

	1	2	3
$x_1 = 0$	$C = C_0$	$vC_0 = vC - D\frac{\partial C}{\partial x}$	$vC_0 = vC - D\frac{\partial C}{\partial x}$
$x_2 = L$	$C = 0$	$C = 0$	$\frac{\partial C}{\partial x} = 0$

The results show that the solutions of the diffusion model for each of the three sets of boundary conditions give essentially the same dispersion coefficient within experimental error. In some cases the three-parameter model gives a better fit of the breakout curve (especially the nonsymmetrical tail). It is not clear how one compares the "dispersion coefficient" for the three-parameter

Table I. Calculated Values of Dispersion (Cm^2/Sec) for Various Values of Parameters of Beta Distribution (Velocity = 0.1 Cm/Sec)

a	b	α	β	$K_L \times 10^3$	$K_T \times 10^4$	K_L/K_T
2	2	0	0	1.77	4.85	3.66
2	2	$-1/2$	$-1/2$	7.28	7.87	9.37
2	2	$-1/2$	1	33.2	16.8	19.8
2	2	2	4	4.82	6.54	7.40
2	2	2	2	3.78	5.87	6.42
2	2	4	2	2.43	4.89	4.96
2	2	1	1	4.54	6.32	7.18
2	4	2	4	3.96	5.15	7.70
4	2	2	4	4.30	5.98	7.20
2	4	1	1	4.78	4.96	9.65
4	2	1	1	7.95	11.6	6.85
2	4	4	2	2.02	4.00	5.05
4	2	4	2	4.26	8.95	4.76
1	1	4	2	1.98	4.28	4.68
1	1	2	4	3.54	5.75	6.16

model with the single parameter. All four solutions fit the breakout data quite well.

Scheidegger (*93*) described dispersion by expressing the motion of a tracer through a porous medium by an equation similar to Kramers and Alberda. He rationalizes the form of the dispersion equation on a statistical basis, using as an assumption the motion at each point is uncorrelated. Thus, by the central limit theorem, the concentration of marked particles is represented by an error function solution and the differential equation whose solution, is the error function is Equation 30. Although Scheidegger (*93*) does not write the equation in tensor form in this paper, in general many authors agree the usual form of the dispersion equation is given by

$$\frac{\partial C}{\partial t} + v_i \frac{\partial C}{\partial x_i} = \frac{\partial}{\partial x_i}\left(D_{jk}\frac{\partial C}{\partial x_k}\right) \tag{31}$$

where D_{jk} is the dispersion tensor.

Rifai *et al.* (*85*) used solutions of Equation 30 for one-dimensional flow to interpret tracer concentration breakout curves resulting from a step input of tracer. Most dispersion coefficients result from interpreting the data with a solution of Equation 30 with simple geometries and boundary conditions.

Scheidegger (*92*) showed there are two possible extremes for the form of the dispersion coefficient: the first directly proportional to velocity, the second proportional to velocity squared, thus

$$D = av \tag{32}$$

$$D = av^2 \tag{33}$$

Equation 32 represents the case where there is enough time for complete mixing of the invading and original fluid in each flow channel. Equation 33 represents the situation where mixing is not complete.

If one adopts a diffusion-type equation as expressed by Equations 30 and 31, there are several models of dispersion that result, depending on whether the dispersion coefficient is allowed to vary with position, whether the dispersion is assumed constant with position, the type of experiment—*i.e.*, point injection or face injections—and whether the experiments are interpreted as finite or infinite in length.

Ogata and Banks (*71*) solved Equation 30 for a homogeneous medium with finite boundary conditions to show these solutions are not symmetrical and found that for most systems, the symmetrical solution is adequate.

Bischoff and Levenspiel (*17*) have investigated the specific forms of Equation 30 and compared the results for various boundary conditions. They present plots showing errors involved in using simplified forms of Equation 30 and for infinite boundary conditions. If we adopt a model such as Equation 30 and fit a solution of this equation to concentration *vs.* time data, it does not necessarily prove the validity of the model. It does show that concentration *vs.* time data can be curve-fit with a solution of Equation 30.

Normally when one investigates longitudinal and transverse dispersion, Equation 30 is written for a cylindrical porous medium with one-dimensional flow

$$\frac{\partial C}{\partial x} = D_L \frac{\partial^2 C}{\partial x^2} + D_T\left(\frac{\partial^2 C}{\partial R^2} + \frac{1}{R}\frac{\partial C}{\partial R}\right) - v_x \frac{\partial C}{\partial x} \tag{34}$$

where D_L and D_T are called the longitudinal and transverse dispersion coefficients, respectively. Most often, experiments are run in such a way that either longitudinal or transverse dispersion is negligible and Equation 34 degenerates to the usual diffusion equation with a convective term. The solution of Equation 34 and the simpler forms are described by Bischoff and Levenspiel (*17*).

Banks and Jerasate (*11*) considered the solution of Equation 30 for longitudinal dispersion where velocity varies as a function of time. They obtain a solution of the equation by assuming the dispersion coefficient is directly proportional to velocity.

Although the discussion to this point applies to uniform, homogeneous, isotropic porous media, the general form of Equation 31 allows the dispersion coefficient to be a tensor so Equation 31 can be applied to nonuniform and anisotropic media by relating the dispersion to the properties of the media and the velocity. Heterogeneity is handled by making the dispersion coefficient a function of position.

Nikolaevskii (*69*) obtained the form of Equation 31 and a relation for the dispersion tensor by analogy to diffusion in homogeneous isotropic turbulent flow. He applies the central limit theorem to marked particles of fluid moving with the mean velocity so that, at an arbitrary instant of time, there exists a three-dimensional probability density of finding particles at some point x_i. If ψ represents the relative concentration of marked particles then

$$\psi(x_i,t) = \frac{(2n)^{3/2}}{\sqrt{\overline{x_1^2}\,\overline{x_2^2}\,\overline{x_3^2}}} \exp\left\{-\frac{1}{2}\sum_{\alpha=1}^{3}\frac{x_\alpha^2}{\overline{x_\alpha^2}}\right\} \tag{35}$$

This is a solution of the diffusion equation of the form

$$\frac{\partial \psi}{\partial t} = \sum_{\alpha=1}^{3}\frac{\partial}{\partial x_\alpha}\left(D_\alpha \frac{\partial \psi}{\partial \psi_\alpha}\right) \tag{36}$$

where D_α is the axial component of dispersion

$$D\alpha = \frac{\overline{X_\alpha^2}}{2T} \tag{37}$$

Since the coefficient of dispersion is different along the axes, D must be a second-order tensor. If we assume D is invariant to rotation about the direction of the mean velocity and mirror reflections relative to planes including the mean velocity vector or perpendicular to this vector, the coefficient will have the form

$$D_{ij} = Au_i^0 u_j^0 + B\delta_{ij} \tag{38}$$

where u_i^0 is the component of the pertubation of mean velocity. Since we assume the fluid motion is represented by Darcy's law, the characteristics of such motion—mean average velocity, fluid viscosity, and a characteristic length of the porous medium—the dispersion coefficient must be proportional to the product of the average velocity and a characteristic length. The characteristic length in general cannot be scalar but must be a tensor. This tensor will have dimensions of length and be of fourth order (the dispersivity tensor) and for an isotropic medium

$$a_{ijkl} = H_1\delta_{ij}\delta_{kl} + H_2\delta_{ik}\delta_{jl} + H_3\delta_{il}\delta_{jk} \tag{39}$$

and in order for Equation 38 to be true

$$D_{ij} = a_{ijkl}\frac{\bar{u}_k\bar{u}_l}{|\bar{u}|} \tag{40}$$

where $H_2 = H_3$ and

$$B = H_1|\bar{\mu}| \tag{41}$$

$$A = 2H_2|\bar{\mu}| \tag{42}$$

The explanation of A and B or H_1 and H_2 depends on the mechanism of the mixing process in the porous medium. At each point in the porous medium the mean velocity vector is modified in a random manner to the local velocity. In general

$$u_i = T_{ij}\bar{u}_j \tag{43}$$

where T_{ij} is the local tensor of the porous medium. He then shows that for one-dimensional flow in an isotropic medium

$$\left.\begin{aligned} D_{11} &= (DT_{11})l\bar{u} = \lambda_1\bar{u} \\ D_{22} &= D_{33} = D(T_{12})l\bar{u} = \lambda_2\bar{u} \\ D_{ij} &= 0 \; i \neq j \end{aligned}\right\} \frac{D_{11}}{D_{22}} = \frac{\lambda_1}{\lambda_2} \tag{44}$$

where l is a "mixing length" for the porous medium. λ_1 and λ_2 are the two components of the dispersivity tensor for an isotropic porous medium. One would estimate l to be about one half of the particle diameter in unconsolidated media or one half of the average of the length of straight lines connecting openings in consolidated media. If one further assumes identical media except for the size of l (particle size in unconsolidated media), then the components of the local tensor T_{ij} should be equal and

$$\frac{D_a}{D_b} = \frac{l_a}{l_b} \tag{45}$$

In other words, he expects the dispersion coefficient to be in ratio to the particle sizes. If we write Equation 36 for a fixed coordinate system

$$\varphi\frac{\partial\psi}{\partial t} = \frac{\partial}{\partial x_1}(\lambda_1 w)\frac{\partial\psi}{\partial x_1} + \frac{\partial}{\partial x_2}(\lambda_2 w)\frac{\partial\psi}{\partial x_2} + \frac{\partial}{\partial x_3}(\lambda_3 w)\frac{\partial\psi}{\partial x_3} - w\frac{\partial\psi}{\partial x_3} \tag{46}$$

where $w = \varphi\bar{u}$ and w is in the direction of x and $\lambda_2 = \lambda_3$. Then, in general,

$$\frac{\partial\psi}{\partial t} = \frac{\partial}{\partial x_j}\left(D_{ij}\frac{\partial\psi}{\partial x_i}\right) - \bar{u}_i\frac{\partial\psi}{\partial x_i} \tag{47}$$

Bear (*12*) obtained a similar result for the form of the dispersion tensor to Nikolaevskii on the basis of geometrical arguments about the motion of marked particles through a porous medium. Bear began on the assumption that the spreading of marked particles from a point system is apparently of the form of a bivariate normal distribution. The resulting distribution depends on a longitudinal and a transverse dispersion coefficient of the porous medium. This assumption includes Equation 31 or 47 as the model for dispersion. The concentration distribution of a point injection of a tracer around an injection point after a displacement $L = vt$ in the direction of flow is

$$C(x,y;\, x_0,y_0) = \frac{c_0}{2\pi\sigma_x\sigma_y}\exp\left\{-\frac{m^2}{2\sigma_x^2} - \frac{n^2}{2\sigma_y^2}\right\} \tag{48}$$

where $m = x - (x_0 + L)$, $n = y - y_0$. The standard deviations are assumed as

$$\left.\begin{aligned} \sigma_x &= (2D_{\mathrm{I}}L)^{1/2} \\ \sigma_y &= (2D_{\mathrm{II}}L)^{1/2} \end{aligned}\right\} \tag{49}$$

Thus the marked particles "spread" as the fluid moves in the x-direction and this "spread" is given by a normal distribution with standard deviation σ_x around the mean position ($x = x_0 + L = x_0 + vt$) in the x-direction and standard deviation σ_y around ($y = y_0 + \Delta y = y_0$) in the y direction. This does not mean there is no movement of particles in the y direction—it means the average motion in the y direction is zero. The problem remains to generalize Equation 49 so that these equations apply to an isotropic medium for an arbitrary coordinate system. Bear proceeds by geometrical arguments to rationalize the form of the tensor which represents the variance of the bivariate normal distribution. In general, the variance transforms as a tensor and we assume

$$(\sigma^2)_{ij} = \tilde{D}_{ijkl}L_{kl} \tag{50}$$

where $\tilde{D}_{ijkl}$ is called the constant of dispersion and L is the tensor of displacement when the mean displacement is $L = v_x t$. In general the mean flow makes an angle β with the x-axis and L_{kl} is given by

$$L_{kl} = \begin{vmatrix} L\cos^2\beta & L\sin\beta\cos\beta \\ L\sin\beta\cos\beta & L\sin^2\beta \end{vmatrix} \tag{51}$$

The variances may be determined in general by geometric arguments by determining the components of $\tilde{D}_{ijkl}$. He rationalized that the components of L_{kl} are in the direction of L and the direction orthogonal to L. The relationship is of the form

$$(\sigma^2)_{ij} = 2\tilde{D}_{\mathrm{II}}L\delta_{ij} + 2(\tilde{D}_{\mathrm{I}} - \tilde{D}_{\mathrm{II}})L_{ij} \tag{52}$$

which means for the tensor $\tilde{D}_{ijkl}$

$$\left.\begin{aligned} \tilde{D}_{1111} &= \tilde{D}_{2222} = \tilde{D}_{\mathrm{I}} \\ \tilde{D}_{1122} &= \tilde{D}_{2211} = \tilde{D}_{\mathrm{II}} \\ \tilde{D}_{1212} &= \tilde{D}_{1221} = \tilde{D}_{2112} = \tilde{D}_{2121} = {}^1/_2(\tilde{D}_{\mathrm{I}} - \tilde{D}_{\mathrm{II}}) \end{aligned}\right\} \tag{53}$$

Thus, Bear defines a fourth-order tensor similar to Nikolaevskii which is a property of the medium and the geometry of the pore-channel system.

Nikolaevskii points out the relation between dispersion and particle size which implies nonuniformity will be handled by superposition. Bear implies the constant of dispersion (dispersivity tensor) is already a function of nonuniformity.

De Josselin De Jong and Bossen (*29*) extend Bear's arguments to construct a differential equation. They did this by assuming dispersion is determined by the three terms of Equation 31. Then applying Bear's (*12*) argument about the form of the dispersion coefficient-velocity relation, they obtained the dispersion coefficient in general must be of the form

$$D_{ij} = a_{ijkl} \frac{v_k v_l}{|v|} \tag{54}$$

where the dispersion tensor is that described by Bear.

Scheidegger (*94*) generalized the results of Bear for an anisotropic porous medium and describes the dispersivity tensor assuming dispersion is given by Equation 31. Based on the assumption that the components of v_i in the direction of v and orthogonal to v are significant, he postulates (similar to Nikolaevskii) Equation 54. The division by $|v|$ is to retain the linearity of the dispersion with Darcy's velocity according to Equation 32. (Though experiments do show dispersion is related to v the relationship is not in general linear; rather v is raised to a power n where n may be as large as 1.2.) The a_{ijkl} is called the dispersivity tensor of the medium and Scheidegger shows it reduces to that defined by Nikolaevskii and Bear for isotropic media. The equation of condition, Equation 31, may be rewritten by inserting Equation 54. The dispersivity tensor is of fourth order and could have as many as 81 components. The number of independent components can be reduced by symmetry arguments.

First we assume

$$a_{ijkl} = a_{ijkl} \tag{55}$$

since the kl terms do not affect the equation of motion˙ Scheidegger (*94*) invokes the Onsager principle for the second symmetry property

$$a_{ijkl} = a_{jikl} \tag{56}$$

since

$$X_j = \frac{\partial \psi}{\partial x_j} \tag{57}$$

is a force and

$$j_i = a_{ijkl} \frac{v_k v_l}{|v|} \frac{\partial \psi}{\partial x_j} \tag{58}$$

is a flux. Then from the force-flux relationship

$$j_i = L_{ij} X_j \tag{59}$$

and the Onsager principle follows:

$$L_{ij} = L_{ji} \tag{60}$$

This is a weak argument on two counts. First, we still have not come to grips with the problem that the "force term" does not cause the flux but results from the flux—it would seem a pressure gradient is the correct force. Second, the Onsager principle is postulated on the strength of microscopic reversibility and we are studying the validity in this purely macroscopic and phenomenological case. (That is, we have adopted a flow picture that predicts results as if the porous medium were not there but represented by superficial velocity and concentration. Since we are being general, it seems more reasonable to accept the first symmetry property but not the second.) However, to proceed with the Scheidegger analysis, let's assume Equation 56 does hold. Scheidegger proceeds to determine the relationship among the various components of the dispersivity tensor assuming there are 36 independent components. He then shows that, to obtain complete symmetry,

$$\left.\begin{aligned} &a_{1111} = a_{2222} = a_{3333} = a_{\mathrm{I}} \\ &a_{1122} = a_{1133} = a_{2233} = a_{2211} = a_{3311} = a_{\mathrm{II}} \\ &a_{1212} = a_{1313} = a_{2323} = a_{2121} = a_{3131} = a_{3232} = \\ &\qquad a_{1221} = a_{1331} = a_{2332} = a_{2112} = a_{3113} = \\ &\qquad\qquad a_{3232} = {}^1/_2(a_{\mathrm{I}} - a_{\mathrm{II}}) \end{aligned}\right\} \tag{61}$$

which is the result obtained by Nikolaevskii and is directly related to the result of Bear. For one-dimensional flow $(v,0,0)$ in a homogeneous medium

$$\frac{\partial \psi}{\partial t} = a_{\mathrm{I}} v_1 \frac{\partial^2 \psi}{\partial x_1^2} + a_{\mathrm{II}} v_2 \frac{\partial^2 \psi}{\partial x_2^2} + a_{\mathrm{II}} v_1 \frac{\partial^2 \psi}{\partial x_3^2} - v_1 \frac{\partial \psi}{\partial x_1} \tag{62}$$

and

$$\left.\begin{aligned} D_{11} &= a_{\mathrm{I}} v_1 \\ D_{22} &= a_{\mathrm{II}} v_2 \\ D_{ij} &= 0 \qquad i \neq j \end{aligned}\right\} \frac{D_{11}}{D_{22}} = \frac{a_{\mathrm{I}}}{a_{\mathrm{II}}} \tag{63}$$

Bachmat and Bear (*9*) adopted a more general form of Equation 31 with velocity inside the space derivative

$$\frac{\partial C}{\partial t} = \frac{\partial}{\partial x_i}\left(D_{ik} \frac{\partial C}{\partial x_k} - v_i C\right) \tag{64}$$

to describe nonuniform flow in homogeneous and isotropic porous media where

$$D_{ik} = a_{iklm} \frac{v_l v_m}{|v|} \tag{65}$$

Since the medium is isotropic, they adopt the model

$$a_{iklm} = \lambda \delta_{ik} \delta_{lm} + 2\mu \delta_{il} \delta_{km} \tag{66}$$

where λ and μ are invariants of the tensor and

$$D_{ik} = \lambda \delta_{ik} v + 2\mu \frac{v_i v_k}{v} \tag{67}$$

If we consider flow along the x-axis $(v,0,0)$, then

$$\left.\begin{aligned} D_{11} &= (\lambda + 2\mu) v \\ D_{22} &= D_{33} = \lambda v \\ D_{ik} &= 0 \qquad i \neq k \end{aligned}\right\} \frac{D_{11}}{D_{22}} = \frac{\lambda + 2\mu}{\lambda} \tag{68}$$

This relation can be related to Scheidegger's if $\lambda + 2\mu = a_{\mathrm{I}}$, $\lambda = a_{\mathrm{II}}$

$$\left.\begin{aligned} D_{11} &= a_I v \\ D_{22} &= D_{33} = a_{II} v \\ D_{ik} &= 0 \qquad i \neq k \end{aligned}\right\} \frac{D_{11}}{D_{22}} = \frac{a_I}{a_{II}} \tag{69}$$

These authors also consider the situation in a generalized frame to determine forms of the coefficients of the dispersion tensor in various coordinate systems.

Poreh (*82*) re-examined the definition of the dispersion tensor given by Equation 54 for isotropic and axisymmetric porous media since experiments indicate dispersion is not directly proportional to velocity. (Unfortunately most dispersion data are reported as coefficients resulting from curve-fitting the error function solutions of Equation 30 for one-dimensional flow. Thus, the results are clouded since one should measure actual point-to-point dispersion to validate the observation.) For an isotropic media the dispersion tensor is contracted with two arbitrary vectors R_i and S_j such that

$$D_{ij}R_iS_j = A_1R_iR_j\delta_{ij} + A_2v_iv_jR_iS_j \tag{70}$$

where A_1 and A_2 are arbitrary functions of v^2 and media properties. Since B_i and S_j are arbitrary

$$D_{ij} = A_1\delta_{ij} + A_2v_iv_j \tag{71}$$

which implies one principal direction of D_{ij} is along the direction of v_i. For the axisymmetric case where the axis of symmetry is given by the unit vector λ the inner product is

$$D_{ij}R_iS_j = B_1\delta_{ij}R_iS_j + B_2v_iv_jR_iS_j + B_3\lambda_i\lambda_jR_iS_j + B_4v_iR_i\lambda_jS_j + B_5v_jS_j\lambda_iR_i \tag{72}$$

Since R_i and S_j are arbitrary

$$D_{ij} = B_1\delta_{ij} + B_2v_iv_j + B_3\lambda_i\lambda_j + B_4v_i\lambda_j + B_5v_j\lambda_i \tag{73}$$

where B_1, B_2, B_3, B_4, and B_5 are arbitrary functions of v^2 and $v_k\lambda_k$. If we assume $D_{ij} = D_{ji}$ then $B_4 = B_5$ and

$$D_{ij} = B_1\delta_{ij} + B_2v_iv_j + B_3\lambda_i\lambda_j + B_4(v_i\lambda_j + v_j\lambda_i) \tag{74}$$

Equation 71 in dimensionless form is

$$\frac{D_{ij}}{D_0} = F_1\delta_{ij} + F_2\left(\frac{l^2}{D_0^2}\right)v_iv_j \tag{75}$$

and F_1 and F_2 are even functions of vl/D_0 and vl/ν. For Equation 71

$$\frac{D_{ij}}{D_0} = G_1\delta_{ij} + G_2\left(\frac{l^2}{D_0^2}\right)v_iv_j + G_3\lambda_i\lambda_j + G_4\left(\frac{l}{D_0}\right)(v_i\lambda_j + v_j\lambda_i) \tag{76}$$

For isotropic media since both F_1 and F_2 are even functions of vl/D_0, then near $vl/D_0 = 0$

$$\frac{D_{ij}}{D_0} = \left(\alpha_1 + \alpha_2\frac{v^2l^2}{D_0^2}\right)\delta_{ij} + \alpha_3l^2v_iv_j/D_0^2 \tag{77}$$

where α_1, α_2, α_3 depend on geometry. In axially symmetric and plane symmetric media

$$\frac{D_{ij}}{D_0} = \left(\beta_1 + \beta_2\frac{v^2l^2}{D_0^2}\right)\delta_{ij} + \beta_3\frac{l^2v_iv_j}{D_0^2} + \left(\beta_4 + \beta_5\frac{v^2l^2}{D_0^2}\right)\lambda_j\lambda_i + \beta_6l^2v(v_i\lambda_j + v_j\lambda_i)/D_0^2 \tag{78}$$

If we choose coordinates $(v_1,0,0)$ from Equation 77

$$\left.\begin{aligned} \frac{D_{11}}{D_0} &= \alpha_1 + (\alpha_2 + \alpha_3)\frac{l^2v_1^2}{D_0^2} \\ \frac{D_{22}}{D_0} &= \alpha_1 + \alpha_2\frac{l^2v_1^2}{D_0^2} \end{aligned}\right\} \frac{D_{11}}{D_{22}} = \frac{\alpha_1 + (\alpha_2 + \alpha_3)\dfrac{l^2v_1^2}{D_0^2}}{\alpha_1 + \alpha_2\dfrac{l^2v_1^2}{D_0^2}} \tag{79}$$

If in addition $\lambda_i = (1,0,0)$ from Equation 78

$$\frac{D_{11}}{D_0} = \beta_5 + (\beta_4 + \beta_6)\frac{l^2v_1^2}{D_0^2} \tag{80}$$

which is similar to the results of Aris (*2*) in circular capillaries.

So far, the equation representing dispersion, Equation 30, and the various models resulting from deciding on a constitutive-type relation for the dispersion tensor and integrating Equation 31 for specific boundaries have been statistical in nature. Basically, the observation has been made that the character of the data for concentration *vs.* time can be fit with solutions related to the error function and thus the differential equation describing the concentration is inferred. While such a statistical approach is reasonable, it also seems reasonable to determine the equation and dispersion tensor relations by direct integration of the equations of change. Whitaker (*105*) volume-averaged the equations of motion and continuity over a finite volume which contains both fluid and solid. For dispersed marked particles, this leads to an equation for the volume-averaged concentration

$$\frac{\partial\bar{C}}{\partial t} + \frac{\partial}{\partial x_i}(\overline{v_iC}) = \frac{\partial}{\partial x_j}\left[\mathfrak{D}\left(\frac{\partial\bar{C}}{\partial x_j} + R\tau_j\right)\right] \tag{81}$$

where the tortuosity vector τ_j is defined as

$$\tau_j = \int_{A(i)}[C]n_jdA \tag{82}$$

The tortuosity vector as defined by Whitaker includes the sinuousness of the pores as well as expansions and contractions in the pores. For "pure" dispersion—*i.e.*, when flow is such that molecular diffusion is negligible compared to dispersion—the tortuosity does not enter the problem. The point velocity and concentrations are considered as the sum of a volume-averaged quantity and a fluctuation quantity as

$$\bar{v}_i = \bar{v}_i + \tilde{v}_i \tag{83}$$

$$C = \bar{C} + \tilde{C} \tag{84}$$

which leads to the definition of the dispersion vector $\bar{\psi}_i$ as

$$\bar{\psi}_i = \overline{\tilde{v}_i \tilde{C}} \quad (85)$$

Equation 81 then becomes

$$\frac{\partial \bar{C}}{\partial t} + \bar{v}_i \frac{\partial \bar{C}}{\partial x_i} + \frac{\partial \bar{\psi}_i}{\partial x_i} = \frac{\partial}{\partial x_j}\left[\mathfrak{D}\left(\frac{\partial \bar{C}}{\partial x_j} + R_{\tau_j}\right)\right] \quad (86)$$

The functional dependence of $\bar{\psi}_i$ is obtained by considering the volume-averaged equation of motion such that

$$\bar{\psi}_i = \bar{\psi}_i\left(l_1\bar{v}_i, \frac{\partial \bar{v}_i}{\partial x_j}, \frac{\partial \bar{c}}{\partial x_i}\right) \quad (87)$$

The following assumptions are made: $\bar{\psi}_i = 0$ when $\bar{v}_i = 0$; the effect of $\partial \bar{v}_i/\partial x_j$ on $\bar{\psi}_i$ is negligible; $\bar{\psi}_i = 0$ when $\partial \bar{C}/\partial x_j = 0$. The first assumption applies for most flow conditions; the second assumption is reasonable for uniform flows; the third assumption is not justifiable. (In fact, in the limit of zero velocity there is no dispersion, just diffusion which results from a concentration gradient in marked particles. However, it seems reasonable the condition is satisfied by the final equations.)

The final result is found by incorporating the assumptions in a Taylor series expansion of $\bar{\psi}_i$ about the point $\bar{v}_i = \partial \bar{C}/\partial x_j = 0$

$$\frac{\partial \bar{C}}{\partial t} + \bar{v}_i \frac{\partial \bar{C}}{\partial x_i} = \frac{\partial}{\partial x_j}\left[\mathfrak{D} \cdot \left(\frac{\partial \bar{C}}{\partial x_j} + R_{T_j}\right)\right] +$$

$$\frac{\partial}{\partial x_i}\left[\left(A^{\mathrm{I}}{}_{jik}\bar{v}_i + A^{\mathrm{II}}{}_{jilk}\bar{v}_i\bar{v}_l + A^{\mathrm{III}}{}_{jilk}v_i \frac{\partial \bar{C}}{\partial x_k}\right)\frac{\partial \bar{C}}{\partial x_k}\right] \quad (88)$$

where the tensors A^{I}, A^{II}, and A^{III} are symmetric and are functions of the structure of the porous medium and the transport properties of the fluid. In general, Equation 85 is not the same as Equation 31 nor is the dispersion tensor as described by Nikolaevskii, Bear, Scheidegger, and others. Since

$$A^{\mathrm{III}}{}_{jikl} = \left.\frac{\partial^3 \bar{\psi}_j}{\partial \bar{v}_i \, \partial\left(\frac{\partial \bar{C}}{\partial x_l}\right)\partial\left(\frac{\partial \bar{C}}{\partial x_k}\right)}\right|_{\bar{v}_i = 0,\ \frac{\partial \bar{C}}{\partial x_l} = 0,\ \frac{\partial \bar{C}}{\partial x_k} = 0} \quad (89)$$

and it may be assumed negligible (Equation 89 involves a second partial of $\bar{\psi}_j$ with respect to concentration gradient which does not affect dispersion greatly), then Equation 89 may be written

$$\frac{\partial \bar{C}}{\partial t} + \bar{v}_i \frac{\partial \bar{C}}{\partial x_i} = \frac{\partial}{\partial y_j}\left[D_{jk} \frac{\partial \bar{C}}{\partial x_L}\right] \quad (90)$$

where the dispersion tensor D_{jk} is defined by

$$D_{jk} = \mathfrak{D}(\delta_{jk} + RB^{\mathrm{I}}{}_{jk}) + A^{\mathrm{I}}{}_{jik}\bar{v}_i + A^{\mathrm{II}}{}_{jilk}\,\bar{v}_i\bar{v}_l \quad (91)$$

Equation 90 is identical to the usual form of the dispersion equation, Equation 31, and thus volume-averaging of the transport equations leads to the usual equation, but no model for the physical phenomenon is assumed. In addition, the average concentration and velocity are clearly defined and the assumptions included are exposed. It is apparent from Equation 91 that at low velocities the diffusion term dominates and

$$D_{jk} \doteq \mathfrak{D}(\delta_{jk} + RB^{\mathrm{I}}{}_{jk}) = \mathfrak{D}_{\mathrm{eff}} \quad (92)$$

The presence of the porous medium causes the molecular diffusion coefficient to be lowered by a factor $(\delta_{jk} + RB^{\mathrm{I}}{}_{jk})$ yielding an "effective" diffusion coefficient. This is a tortuosity effect reflected in the tensor $B^{\mathrm{I}}{}_{jk}$, and media of different tortuosities yield different effective diffusion coefficients substantiated by experiment (*75*). At higher velocities the diffusion term is less important and

$$D_{jk} \doteq A^{\mathrm{I}}{}_{jik}\bar{v}_i + A^{\mathrm{II}}{}_{jilk}\bar{v}_i\bar{v}_l \quad (93)$$

Note this relation for the dispersion tensor is different from Nikolaevskii, Bear, Scheidegger, and others (Nikolaevskii considered a form such as Equation 93 but eliminated A^{I} since he considered only isotropic media) in that there is both a third- and fourth-order symmetric tensor associated with velocity. Also Equation 93 predicts no direct influence of tortuosity on the dispersion tensor. The effect of the structure of the medium is accounted for by the tensors A^{I} and A^{II}.

For an isotropic medium $A^{\mathrm{I}} = 0$ and

$$D_{jk} = A^{\mathrm{II}}(2v_jv_k + \delta_{jk}|v|^2) \quad (94)$$

With flow along the x-axis $(v,0,0)$

$$\left.\begin{aligned} D_{11} &= 3A^{\mathrm{II}}v^2 \\ D_{22} &= D_{33} = A^{\mathrm{II}}v^2 \\ D_{jk} &= 0 \qquad j \neq k \end{aligned}\right\} \quad \frac{D_{11}}{D_{22}} = 3 \quad (95)$$

Data

Longitudinal and transverse dispersion. Data on longitudinal and transverse dispersion in both consolidated and unconsolidated porous media were obtained by Grane and Gardner (*42*). The consolidated porous media were a Berea sandstone with porosity approximately 20%, average grain diameter about 180μ, and average pore size about 15μ. The unconsolidated media were packs of glass beads of uniform size and variously 1.5, 0.25, and 0.074 mm in diameter giving respective porosities of 37.5, 41, and 38%. The statistical model of Haring and Greenkorn (*46*) in Figures 8 and 9 is fitted to this data. In each case, a regime where molecular diffusion controls and a regime where dispersion controls can be distinguished. In fitting the statistical model to the longitudinal and transverse dispersion for unconsolidated media (0.025-cm glass beads), a parametric distribution was used where the parameters in the beta distribution were $a = 2$ and $b = 2$, yielding a probability density function with shape not unlike that

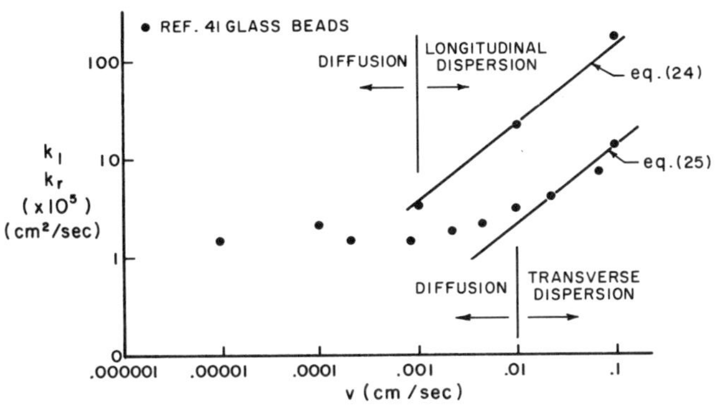

Figure 8. Statistical model of Haring and Greenkorn fitted to data for unconsolidated media

of a normal distribution. An average length of 0.0062 cm was assumed, a value which yields the largest pore of approximately one-half the bead diameter. The largest pore does not approach the bead diameter, because it is assumed that the smaller beads will tend to fall between the larger beads and thus reduce the maximum pore diameter. The radius distribution was used with $\alpha = 2$ and $\beta = 4$. The calculated ratio of longitudinal to transverse dispersion coefficients shown is 10.9, which falls at the high end of the range expected for unconsolidated porous media based on the statistical model (uniform porous media yield a ratio of longitudinal to transverse dispersion approaching 3 in the statistical model). The longitudinal and transverse dispersion coefficients for the Berea sandstone were fitted by assuming a length distribution with $a = 2$, $b = 4$, and with a radius distribution with parameters $\alpha = -1/2$ and $\beta = 6$. For actual consolidated media, the length distribution is moderately skewed, and the radius distribution somewhat resembles the gamma function. The probability density functions selected exhibit these characteristics. The average length used was 0.00188 cm. Here the calculated value of the ratio of longitudinal to transverse dispersion coefficient is 61.5.

Heterogeneity. Harleman *et al.* (*47*) have taken data on some 10 different media of uniform size and either spherical or angular (sand grains) shape. From these data they developed a correlation for the dimensionless ratio of the longitudinal coefficient of dispersion to the kinematic viscosity in terms of a grain size, Reynolds number, ud_{50}/ν, which incorporates the seepage velocity, the 50% grain size, and the intrinsic permeability; and in terms of a permeability Reynolds number, $u\sqrt{k}/\nu$, which incorporates the same velocity and permeability (Figure 10). This correlation is used in a later paper by Shamir and Harleman (*97*) in which a specific type of nonhomogeneous media is studied—*i.e.*, layered media. Figure 11 depicts longitudinal dispersion behavior in flow perpendicular to the layers. The results are shown in Figure 12 for dispersion based on a calculation incorporating an equivalent dispersion (measured) and the dispersions calculated by estimating

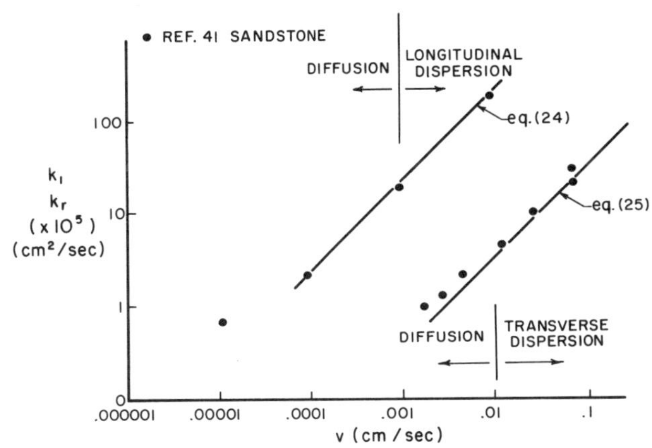

Figure 9. Statistical model of Haring and Greenkorn fitted to data for consolidated media

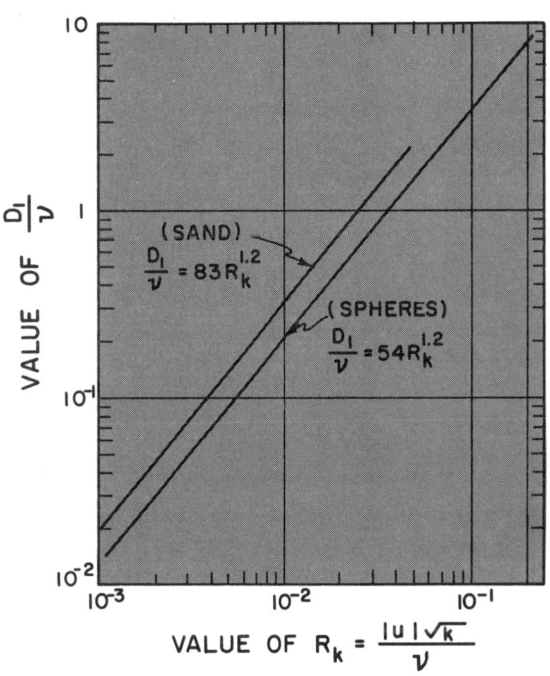

Figure 10. Correlation for longitudinal dispersion coefficient by Harleman, Melhorn, and Rumer

From Journal of the Hydraulics Division: Proceedings of the American Society of Civil Engineers, page 67 (March 1963)

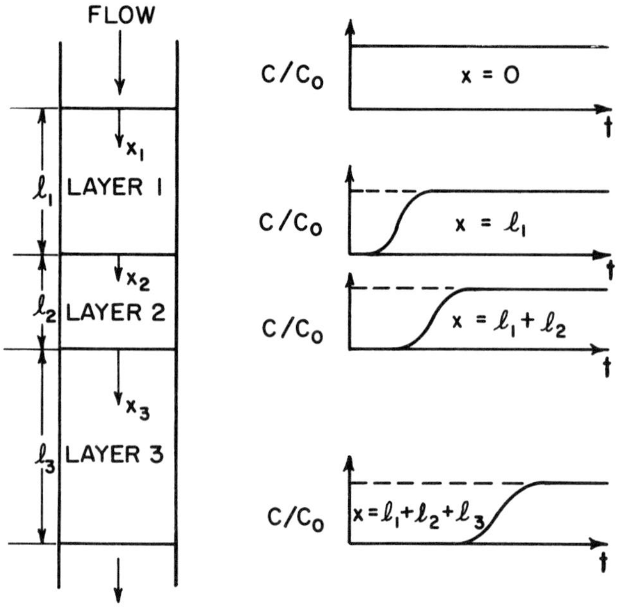

Figure 11. Longitudinal dispersion behavior in flow perpendicular to layers

From Shamir and Harleman, Journal of the Hydraulics Division: Proceedings of the American Society of Civil Engineers, page 237 (September 1967)

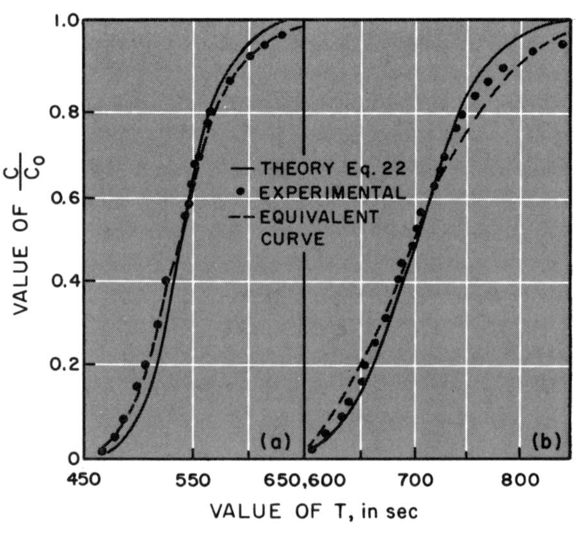

Figure 12. Longitudinal dispersion in layered media

From Shamir and Harleman, Journal of the Hydraulics Division: Proceedings of the American Society of Civil Engineers, page 237 (September 1967). Equation numbers refer to original Journal article

dispersion from the permeabilities of the individual media in the separate layers and superimposing solutions for each layer. These results incorporate the dispersion-permeability correlation developed by Harleman *et al.* mentioned above (shown in Figure 10). In Figure 13 is shown a depiction of a lateral dispersion in a layered media (two layers only) and in Figure 14 the results are shown in which the experimental data were described in terms of a model for an "equivalent" medium which is homogeneous and isotropic.

Pleshek (*80*) has measured dispersion in linear heterogeneous porous media and found a substantial difference in dispersion coefficients caused by bead size or permeability. Pleshek investigated reciprocity, that is, whether or not taking precisely the same layers and letting fluid flow through them in a different order made any difference in the dispersion obtained. The results in Figure 15 show a difference in dispersion depending on the order in which the fluid flows through the various layers. The reason for this is not entirely clear at the moment; however, it appears that difficulty in experimental equipment is not the cause. Pleshek had little success in attempting to correlate flow in porous media using some average dispersion coefficient obtained by combining the dispersion coefficients for the various media layers (even though the composite permeability could be accurately predicted from the individual data). Pleshek also observed a significant effect of particle size or permeability on the exponent n in the equation

$$D = av^n \tag{96}$$

Nonuniformity. Niemann (*68*) has taken data on dispersion for flow in nonuniform heterogeneous porous media, as shown in Figures 16 and 17. Niemann's media were glass beads with an approximately logarithmic distribution of diameters. Niemann found an effective bead diameter for permeability calculations should be somewhat greater than the value at the 10% diameter (at this value average permeability of his models is about the same—even though the variance in bead diameter is different). Niemann found correlation for dispersion coefficient based on D (*68*) showed that the dispersion coefficient could be correlated well with permeability according to the equation: $D/\nu = \Gamma R_k{}^m$. An interesting result was that the slope of the log normal number distribution plot of bead size—*i.e.*, the variance—had a significant effect upon the dispersion coefficient. Again, attempts at using an averaged dispersion coefficient based on homogeneous models was unsuccessful.

Anisotropy. Lenth (*60*) has studied dispersion in anisotropic media as shown in Figure 18. Permeability was a function of direction as shown in Figure 19. The magnitude of the permeability was higher than values obtained from linear models incorporating similar

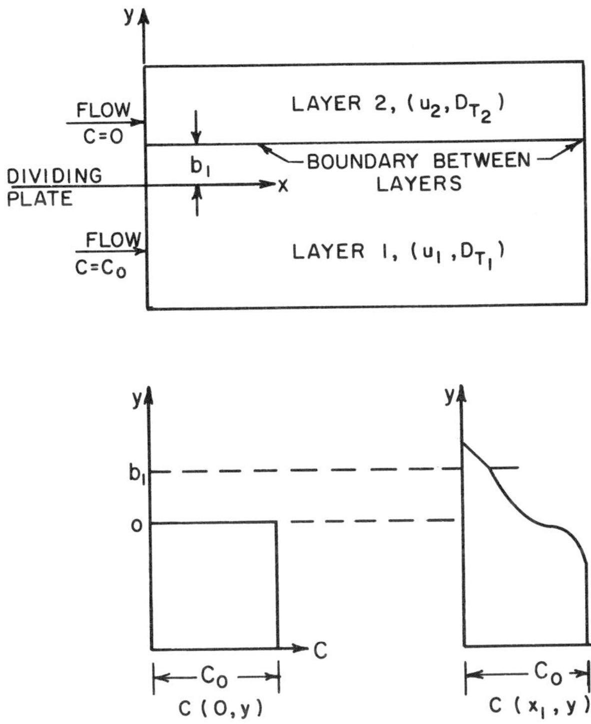

Figure 13. Lateral dispersion in layered media

From Shamir and Harleman, Journal of the Hydraulics Division: Proceedings of the American Society of Civil Engineers, page 237 (September 1967)

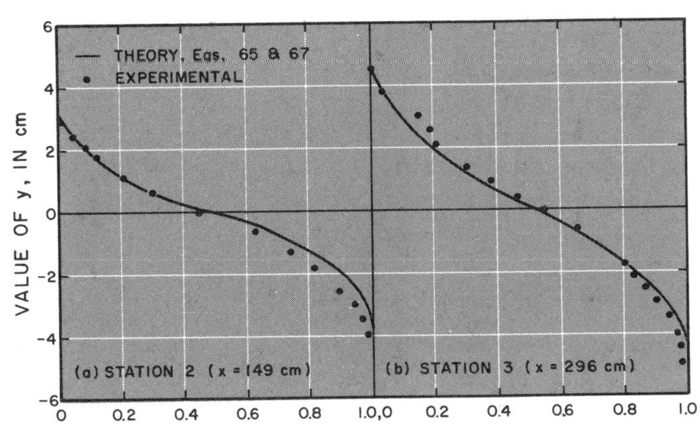

Figure 14. Lateral dispersion in layered media

From Shamir and Harleman, Journal of the Hydraulics Division: Proceedings of the American Society of Civil Engineers, page 237 (September 1967). Equation numbers refer to original journal article

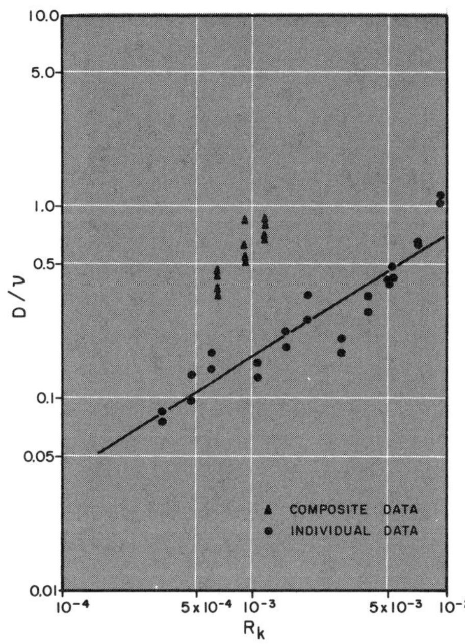

Figure 15. Pleshek data on reciprocity

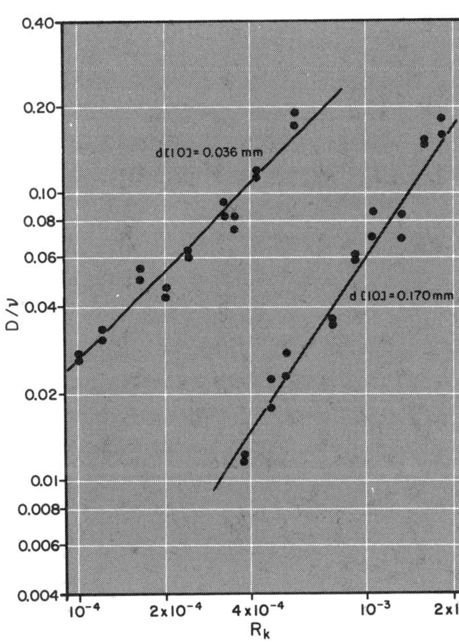

Figure 16. Niemann data in unconsolidated media made up of glass beads with log-normal diameter distribution

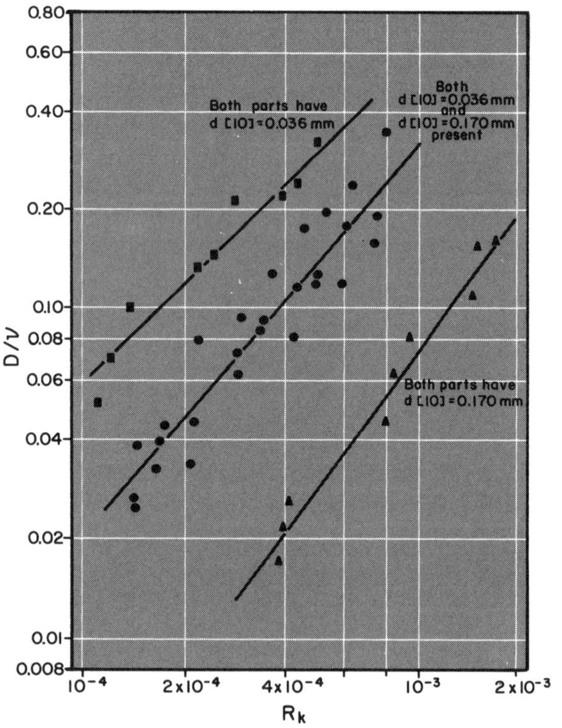

Figure 17. Niemann data on reciprocity for combining sections of the media from Figure 16

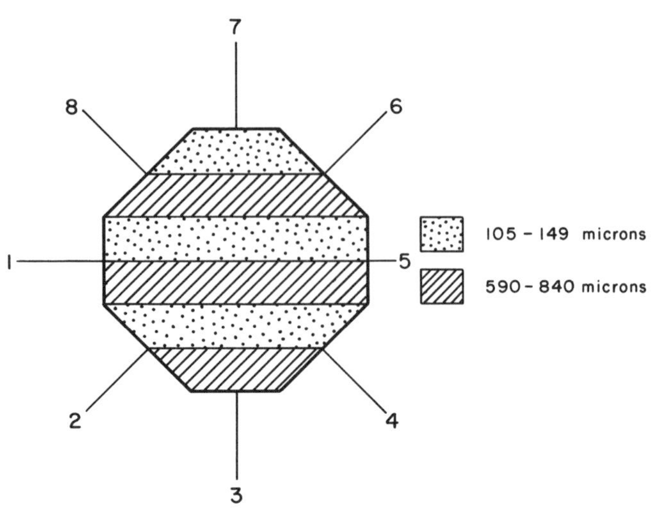

Figure 18. Lenth apparatus for study of anisotropy

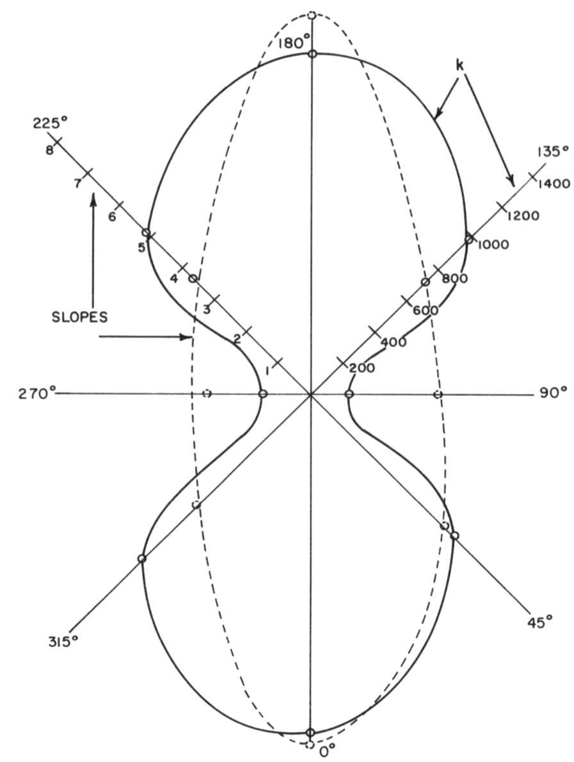

Figure 19. Lenth data for permeability in anisotropic media

Glass spheres as shown in Figure 18

sizes of glass spheres. (One problem which may contribute to this effect is the fact that the cross-sectional area is not constant with the experimental model used and therefore, it was necessary to choose a value of area to use in Darcy's law.)

One interesting effect was that the permeability observed by simply reversing the flow direction through the model was not the same as that for the original direction of flow.

Longitudinal dispersion coefficients are not presented but the slope of the breakout curve is given. The effects of both direction and flow rate are significant, and there was some effect of interaction between the flow rate and direction.

Nomenclature

NOTE: The authors have retained the nomenclature of the original papers to facilitate use of the original references—this causes some unfortunate duplication in use of symbols, but an attempt has been made to define all symbols in context as well as below.

a = constant, parameter of beta distribution
a_{ijkl} = dispersivity tensor
A = area
$A(i)$ = area of medium
A_i = function of v^2, media properties
A_{ijk} = tensor which is function of properties of medium structure and fluid properties
b = parameter of beta distribution
B_i = arbitrary functions of v^2 and $v_k\lambda_k$
B_{ik} = tensor
C = concentration

C_0 = inlet concentration
d_n = size of nth% particle of percentage distribution of sizes
$\mathfrak{D}$ = diffusion coefficient
D = dispersion coefficient
D_L = longitudinal dispersion coefficient
D_T = transverse dispersion coefficient
D_{jk} = dispersion tensor
$\tilde{D}_{ijkl}$ = constant of dispersion
E = probability of existence of pore with specified properties
F = undetermined function of medium
G = undetermined function of medium
j_i = flux
$J = \dfrac{(\alpha + 1)(\alpha + 2)(\alpha + \beta + 4)(\alpha + \beta + 5)}{(\alpha + 3)(\alpha + 4)(\alpha + \beta + 2)(\alpha + \beta + 3)}$
k = permeability
l = characteristic length
l^* = l/L
l_i = "mixing length" for porous medium
L = length of medium, length of longest pore
L_{ij} = phenomenological coefficient, tensor of displacement
M = normalization constant
n_j = outward normal
p = pressure
$P(N)$ = probability of N
q = volumetric flow rate
r = radius
r^* = r/R
R = coefficient, largest pore radius
R_i = vector
R_k = mean Reynolds number based on permeability

SUMMARY

In general, the most effort for real media seems to be in the continuum theories. There is relatively less on the statistical models reported, and meager amounts of data are interpreted in nonideal media.

If one accepts the form of Equation 31, the effect of heterogeneity can be incorporated by defining D_{jk} as a function of position. Any useful predicted results will probably have to be made using numerical techniques for solving Equation 31.

The effect of nonuniformity is incorporated in the form of dispersion tensor and is summarized in terms of the ratio of the longitudinal to transverse dispersion coefficients for one-dimensional flow in Table II.

These models and the data can be made to agree, except in the case of Whitaker's results which seem to show that for the assumptions made to get to a comparable result we must assume the medium to be uniform. Otherwise, the form of the ratio from Whitaker is a complex function of several tensors.

To attack anisotropy, we need more experimental data, extension of the statistical model for this case, and further investigation of the forms of the dispersion tensor. Empirically, it is appealing to adopt Bachmat and Bear's model of the tensor and find λ and μ as functions of angle.

There are three major questions to be answered after we get a meaningful amount of data and interpretation for nonideal system.

(1) What is the "correct" model for flow in porous media in the sense of giving sufficient accuracy? Do we need anything in addition to curve fitting using the normal distribution (error function) of Equation 2 and empirical corrections to it?

(2) Can one characterize a porous medium with a reasonable number of parameters—*i.e.*, at what level do heterogeneity, nonuniformity, and anisotropy become important?

(3) What is the limiting step in our analysis of the problem: the mathematical model, the computing techniques, or the experimental data?

Table II. Ratio of Longitudinal to Transverse Dispersion Coefficients Effect of Nonuniformity

Experimental data

$$3 \leq \frac{D_L}{D_T} \leq f$$

Statistical model

Haring and Greenkorn

$$\frac{D_L}{D_T} = \frac{4}{95} \ln \left[\frac{27}{2} \frac{(a + b + 2)^2}{(a + 1)J^2} \frac{vT}{\langle l \rangle} \right]$$

Continuum models

Nikolaevskii

$$\frac{D_L}{D_T} = \frac{\lambda_1}{\lambda_2}$$

Bear, Scheidegger, de Josselin de Jong

$$\frac{D_L}{D_T} = \frac{A_I}{A_{II}}$$

Bachmat and Bear

$$\frac{D_L}{D_T} = \frac{\lambda + 2\mu}{\lambda}$$

Poreh

$$\frac{D_L}{D_T} = \frac{\alpha_1 + (\alpha_2 + \alpha_3)\dfrac{l^2 v_1^2}{D_0^2}}{\alpha_1 + \alpha_2 \dfrac{l^2 v_1^2}{D_0^2}}$$

Whitaker

$$\frac{D_L}{D_T} = 3$$

S_i = vector
S_j = vector
t = time
T = total time
T_{ij} = local tensor of medium
u = velocity
u_i^0 = component of perturbation of mean velocity
v = velocity
V = volume throughput of liquid
V_0 = pore volume
w = rectangular coordinate
x = rectangular coordinate
X = generalized force
y = rectangular coordinate
z = rectangular coordinate

Greek

α = parameter of beta distribution
α_i = coefficient
β = parameter of beta distribution
Γ = coefficient
δ = dirac delta function
δ_{ij} = unit tensor
ϵ = radius of elemental volume
θ = spherical coordinate
λ = invariant of tensor
λ_i = component of dispersivity tensor for isotropic porous medium
μ = viscosity of fluid, invariant of tensor
ν = kinematic viscosity
ξ_i = constant
σ^2 = variance
τ_j = tortuosity vector
ψ = relative concentration, spherical coordinate

Overscores

bar = average
tilde = for particle following most probable path, fluctuation

BIBLIOGRAPHY

(1) Aranow, R. H., *Phys. Fluids*, **9** (9), 1721 (1966).
(2) Aris, R., *Proc. Roy. Soc. London, Ser. A*, **235**, p 67 (1956).
(3) Aris, R., and Amundson, N. R., *A.I.Ch.E. J.*, **3** (2), 280 (1957).
(4) Aronofsky, J. S., and Heller, J. P., *Petrol. Trans., AIME*, **210**, 345 (1957).
(5) Ashcroft, G. L., Ph.D. thesis, Oregon State University, Corvallis, 1962.
(6) Babbitt, J. D., *Can. J. Res.*, **28**, 449 (1950).
(7) Babcock, R. E., Green, D. W., and Perry, R. H., *A.I.Ch.E. J.*, **12** (5), 923 (1966).
(8) Babcock, R. E., Perry, R. H., and Crosser, O. K., *Chem. Eng. Prog. Symp. Ser.*, **63** (77), 102 (1967).
(9) Bachmat, Y., and Bear, J., *J. Geophys. Res.*, **69** (12), 2561 (1964).
(10) Banks, R. B., and Ali, I., *J. Hydraul. Div.: Proc., Am. Soc. Civil Eng.*, p 13 (1964).
(11) Banks, R. B., and Jerasate, S., *ibid.*, **3**, 3109 (1962).
(12) Bear, J. *J. Geophys. Res.* **66** (4), 1185 (1961).
(13) Bear, J., *J. Hydraul. Div.: Proc. Am. Soc. Civil Eng.*, 11 (1960).
(14) Bear, J., *J. Geophys. Res.*, **66** (8), 2455 (1961).
(15) Biggar, J. W., and Nielsen, D. R., *ibid.*, **65** (9), 2887 (1960).
(16) Biggar, J. W., and Nielsen, B. R., *ibid.*, **67** (9), 3636 (1962).
(17) Bischoff, K. B., and Levenspiel, O., *Chem. Eng. Sci.*, **17**, 245 (1962), *ibid.*, p. 257.
(18) Blackwell, R. J., *Soc. Petrol. Eng. J.*, **2** (1), **1** (1962).
(19) Blackwell, R. J., Terry, W. M., and Rayne, J. R., *TP8047 Trans. AIME*, **216**, 1 (1959).
(20) Cairns, E. J., and Prausnitz, J. M., *Chem. Eng. Sci.*, **12**, 20 (1960).
(21) Carberry, J. J., and Bretton, R. H., *A.I.Ch.E. J.*, **4** (3), 367 (1958).
(22) Chandari, N. M., and Scheidegger, A. E., *Can. J. Phys.*, **43**, 1776 (1965).
(23) Chao, R., and Hoelscher, H. E., *A.I.Ch.E. J.*, **12** (2), 271 (1966).
(24) Chung, S. F., and Wen, C. Y., *ibid.*, **14**, 857 (1968).
(25) Coats, K. H., and Smith, B. D., *Soc. Petrol. Eng. J.*, p. 73 (March, 1964).
(26) Dagan, G., *J. Geophys. Res.*, **72** (16), 4075 (1967).
(27) Day, P. R., *Trans. Am. Geophys. Union*, 595 (1956).
(28) De Josselin De Jong, G., *ibid.*, **39** (1), 67 (1958).
(29) De Josselin De Jong, G., and Bossen, M. J., *J. Geophys. Res.*, **66** (10), 3623 (1961).
(30) Deans, H. A., *Soc. Petrol. Eng. J.*, 49 (1963).
(31) Dudgeon, C. R., *J. Hydraul. Div.: Proc. Am. Soc. Civil Engrs.*, p 137 (Sept 1967).
(32) Ebach, E. A., and White, R. R., *AIChE J.*, **4** (2), 161 (1958).
(33) Edwards, M. F., and Richardson, J. F., *Chem. Eng. Sci.*, **23**, 109 (1968).
(34) Elrick, D. E., Erh, K. T., and Krupp, H. K., *Water Resources Res.*, **2** (4), 717 (1966).
(35) Evans, E. V., and Kenney, C. N., *Trans. Inst. Chem. Eng.*, **44**, T-189 (1966).
(36) Fatt, I., *Petrol. Trans. AIME*, **207**, 144 (1956).
(37) Fatt, I., *ibid.*, p 160.
(38) Frisch, H. L., *Trans. Soc. Rheol.*, **9** (1), 293 (1965).
(39) Glaser, M. B., and Litt, M., *A.I.Ch.E. J.*, **9** (1), 103 (1963).
(40) Goddard, R. R., *Soc. Petrol. Eng. J.*, **6** (2), 143 (1966).
(41) Gottschlich, C. F., *A.I.Ch.E. J.*, **9** (1), 88 (1963).
(42) Grane, F. E., and Gardner, G. H. F., *J. Chem. Eng. Data*, **6** (2), 283 (1961).
(43) Greenkorn, R. A., *J. Petrol. Technol.*, 87 (1962).
(44) Greenkorn, R. A., Johnson, C. R., and Shallenberger, L. K., *ibid.*, p 124 (1964).
(45) Haring, R. E., M.S. thesis, University of Tulsa, Tulsa, Okla., 1961.
(46) Haring, R. E., and Greenkorn, R. A., *A.I.Ch.E. J.*, *16*, 477 (1970).
(47) Harleman, D. R. F., Melborn, P. F., Rumer, R. R., Jr., *J. Hydraul. Div.: Proc. Am. Soc. Civil Eng.*, 67 (1963).
(48) Harleman, D. R. F., and Rumer, R. R., *J. Fluid Mech.*, **16** (3), 385 (1963).
(49) Harleman, R. F., and Shamir, U., *J. Hydraul. Div., Proc. Am. Soc. Civil Eng.*, 330 (1965).
(50) Hassinger, R. C., and Von Rosenburg, D. U., *Soc. Petrol. Eng. J.*, **8** (2), 195 (1968).
(51) Hoopes, J. A., and Harleman, D. R. F., *J. Geophys. Res.*, **72** (14), 3595 (1967).
(52) Hoopes, J. A., and Harleman, D. R. F., *J. Hydraul. Div.: Proc. Am. Soc. Civil Eng.*, 51 (1967).
(53) Kaufmann, W. J., Ewing, B. B., Kerrigan, J. V., and Inoue, Y., *J. Water Pollution Control Fed.*, p. 73 (Jan 1961).
(54) Kirwaan, A. D., Jr., *J. Geophys. Res.*, **70** (12), 2923 (1965).
(55) Klinkenberg, I. A., *Trans. Inst. Chem. Eng.*, **43**, T141 (1965).
(56) Kramers, H., and Alberda, G., *Chem. Eng. Sci.*, **2**, 173 (1953).
(57) Krupp, H. K., and Elrick, D. E., *Water Resources Res.*, **4** (4), 809 (1968).
(58) Lamb, D. E., and Wilhem, R. H., Ind. Eng. Chem. Fundam., **2** (3), 173 (1963).
(59) Legatski, M. W., and Katz, D. L., *Soc. Petrol. Eng. J.*, **7** (1), 43 (1967).
(60) Lenth, E. G., M.S. thesis, Purdue University, Lafayette, Ind. (1968).
(61) Levich, V. G., Markin, V. S., and Chismadzhez, Y. A., *Chem. Eng. Sci.*, **22**, 1357 (1967).
(62) Li, W. H., and Canale, R. P., *J. Hydraul. Div.: Proc. Am. Soc. Civil Eng.*, 65 (1967).
(63) Li, W. H., and Yeh, G. T., *Water Resources Res.*, **4** (2), 369 (1968).
(64) Litwiniszyn, J., *Chem. Eng. Sci.*, **22**, 1315 (1967).
(65) Lusczynski, N. J., and Swarzenski, W. V., *J. Hydraul. Div.: Proc. Am. Soc. Civil Eng.*, 173 (1962).
(66) Miller, S. F., and King, C. J., *A.I.Ch.E. J.*, **12** (4), 767 (1966).
(67) Miyauchi, T., and Vermeulen, T., Ind. Eng. Chem. Fundam., **2** (2), 113 (1963).
(68) Niemann, E. H., M.S. thesis, Purdue University, Lafayette, Ind. (1969).
(69) Nikolaevskii, V. N., *Prikl. Math. Mech.*, **23** (6), 1042 (1959).
(70) Ogata, A., Ph.D. thesis, Northwestern University, Evanston, Ill. (1958).
(71) Ogata, A. N., and Banks, R. B., Geological Survey Professional Paper **411-A**, U. S. Govt. Printing Office, Washington, D. C. (1961).
(72) Olbrich, W. E., Agnew, J. B., and Potter, O. E., *Trans. Inst. Chem. Eng.*, **44**, T207 (1966).
(73) Pakula, R. J., M.S. thesis, Purdue University, Lafayette, Ind. (1968).
(74) Patel, R. D., and Greenkorn, R. A., *A.I.Ch.E. J.*, *16*, 332 (1970).
(75) Perkins, T. K., and Johnston, O. C., *Soc. Petrol. Eng. J.*, p. 70 (1963).
(76) Perkins, T. K., and Johnston, O. C., *ibid.*, **9**, (1), 39 (1969).
(77) Perkins, T. K., Johnston, O. C., and Hoffman, R. N., *ibid.*, **5** (4), 301 (1965).
(78) Perrine, R. L., *ibid.*, **3**, 205 (1963).
(79) Pis'men, L. M., Kuchanov, S. I., and Levich, V. G., *Dokl. Akad. Nauk, USSR*, **174** (3), 650 (1967).
(80) Pleshek, R. C., M.S. thesis, Purdue University, Lafayette, Ind., 1968.
(81) Poreh, M., *J. Geophys. Res.*, **70** (16), 3909 (1965).
(82) Pozzi, A. L. and Blackwell, R. J., *Soc. Petrol. Eng. J.*, **3** 28–40 (1963).
(83) Raats, P. A. C., and Scotter, D. R., *Water Resources Res.*, **4** (3), 561 (1968).
(84) Raimondi, P., Gardner, G. H. F., and Petrick, C. B., *Preprint* **43** AIChE-SPE Joint Symp., San Francisco, December 1959.
(85) Rifai, M. N. E., Kaufman, W. J., and Tood, D. K., Sanitary Res. Lab.: Div. of Civil Eng., University of California, Berkeley, Report No. **3**, IER Series 90 (1956).
(86) Rose, W., *J. Geophys. Res.*, **69** (24), 5432 (1964).
(87) Rumer, R. R., *J. Hydraul. Div.: Proc. Am. Soc. Civil Eng.*, 147 (1962).
(88) Rumer, R. R., and Harleman, D. R. F., *ibid.*, 193 (1963).
(89) Saffman, P. G., *Fluid Mech.*, **6**, 21 (1959).
(90) Scheidegger, A. E., *Can. J. Phys.*, **45**, 1783 (1967).
(91) Scheidegger, A. E., *Can. Mining Metallurg. Bull.*, 1 (1959).
(92) Scheidegger, A. E., *Compt. Rend. Assoc. Geo., Toronto, Assoc. Intern. Hydrol. Scient.*, **2**, 236 (1957).
(93) Scheidegger, A. E., *J. Appl. Phys.*, **25** (8), 994 (1954).
(94) Scheidegger, A. E., *J. Geophys. Res.*, **66** (10), 327 (1961).
(95) Scheidegger, A. E., *Trans. Soc. Rheol.*, **9** (1), 313 (1965).
(96) Scotter, D. R., and Raats, P. A. C., *Water Resources Res.*, **4** (6), 1201 (1968).
(97) Shamir, U. Y., and Harleman, D. R. F., *J. Hydraul. Div.: Proc. Am. Soc. Civil Eng.*, 237 (1957).
(98) Shamir, U. Y., and Harleman, D. R. F., *Water Resources Res.*, **3** (2), 557 (1967).
(99) Sinclair, R. J., and Potter, O. E., *Trans. Inst. Chem. Eng.*, **43**, T3 (1965).
(100) Sing, R., *Hydraul. Div.: Proc. Am. Soc. Civil Eng.*, 43 (1967).
(101) Stahel, E. P., and Geankoplis, C. J., *A.I.Ch.E. J.*, **10** (2), 174 (1964).
(102) Szabo, B. A., "Permeability of Orthotropic Porous Mediums," *Water Resources Res.*, **4** (4), 801 (1968).
(103) Von Rosenberg, D. U., *AIChE J.*, **2** (1), 55 (1956).
(104) Warren, J. E., and Skiba, F. F., *Soc. Petrol. Eng. J.*, **4** (3), 215 (1964).
(105) Whitaker, S., *A.I.Ch.E. J.*, **13** (3), 420 (1967).

9

Mechanisms Affecting Dispersion and Miscible Displacement

RICHARD J. NUNGE
WILLIAM N. GILL

Mechanisms Affecting Dispersion and Miscible Displacement

Mechanisms influencing dispersion of solute material in another miscible phase in porous media consisting of several microstructure models

The nature of dispersion or miscible displacement in porous media may be examined from both an overall, or macroscopic, and a local, or microscopic, viewpoint. Because of the complexities of the structure of such media, there are difficulties associated with both approaches. Overall behavior, such as the average concentration distribution at the system outlet as a function of time, can be observed in sets of experiments, as is frequently done in practice, and the results can be correlated with the variables investigated on empirical or semiempirical bases. Such studies make up a large fraction of the existing work on miscible displacements in the literature. However, the generality of such results is questionable unless accompanied by a unifying theory.

To date, no theory which accounts for all the effects present in porous media has been established, and perhaps not even all the effects have been delineated. To construct macroscopically correct laws for describing miscible displacements, a knowledge of local behavior is required, since integral results which reflect overall behavior are obtained from integrating the differential equations which are assumed to describe local behavior. At this point in the development, it is assumed that the microscopic equations of change are accurate enough to describe phenomena as they occur in a porous medium. However, the application of these equations to porous media is difficult. Consider the flow boundaries in even the most orderly packed bed, and the geometric complexities are clear.

Perhaps the theory of transport in porous media will never reach a predictive level and will always remain closely linked with experiment for final parametric evaluation, just as the theory of turbulence relies heavily on experimental results. To improve the present situation, the need for more physically realistic models and more refined experimental evidence is clear.

This review describes the progress made in one attempt to examine the mechanisms influencing the dispersion of solute material in another miscible phase in porous media. The medium is considered to consist of a microstructure made up of the pores and void spaces in and between the solid materials. Because the geometry of the microstructure is generally too complex to be described by a single model, it is broken into several different types of models. This requires that the important mechanisms controlling the transport be identified through experiment so that realistic microstructure models may be developed.

Finally, the microscopic results need to be combined and averaged so that a comparison with macroscopic behavior can be attained. At this crucial point, the theory is intimately related to experiment since, like turbulence, it is necessary to combine experimental measurements with theoretical formulations to find the values of undetermined parameters. One of the main difficulties associated with this final step is that the physical structure of the media may contribute effects which are not easily included in the microscopic analysis and yet which contribute in a significant manner to the overall process. These include channeling and backmixing for example. Hence, most of the material discussed here relates to the microstructure models and cannot be applied to porous media without modification.

General Comments about Dispersion

When two miscible liquids are contacted and mixed as in a miscible displacement process, there can be no equilibrium except that of one phase uniformly distributed throughout the second. Hence, all of the phenomena of interest are transient. The process of distributing one phase in the second may be termed dispersion. In a forced flow, many modes of transfer contribute to the overall process. These include longitudinal and transverse molecular diffusion and/or Knudsen diffusion, eddy diffusion if the flow is turbulent, free and forced convection, and surface tension–driven flows. In addition, the nature of the dispersion may be affected by the boundary conditions, flow pulsations, and, most significantly in a porous medium, geometrical configuration.

Here we shall be primarily concerned with the influence of these factors in simple configurations, but reference will be made to similar behavior in porous media, and some of the implications of the results for the simplified models toward studies in porous media will be pointed out. In all cases, the bounding material—*e.g.*, the solid phase making up the porous medium—is considered stationary and inert.

We shall restrict our discussion to binary systems that can be described by a continuum model and, in this case, the general partial differential equation which describes the dispersion process is

$$\frac{\partial C_A}{\partial t} + \nabla \cdot \bar{n}_A = r_A \tag{1}$$

If the flow is laminar this can be written as

$$\frac{\partial C_A}{\partial t} + (\nabla \cdot C_A\bar{v}) = \nabla \cdot \rho D \nabla \omega_A + r_A \tag{2}$$

where

C_A = mass concentration of component A
$\bar{v}$ = vector representing mass average velocity
ρ = density of the system
D = molecular diffusion coefficient
ω_A = mass fraction of A
r_A = reaction rate at which A is produced
$\bar{n}_A$ = vector representing mass flux of A
∇ = del operator

This equation applies to the local behavior of component A, and its solution provides one with a complete map of the mass concentration distribution of A throughout the system. However, even for dispersion in simple, well-defined systems, such as fully developed laminar flow in circular tubes, Equation 1 is difficult to solve analytically.

For systems of complex configuration such as porous media, Equation 1 is probably impossible to solve directly. Therefore, it is desirable to consider an alternate approach in which one settles for less information, which is also less exact, than that provided by a direct solution of Equation 1. Thus, appeal is made to the dispersion model which provides a good approximation to the cross-sectional area average of C, C_m, if the system is sufficiently long so that the residence time is large enough—*i.e.*, the dispersion model is asymptotic in time and applies only after sufficient time has passed since the process was started.

Complex geometrical configurations complicate the solution of Equation 1 in two important ways. First, it is difficult or perhaps impossible to determine the velocity field, $\bar{v}$, accurately. Second, boundary conditions become complex and difficult to deal with.

Much experimental work on miscible displacement in porous media has been interpreted and correlated on the basis of an equation identical in form to the one-dimensional unsteady-state diffusion equation:

$$\frac{\partial C_m}{\partial t} = E\frac{\partial^2 C_m}{\partial x_1^2} \tag{3}$$

with the molecular diffusion coefficient replaced by an effective longitudinal diffusion coefficient or dispersion coefficient E; x_1 is a space coordinate in the direction of main flow which moves with the mean interstitial velocity. The solution of Equation 3 with a step change in the inlet concentration, C_0, is well known to be

$$C_m/C_0 = 1/2\left[1 - \text{erf}\left(\frac{x_1}{2\sqrt{Et}}\right) + \exp\left(\frac{u_m(x_1 + u_m t)}{E}\right)\left(1 - \text{erf}\,\frac{x_1 + 2u_m t}{2\sqrt{Et}}\right)\right] \tag{4}$$

The exponential term in Equation 4 decays rapidly and hence, for sufficiently large times, Equation 4 plots as a straight line on arithmetic probability paper. Since E is usually an unknown, it is necessary to perform experimental measurements to determine its value. Available experimental evidence has been obtained mostly by measuring the effluent concentration as a function of time. This is then plotted as suggested by Equation 4 and the dispersion coefficient is determined from the following relationship (*67*).

$$E = \frac{1}{t}\left[\frac{L'}{3.625}\right]^2 \tag{5}$$

Here L' is the mixing length, or as defined by Taylor (*67*), the axial distance over which the average concentration of the displacing fluid changes from 10% to 90% of its inlet value. After E has been determined for several systems, it is assumed that this parameter is of value for extrapolation to similar situations.

One of the main deficiencies of this procedure is that experimental results have seldom followed the straight line on probability paper predicted by Equation 4, although this is generally ignored, perhaps because deviations may be ascribed to experimental error.

The lumped parameter model described above leaves much to be desired from a mechanistic point of view. The requirements to be imposed on a complex system before Equation 3 can be used with confidence have not been established on a firm theoretical basis. Much

---

AUTHORS *R. J. Nunge and W. N. Gill are members of the faculty of Clarkson College of Technology, Potsdam, N. Y. 13676, where Dr. Gill is Professor and Dr. Nunge is Associate Professor of Chemical Engineering.* *The work for this paper was supported by the Office of Saline Water.*

of the work reviewed here is addressed to this problem in simple geometries to establish such a basis under many conditions similar to those encountered in porous media.

Laminar Dispersion in Capillary Tubes

Perhaps the simplest model for the microstructure is the straight tube. Bundles of straight capillaries have long been used to model flow through porous media (*12, 61*), and assemblages of randomly oriented straight pores or capillaries, where it is assumed that the path of a marked element consists of a sequence of statistically independent steps, direction and duration of which vary in a random manner, have been suggested by de Josselin de Jong (*22*), Saffman (*57–59*), and Scheidegger (*61*). Here the results of capillary dispersion are considered for the purpose of illustrating the interactions of some of the factors influencing miscible displacements in porous media. The straight tube provides a well-defined hydrodynamical system where the dispersion process is most easily described while still retaining many of the main features of the same process in porous media.

Consider first the dispersion of two fluids of the same constant physical properties in a straight capillary tube with steady fully developed laminar flow prevailing. An early experimental work which demonstrated the essence of the process without mathematical treatment was reported by Griffiths (*35*) in 1911, who observed that a tracer injected into a stream of water spreads out in a symmetrical manner about a plane in the cross section which moves with the mean speed of flow. As Taylor (*66, 67*) pointed out, this is a rather startling result for two reasons. First, since the water near the center of the tube moves with twice the mean speed of flow and the tracer at the mean speed, the water near the center must approach the column of tracer, absorb the tracer as it passes through the column, and then reject the tracer as it leaves on the other side of the column. Second, although the velocity is unsymmetrical about the plane moving at the mean speed, the column of tracer spreads out symmetrically. The concentration of the tracer material is described by the two-dimensional unsteady convective diffusion equation, Equation 6.

$$\frac{\partial C}{\partial t} + u(r)\frac{\partial C}{\partial x} = D\left(\frac{\partial^2 C}{\partial x^2} + \frac{1}{r}\frac{\partial}{\partial r}r\frac{\partial C}{\partial r}\right) \tag{6}$$

Here C is the point concentration, u the parabolic laminar velocity profile, and x and r the axial and radial coordinates, respectively. For large values of time, Taylor showed that the process could be described by a one-dimensional dispersion model, as given in Equation 7.

$$\frac{\partial C_m}{\partial t} + u_m\frac{\partial C_m}{\partial x} = k\frac{\partial^2 C_m}{\partial x^2} \tag{7}$$

Upon defining a coordinate which moves with the mean speed of flow as

$$x_1 = x - u_m t \tag{8}$$

Equation 7 becomes

$$\frac{\partial C_m}{\partial t} = k\frac{\partial^2 C_m}{\partial x_1^2} \tag{9}$$

and therefore Equation 9 is simply the one-dimensional unsteady diffusion equation to which solutions are readily available under a variety of conditions. The molecular diffusion coefficient has been replaced by an effective axial diffusion coefficient or dispersion coefficient which, in the absence of axial molecular diffusion, Taylor showed to be

$$k = \frac{4a^2u_m^2}{192D} \tag{10}$$

where a is the tube radius. This simple equation provides a great deal of physical insight into the nature of the dispersion process if we interpret the numerator to be a measure of axial convection and the denominator to reflect the intensity of transverse mixing rather than just transverse molecular diffusion.

In this event, we see that the dispersion coefficient, which is a measure of the rate at which material will spread out axially in the system, is enhanced by having large differences in velocity exist across the flow and by taking place in equipment with large transverse dimensions. In contrast, any mechanism which increases transverse mixing, such as turbulence or transverse convection currents, reduces the dispersion coefficient. These arguments apply, in a qualitative way, to porous media as well as to simple configurations. That is, in porous media, dispersion is created by both the microscopic differences in velocity which exist in the interstices between particles and by large-scale or macroscopic effects such as channeling.

Aris (*4*) later extended the analysis to include axial molecular diffusion and demonstrated that the dispersion coefficient for this case contains Taylor's result with an additive term due to diffusion:

$$k = D + \frac{4a^2u_m^2}{192D} \tag{11}$$

This combined result will be referred to as the Taylor-Aris theory.

It is important to note the restriction placed on the Taylor-Aris theory with respect to time, since, unless this time is exceeded, the mean concentration distribution is not described by a dispersion model. For porous media, the dimension in the flow direction must be large enough to ensure sufficient residence time for a dispersion model to apply. Otherwise, measured dispersion coefficients, together with the solution to Equation 3, could not be used to predict the average concentration distribution.

Gill (*28, 29*) has developed an analytical solution of the time-dependent convective diffusion equation which can be used to predict local concentrations at times large enough so that the dispersion model holds—*i.e.*, the solution is asymptotic in time in the same sense as the

Taylor-Aris theory. The basis for computing the local concentration is an expansion of the point concentration in a Taylor series about the mean concentration as given in Equation 12.

$$C_A = C_m + \sum_{n=1}^{\infty} f_n(\tau,y) \frac{\partial^n C_m}{\partial x_1{}^n} \tag{12}$$

The dispersion model, Equation 9, and the nth derivative of Equation 9 with respect to x_1,

$$\frac{\partial^{n+1} C_m}{\partial x_1{}^n \partial t} = k \frac{\partial^{n+2} C_m}{\partial x_1{}^{n+2}} \tag{13}$$

are used along with Equation 12 in the full convective diffusion equation, Equation 6 for example, to generate an infinite set of equations for the f_n functions. The dispersion coefficient is determined from the first-order function and appears in the problem as an eigenvalue. Note that Equation 13 assumes that k is not a function of axial distance. For dispersion in straight tubes, the dispersion coefficient obtained by this procedure agrees with the Taylor-Aris theory (Equation 11). The first three terms in Equation 12 are needed to give good agreement with the finite difference experiments (*1*) in the mixing zone $0.3 < x/(2u_m t) < 0.7$, while the higher order terms correct the local concentration in the extremes of the mixing zone. This procedure has also been generalized to include time-dependent velocity fields and has been employed in a variety of miscible displacement problems, as will be discussed later.

A complete description of the mechanisms controlling dispersion in a capillary tube as a function of dimensionless time and Peclet number is available from the finite difference solution of the full convective diffusion equation obtained by Ananthakrishnan *et al.* (*1*) for a step change in inlet concentration and by Gill and Ananthakrishnan (*31*) for the slug stimulus. Consider, for illustrative purposes, a slug of solvent A injected at time zero into a fully developed laminar flow of B. The initial configuration is shown in Figure 1*a*.

Different mechanisms dominate the transport process depending on the amount of time elapsed since it was initiated—*i.e.*, four-time regimes exist in which different mechanisms are most important. These are: (*a*) at very short times after injection, except at high flow rates, the dispersion of A into B takes place by pure axial molecular diffusion because of the high axial concentration gradients established. Thus, for a very short time, which is usually too brief to be of practical interest, the dispersion takes place as if the two phases were stagnant.

Since diffusive length is proportional to the square root of time and convective length to the first power, (*b*) at slightly higher times axial convection participates significantly in the dispersion. At high flow rates, axial convection is the controlling mechanism, and Figure 1*b* shows the distortion of the solvent slug into a parabolic shape when convection controls. From this it is clear that the action of axial convection is to

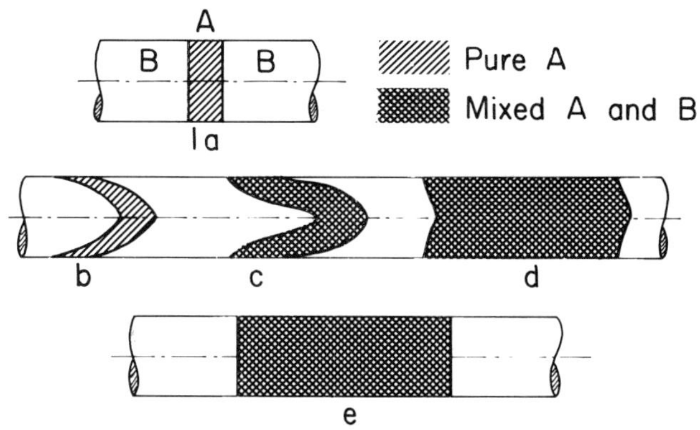

Figure 1. Schematic representation of the mixed zone for a slug of A dispersing in B as a function of time, A-B completely miscible

establish large radial concentration gradients and hence (*c*) at slightly higher times, radial molecular diffusion contributes to the dispersion process. Note that A is now transported radially toward the wall at the front of the slug and toward the centerline at the rear of the slug, the result being a mixed zone of varying concentration as shown in Figure 1*c*.

As time goes on, the action of radial diffusion continues to inhibit axial dispersion by diffusion and convection and makes the mixed zone more uniform as shown in Figure 1*d*. Finally, (*d*) at higher times, a pseudoequilibrium is established, Figure 1*e*, in which all three mechanisms contribute to the dispersion with the net effect appearing as if the fluid were in plug flow, whereas in fact the velocity is radially distributed. For this and higher times, the dispersion model results of the Taylor-Aris theory apply. As time increases further, the only effect is an increase in the length of the mixed zone.

The mechanisms are the same for the step change in inlet concentration (which occurs frequently in miscible displacements) as they are for a slug input; the mixed zones shown in Figure 1*c* through 1*e* then have concentrations varying between 1% and 99% A with all material to the left of the rear contour being pure A. Slight differences in the time required to reach the pseudoequilibrium region exist for the two types of experiments described above, as will be discussed later.

A graphical summary of the regions of applicability of various analytical solutions for dispersion for the step change in inlet concentration is given in Figure 2 with the dimensionless time and the Peclet number as parameters. Information on the intermediate range of parameters is provided (*1*). An important point relating to the dispersion model is that one finds that the dimensionless time $\tau = tD/a^2$ must be sufficiently large for it to apply—say 0.8 for fully developed laminar flow in tubes. Consequently the real time required is

$$t \approx \frac{0.8a^2}{D}$$

This simple formula provides us with a fairly general qualitative picture of when the dispersion model is likely to apply if we interpret "a" as a typical transverse dimension of a system and D to be a measure of transverse mixing rates which may be caused by turbulence or convection in addition to molecular diffusion. In this event, we see that the real time required for the dispersion model to apply increases rapidly with an increase in the transverse dimension and is decreased by greater transverse mixing. Therefore, for example, we could expect the dispersion model to apply more generally to turbulent than to laminar systems.

The mixing length, L', the axial distance required for the average concentration of displacing fluid to change from 10% to 90% of its inlet value, is of critical importance in determining the usefulness of the displacement since, in general, it is desirable to keep the two phases as separate as possible (minimize axial dispersion). For large enough values of time, Taylor's analysis and the finite difference results of Ananthakrishnan *et al.* yield

$$L = \frac{DL'}{2a^2u_m} = 0.26\left(\frac{tD}{a^2}\right)^{1/2}, \quad \tau = \frac{tD}{a^2} > 0.8$$

Bailey and Gogarty (*7*, *8*) report a different exponent for the dimensionless time, but this evidently can be associated with inaccuracies in their numerical procedure (*1*). For small values of time, the mixing length results are correlated by

$$L = 0.8\ \tau, \quad \tau < 0.01$$

For small Peclet numbers the numerical results of reference 1 are correlated by

$$L = 2.5N_{Pe}^{-0.75}\ \tau^{0.55}, \quad \tau > 0.05 \qquad (14)$$

$$L = 2.5N_{Pe}^{-0.75}\ \tau^{0.60}, \quad \tau < 0.05 \qquad (15)$$

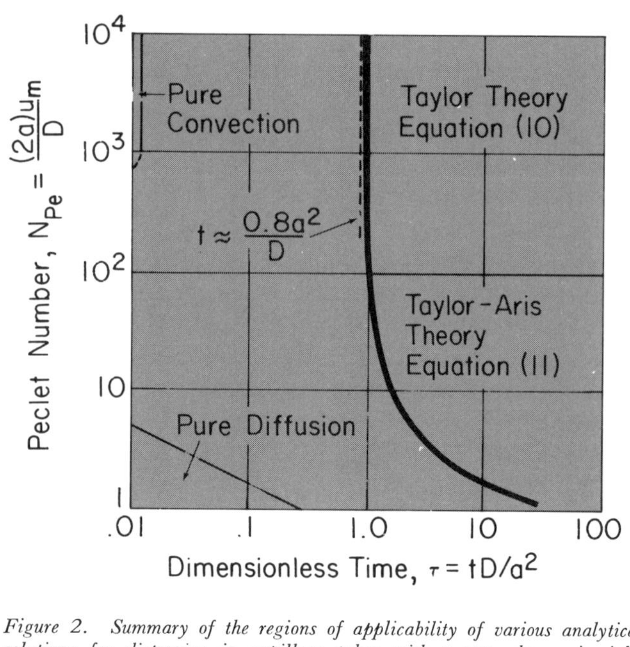

Figure 2. Summary of the regions of applicability of various analytical solutions for dispersion in capillary tubes with a step change in inlet concentration as a function of τ and N_{Pe} (1)

Table I. Minimum Dimensionless Time Required for Taylor-Aris Theory to Apply to Dispersion in a Tube as a Function of Inlet Boundary Condition and N_{Pe} (30)

N_{Pe}	τ_{min} (a)	(b)	(c)
1000	0.80	0.80	0.80
25	1.25	1.25	1.25
10	1.75	0.50	0.50
1	>4.50	4.50	>4.50

There is no equivalent axial dispersion coefficient that can be calculated from these mixing lengths unless the dispersion model region has been reached. Only then can the complete average concentration distribution be computed from a minimum of data.

Effect of inlet boundary condition. The extent to which the nature of the inlet boundary condition influences dispersion was investigated by Gill and Ananthakrishnan (*30*) using finite difference solutions to the convective-diffusion equation for laminar flow in tubes. They considered (*a*) the step change at the inlet ($x = 0$), (*b*) a step change at $x = -\infty$, and (*c*) the Danckwerts' boundary condition at the inlet. The step change at $-\infty$ is exact when two identical infinitely long capillaries, one filled with A and the other with B, are joined at $x = 0$ and the flow is started simultaneously. Danckwerts' boundary condition is an approximation to this situation. The important conclusions reached may be summarized as: (1) for $N_{Pe} > 100$, the dispersion is essentially independent of the inlet condition and (2) for low N_{Pe}, the Taylor-Aris theory applies over significantly larger ranges of the dimensionless time for (*b*) than for (*a*). This last conclusion is significant for small Peclet number experiments if the results are to be interpreted by a dispersion model, since inlet condition (*b*) should be simulated. Table I gives the minimum time for case (*b*) to follow the Taylor-Aris result as a function of the Peclet number.

Slug stimulus. The slug stimulus is an interesting case for study since stimulus response techniques are widely used for determining the flow characteristics of process equipment including porous media. A summary of the work done in this area is available (*44*). Since in real systems a true pulse is difficult to achieve, finite slugs are more generally the case. The main result of

Table II. Minimum Dimensionless Time Required for Dispersion in a Tube with Slug Stimulus to Follow Taylor-Aris Theory as a Function of N_{Pe} (30)

N_{Pe}	τ_{min}
>100	0.60
25	0.50
10	0.30

a finite difference treatment of this problem (*30*) is that the time required for a dispersion model of the Taylor-Aris type to apply is less than that for inlet condition (*b*) above, as is clear if the results summarized in Table II for the slug stimulus are compared with those in Table I. Within the ranges studied, the slug length had little effect on the minimum time requirement, but since condition (*b*) represents an infinitely long slug, and since the results for this case differed from those obtained from the set of small slugs studied, the slug length will have some influence as it becomes longer. For the slug stimulus an approximate criterion for the time required for the dispersion model to apply is

$$\tau_{\min} \simeq \frac{0.6\, N_{Pe}^2}{N_{Pe}^2 + 192}$$

Developing velocity fields and pulsating flow. Two types of problems have been investigated by finite difference solutions of the convective diffusion equation for determining the effects of developing velocity fields on dispersion in tubes (*32*). These include the velocity entrance region wherein the velocity distribution depends on axial distance and the velocity distribution which develops from rest and therefore depends on time. Each of these problems is most likely to be significant in gaseous systems with small Schmidt numbers. For high Schmidt number systems, the velocity entrance region comprises only a small portion of the mixed zone between displacing and resident fluids, and therefore this region is relatively unimportant even at small dimensionless times. For small Schmidt number systems, a relatively large time is required to reach a mixed length equivalent to the velocity entrance length. For velocities developing from rest, the dimensionless time required to reach the fully developed state is inversely proportional to the Schmidt number, such that the time-dependent nature of the velocity is unimportant except for small Schmidt number systems.

In performing experiments, the velocity entrance region would most often be important for the step-change inlet condition while velocity fields developing from rest are more likely to occur in slug stimulus experiments since it is difficult to introduce a well-defined slug of tracer into a flowing stream.

The analytical procedure involving the dispersion model and Equation 12 was also used to study these two situations, so that the applicability of the dispersion model might be determined. For time-dependent velocities, one obtains a dispersion coefficient which is time-dependent (only because the velocity is time-dependent) and the concentration distributions predicted compare favorably with the finite difference results. However, the approach fails for the velocity entrance region (*32*) because of the axial dependence of the velocity which causes the dispersion coefficient to depend on x, in contradiction to the assumption given in Equation 13—*i.e.*, the dispersion model given by Equation 9 is valid only when the velocity does not depend on x. This conclusion was also reached by a different approach by Batchelor (*9*).

The finite difference results, for dispersion with a velocity entrance region, again produced a mean concentration described by an error function at high times. However, the apparent dispersion coefficient determined from this behavior is an unpredictable function of dimensionless time and approached the fully developed value only at large dimensionless times. The apparent dispersion coefficients for this case are smaller than those for fully developed flow. This occurs because the velocity profile is flatter in entrance region flows.

The dispersion coefficient for flows developing from rest is also significantly smaller than the steady-state value and approaches the steady value asymptotically. However, practical use of the dispersion model for this case is possible because error function behavior of the mean concentration is achieved in the finite difference experiments at dimensionless times of approximately 0.6 for large Peclet numbers, which is in close agreement with the minimum time (*1*) required for the dispersion model to apply in steady-state flows.

The effect of the developing velocity field on the extent of dispersion in each case was a decrease in the extent of dispersion. This occurs because nonuniformities in the velocity field in a cross section are greatest when the flow is fully developed.

Aris (*5*), again using the method of moments, investigated the dispersion problem in a tube due to a pulsating pressure gradient in laminar flow. After averaging over one complete period, it was found that if the fluctuations in the pressure gradient are smaller than the mean pressure gradient, the effect of the periodic flow will contribute less than 1% of the total dispersion in the dispersion model regime.

Transition and turbulent flow dispersion. Fortunately, the one-dimensional unsteady dispersion model is more generally applicable to turbulent than laminar flows—*i.e.*, the time required for this model to describe turbulent systems accurately is much less than it is in laminar systems. However, to predict the dispersion coefficient for turbulent flow requires the use of an expression for the velocity distribution and the eddy diffusivity of mass, ϵ_D, which is usually taken to be equal to the eddy diffusivity of momentum, ϵ_m—*i.e.*, since the shear stress, τ_s, is given by

$$\tau_s = -(\nu + \epsilon_M)\rho \frac{\partial u}{\partial r}$$

and the mass flux, n, is given by

$$n = -(D + \epsilon_D)\frac{\partial C}{\partial r}$$

it follows that

$$\epsilon_M = \epsilon_D = \frac{-\tau_s}{\rho \dfrac{\partial u}{\partial r}} - \nu = \frac{-n}{\dfrac{\partial c}{\partial r}} - D$$

if one accepts the Reynolds analogy that $\epsilon_D = \epsilon_M$. This is useful since a great deal more empirical information is available in the literature regarding ϵ_M than ϵ_D. On this basis, the behavior of ϵ_D is known to a reasonable approximation and therefore one can use this expression for n, together with Equation 1, to describe the concentration distribution.

Once the velocity distribution and eddy diffusivity functions are known, one can proceed to determine the dispersion coefficient for turbulent flow in essentially the same way as for laminar flow—*i.e.*, one uses Equation 6 with D replaced by $D + \epsilon_D$, together with Equations 9 and 12. After Equations 9 and 12 are substituted into the modified form of Equation 6 and one collects coefficients of $\partial C_m/\partial x_1$, the significant part of the f_1 function is given by

$$\frac{1}{y}\frac{d}{dy}y\left(1 + \frac{\epsilon_D}{D}\right)\frac{df_{1s}}{dy} = \frac{u}{u_m} - 1 \qquad (16)$$

and the boundary conditions for f_{1s}, the steady-state part of the f_1 function, are

$$\frac{df_{1s}}{dy}(0) = \frac{df_{1s}}{dy}(1) = 0 \qquad (17)$$

so that f_{1s} is given by

$$f_{1s} = f_{1s}(0) + \int_0^y \frac{1}{\xi\left(1 + \frac{\epsilon_D}{D}\right)} \int_0^\xi \sigma\left[\frac{u(\sigma)}{u_m} - 1\right] d\sigma d\xi \qquad (18)$$

with

$$f_{1s}(0) = -2\int_0^y y \int_0^y \frac{1}{\xi\left(1 + \frac{\epsilon_D}{D}\right)} \int_0^\xi \sigma\left[\frac{u(\sigma)}{u_m} - 1\right] d\sigma d\xi dy \qquad (19)$$

The expression for the dispersion coefficient, neglecting axial molecular or turbulent diffusion, is then

$$k = -\frac{2u_m^2a^2}{D}\int_0^1 y\left[\frac{u}{u_m} - 1\right] f_{1s} dy \qquad (20)$$

where

u_m = mean velocity
a = tube radius
D = molecular diffusion coefficient

By assuming $\epsilon_D \gg D$, which is true except very close to the wall, Taylor was able to obtain the very simple expression for k,

$$k = 10.1\ av^* \qquad (21)$$

where v^* is the friction velocity defined by

$$\tau_w = \rho v^{*2} \qquad (22)$$

This can be related to the Reynolds number through the friction factor, λ, by

$$v^* = \sqrt{\frac{\tau_w}{\rho}} = \sqrt{\frac{\lambda\left(\frac{1}{2}\rho u_m^2\right)}{\rho}} = u_m\sqrt{\lambda/2} \qquad (23)$$

Consequently

$$k = 7.14\ au_m\sqrt{\lambda} \qquad (24)$$

and a variety of formulas are available which express the friction factor as a function of Reynolds number. Presumably Equation 24 is valid for rough as well as smooth pipes.

This result was supported with limited experimental data. However, Taylor used an expression for the velocity profile valid only for fully developed turbulence and did not include axial molecular diffusion since he was concerned with high Peclet number systems. As pointed out later by Tichacek *et al.* (*72*), the accuracy of the predicted dispersion coefficient is very sensitive to the accuracy of the velocity distribution. At a Reynolds number of 42,000, two sets of experimental data for the velocity distribution were employed to compute the dispersion coefficient and, although the the maximum deviation between the two was about 3%, the resulting dispersion coefficients differed by 50%. Tichacek *et al.* finally used the smoothed data of several studies to investigate the dispersion behavior. The difference in the results due to the assumed form of the velocity distribution is again evident in the work of Flint and Eisenklam (*27*) who used other smoothed data and compared their result with the previous theories. Both of these later papers noted that the coefficient in Equation 21 cannot remain constant over the entire turbulent regime from the onset of turbulence.

The interpretation of experimental data for dispersion in turbulent flow has broken down into studies of liquid systems with high Schmidt numbers and gaseous systems with Schmidt numbers near unity. The reason for this, as put forward by Tichacek *et al.*, is that in computing the dispersion coefficient, the changes brought about in gaseous systems by having the Schmidt number different from unity are so small as to be within the error involved in determining the coefficient experimentally. Since the total effective radial diffusion coefficient (using the analogy with momentum transfer) may be written as,

$$\epsilon_D + D = -\frac{\tau_s}{\rho\partial u/\partial r} - \nu\left(1 - \frac{1}{N_{Sc}}\right) \qquad (25)$$

assuming the Schmidt number N_{Sc} to be unity, eliminates a term from the expression and the Schmidt number as a parameter. Hence, available experimental data for turbulent flow were correlated separately for gases and liquids. Although Flint and Eisenklam retained the Schmidt number as a parameter, they compared and correlated their own data and theoretical expression only with previous data on gaseous systems. These two papers summarize most of the existing data on turbulent flow dispersion although recent results are reported in (*62A*) and some other interesting results (*70, 71*) will be described in connection with the effects of flow pulsations on dispersion.

The data of Flint and Eisenklam for gaseous horizontal dispersion with a pulse tracer are of interest because they bridge the gap between laminar and turbulent flow in the Reynolds number range 3×10^2 to 10^4 with Schmidt numbers between 0.27 and 1.0. The characteristic shape of a plot of the dispersion coefficient *vs.* the Reynolds number as given by these authors is shown in Figure 3 along with the theoretical result. This is instructive since it demonstrates the effect of turbulence on the dispersion process. [The plot of laminar flow results

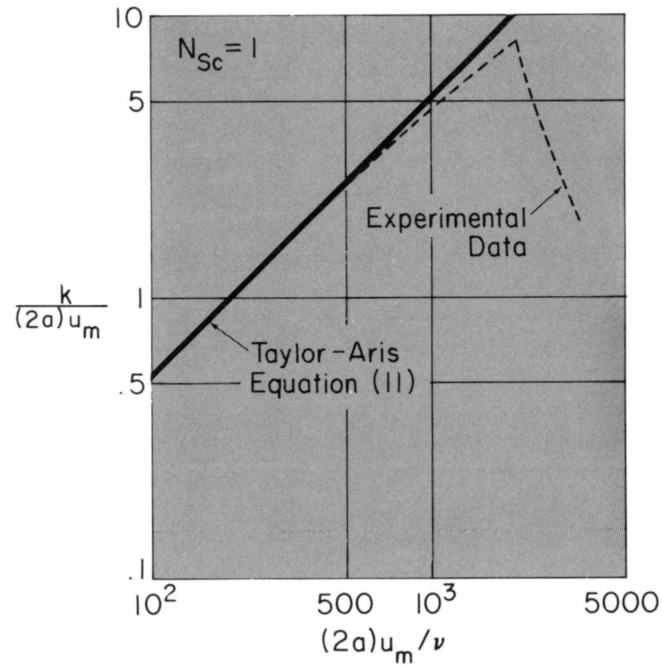

Figure 3. Schematic representation of the effect of transition from laminar to turbulent flow on k. Data in the transition region are available (27)

in Figure 3 as used by Flint and Eisenklam is misleading since it implies that the dispersion coefficient for laminar flow is dependent on the Schmidt number, even though in their experiments the contribution of molecular diffusion was less than 1% (smallest Peclet number about 81). In this situation Equation 33 indicates that the proper dimensionless coefficient is a constant. Although Flint and Eisenklam report deviations from the Taylor result at values of the Reynolds number considerably below the critical value, these deviations do not indicate a discrepancy in the basic theory since they are ascribed to the formation of turbulent spots. Furthermore, the discrepancies occur at different Reynolds numbers for different Schmidt numbers and do not seem to occur to any considerable degree at a Schmidt number of 0.27 (see their Figure 6).]

Over the range of Reynolds numbers 10^2 to 10^3 the experimental data follow the Taylor results and peak at approximately the critical Reynolds number followed by a rapid decrease as the Reynolds number increases further. In turbulent flow as compared to laminar, material is dispersed by the additional mechanism of eddy diffusion. In the radial direction, this enhances the action of radial molecular diffusion and hence the dispersion coefficient decreases. Tichacek *et al.* and Taylor have presented arguments which indicate that turbulent axial diffusion has negligible effects on the dispersion because its contribution is small compared to mixing caused by radial differences in the velocity.

Axial dispersion in concentric annuli in fully developed turbulent flow was studied by Smith (*65*) both experimentally and with a Taylor dispersion model. Again, an extreme sensitivity of the predicted dispersion coefficient to the velocity distribution used was noted. Experimental results for pipes and annuli correlated well if the hydraulic diameter was used as the length parameter.

Unsteady turbulent flow due to a sinuosidal pressure gradient in a liquid system was studied experimentally by Taylor and Leonard. Although the initial results (*70*) indicated a large increase in the dispersion due to small amplitude pulsations, later experiments (*71*) failed to reproduce the initial results and demonstrated only small changes in the dispersion coefficient due to small flow pulsations. An analytical treatment of this problem is now in progress (*47*), and preliminary results indicate that the trends of the experimental findings may be predicted using an empirical eddy diffusivity in the dispersion model.

Interesting steady turbulent flow dispersion results

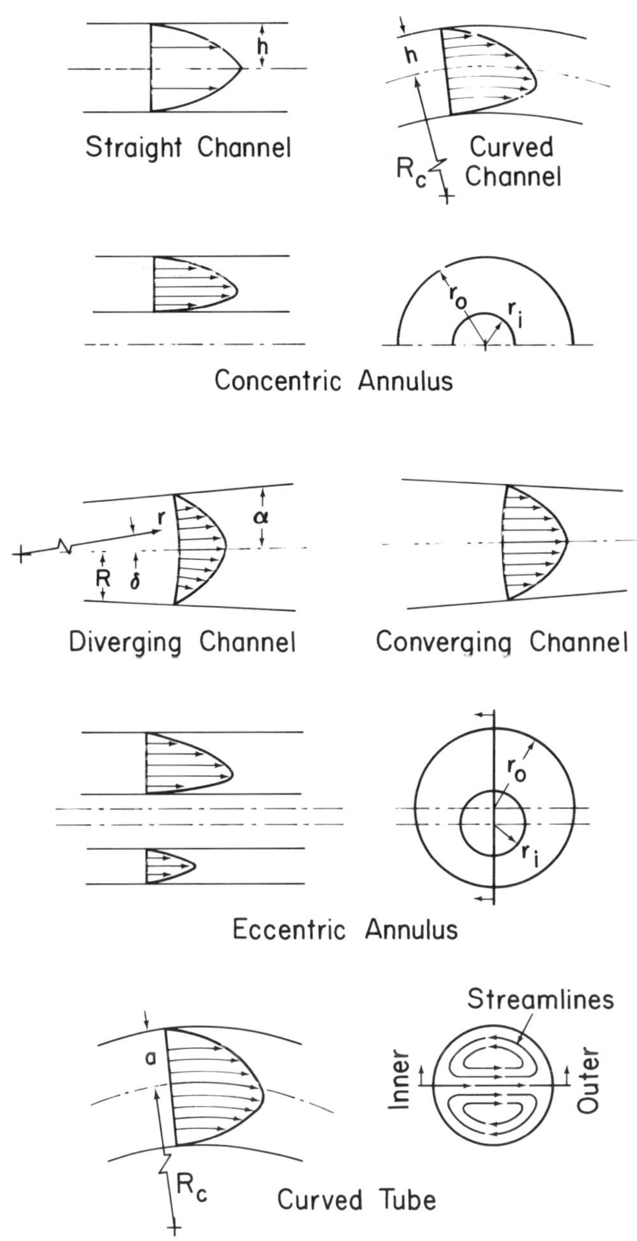

Figure 4. Diagrams of various microstructure units indicating the nature of the velocity profiles

on a liquid system are reported by Taylor (*71*) for comparison with the pulsating flow result. Two different tube sizes of commercially "drawn" tubing were used in experiments over a Reynolds number range of 3500 to 26,000, namely 0.242 in. and 0.745 in. i.d. For the larger tube, the results are in reasonably good agreement with the correlation suggested by Tichacek *et al.* For the smaller tube, the results are significantly higher (by about a factor of 2). One explanation for this may be the tube roughness. On the assumption that the two tubes had the same size roughness elements, it is possible that the roughness affected the results in the small tube but not in the large one. The friction factor in Equation 24 is a function of only the Reynolds number if the tube is hydraulically smooth. If, however, the roughness elements are large enough so that the system cannot be considered smooth, then λ increases as the relative roughness, the ratio of the height of a roughness element to the tube radius, increases. Hence, Equation 24 predicts trends in agreement with the experimental results if the small tube is assumed to have a larger relative roughness than the larger tube. As noted by Tichacek *et al.*, roughness tends to make the velocity less like plug flow than in systems which may be considered hydraulically smooth, and hence increases the dispersion. However, studies of the friction factor variation with Reynolds number by Taylor failed to show any significant dependence on tube size.

Other Microstructure Units

A number of microstructure units other than straight capillary tubes are of interest in formulating a description of possible geometries making up the macrostructure. Among those which have significant effects on the nature of the dispersion and which have been investigated using either the finite difference solution of the convective diffusion equation and/or the dispersion model procedure are: (1) the straight channel; (2) the straight annulus, both concentric and eccentric, wherein the effects of an asymmetrical velocity profile may be isolated; (3) the curved channel and tube which introduce the effect of curvature and secondary flow in the tube case; and (4) converging and diverging channels which have the important property of nonparallel boundaries. A schematic representation of the various units is given in Figure 4. In addition, the mixing cell may form one of the important units in the microstructure, and a series of stirred tank arrangements is frequently used in the literature (*43, 50, 75*) as a model of porous media.

It is important to investigate the relationship between the mixing cell model and an actual porous medium to determine if such a model is physically realistic in the ranges of flow rates over which it has been used. The correlation of much of the experimental data existing in the literature for unconsolidated media, such as sand packs by Perkins and Johnston (*50*), demonstrates that an effective axial dispersion coefficient correlates well with the first power of the mean interstitial velocity over a fairly large range of velocity, although to cover very large ranges of velocity a power series in a Peclet number based on pore dimensions (*39*) may be required. The mixing cell model can be used to predict that the dispersion coefficient is proportional to the velocity whereas in a single capillary, the dispersion coefficient is proportional to the square of the velocity. However, the mechanism which one would consider as forming a mixing cell at the junctures of solid particles—*i.e.*, turbulent mixing—has not been observed in such media and furthermore—if the velocity is increased, experimental dispersion coefficients become proportional to a higher power of the velocity. Visual observations by von Rosenberg (*78*) and Blackwell (*11*) suggest that individual streams retain their identity at a juncture in a sand pack and that transport occurs between them solely by molecular diffusion. Hence, the mixing cell in this case really approximates a situation which occurs at relatively low flow rates such that the fluid streams from different pores leading to a juncture remain in the junction sufficiently long enough to have their concentrations equalized by molecular diffusion and not turbulent mixing. The situation is different in other types of packing. For example, Shulman and Mellish (*62*) observed the flow patterns in a column packed with 1-in. Raschig rings and found that although laminar flow predominated in the flow over packing, turbulent mixing occurred at the junctions between pieces of packing.

Straight channel. The dispersion coefficient for a parallel plate duct with fully developed laminar flow can be shown to be (*54, 74*):

$$k_{f.p.} = D + \frac{8}{945}\frac{u_{\max}{}^2h^2}{D} \tag{26}$$

where h is the half-width of the channel and $u_{\max}$ is the centerline velocity. Subject to a step change in inlet concentration, the mean concentration in the region of applicability of the dispersion model is

$$C_m/C_0 = 1/2\left\{\operatorname{erfc}\left(\frac{x_1}{2\sqrt{k_{f.p.}}}\right) + \exp\left[\frac{u_m(x_1 + u_mt)}{k_{f.p.}}\right]\operatorname{erfc}\left(\frac{x_1 + 2u_mt}{2\sqrt{k_{f.p.}t}}\right)\right\} \tag{27}$$

The full convective diffusion equation has been solved by finite difference (*33*) and the regions of validity of analytical solutions are shown in Figure 5. Note that for the dispersion model to apply requires a dimensionless time $\tau = tD/h^2$ of approximately unity. At high Peclet numbers with the same fluids, a straight channel having the same equivalent diameter (four times the hydraulic radius) as a tube requires a minimum real time of approximately 1/3 of that for the tube before the dispersion model applies, thus indicating that the minimum real time requirement is a strong function of the geometry of the system.

Concentric and eccentric annuli. The concentric annulus represents a generalization of the straight channel and straight tube cases and demonstrates the effect of an asymmetric velocity distribution of dispersion. The dispersion coefficient for this situation (*33*) is a function of the radius ratio, $1/\kappa$, the ratio of outside to inside radii, r_o/r_i. For small values of $1/\kappa$ (narrow gap) ≤ 1.5, the flat plate result given in Equation 26 may be written as

$$k = D + \frac{8}{945}\left[\frac{r_o^2 u_{\max}^2 (1 - \kappa^2)}{4D}\right] \quad (28)$$

with r_o the radius of the outer cylinder; it provides an excellent approximation to the annulus result. As the radius ratio increases, the dispersion coefficient becomes larger than predicted by Equation 28, thus indicating that the effect of an asymmetrical velocity is to increase the dispersion relative to the symmetrical case of the straight channel. As $1/\kappa$ approaches 100, the dispersion coefficient approaches that for a tube given in Equation 11. The limits of applicability of the dispersion model with respect to time have not yet been established but for the limiting cases (and probably in between), they may be inferred from the results for tubes and channels given previously.

Dispersion in eccentric annuli is most markedly affected by the degree of eccentricity when the ratio of the outer to inner radius approaches unity so that the system approximates nonparallel plates with a constant cross-sectional area along the direction of flow (*60*). Over the range of radius ratios and eccentricities studied, the dispersion coefficient can vary by more than two orders of magnitude with changes in eccentricity.

Curved systems. The curved tube with fully developed laminar flow has the additional complicating factor of a secondary circulatory motion in a cross section superimposed on an asymmetrical axial velocity profile. Jones (*40B*) and Wright (*81*) have recently presented evidence that such circulations are the reason for departures from Darcy's law in granular beds. For large values of the ratio of the radius of curvature to the tube radius RC/a, Erdogan and Chatwin (*25*), using an approach similar to Taylor's (*67*), demonstrated that the effect of curvature is to decrease the dispersion coefficient over that in a straight tube if the Schmidt number is >0.124. Later it was demonstrated (*46*), by relaxing the assumption with regard to curvature ratio, that the competing effects of transverse mixing caused by the secondary flow and the asymmetrical axial velocity cause the dispersion coefficient first to increase above and then decrease below the straight tube result as the Reynolds number increases, these effects being most pronounced at small values of the curvature ratio. The applicability of the result with respect to the minimum time requirement is not yet available although approximate minimum times have been estimated (*47A*).

In the curved channel, no secondary flow exists and hence the dispersion coefficient is increased over that in a straight channel under the same conditions. The effect of molecular diffusion on dispersion is modified by the curvature. For example, in the curved channel, for a given Peclet number, molecular diffusion contributes a smaller fraction of the total dispersion effect than it does in straight channels.

Experimental studies on miscible displacements in

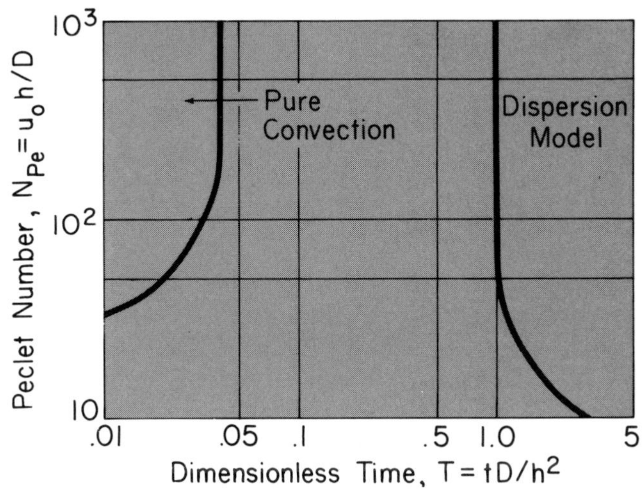

Figure 5. Summary of the regions of applicability of analytical solutions for dispersion in straight channels with a step change in inlet concentration as a function of τ and N_{Pe} (33)

systems with curvature have mainly involved turbulent flow, and conflicting results concerning the effect on the dispersion have been reported (*6, 16, 17, 41*). These discrepancies may be due to differences in the experimental apparatus used or to differences in the turbulence characteristics of the flow, but it appears that more work is needed in this area.

Converging and diverging channels. Flow in the interstices of porous media experiences local accelerations and decelerations because of changes in cross-sectional area. Some indication of the effect of variable area passages on the dispersion process has been provided by a recent study of Jeffrey-Hamel flows between nonparallel flat planes (*33, 36*). Even small angles of divergence can markedly affect the dispersion process, particularly when the Peclet number is large. Perhaps the most striking result of this study is that the pure convection solution is valid for dispersion in flow between nonparallel walls for much larger values of dimensionless time than for parallel plate or capillary systems. The implication is that both transverse and axial molecular diffusion play a minor role in dispersion in diverging channels at high Peclet number.

Some physical insight into the reasons why molecular diffusion is less effective can be gained by considering the velocity distribution. For stable, fully developed flow in diverging channels the velocity field can be written in terms of Jacobian elliptic functions which can be simplified considerably for small Reynolds numbers and angles of divergence to the form

$$u = \frac{U_{\text{ent}} R}{\tan \alpha}\left(\frac{1}{r}\right)(1 - \theta^2)$$

where

U_{ent} = velocity at entrance to system
R = halfspacing at entrance to system
α = angle of divergence
r = axial coordinate
θ = dimensionless angular coordinate (δ/α)

If one plots this distribution, it becomes flatter as one moves downstream. This reduces the surface area across which a steep radial concentration exists, compared to that in parallel plates, and therefore decreases the effect of transverse molecular diffusion. As the velocity gets flatter, in the limit, the surface area across which molecular diffusion occurs degenerates to a plane, the outer normal of which is parallel to the direction of main flow. In contrast, with tubes or parallel plates the velocity profile is invariant with axial position and as time increases, convection creates an increasingly large paraboloid surface, across which transverse concentration gradients are very large.

Capacitance Effect

The existence of stagnant pockets of fluid in porous media due to dead-end passages along the flow path is one consequence of the geometry of the medium which may be incorporated relatively easily into the microstructure capillary model. The effect of system capacitance is to increase the length of the mixing zone. Capacitance models for porous media have been suggested and employed for the interpretation of experiments by several authors (*18, 21, 34, 37, 38, 45, 50, 73, 74*). To construct the microstructure model, some physical picture of the nature of the stagnant regions must be assumed. One extreme of the types of stagnant regions possible is the "bottleneck" (*18, 34*) shown in Figure 6*a* wherein the concentration of the main volume of stagnant fluid is assumed to change with time, due to a steady-state diffusion process in the pores leading to the main volume. Turner (*73*) visualized a porous medium as consisting of rectangular dead-end pockets communicating with the main flow channels as shown in Figure 6*b*. Material passes into or out of these pockets only by molecular diffusion. In the main channels, dispersion takes place by convection and diffusion, while the concentration in the stagnant volume changes according to

$$\frac{\partial C}{\partial t} = D\frac{\partial^2 C}{\partial y^2} \qquad (29)$$

Later Gill and Ananthakrishnan (*30*) used a continuous stagnant pocket or Turner structure as shown in Figure 6*c*. The point concentration in the stagnant region varies according to the two-dimensional unsteady diffusion equation. The continuous Turner structure retains the elements of the other representations and is simplest to deal with numerically. Hence, we shall consider in some detail the analysis and results for the Turner model microstructure unit.

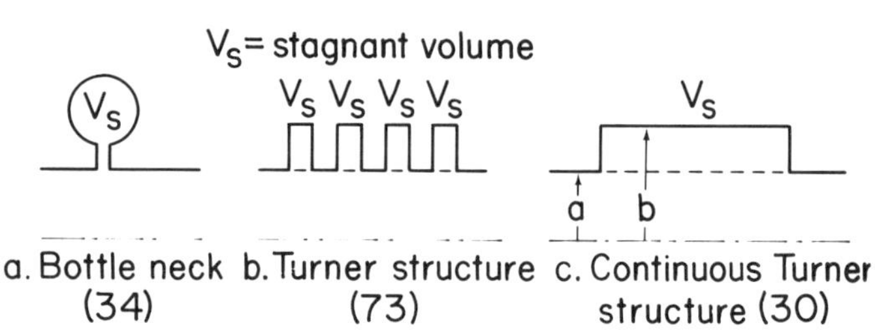

Figure 6. Diagrams of different capacitance models

Turner (*73*) and later Aris (*2*), who generalized the Turner model, were concerned with the frequency response of such models and did not discuss the applicability of a dispersion model, although Turner (*74*) interpreted some experimental results for a flat plate system to which were connected dead-end capillaries by using a Taylor dispersion coefficient based on the distance between plates. However, Aris (*3*) pointed out that the Taylor-Aris result given in Equation 11 is not the correct expression for the dispersion coefficient in Turner-type structures. Using the method of moments it was determined that the Taylor dispersion coefficient could be anywhere from one to 11 times the value it has for a straight capillary without the capacitance effect. The low and high extremes occur when there are no stagnant regions and when the ratio of stagnant volume to pore volume become infinite, respectively.

The applicability of a dispersion model of the type postulated in the Taylor-Aris theory was investigated with numerical experiments by Gill and Ananthakrishnan (*30*) on the continuous Turner structure for various inlet boundary conditions, Peclet numbers and stagnant to pore-volume ratios. For high Peclet numbers ($N_{Pe} \gtrsim 100$) where the effects of inlet boundary conditions and axial molecular diffusion are unimportant, the mean concentration becomes an error function at sufficiently high dimensionless time. A comparison of the result given in Equation 30

$$C_m/C_0 = 0.5\left[1 - \text{erf}\left\{\frac{x - u_{max}t/2\gamma^2}{\sqrt{4kt/\gamma^2}}\right\}\right] \qquad (30)$$

where γ is the ratio b/a in Figure 6*c*, with the Taylor theory demonstrates that dispersion in Turner structures must be considered to take place in a tube of sufficient radius to include the stagnant pore volume rather than just the main channel volume alone; then the dispersion coefficient obtained by Aris (*3*)

$$k = \left[\frac{8(1+\beta) + 12(1+\beta)^2 \ln(1+\beta) - 7(1+\beta)^2}{(1+\beta)^2}\right] \times \frac{u_{max}^2 a^2}{192D} \qquad (31)$$

where

$$\beta = \gamma^2 - 1$$

Table III. Minimum Values of Dimensionless Time Necessary for Dispersion Model to Apply to Continuous Turner Structure Capacitance Model at $N_{Pe} \gtrsim 100$ as a Function of Ratio, Total Volume to Pore Volume, γ (30)

γ	τ_{min}
1.0	0.80
1.1	1.25
1.2	2.00
1.3	4.00
2.0	>15.00

can be used to predict the mean concentration. This result implies that for complex systems, such as porous media, the entire void volume should be used in determining flow parameters.

The numerical experiments (*30*) demonstrated that for high Peclet number systems, the limiting value of the dimensionless time necessary for Equations 30 and 31 to predict the mean concentration distribution was a function of γ as shown in Table III. It is clear from Table III that systems having a large capacitance require significantly longer times before a dispersion model applies. This implies that for porous media with large amounts of dead space a long length is required to ensure sufficient residence time for the effluent concentration to obey the error function description.

For low values of the Peclet number, the effects of axial molecular diffusion and inlet boundary conditions create a more complex situation. At very low Peclet numbers, pure molecular diffusion is the controlling mechanism if

$$N_{\mathrm{Pe}}\sqrt{\tau}\begin{cases}<0.4 \text{ with } \gamma = 1.2\\<0.5 \text{ with } \gamma = 1.0\end{cases} \tag{32}$$

For large values of the dimensionless time, the mean concentration again becomes an error function; however, in this situation, the minimum time is strongly dependent on Peclet number and the nature of the inlet boundary condition. Interestingly, the trend of behavior is not the same as that for tubes without capacitance for inlet conditions of the types *b* and *c* described previously. In fact, the minimum time required for case *b* to be described by an error function, as given in Table IV, is reduced from 2.75 to 1.0 by changing the Peclet number from 25 to 10. Note that this lower value is less than that required at high Peclet numbers. The minimum values of time listed in Table IV reinforce the conclusion drawn previously that if miscible displacements are to be interpreted by using a dispersion model they should simulate inlet boundary condition *b* since this will reduce markedly the system length necessary.

It is significant to note that using an analysis similar to that used originally by Taylor (*67*, *69*), Gill and Ananthakrishnan were able to predict local concentration distributions which agreed with the finite difference experiments at sufficiently large values of time using the entire void volume of the system. This implies that local as well as average behavior may be predicted in complex systems if only the local velocity distribution is known.

Dayan and Levenspiel (*20A*) further generalized the Turner model to include adsorption on the solid surface including that inside the dead end pores. Thus they were able to simulate the separate effects of holdup in the pores and adsorption on the surface. They employed the method of moments developed by Aris. For the case of a linear equilibrium relation between phases and assuming equilibrium always exists between the interfacial liquid and the adsorbed phase, they derived an expression for the dispersion coefficient and showed that adsorption increases the dispersion coefficient. Furthermore they showed that pore holdup and adsorption are not simply additive contributions to the dispersion coefficient but are combined in a complex fashion.

Aris (*4A*) also applied his moment analysis to the case of two phases flowing contiguously with heat or mass transfer across the interface between them. This interesting analysis provides the basis for analyzing both cocurrent and countercurrent systems. Gill (*33A*) using a generalized version of the series expansion in Equation 12, later extended the two phase problem to include the effect of time variable flow. The dispersion coefficient obtained from these analyses should be used in conjunction with a dispersion model based on the average velocity for the entire system (*33A*).

Buoyancy Effects in Tubes

From a dimensional analysis of the general equations of change for dispersion in tubes, it is possible to demonstrate that the dispersion process depends on the independent parameters

$$N_{\mathrm{Pe}} = \frac{au_m}{D},\ N_{\mathrm{Re}} = \frac{au_m\rho_0}{\mu_0},\ N_{\mathrm{Gr}} = \frac{a^3g(\Delta\rho/\rho_0)}{\nu_0^2},\ N_{\mathrm{Fr}} = \frac{u_m^2}{ag}$$

these being, respectively, the Peclet, Reynolds, Grashof, and Froude numbers. Thus far we have considered laminar systems in which density and viscosity differences between resident and invading fluids are negligible and no free surface exists. Hence, it is easily demonstrated from Equation 11 for straight capillary tubes that the dimensionless dispersion coefficient, K, depends only on the Peclet number, and for large N_{Pe}, K is independent of the Peclet number.

$$K = \frac{k}{DN_{\mathrm{Pe}}^2} = \frac{1}{N_{\mathrm{Pe}}^2} + \frac{1}{48} \tag{33}$$

In practice, density and viscosity differences give rise to significant changes in the nature of dispersion in both capillary tubes and porous media. In a completely submerged system in which density is constant, the effect of gravitational force is eliminated by the proper choice of static pressure and hence, the Froude number which represents the ratio of inertial to gravitational forces is not a parameter. In displacement systems with density differences, however, it is not immediately evident that the initial interface established between resident and invading phases does not have some of the properties of a free surface. Surface tension–driven convection can also play a role in determining the nature of the process which occurs at this interface. Some of the effects which arise in horizontal and vertical

Table IV. Minimum Values of Dimensionless Time Necessary for Dispersion Model to Apply to Continuous Turner Structure Capacitance Model as Function of N_{Pe} with $\gamma = 1.2$ (*30*)

N_{Pe}	τ_{min}
1000	2.00
25	2.75
10	1.00

systems due to natural convection caused by concentration gradients are considered here.

A number of studies of miscible displacements in tubes (*13*, *26*, *55*, *56*, *67*, *76*) have verified the Taylor theory for large Peclet numbers and dimensionless times near or greater than unity. A majority of these results are for systems wherein the two fluids have approximately the same density and viscosity. More difficulty has been experienced with experimental verification of the Aris modification for small Peclet numbers, although the work with gases of Evans and Kenney (*26*) for large values of the dimensionless time and the experiments with liquids of Rejhsinghani *et al.* (*56*) for small values of time and small bore tubes show excellent agreement when effects not included in the Taylor-Aris theory are not present. The main factor associated with deviations from this theory at low values of the Peclet number appears to be natural convection currents established by concentration gradients, although instabilities in the flow may contribute.

Significant deviations from the Taylor-Aris result for higher times and from the numerical experiments (*1*) at lower times have been noted in horizontal systems with density differences $\Delta\rho/\rho$ as small as 1.2×10^{-4} in aqueous solutions (*56*). Bournia *et al.* (*13*), working with gases having the same molecular weight in vertical downward flow, found large discrepancies between theory and experiment in the low Peclet number range. Density differences in such gaseous pairs are due to differences in the compressibility factors. It has been estimated (*56*) that the Grashof number for their experiments was on the order of magnitude of 10,000, thus indicating that natural convection effects were probably present.

From first considerations, natural convection seems to have a different influence on the extent of dispersion in horizontal and vertical systems since, in the first, the action of gravity is normal to the main flow and, in the second, parallel. We shall now explore the modes by which natural convection influences the dispersion.

Assuming a step change in the inlet concentration at time zero, at small values of time, axial convection dominates and a parabolic finger of the displacing fluid invades the resident fluid. This situation prevails in both horizontal and vertical systems and whether the light fluid displaces the heavier or vice versa.

In horizontal systems, if the heavier fluid displaces the lighter, at low flow rates, the core of heavier fluid tends to sink to the tube wall displacing the less dense fluid and creating a transverse movement in the cross section. If the lighter fluid is used as the displacing medium, the core tends to rise, again creating a transverse motion. This secondary movement increases radial mixing—*i.e.*, it enhances the action of radial molecular diffusion and thus tends to inhibit axial dispersion. However, there is a second effect in systems with natural convection which must be considered.

Axial density gradients cause a variation in the axial pressure gradient and thus axial movement. For the case of a denser fluid displacing a lighter one, this effect enhances axial movement in the flow direction, thus enforcing the effects of axial convection and increasing the amount of axial dispersion. These two effects of natural convection in horizontal systems are competitive, one tending to increase and the other to decrease the axial dispersion. The experimental results (*56*) agree with theory which predicts that in the low Peclet number range, natural convection enhances axial dispersion at lower Peclet numbers and depresses it at higher Peclet numbers. For intermediate values, where the two competing effects cancel, and for very large Peclet numbers, where natural convection is negligible, the experimental results approach the numerical solution ignoring natural convection. This was first explained qualitatively on the basis of an analogy to heat transfer (*56*) and later explained quantitatively by the analysis of Erdogan and Chatwin (*25*).

Buoyancy effects in vertical upward displacements can either increase or decrease the extent of axial mixing depending upon whether the heavy or light fluid is the displacing medium. The contributing effects can again be explored if one considers the small time behavior of the system. At small times, a paraboloid of displacing fluid invades the resident fluid as shown in Figure 7*a*. If we ignore the effects of molecular diffusion, as time increases, if the heavy fluid is the displacing fluid, in a cross section, the centrally located fluid is denser than that near the wall, and therefore the action of gravity is to retard the movement near the centerline and accelerate it near the wall. The flattened profile in Figure 7*b* results.

If the heavier phase is displaced, the action of gravity is to accelerate the central core relative to the fluid near the wall, and the elongated profile of Figure 7*c* occurs. Comparison of the relative lengths of the zones over which the invading fluid is present in Figures 7*b* and 7*c* shows that the axial dispersion is depressed by natural convection in case *b* and enhanced in case *c*. An approximate analytical analysis which predicts these trends is given by Reejhsinghani *et al.* (*55*).

When the initial interface is created, if the lighter phase is on the bottom and displaces the heavier one, we have an unstable hydrodynamic system which is analogous to the classical problem of a fluid layer heated from below. This instability has a profound effect on the process as evidenced by the experiments (*55*). In

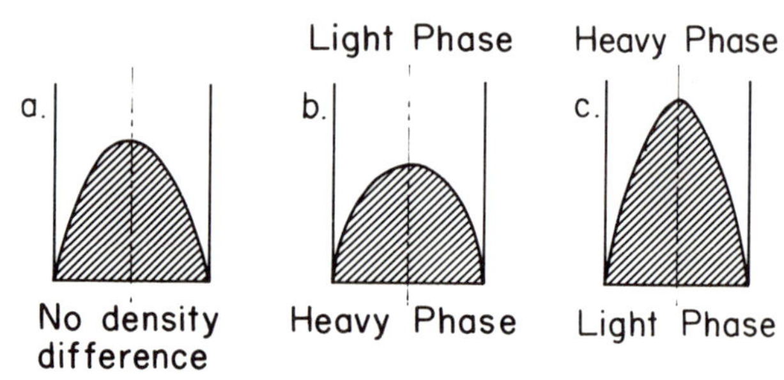

Figure 7. Schematic diagrams illustrating the effect of buoyancy on dispersion in vertical upward displacements in tubes

particular, for such unstable cases, the approximate theory which ignores this effect grossly underestimates the value of the dispersion coefficient.

In downward flow, the same arguments applied to upward flow can be used to predict that a heavier fluid displacing the lighter should lead to increased axial mixing and a lighter displacing the heavier to decreased axial mixing. These mechanisms would seem to apply as long as the system is in stable laminar flow.

The vertical downward experiments of Bournia *et al.* (*13*) are more difficult to interpret since slugs of tracer gas were used rather than the step change in inlet concentration. The small density differences in these experiments were noted previously. Each of the gases was used as the slug with the other as the surrounding medium. Hence, both a light slug in a heavy carrier phase and a heavy slug in a light carrier stream were investigated. The configurations which exist at small times are shown in Figures 8*a* and 8*b*. By the previous arguments, the two slugs will be distorted by gravity into the shapes shown in Figures 8*c* and 8*d*. In the case of the heavy slug, the forward portion is elongated and the rear flattened. If the flow is stable then clearly the mixing length is increased. For the light slug the situation is not clarified by this simple analysis since the rear portion of the slug will tend to catch up with the front portion. The experiments of Bournia *et al.* yielded large increases in the amount of dispersion for both types of experiments at low velocities. In the case of the heavy slug, the trend is predicted by our simple analysis while for the light slug the analysis fails. Perhaps in both cases the stability of the flow should be examined since a heavy fluid on top of a lighter one in a vertical system is an unstable state; clearly, with a slug of different density from its surroundings, one of its two interfaces is always unstable.

Most experimental data available on vertical displacements in porous media have been performed with a favorable density difference—*i.e.*, the less dense fluid above the more dense, since viscosity instabilities are lessened by favorable density gradients, as will be described later. Hence, existing analytical studies of the stability problem have focused mainly on the description of viscous fingering and assumed matched densities or favorable density gradients. It is important to note that gravity effects in porous media even for fluids of apparently matched densities (matched to two significant figures, for example) may have significant effects and, by analogy to capillary tubes, render a dispersion model useless for interpreting results.

Viscosity and Density Effects in Porous Media

Since density effects have been found to influence significantly the nature of dispersion in capillary tubes, it is pertinent to examine the literature pertaining to porous media to determine if similar effects are noted. Viscosity and/or density differences can lead to unstable displacements—*i.e.*, the formation of viscous "fingers" or a gravity "tongue" of displacing fluid in the resident phase. These effects, which are described further below,

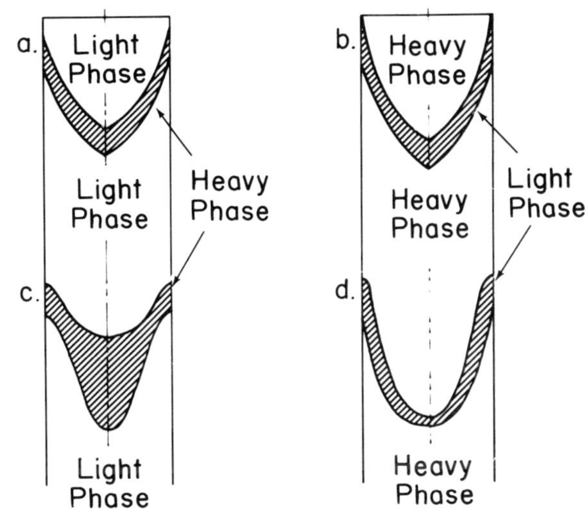

Figure 8. Schematic diagrams illustrating the effect of buoyancy in downward flow with a slug stimulus

tend to lengthen the mixed zone between phases; thus, they have a deleterious effect in most practical miscible displacements. Several different possibilities exist for the interaction of viscosity and density effects. One may have either a favorable viscosity ratio (the displacing fluid is more viscous than the displaced phase), or an unfavorable viscosity ratio coupled with either a favorable (the upper fluid is less dense than the lower in a vertical system), or an unfavorable density difference. Viscosity ratios and density differences are favorable if they result in a shorter mixed zone.

Consider first a system with matched densities. Viscous fingering results if the displacing fluid penetrates at different rates in a cross section of the porous medium. This initial preferential penetration is evidently due to local heterogeneities in the permeability, since it occurs (*12*) in what would normally be considered homogeneous media (*e.g.*, sand packs with a small range of particle size). Under an unfavorable viscosity ratio, these initial fingers grow in length and width with distance penetrated. Such fingers have been observed by several workers (*12*, *39*, *64*) and resemble those shown in Figures 9*a* and *b* for very high and smaller unfavorable viscosity ratios in horizontal systems. At low flow rates, transverse diffusion has been observed (*64*) to cause the formation of one large finger. As indicated in Figure 9, viscous fingering can lead to bypassed regions and extremely long mixed zones as the viscosity ratio becomes more unfavorable. Although this is an extremely complex phenomenon, Peaceman and Rachford (*49*) were able to solve a mathematical model which yielded qualitative behavior similar to that shown in Figure 9. To cause fingers to form spontaneously in the mathematical model, it was necessary to use small random variations of permeability with position. Other mathematical analyses of horizontal systems have been given (*23*, *42*).

Matched density experiments in vertical upward flow with both favorable and unfavorable viscosity ratios have been reported by Brigham *et al.* (*14*). Results showed: (*a*) for favorable viscosity ratios of 0.175 and 0.998 and unfavorable ratios of 1.002 and 5.71, the mixing zone length increases as the viscosity ratio increases; (*b*) a change from a favorable viscosity ratio of 0.998 to 1.002 unfavorable brought about a disproportionately large increase in the mixing zone length; and (*c*) for unfavorable viscosity ratios, the mean concentration at the outlet did not follow the error function behavior of Equation 4. Point (*b*) has been discussed (*40*) with regard to the prediction of stability theories, while von Rosenberg (*78*) has mentioned an effect similar to that in point (*c*), and the downward flow matched density experiments of Slobod and Thomas (*64*) for unfavorable viscosity ratios do not appear to have the characteristic error function "S" shape on a plot of mean concentration *vs.* pore volumes injected although these data were not plotted on probability paper to test this.

Gravity tonguing occurs in horizontal systems as the less dense displacing phase (most experiments available have used this arrangement) tends to flow over the more dense resident phase, thus increasing the mixing length. Experimental evidence for this effect has been reported (*68, 69, 77*). In slow flows with large gravity effects, the tendency is to form one large finger (*20*) of displacing fluid rather than several smaller fingers. Transverse mixing by molecular diffusion is probably enhanced by natural convection in this situation by a mechanism similar to that described for empty tubes previously, so that the displacement front is smoothed out compared to that shown in Figure 9. Density effects are also considered in mathematical models (*23, 49*).

In vertical systems, gravity effects may be used to inhibit viscous fingering and may effectively prevent the formation of fingers under certain conditions. Consider the schematic diagram in Figure 10 of a vertical downward displacement with a lighter, less viscous fluid invading the resident fluid. If viscous fingering occurs, then the shaded area represents the region occupied by the displacing fluid. A cross-sectional cut of the medium is denoted by the two horizontal lines. In the heavily shaded regions the material is lighter than the unshaded areas because of the density difference. Hence, the action of gravity will be to accelerate flow through the unshaded regions and to retard flow in the shaded regions as indicated by the direction of the arrows. Thus, the gravity effect is clearly one of inhibiting the growth of viscous fingers and reducing the length of the mixed zone. If the positions of the heavy and light fluids are reversed but the unfavorable viscosity ratio is maintained, then the growth of viscous fingers is accelerated by density effects. This reasoning is analogous to that used to explain the effect of gravity on the velocity distribution for displacements in tubes.

Some of the most extensive experimental work on the length of the mixed zone under all four combinations of viscosity and density differences in vertical downward displacements was carried out by Slobod and Howlett (*63*). For a less viscous, less dense displacing phase, a simple stability analysis which assumes no mixed zone

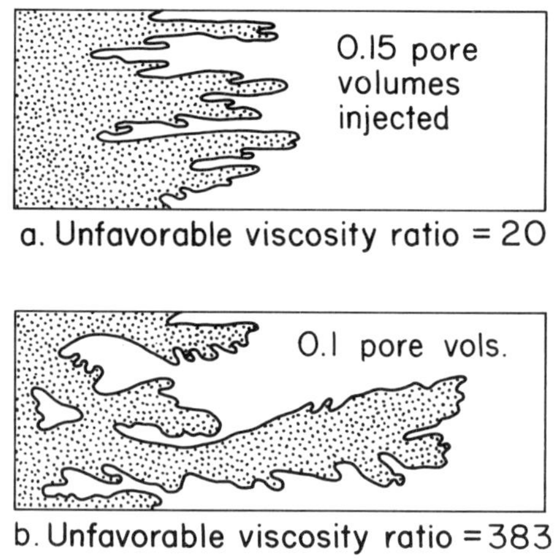

Figure 9. Schematic representation of viscous fingering in a horizontal porous medium for two unfavorable viscosity ratios (12)

and uses Darcy's law with gravity effects included has indicated that a critical flow rate exists below which the displacement is stable (*24, 51–53, 79, 80*). If one considers a horizontal, sharp interface between the two phases in which a small dimple of invading fluid has formed in the resident phase, then if the pressure gradient in the displacing phase is larger than the pressure gradient in the resident phase, this protrusion will grow and the displacement is considered unstable. If the pressure gradient in the displacing phase is less than that in the resident phase, the protrusion is suppressed and the displacement stable. The criterion for stability is thus

$$\left(\frac{dp}{dx}\right)_d < \left(\frac{dp}{dx}\right)_r$$

or using Darcy's law

$$\frac{-q\mu_d}{P} + \rho_d g < \frac{-q\mu_r}{P} + \rho_r g$$

Hence the critical volumetric flow rate is given by

$$q_c = Pg\frac{(\rho_r - \rho_d)}{(\mu_r - \mu_d)} \tag{34}$$

In Equation 34, P is the permeability and the subscripts d and r denote the displacing and resident phases, respectively. The ratio of the experimental flow rate q to the critical q_c was used by Slobod and Howlett to correlate their results for various flow rates, viscosity ratios, and density differences within one class of experiments—*e.g.*, favorable viscosity and density ratios. The ratio q/q_c represents essentially the ratio of viscous to gravity forces or, in dimensionless groups, the ratio of the Reynolds number to the Grashof number.

A summary of the experiments using fluids with no density differences as the reference cases is given in Table V. These results clearly demonstrate that the effect of gravity on the mixing zone length is explained qualitatively by the elementary physical arguments advanced previously. Furthermore, rather large changes in the mixing zone length were affected in the cases where flow rates were small enough for gravity forces to become important. These effects were most pronounced for unfavorable viscosity ratios but also strongly influenced the mixing zone length for $q/q_2 < 1$ with favorable viscosity ratio but unfavorable density ratio.

Dumore (*24*) offers a treatment of stability based on the same criterion regarding the pressure gradient but assuming monotonic concentration changes in the mixed zone and a monotonic dependence of the density and viscosity on the concetration in the mixed zone. Perrine (*51–53*) has applied the theory of hydrodynamic stability using Darcy's equation as the equation of motion, although some of this work has been questioned (*48*). For more recent and extensive treatments of the stability problem for vertical systems in the presence of viscosity and density differences and including the effects of a mixed zone between resident and displacing fluids, the works of Schowalter (*61A*) and Heller (*40A*) are recommended.

Summary

We have tried to point out how the mechanisms of dispersion in simple geometries reflect on the more complex problem of dispersion in porous media, and in particular we have emphasized the restrictions necessary for a diffusion-type equation—*e.g.*, Equation 3—to describe miscible displacements. The most definitive study of this question is presented by Bischoff and Levenspiel (*10*). By considering the time involved in the process to be approximately the mean residence

Table V. Summary of Effects of Density Differences on Mixing Zone Length in Vertical Downward Displacements in Porous Medium with Both Favorable and Unfavorable Viscosity Ratios (63)

Experimental results compared to reference cases with no density differences

Viscosity ratio	Density difference	Comments
Favorable	Favorable	Mixing zone length shows only minor decreases over the reference case
Favorable	Unfavorable	Mixing zone length shows a rapid and large increase as q/q_c becomes <1, indicating viscous fingering. Close agreement with reference case for $q/q_c \geqslant 2$
Unfavorable	Favorable	Mixing zone length shortened significantly and approaches the reference case as $q/q_c \rightarrow 7$
Unfavorable	Unfavorable	Mixing zone length shows large increases for $q/q_c \geqslant -3$. The reference case is approached near $q/q_c = -7$

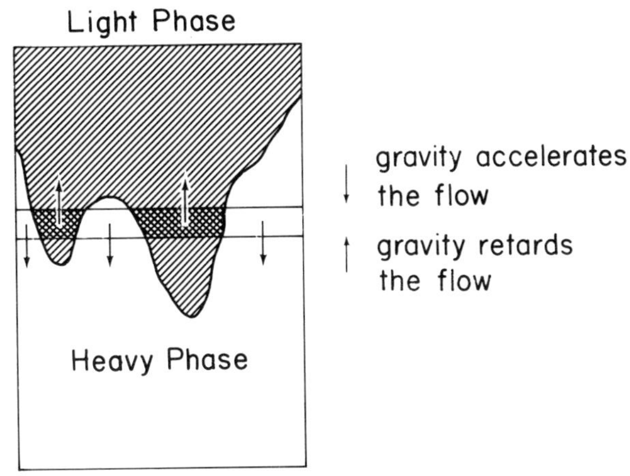

Figure 10. Diagram illustrating the tendency of a favorable density difference to smooth out the displacement front in a porous medium

times, L_T/u_m, where L_T is the length of the system in the main flow direction, these investigators concluded that

$$\frac{L_T}{(2a)} > 0.04 \frac{(2a)u_m}{D} \tag{35}$$

was a criterion for the applicability of the dispersion model. By introducing an effective or hydraulic diameter for packed tubes and the particle diameter, this same criterion was used to establish order of magnitude estimates of the linear dimensions of packed beds necessary to apply the dispersion model.

For miscible displacements in tubes, the minimum time criterion (*1*)

$$\tau_{\min} \geq 0.8 \tag{36}$$

may also be written in terms of length-to-diameter ratio by taking the real time to be the residence time. Then

$$\frac{L_T}{2a} \geq 0.2 \frac{(2a)u_m}{D} \tag{37}$$

is the criterion. If one compares this result to Equation 35, the true criterion is more restrictive than that estimated by Equation 35 and hence, the order of magnitude estimate provided by Bischoff and Levenspiel for packed beds must be used with caution. However, the minimum real time requirement for straight channel systems based on the same equivalent diameter is much smaller than that for tubes and yields a length-to-diameter ratio in much closer agreement with Equation 35.

We have also described in some detail two important effects in porous media—namely, bed capacitance and the instabilities caused by differences in viscosities and densities between displacing and resident phases. From an analysis of the capacitance model, it is clear that systems with large amounts of dead space require significantly larger dimensionless time (or length-to-diameter ratios) for the dispersion model to apply. The effect of viscosity and density differences in unstable situations are such that a dispersion model cannot be used to interpret and extrapolate the results. In capillary tubes, small density differences can have profound effects on the nature of the dispersion and hence, matched density experiments in porous media must be closely examined for the influence of natural convection.

NOMENCLATURE

a = tube radius
b = radius of a tube including stagnant region in Figure 6c
C = mass concentration
D = molecular diffusion coefficient
E = effective axial diffusion coefficient (dispersion coefficient) defined by Equation 3
f_n = nth expansion function in Equation 12
g = acceleration of gravity
h = half width of a parallel plate channel
K = dimensionless dispersion coefficient, $k/(DN_{Pe})$
k = dispersion coefficient
L' = mixing length defined below Equation 5
L = dimensionless mixing length, $DL'/(2a^2u_m)$ for tubes
L_T = length of the system in the main flow direction
N_{Fr} = Froude number
N_{Gr} = Grashof number
N_{Pe} = Peclet number
N_{Re} = Reynolds number
N_{Sc} = Schmidt number
$\bar{n}$ = vector representing the mass flux
n = component of the mass flux vector
P = permeability
p = pressure
q = volumetric flow rate
R = half spacing at the entrance to the diverging channel
R_C = radius of curvature
r = radius radial coordinate
r_A = reaction rate at which A is produced
t = time
U_{ent} = velocity at the entrance to the diverging channel
u = axial velocity component
$\bar{v}$ = vector representing the mass average velocity
v^* = friction velocity defined by Equation 23
x = axial coordinate
x_1 = axial coordinate moving with the mean speed of flow

Greek Letters

α = angle of divergence of the diverging channel
β = $\gamma^2 - 1$
γ = b/a
δ = angular coordinate
ϵ = eddy diffusivity
θ = dimensionless angular coordinate, δ/α
κ = r_i/r_o
λ = friction factor
μ = viscosity
ν = kinematic viscosity, μ/ρ
ρ = density
τ = dimensionless time, tD/a^2 for tubes
τ_s = shear stress
τ_w = shear stress at the wall
ω = mass fraction

Subscript

A = component A
c = critical value
D = mass
d = displacing
$f.p.$ = flat plate
i = inner
M = momentum
m = average
max = maximum
o = outer
r = resident
0 = reference quantity

REFERENCES

(1) Ananthakrishnan, V., Gill, W. N., and Barduhn, A. J., *A.I.Ch.E. J.*, **11**, 1063 (1965).
(2) Aris, R., *Chem. Eng. Sci.*, **10**, 80 (1959).
(3) Aris, R., *ibid.*, **11**, 194 (1959).
(4) Aris, R., *Proc. Roy. Soc.*, **A235**, 67 (1956).
(4A) Aris, R., *ibid.*, A252, 538 (1959).
(5) Aris, R., *ibid.*, **A259**, 370 (1960).
(6) Aunicky, Z., *Can. J. Chem. Eng.*, **46**, 27 (1968).
(7) Bailey, H. R., and Gogarty, W. B., *Proc. Roy. Soc.*, **A259**, 352 (1962).
(8) Bailey, H. R., and Gogarty, W. B., *Soc. Petrol. Eng. J.*, **3**, 256 (1963).
(9) Batchelor, G. K., Proc. 2nd Australasian Conference on Hydraulics and Fluid Mechanics, Auckland, Dec. 1965.
(10) Bischoff, K. B., and Levenspiel, O., *Chem. Eng. Sci.*, **17**, 257 (1962).
(11) Blackwell, R. J., *Soc. Petrol. Eng. J.*, **2**, 1 (1962).
(12) Blackwell, R. J., Rayne, J. R., and Terry, W. M., *Trans. AIME*, **216**, 1 (1959).
(13) Bournia, A., Coull, J., and Houghton, G., *Proc. Roy. Soc.*, **A261**, 227 (1961).
(14) Brigham, W. E., Reed, P. W., and Dew, J. N., *Soc. Petrol. Eng. J.*, **1**, 1 (1961).
(15) Carman, P. C., "Flow of Gases through Porous Media," Academic Press, New York, N. Y., 1955.
(16) Carter, D., and Bir, W. G., *Chem. Eng. Progr.*, **58**, 40 (1962).
(17) Cassell, R. E., and Perona, J. J., *A.I.Ch.E. J.*, **15**, 81 (1969).
(18) Coats, K. H., and Smith, B. D., *Soc. Petrol. Eng. J.*, **4**, 73 (1964).
(19) Craig, F. F., Sanderlin, J. L., Moore, D. W., and Geffen, T. M., *Trans. AIME*, **210**, 275 (1957).
(20) Crane, F. E., Kendall, H. A., and Gardner, G. H. F., *Soc. Petrol. Eng. J.*, **3**, 277 (1963).
(20A) Dayan, J., and Levenspiel, O., *Chem. Eng. Sci.*, **23**, 1327 (1968).
(21) Deans, H. A., *Soc. Petrol. Eng. J.*, **3**, 49 (1963).
(22) De Josselin de Jong, A., *Trans. Am. Geophys. Union*, **39**, 67 (1958).
(23) Dougherty, E. L., *Soc. Petrol. Eng. J.*, **3**, 155 (1963).
(24) Dumore, J. M., *ibid.*, **4**, 356 (1964).
(25) Erdogan, M. E., and Chatwin, P. C., *J. Fluid Mech.*, **29**, 465 (1967).
(26) Evans, E. V., and Kenney, C. N., *Proc. Roy. Soc.*, **A284**, 540 (1965).
(27) Flint, L. F., and Eisenklam, P., *Can. J. Chem. Eng.*, **47**, 101 (1969).
(28) Gill, W. N., *Chem. Eng. Sci.*, **22**, 1013 (1967).
(29) Gill, W. N., *Proc. Roy. Soc.*, **A298**, 335 (1967).
(30) Gill, W. N., and Ananthakrishnan, V., *A.I.Ch.E. J.*, **12**, 906 (1966).
(31) Gill, W. N., and Ananthakrishnan, V., *ibid.*, **13**, 801 (1967).
(32) Gill, W. N., Ananthakrishnan, V., and Nunge, R. J., *ibid.*, **14**, 939 (1968).
(33) Gill, W. N., Güceri, Ü., and Nunge, R. J., Office of Saline Water, Research and Development Report No. **443**, June 1969.
(33A) Gill, W. N., "Axial Dispersion with Time Variable Flow in Multiphase Systems," *A.I.Ch.E.J.*, in press, 1969.
(34) Goodknight, R. C., Klikoff, W. A., and Fatt, I., *J. Phys. Chem.*, **64**, 1162 (1960).
(35) Griffiths, A., *Proc. Phys. Soc. London*, **23**, 190 (1911).
(36) Güceri, Ü., "Dispersion in Jeffery-Hamel Flows," M.S. thesis, Clarkson College of Technology, Potsdam, N. Y., June 1968.
(37) Han, C. D., *Chem. Eng. Sci.*, **22**, 837 (1967).
(38) Han, C. D., and Bixler, H. J., *A.I.Ch.E. J.*, **13**, 1058 (1967).
(39) Heller, J. P., *ibid.*, **9**, 452 (1963).
(40) Heller, J. P., *Soc. Petrol. Eng. J.*, **1**, 213 (1961).
(40A) Heller, J. P., *J. Appl. Phys.*, **31**, 1566 (1966).
(40B) Jones, W. M., *Brit. J. Appl. Phys.*, Ser. 2, **1**, 1559 (1968).
(41) Koutsky, J. A., and Adler, R. A., *Can. J. Chem. Eng.*, **42**, 239 (1964).
(42) Koval, E. J., *Soc. Petrol. Eng. J.*, **3**, 145 (1963).
(43) Kramers, H., and Westerterp, K. R., "Elements of Chemical Reactor Design and Operation," Chapman and Hall, London, 1963.
(44) Levenspiel, O., and Bischoff, K. B., "Advances in Chemical Engineering," Vol. 4, pp 98–198, Academic Press, New York, N. Y., 1963.
(45) Levich, V. G., Markin, V. S., and Chismadzhev, Yu. A., *Chem. Eng. Sci.*, **22**, 1351 (1967).
(46) Lin, T.-S., M.S. thesis, Clarkson College of Technology, Potsdam, N. Y., June 1969.
(47) Lu, S.-Z., *ibid.*, in progress.
(47A) Nunge, R. J., Lin, T.-S., and Gill, W. N., "Laminar Dispersion in Curved Tubes and Channels," paper presented at A.I.Ch.E. 62nd Annual Meeting, Washington, D. C., Nov. 1969.
(48) Outmans, H. D., *Soc. Petrol. Eng. J.*, **1**, 298 (1961).
(49) Peaceman, D. W., and Rachford, H. H., *Soc. Petrol. Eng. J.*, **2**, 327 (1962).
(50) Perkins, T. K. and Johnston, O. C., *ibid.*, **3**, 70 (1963).
(51) Perrine, R. L., *ibid.*, **1**, 9 (1961).
(52) Perrine, R. L., *ibid.*, p 17.
(53) Perrine, R. L., *ibid.*, **3**, 205 (1963).
(54) Philip, J. R., *Australian J. Phys.*, **16**, 287 (1963).
(55) Reejhsinghani, N. S., Barduhn, A. J., and Gill, W. N., *A.I.Ch.E. J.*, **14**, 100 (1968).
(56) Reejhsinghani, N. S., Gill, W. N., and Barduhn, A. J., *ibid.*, **12**, 916 (1966).
(57) Saffman, P. A., *Chem. Eng. Sci.*, **11**, 125 (1959).
(58) Saffman, P. A., *J. Fluid Mech.*, **6**, 321 (1959).
(59) Saffman, P. A., *ibid.*, **7**, 194 (1960).
(60) Sankarasubramanian, R., M.S. thesis, Clarkson College of Technology, Potsdam, N. Y., June 1969.
(61) Scheidegger, A. E., "The Physics of Flow Through Porous Media," Macmillan, New York, N. Y., 1960.
(61A) Schowalter, W. R., *A.I.Ch.E.J.*, **11**, 99 (1965).
(62) Shulman, H. L., and Mellish, W. G., *A.I.Ch.E. J.*, **13**, 1137 (1967).
(62A) Sittel, C. N., Threadgill, W. D., and Schnelle, K. B., IND. ENG. CHEM. FUNDAMENTALS, **7**, 39 (1968).
(63) Slobod, R. L., and Howelett, W. E., *Soc. Petrol. Eng. J.*, **4**, 1 (1964).
(64) Slobod, R. L., and Thomas, R. A., *ibid.*, **3**, 9 (1963).
(65) Smith, J. V., M.S. thesis, Univ. of Tennessee, Knoxville, Tenn., June 1967.
(66) Taylor, G. I., *Proc. Phys. Soc. London*, **67**, 857 (1954).
(67) Taylor, G. I., *Proc. Roy. Soc.*, **A219**, 186 (1953).
(68) Taylor, G. I., *ibid.*, **A223**, 446 (1954).
(69) Taylor, G. I., *ibid.*, **A225**, 473 (1954).
(70) Taylor, H. M., and Leonard, E. F., *A.I.Ch.E. J.*, **11**, 686 (1965).
(71) Taylor, H. M., Ph.D. thesis, Columbia University, New York, N. Y., 1967.
(72) Tichacek, L. J., Barkelew, C. H., and Baron, T., *A.I.Ch.E. J.*, **3**, 439 (1957).
(73) Turner, G. A., *Chem. Eng. Sci.*, **7**, 156 (1958).
(74) Turner, G. A., *ibid.*, **10**, 14 (1959).
(75) Turner, J. C. R., *British Chem. Eng.*, **9**, 377 (1964).
(76) Van Deemter, J. J., Broeder, J. J., and Lauwerier, H. A., *Appl. Sci. Res.*, **5A**, 374 (1956).
(77) Van der Poel, C., *Soc. Petrol. Eng. J.*, **2**, 317 (1962).
(78) Von Rosenberg, D. U., *A.I.Ch.E. J.*, **2**, 55 (1956).
(79) Wooding, R. A., *Proc. Roy. Soc.*, **A252**, 120 (1959).
(80) Wooding, R. A., *ZAMP*, **13**, 256 (1962).
(81) Wright, D. E., *J. Hydral. Div., Proc. ASCE*, **94**, 851 (1968).

10

Progress in Induced Coalescence and a New Theoretical Framework for Coalescence by Porous Media

Progress in Induced Coalescence and a New Theoretical Framework for Coalescence by Porous Media

The problem of separating two immiscible liquids when one is finely dispersed within the other is frequently encountered in engineering practice. Situations of industrial importance include liquid-liquid extractions (especially when product contamination occurs), direct-contact heat transfer as applied to seawater conversion, desalting and dewatering of crude oils, caustic washing of light distillates, and water haze removal from aviation fuels. Aqueous wastes from petroleum refineries, steel mills, ships, and other sources often carry finely divided oils that adversely affect the quality of the water into which these wastes are discharged.

For such dispersions, gravity settling, though normally effective in achieving an initial or primary separation that is adequate for many purposes, is often insufficient for eliminating the residual secondary haze that results when the dispersed phase consists of micron and smaller droplets that constitute a relatively small fraction of the total volume. Prohibitively long separator holding times may then be required for Brownian- and velocity gradient-induced coalescence to produce globules large enough to settle out.

Accordingly, various auxiliary procedures have been proposed for use in conjunction with gravity settling to promote more rapid coalescence and separation of the dispersed phase. A number of these have been reviewed by Treybal (*60*) and Clayton (*5*) and include among others, the use of alternating electrical fields, centrifugation, the addition of chemical coagulants, and coalescence by flow through porous solids. Certain procedures are often too specialized, expensive, or inefficient for many applications.

Among the various auxiliary procedures, coalescence induced by flow through appropriate porous solids is especially attractive, as it can be conducted continuously with high effectiveness of separation under favorable conditions. Micron and smaller droplets can thus be coalesced into millimeter and larger globules that are readily separated through subsequent gravity settling.

Summary of Studies of Coalescence by Flow Through Porous Media

A wide variety of materials have been examined for use as coalescing media. Natural fibers such as cotton and wool, as well as synthetic fibers such as glass, viscose, and Teflon have been tried. Metal screens, steel wool, membranes, and granular beds of gravel, pebbles, diatomaceous earth, and other packings have been used. The design of coalescing devices employing porous solid contact has been largely empirical and such devices have incorporated many processing arrangements; these include packed beds, cylindrical cartridges, staged arrangements including settling chambers, and sequential layers of media having different properties.

Burtis and Kirkbride (*2*) investigated the use of beds of glass fibers for desalting crude oils. They varied bed density, bed thickness, amount of fresh water added to the crude oil, flow rate and temperature, measuring effluent salt content and pressure drop. They observed that effluent salt content increases with increasing flow velocity, decreasing temperature, lower inlet water-to-oil ratio, decreasing bed thickness, and decreasing packing density. Hayes *et al.* (*22*) did a full-scale study of desalting by contact with glass-fiber

LLOYD A. SPIELMAN
SIMON L. GOREN

A review of progress in the coalescence of liquid-liquid suspensions and a new theoretical framework for coalescence by porous media are presented

beds extending over 58 weeks. They observed sporadic behavior; however, it could be made steadier through addition of a demulsifying agent. Hayes *et al.* and Burtis and Kirkbride concluded desalting by glass-fiber beds to be commercially attractive.

Voyutskii *et al.* (*62*) studied water-in-oil emulsion filtration using fibrous materials. They claim separation occurred only below a certain critical velocity and below that velocity coalescence depended on total fiber contacting surface rather than on pore size. By varying relative wettability, they observed that an intermediate wettability gave the most effective separation, and concluded that for the best performance the filter should be sufficiently water-wetted to coalesce the water, but not so saturated as to produce excessive clogging by accumulated water. In a second study, Voyutskii *et al.* (*63*) observed that mixtures of viscose and wool fibers resulted in improved coalescence performance over single fibrous materials.

Graham (*19*) and Sweeney (*59*) investigated coalescence induced by beds of glass fibers with subsequent gravity settling of water-in-Aroclor dispersions prepared by turbulent mixing. Fiber diameter and bed depth were varied and the total water content in the effluent was determined by titration. Effluent water content steadily decreased with increased flow rate and increased total surface.

Kintner's laboratory has produced a number of experimental studies that have provided a good deal of information about fibrous coalescence. Rose (*46*) studied the coalescence of different water-organic suspensions made by turbulent mixing. Both fiber diameter and fiber material were varied, but independent variation was apparently prevented by the unavailability of materials. Separation was judged visually, and accompanying pressure drops were measured. Rose found a significant effect in varying interfacial tension—coalescence improving with higher interfacial tension—but suggested that the inlet drop size could have been influenced by varied interfacial tension during drop formation. Gudesen (*20*) investigated the coalescence of petroleum fractions dispersed in water by turbulent mixing, using mixed fibrous beds of cotton and glass wool. Coalescence performance was again judged visually and pressure drop measured. Gudesen, like Voyutskii, found a flow velocity below which effectively complete separation was observed. This velocity was observed to pass through a maximum with changing bed depth, and a minimum with changing bed composition. Sareen *et al.* (*49*) investigated the use of mixed cotton and supporting fiber (Teflon, glass, Dynel) in coalescing oil-in-water dispersions prepared by turbulent mixing. They present further experimental results for surfactant addition and variation of oil viscosity. Surfactants reduced the interfacial tension and inhibited coalescence. The "critical separation velocity" decreased with increased oil viscosity, appearing to become constant at higher oil viscosities.

Vinson (*61*) investigated the coalescence of dispersions of organic-in-aqueous dispersions, using fine mesh screens in an effort to simulate layers of fibers. Average droplet diameters were 2 and 4 μ, prepared by an homogenizing device. Droplet size distributions were estimated by using a light scattering technique. The important variables affecting coalescence were interstitial velocity, screen filament width, and aqueous-phase viscosity, the extent of coalescence decreasing as any of these variables was increased. Only a slight increase in coalescence performance was observed over a 30-fold increase in droplet viscosity. Varying the interfacial tension above 30 dyn/cm was observed to have little effect, but when decreased below 11 dyn/cm, coalescence performance was found to decrease.

Shackleton *et al.* (*51*) and Douglas and Elliot (*11*) investigated separation of oil-in-water dispersions, using pebble beds for possible application to ballast and bilge water treatment before discharge. Attention was given to drop sizes formed during pumping. Farley and Valentin (*12*) investigated the coalescence of high-viscosity oils (3500 to 20,000 cP) dispersed in water by pumping with application to ballast water treatment. Beds of gravel, steel gauze, and glass fiber were used with suspended oil content determined by optical methods. Effluent oil content increased with increasing flow rate, decreasing the number of gauze layers and decreasing oil viscosity. The Permutit Company (*39*), who undertook research and development of a practical package unit for shipboard separation of oil from ballast water, found that influent oil present in concentrations of less than one per cent was reduced to concentrations of 100 to 200 ppm when coalescence was induced by flow through a series of woven metal screens. Further reduction was achieved through the use of a replaceable fiberglass cartridge for final polishing.

Many kinds of porous materials have been examined as possible coalescing media. Performance depends on internal geometric and surface properties of the solid matrix, physical and chemical properties of the liquids, flow rate, drop size, and bed depth. The qualitative trends that seem to be generally applicable are: coalescence performance decreases with increased flow rate; performance increases with larger internal surface of the porous solid; performance increases as bed thickness is increased; coarse suspensions are coalesced more readily than fine suspensions; chemical agents that promote or prevent coalescence through coagulation have similar influences for induced coalescence. However, even these seemingly obvious generalizations should be accepted with some caution. Indeed one discouraging aspect of porous coalescence seems to be the uncertainty of design associated with a particular application and the need for extensive pilot testing. Increased understanding of the mechanisms describing coalescer performance could contribute very significantly to the design and operation of porous coalescers.

Scope of the Theoretical Framework

The present effort is aimed at development of a general mathematical description of porous coalescer performance. This description provides rational criteria for coalescer design and scale-up, as well as outlining relevant measurables, thus giving a basis for future experimental investigation. The conclusions presented here were drawn from a combined experimental and theoretical program employing glass fibrous mats to coalesce oil-in-aqueous suspensions under controlled conditions. However, only those limited aspects of the experiments important to the development of the theoretical scheme are presently discussed. A thorough treatment of the experimental program, results, and interpretation through mathematical solutions are given elsewhere [Spielman (*53*); Spielman and Goren (*55*)].

In the design of coalescers, knowledge of steady-state separation performance and pressure drop as they depend on system variables such as drop size, flow rate, fluid, and medium properties, is of primary importance. Therefore, special emphasis has been placed on incorporation of those quantities into the theory. Our initial approach employed models successfully used to describe aerosol collection by fibrous filters, but as experimental work progressed it became apparent that the steady-state coalescence process differed in important ways from the filtration of aerosols. The development of a model logically incorporating these essential differences has been made possible by exploiting other areas of study, namely water filtration, soil science, and petroleum reservoir engineering, the latter two having evolved a rather sophisticated theory of multiphase capillary flow in porous solids. Information from these areas, supplemented by simple assumptions supported by experimental observations, has made possible the formulation of a theoretical framework for the description of coalescence by porous media.

Though the theoretical scheme proves useful for understanding certain observations, and for correlating and scaling up data for pressure drop and coalescence efficiency, its utility is greatly enhanced through deeper insight into four fundamental underlying quantities. These quantities appear in the theory as coefficients depending on coalesced phase holdup and other system variables. The coefficients are the relative permeabilities for the two phases, the capillary pressure, and the filter coefficient. A brief review of the literature pertinent to evaluating these coefficients is finally presented. This review is by no means exhaustive, but serves to summarize important information that might be of value, and to indicate the direction of further investigation.

Development of the Theoretical Framework

When liquid-liquid dispersions pass through porous solids, suspended droplets are transported to solid and entrapped liquid interfaces where they are captured and coalesced into the bulk of previously captured liquid. Under the action of hydrodynamic, gravitational, and capillary forces, coalesced liquid drains through the solid matrix. At steady state, coalesced liquid leaves the porous solid at the same volumetric flow rate as suspended droplets are coalesced within the matrix. Water and air filtration contribute models describing particle capture by porous media. Soil physics and petroleum reservoir engineering provide equations describing the flow of coalesced liquid within the solid.

Our experiments (*53, 55*) coalescing dilute oil in aqueous suspension by flow through fibrous mats clearly indicate the importance of accounting for the presence of coalesced oil held up within the void spaces of the mats, since both pressure drop and coalescence efficiency are observed to increase significantly as deposited oil accumulates in approaching steady state from an initial oil-free condition. The importance of retained oil is further indicated by the observation that at steady state typically 30 to 40% of the medium voids are occupied by oil, while fibers themselves occupy about 5% of the total medium volume. Further indication is provided by microscopic visual observations of Cordes (*7*).

Capillary pressure. To predict ground-water movement in partially saturated soils, and to describe petroleum production processes wherein immiscible fluids flow through porous rock, soil physicists and petroleum engineers have evolved a theoretical model in which each of the immiscible fluids is considered to flow within more or less fixed microscopic channels, the boundaries of which consist of fluid-fluid interfaces, as well as solid-fluid interfaces. An important simplification results through the hypothesis that two-phase capillary equilibrium governs the local microscopic distribution of the fluids, even though they are in a flowing state. This last assumption is justified when fluid-fluid interfacial forces are large enough to resist significant interface deformation by local hydrodynamic forces; it is normally satisfied if the flow rates are not too high [Rapoport and Leas (*43*); Slattery (*52*)].

Capillary equilibrium can be pictured by means of an experiment in which a (noncompressible) porous medium is preferentially wetted by and initially saturated with a given fluid. If a relatively nonwetting immiscible fluid is now forced under pressure into the medium, a certain amount of wetting fluid is displaced, and after sufficient time the two phases distribute throughout the medium in a macroscopically uniform volumetric proportion, depending only on the applied pressure and the nature of the fluids and porous solid. As the pressure is increased, more wetting fluid is displaced and eventually a new equilibrium distribution of the fluids occurs. The applied pressure corresponds to the pressure difference across fluid-phase interfaces within the medium and is called the capillary pressure p_c. As nonwetting fluid is made to occupy successively smaller pores, the radii of curvature of microscopic liquid interfaces become smaller and the capillary pressure increases according to the relation

$$p_c = \gamma\left(\frac{1}{R_1} + \frac{1}{R_2}\right) \tag{1}$$

where γ is the interfacial tension and R_1, R_2 are the principal radii of curvature of the fluid interfaces.

The relationship between the capillary pressure and the corresponding fraction S of the voids occupied by the nonwetting fluid constitutes the capillary pressure function

$$p_c = p_c\,(S) \tag{2}$$

Displacement of the wetting phase as described above is referred to as "drainage." A typical drainage capillary pressure curve is shown schematically in Figure 1.

Two-phase flow. During two-phase flow within the voids, each phase flows at its own local pressure, the local pressure difference between the phases being the capillary pressure, which is related to the local oil saturation according to Equation 2, *i.e.*

$$p_2 - p_1 = p_c\,(S_2) \tag{3}$$

Here p_1 is the local pressure of flowing water, and p_2 is the local pressure of flowing oil. The establishment of fixed microscopic flow channels through local capillary equilibrium, in conjunction with the predominance of viscous over inertial hydrodynamic forces, as normally prevails for low-speed flow through porous media, implies the validity of Darcy's law appropriately extended to describe the capillary-conduction of the two fluids within the medium. The appropriate extensions for one-dimensional, gravity-free flow are

$$q_1 = -\frac{k_1(S_2)}{\mu_1}\frac{dp_1}{dx} \tag{4}$$

$$q_2 = -\frac{k_2(S_2)}{\mu_2}\frac{dp_2}{dx} \tag{5}$$

where 1 equals the capillary-conducted aqueous phase, 2 equals the capillary-conducted oil phase, q_i equals superficial velocity of phase i, μ_i equals viscosity of

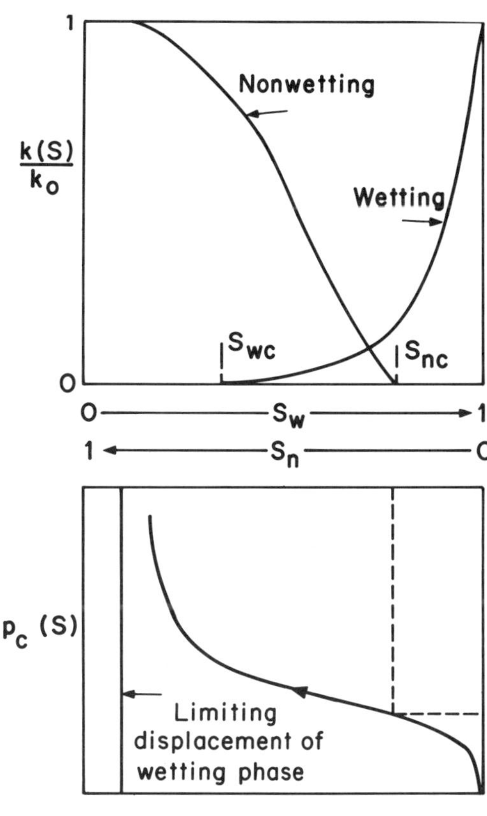

Figure 1. Capillary pressure vs. saturation (lower); relative permeability vs. saturation (upper)

phase i, x equals position in porous medium, and k_i equals Darcy permeability to phase i.

The permeabilities depend only on the local geometric microstructure of the fluid-fluid-solid configuration. However, since the detailed geometry is taken to be governed by capillary equilibrium, the permeabilities are completely determined for a given porous solid and fluid pair through specification of the local oil saturation. This is stressed by explicitly designating the dependence of the permeabilities on oil saturation.

The basic ideas leading to the mathematical formulation of multiphase flow incorporating capillary pressure effects were first set forth by Richards (*44*). The extension of Darcy's law to two flowing phases was suggested by Muskat and Meres (*34*), but the ideas have since been refined and expanded [Miller and Miller (*31*); Slattery (*52*)]. Standard texts dealing with multiphase flow through porous media include Muskat (*33*), Scheidegger (*50*), and Collins (*6*).

It follows from dimensional analysis that the ratio $k_i(S)/k_0$ is a function of saturation alone for geometrically similar porous media having the same wetting angle (or distribution of wetting angles for mixtures of solids). The quantity k_0 is the permeability of the medium to a single phase and depends only on medium geometry. The ratio $k_i(S)/k_0$ is called the relative permeability to phase i. Typical qualitative behavior of the relative permeabilities is shown in Figure 1. Further detail concerning the nature of the relative permeabilities is reserved for subsequent discussion, but one important general feature to be recognized is the existence of a limiting saturation of nonwetting phase below which its permeability effectively vanishes, implying that no flow of nonwetting phase can occur in that saturation range. This results from the inability of the phase to form a sufficiently continuous network to support its flow when it occupies too little of the total pore space. The wetting phase may exhibit a similar limiting saturation, although some workers [Fatt (*13*); Jones (*26*)] suggest that for strongly wetting fluids the limiting wetting saturation is zero. Another important feature is that the sum of the relative permeabilities is generally less than unity, which indicates that the presence of one phase markedly inhibits the flow of the other rather than simply decreasing the area available for flow.

Capture and coalescence. For concreteness, in the subsequent discussion the coalescing dispersed phase *arbitrarily* is designated as oil and the continuous phase as water. Clearly, the theory describing capillary-conduction of two phases in a porous solid alone is unable to describe the most important aspects of coalescence by flow through porous solids. Experimental observations of the coalescence process support the hypothesis that two predominantly discernible interstitial regimes of the (coalescing) oil exist simultaneously within the voids. One oil regime is very finely dispersed within a capillary-conducted aqueous phase and consists of droplets considerably smaller than the pores. The other oil regime is capillary-conducted and is maintained by the steady deposition of the finely dispersed oil at solid-liquid and liquid-liquid interfaces. Figure 2 depicts the interstitial regimes assumed.

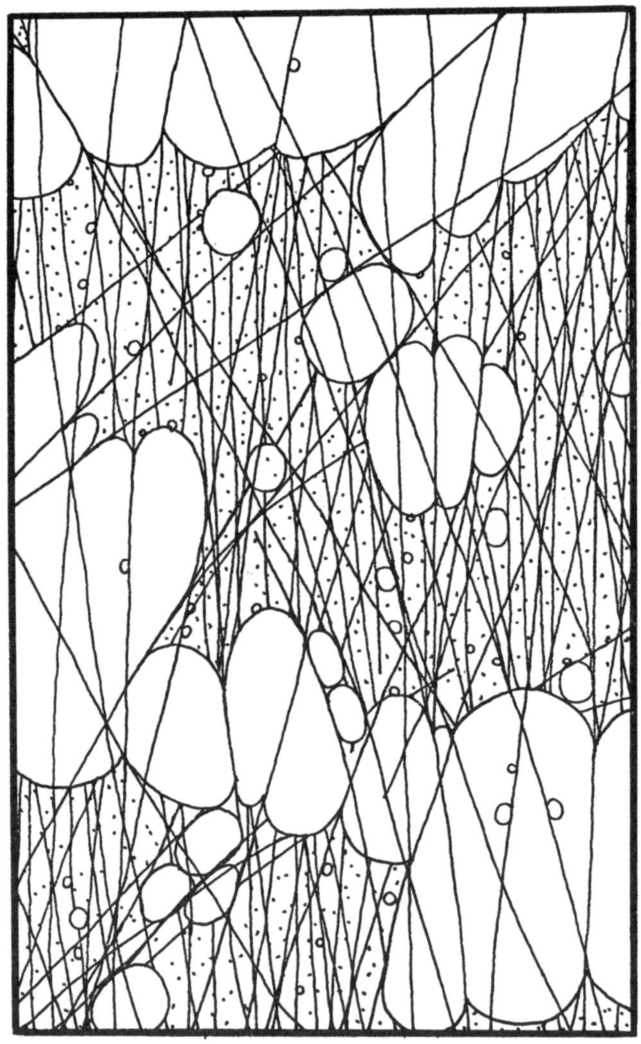

Figure 2. Depiction of suspension flowing in fibrous medium (oil nonwetting)

The existence of two such oil regimes within the voids is supported by the observation of two readily distinguishable oil regimes leaving the filter medium. Macroscopic globules of coalesced oil always larger than a fraction of a millimeter in diameter leave the medium from more or less fixed points on the outlet face, if the oil is nonwetting; if the medium is preferentially wet by the oil, deposited oil is observed to accumulate in the form of large globules adhering to the outlet face. On the other hand, the effluent aqueous regime remains residually turbid, though considerably less turbid than the incoming suspension. The diameter of suspended oil droplets producing the effluent turbidity is less than several microns and is always comparable to that producing the influent turbidity. Further evidence of the two principle oil regimes is obtained from microscopic observation of coalescence by flow through thin fiber mats [Cordes (7)].

It is assumed then that Equation 4 describes the capillary-conduction of the aqueous suspension, Equation 5 describes the capillary-conduction of the coalesced oil regime, and Equation 3 gives the local pressure difference across the interface between the coalesced oil regime and the continuous phase of the aqueous suspension, where S_2 is the local coalesced oil saturation and does not include any finely suspended oil.

At this point it is appropriate to introduce an equation describing the local rate of deposition of finely dispersed oil from the aqueous suspended regime into the coalesced oil regime. The expression is

$$\frac{d(q_1 n)}{dx} = -\lambda(S_2, q_1) \cdot (q_1 n) \qquad (6)$$

$n(a_p)\,da_p$ = number of finely dispersed oil droplets having radii on the interval $(a, a_p + da_p)$, per unit volume of aqueous suspension

$\lambda(S_2, q_1)$ = local filter coefficient or deposition coefficient

The basic premise underlying Equation 6 is that the local rate of capture of particles of given size and chemical nature is directly proportional to the number concentration of those particles locally suspended. At volume fractions below a few per cent suspended particle-particle encounters resulting from velocity gradients are comparatively infrequent, and removal results primarily from individual particle encounters with static interfaces. No provision is made for reentrainment of microscopic particles once captured.

A first-order deposition equation was first used by Iwasaki (25) to describe the filtration of suspended solids from water using sand filters. Such equations have since found extensive application in water and air filtration. A primary goal of filtration theory has been the prediction of the filter coefficient, as it depends on particle size, medium geometry, flow rate, surface chemistry, etc., when various mechanisms are operative to assist or impair deposition. The feature common to all such theoretical treatments is the first-order dependence of deposition on the local number concentration of particles at a given depth in the filter. Especially in aerosol filtration, significant advances have been made and incorporated into recent texts such as Fuchs (17) and Davies (9). The principal mechanism of deposition believed to be operative under the conditions of the experiments performed in the present study is that of convective transport of suspended oil droplets to the vicinity of solid and liquid interfaces forming the matrix. There molecular adhesive forces remove and hold particles at the fiber and held up oil surfaces until they are eventually coalesced into the capillary-conducted oil regime.

The filter coefficient λ in general depends on particle size, local suspension flow rate, surface conditions, chemical environment, and local geometry of the capturing matrix. The local geometric configuration of the solid-liquid-liquid matrix, however, is assumed to be determined by the capillary equilibrium distribution of the aqueous suspension and coalesced oil. For a given porous solid and preferential wettability, the geometric dependence of λ is fixed in a manner quite analogous to that for the capillary pressure and permeabilities once the saturation is specified. This has been stressed by explicitly indicating the saturation dependence in Equation 7. In fact, in sand filtration of solids from water a common procedure has been to assume the local filter coefficient is fixed for a given local total solids deposit, even though the basis for such an approximation does not seem nearly so well justified as for the coalescing liquid situation.

Oil captured at interfaces presumably exists distinct from capillary-conducted oil until its eventual coalescence into that regime. If coalescence proceeds sufficiently fast compared to the rate of capture, then the volume of oil captured but not yet coalesced into the conducted regime will be small when compared with the total oil saturation. The assumption of local capillary equilibrium should therefore be best justified for relatively unstable suspensions that coalesce readily upon capture. On the other hand, sufficiently well stabilized suspensions could permit significant volumes of captured but uncoalesced oil to accumulate, making the approximation of capillary equilibrium doubtful. To behave in this way oil drops would have to exhibit a strong tendency for adherence at interfaces with little tendency to spread or coalesce. Indeed, if the droplets were completely unable to overcome barriers to coalescence, they should accumulate as do solid particles in water or aerosol filtration. For a given suspension, capillary equilibrium should become better established with greater distances downstream, because of the decreasing deposition rate as the number of suspended droplets is steadily decreased.

The preceding discussion suggests a possible modification of the present treatment to incorporate effects of captured but uncoalesced oil. However, present theoretical understanding does not permit so rigorous a treatment. The approximations of capture rate control and local capillary equilibrium are therefore made throughout this analysis but with cognizance of the restrictions discussed above.

The dependence of the filter coefficient of the local suspension flow rate q_1 is also shown explicitly in Equation 6, since q_1 is, in general, a function of position in the medium to be determined by solution of the scheme.

The dependences of the filter coefficient on quantities other than the local suspension flow rate and coalesced oil saturation are parametric and therefore not explicitly indicated.

Material balances. Additional equations to complete the scheme are obtained through steady-state material balances. For the oil,

$$\frac{dq_2}{dx} = -\frac{d(q_1 \mathfrak{f})}{dx} \tag{7}$$

and for the total aqueous suspension,

$$\frac{dq_1}{dx} = \frac{d(q_1 \mathfrak{f})}{dx} \tag{8}$$

where

$$\mathfrak{f} \equiv \int_0^\infty \frac{4}{3} \pi a_p^3 n(a_p) da_p \tag{9}$$

is the total volume fraction of finely dispersed oil in aqueous suspension at a given depth of the medium.

It is important to recognize that $\mathfrak{f}$ accounts only for the volume of suspended oil and excludes the coalesced oil regime, whereas S_2 includes only coalesced capillary-conducted oil, excluding any oil droplets in aqueous suspension.

Equations 3 through 9 constitute seven coupled relations in the seven unknown functions of x: p_1, p_2, q_1, q_2, S_2, n, and $\mathfrak{f}$. A sufficient set of equations describing the steady-state coalescence process is therefore obtained through specification of the detailed mathematical forms of the capillary pressure, permeabilities, and filter coefficient, giving their dependences on the coalesced oil saturation and other system variables. Influences of relative wettability are implicitly accounted for through its effect on the coefficients. Phenomena occurring at the medium boundaries must also be described to establish boundary conditions for the unknowns, and to predict coalesced globule dimensions. These are treated subsequently.

Simplification of the Basic Scheme

Integrating Equations 7 and 8 from $x = 0$, to $x = x$, and solving for the velocities,

$$q_1 = q_{10} (1 - \mathfrak{f}_0)/(1 - \mathfrak{f}) \tag{10}$$

and

$$q_2 = q_{20} + q_{10} (\mathfrak{f}_0 - \mathfrak{f})/(1 - \mathfrak{f}) \tag{11}$$

In Equations 10 and 11 a zero subscript denotes inlet values. The inlet oil rate q_{20} is as yet unspecified, but the other inlet quantities q_{10} and $n_0(a_p)$ (hence $\mathfrak{f}_0$) are presumed specified. Formal integration of Equation 6 between the same limits, and elimination of q_1 by substitution from Equation 10 gives

$$n(a_p) = \frac{n_0(a_p) \cdot \exp\left\{-\int_0^x \lambda[S_2, q_{10}(1 - \mathfrak{f}_0)/(1 - \mathfrak{f})]dx\right\}}{(1 - \mathfrak{f}_0)/(1 - \mathfrak{f})} \tag{12}$$

Differentiating Equation 3 with respect to x, eliminating the pressure gradients by substitution from Equations 4 and 5 and the velocities by means of Equations 10 and 11, gives the important equation

$$\frac{k_1(S_2)p_c'(S_2)}{\mu_1 q_{10}} \frac{dS_2}{dx} - (1 - \mathfrak{f}) = -\frac{\mu_2 \mathfrak{f}_0}{\mu_1} \frac{k_1(S_2)}{k_2(S_2)} \left(\frac{q_{20}}{q_{10}\mathfrak{f}_0} + 1 - \frac{\mathfrak{f}}{\mathfrak{f}_0}\right) \tag{13}$$

Equations 12 and 13 [with $\mathfrak{f}$ given in terms of $n(a_p)$ through the identity Equation 9] are two independent coupled equations in the unknown coalesced oil saturation S_2, and suspended drop size distribution $n(a_p)$. Once these are obtained as functions of position in the solid, the velocities can be straightforwardly computed from Equations 10 and 11 and the pressure distributions from Equations 4 and 5 (the latter equations separating variables on substitution of S_2, q_1, and q_2 as known functions of x). The problem has thus been effectively reduced to that of solving Equations 12 and 13 simultaneously for S_2 and $n(a_p)$ under appropriate boundary conditions.

Boundary conditions. The occurrence of S_2 through derivatives of first order in Equations 12 and 13, suggests the need for one boundary condition on S_2. However, because q_{20}, which appears in Equation 13, was left unspecified in the integration leading to Equation 11, a second condition is required for complete determinacy. Depending on experimental conditions, this second condition may either be placed explicitly on q_{20}, or given as a boundary condition on S_2.

One boundary condition to be extensively used is that for which a given face of the porous solid is in immediate contact with effectively pure wetting fluid. In keeping with the assumption of local capillary equilibrium inside the solid, the saturation just within the solid boundary corresponds to capillary equilibrium of the solid in contact with bulk wetting fluid. This implies the largest wetting phase saturation possible, and that wetting fluid contacting the face continuously connects with wetting fluid within the solid. The equilibrium is, however, subject to the constraint of nonzero flow of the nonwetting phase; therefore the wetting phase saturation cannot be so large that the permeability to the nonwetting phase completely vanishes over a finite extent of the solid. The saturation at the

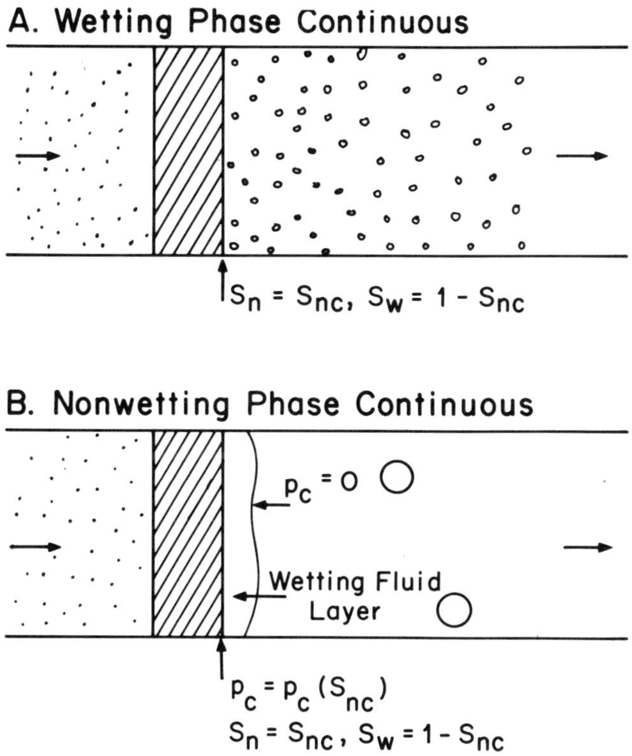

Figure 3. Outlet boundary conditions, $S_n = S_{nc}$

boundary therefore corresponds to

$$S_w = 1 - S_{nc} \tag{14}$$

this being the largest possible saturation of wetting phase consistent with finite nonwetting phase flow. (See Figure 1.)

The outlet face under all experimental conditions was observed to be in immediate contact with wetting fluid, thus Equation 14 applies to the outlet. For dispersed oil nonwetting, the fluid adjacent to the outlet is just the wetting aqueous phase as shown in Figure 3A, and Equation 14 thus gives

$$S_2 = S_{2c} \text{ at } x = L \tag{15}$$

For dispersed oil preferentially wetting, the fluid in immediate contact with the outlet face is oil, which occurs as an adhering layer or sheath of finite extent (see Figure 3B). Then according to Equation 14 the outlet condition is

$$S_2 = 1 - S_{1c} \text{ at } x = L \tag{16}$$

The presence of an adhering layer of wetting oil at the outlet produces an additional mechanical resistance to flow of nonwetting fluid. The pressure of the nonwetting aqueous phase just within the solid exceeds that of the oil there by the amount $|p_c(1 - S_{1c})|$ (Equation 14), while at the outermost edge of the adhering layer the capillary pressure is approximately zero because the interfacial curvature becomes large. The nonwetting aqueous pressure must therefore decrease by the amount $|p_c(1 - S_{1c})|$ as the aqueous phase flows past the layer, because the oil pressure remains approximately constant interior to the nearly stagnant layer. This additional resistance to flow, which results when the continuous aqueous phase is nonwetting, should be taken into account when evaluating the total aqueous pressure drop across the system.

The outlet condition (Equation 14) is also supported by the findings of Richardson *et al.* (*45*), who measured pressure and saturation profiles for steady-state flow of preferentially wetting oil and nonwetting gas through porous rock into the open receiving vessel. Over a wide range of flow conditions those workers found the pressure and saturation profiles to be accurately described by Equations 3, 4, and 5, with outlet saturation S_{nc} corresponding to that where the permeability to the nonwetting phase just vanishes. They also observed the added resistance to gas flow caused by wetting liquid at the outlet face.

As already mentioned, the second condition may alternatively be placed explicitly on q_{20} or expressed as an inlet boundary condition on S_2, permitting q_{20} to be determined. Because all oil introduced into the medium in the coalescence application is in fine suspension, the inlet oil rate q_{20} must satisfy the inequality

$$-1 \leq \frac{q_{20}}{q_{10}f_0} \leq 0 \tag{17}$$

That is, at steady state any nonzero flow of capillary-conducted oil at the inlet face must originate from oil deposited within the medium and must therefore flow in the negative x-direction (opposite in sign to the flow q_{10} of inlet suspension and out of the inlet face), not exceeding the total inflow of suspended oil. A formulation allowing for both zero and negative oil flow at the inlet is motivated by the observation of both situations under different experimental conditions. This formulation is believed relevant in describing so-called "interfacial tension separation" by porous membranes, for which captured dispersed phase is retained at the inlet face [Jordan (*27*)]. If q_{20} equals 0, all finely suspended oil entering the inlet face must leave the outlet face either in residual suspension or as coalesced globules that originate from capillary-conducted oil released at the outlet. With finite backflow, nonwetting oil accumulates in front of the inlet face in the form of macroscopic globules that are separated from capillary-conducted oil in the solid by preferentially wetting aqueous phase, and thus not continuously connected with oil in the solid. The oil flowing back from the inlet and the accumulated oil must therefore coalesce as the former joins the latter. Since the aqueous phase in immediate contact with the inlet face preferentially wets the solid, Equation 14 applies and the inlet condition is

Table I. Summary of Boundary Conditions Oil Dispersed in Water[b]

Location	Oil nonwetting: Zero back flow	Oil nonwetting: Finite back flow	Oil wetting[a]: Zero back flow
Inlet $x = 0$	$q_{20} = 0$ (above critical q_{10}, $S_2 = S_{2c}$)	$S_2 = S_{2c}$ (above critical q_{10}, $q_{20} = 0$)	$q_{20} = 0$
Outlet $x = L$	$S_2 = S_{2c}$	$S_2 = S_{2c}$	$S_2 = 1 - S_{1c}$

[a] *For oil wetting, finite back flow cannot occur.*
[b] *For water dispersed in oil: interchange oil with water everywhere.*

$$S_2 = S_{2c} \text{ at } x = 0 \tag{18}$$

The oil accumulating in front of the inlet face eventually mechanically inhibits flow of the aqueous phase, producing an additional aqueous pressure drop that should be included when computing the total aqueous pressure drop across the system. As the inlet oil accumulates in volume, its resistance to aqueous flow increases until the additional pressure drop becomes large enough to force the accumulated oil into the pores where it then continuously connects with capillary-conducted oil in the solid. This occurs when the aqueous pressure drop across the accumulated oil just equals the capillary pressure immediately within the solid. Backflow of oil then ceases and all captured oil is conducted in the downstream direction. No further increase in the quantity of accumulated oil can occur, and the pressure drop corresponding to its resistance to aqueous flow just balances the capillary pressure immediately within the solid. This steady condition is described mathematically by explicitly requiring zero oil backflow—*i.e.*

$$q_{20} = 0 \tag{19}$$

Equation 19 is, in this case, the second condition used in conjunction with the outlet condition (Equation 15). Applying Equations 12, 13, 15, and 19, the inlet face saturation becomes determined implicitly and the corresponding value of the inlet capillary pressure $p_c(x = 0)$ gives the steady-state aqueous pressure drop across the accumulated oil at the inlet. If, however, some means were provided for continually removing the back flowing oil so that it could not accumulate to satisfy Equation 19, then another steady state should be reached where the backflow persists indefinitely. While the situation during oil accumulation at the inlet is not truly at steady state, its unsteady character results only from the inlet oil accumulation. Accumulation can be slow relative to approach to steady state within the solid, and the steady-state Equations 3 through 9 will still apply at any instant; the accumulating condition is thus a "pseudo" steady state.

Another interesting possibility occurs if the oil is nonwetting. From the steady-state equations, it can be shown that for aqueous inlet velocities above a definite value (which depends on particular experimental conditions), specifying outlet Equation 15 and either of Equations 18 or 19 implies a solution that automatically satisfies the remaining of the two latter conditions. Physically, this means that backflow of oil theoretically cannot spontaneously occur if the aqueous flow rate exceeds a certain value, and all captured oil is conducted downstream.

If the suspended oil preferentially wets the solid, backflow should never occur, since oil would then be required to flow from a state of low pressure within the solid (negative capillary pressure) to a state of higher pressure in front of the solid (zero capillary pressure). The appropriate boundary conditions for oil wetting are therefore Equations 16 and 19. The different possible boundary conditions are summarized in Table I.

Estimation of the size of the coalesced oil globules in the effluent. According to the presented description of the coalescence-filtration process, the effluent suspension contains two regimes of oil. One regime consists of the residual minute uncaptured droplets in suspension, while the other results from the release of coalesced, capillary-conducted oil at the outlet. The basic equations provide only a description of the drop size distribution of the finely dispersed oil remaining in the effluent. A description of the entire effluent oil drop size distribution should also include a prediction of the size of globules that are formed when the captured and coalesced oil regime is released into the effluent. To estimate the size of the effluent coalesced oil globules, it is assumed that the globules are formed at the same capillary pressure that exists at the point of their formation.

If the oil is nonwetting, the globules are taken to be formed at the capillary pressure that exists at the outlet face of the medium. According to the outlet boundary condition established as Equation 15, the spherical oil globule radius R_G is given by

AUTHORS *Lloyd A. Spielman is with the Division of Engineering and Applied Physics, Harvard University, Cambridge, Mass. 02138, and Simon L. Goren is with the Department of Chemical Engineering, University of California, Berkeley, Calif. 94720.*

$$\frac{2\gamma}{R_G} = p_c(S_{2c})$$

or (20)

$$R_G = \frac{2\gamma}{p_c(S_{2c})}$$

This estimate of globule size is based on capillary-statics arguments, and is independent of flow rate; it is clear that some typical pore dimension characterizes the effluent globule size. This is particularly important for fibrous media whose pore dimensions can be considerably greater than the fiber diameters. It also follows from Equation 20 that scaling criteria for the capillary pressure also imply scaling criteria for the globule size R_G. As a special case for fibrous media (at the same porosities) similarly formed from fibers having uniform diameters, the nonwetting effluent oil globules are scaled in proportion to the fiber diameter (see Equations 21–23). Thus it might be advantageous to place a layer of coarser fibers at the outlet to produce large globules.

If the oil preferentially wets the solid, the globules are hypothetically formed at zero capillary pressure since this is implied as a result of their formation from the stagnant oil layer adhering to the outlet face. The interface at the outermost portion of the oil layer is approximately planar, and globules formed by breaking away at this point should therefore be indefinitely large. In reality, however, occasional effluent oil globules occur when the nearly stagnant oil sheath adhering to the outlet face breaks up as a result of disturbances; primarily, by the aqueous flow in the system. In this case the globules are of the same order as the thickness of the sheath and typically in the centimeter range. Figure 3 depicts globule formation for both oil wetted and nonwetted media.

Background for Correlation and Prediction of the Coefficients Underlying the Theoretical Framework

This section surveys previous work on two-phase flow and filtration relevant in obtaining mathematical descriptions of the capillary pressure, relative permeabilities, and the filter coefficient. Since these quantities are presumed known within the theoretical scheme, it is essential to assess their behavior with system variables by independent experimental or theoretical means to apply the above theoretical framework to coalescence in porous solids.

Since the majority of two-phase flow applications are to soils and porous-rock reservoirs, most available data in this area apply to granular rather than fibrous solids. Most water filtration is performed using beds of sand or other unconsolidated solids, so again available data are concerned predominantly with granular media. Porosities in granular solids commonly fall in the vicinity of 40%, while porosities of 90% and higher are typical for fibrous materials. The difference results from different packing characteristics of the respective materials. Caution must therefore be exercised when attempting to apply these data to fibrous media.

Hysteresis phenomena in two-phase flow. A difficulty so far ignored is the possible role of hysteresis in the coalescence process. Indeed, capillary pressure and relative permeabilities are known to depend not only on phase saturation but also on the saturation history leading to a given state. Displacement of the wetting phase by the nonwetting phase (drainage) or, conversely, displacement of the nonwetting phase by the wetting phase (imbibition) provide extremes giving somewhat different values of capillary pressure and relative permeabilities for the same final saturation. The importance of hysteresis in fibrous media is a matter open to experimentation, as it appears to have received little attention.

Hysteresis effects in more typical two-phase flow applications are dealt with by considering whether the particular application more closely resembles a drainage or imbibition process, and then using the corresponding functions of saturation. One might speculate that similar considerations be given for the selection of the appropriate functions in the coalescence process. If coalescence of a nonwetting oil is accomplished by approach to steady state from the initially oil-free condition, then the depositing nonwetting oil displaces water as oil accumulates in the medium, indicating a drainage process. If the coalescing oil preferentially wets the solid, it is often convenient to begin operation with the medium fully saturated with oil, since preliminary saturation with water does not normally proceed spontaneously in the presence of air. With media initially oil-saturated, the nonwetting aqueous phase must displace the wetting oil, again suggesting a drainage process. It must be concluded, however, that because the porous coalescence process is a considerably different situation of two-phase flow from those typically encountered in soil science and petroleum production, the role of hysteresis, as well as other aspects, should be more thoroughly investigated.

A good discussion of the origin of hysteresis effects can be found in the work of Miller and Miller (*31*). Recent experimental papers that are concerned with hysteresis under static and flow conditions are those of Naar *et al.* (*35*), Raimondi and Torcaso (*42*), and Morrow and Harris (*32*).

Capillary pressure. The capillary pressure function depends on the microscopic geometry of the porous medium, as well as the interfacial tension and relative wettability of the fluids. A drainage capillary pressure curve is obtained by slowly displacing the wetting fluid with the nonwetting fluid through applied pressure and measuring the displaced fluid volumes. Measurement techniques are reviewed by Scheidegger (*50*). Capillary pressure is often associated with pore size distribution, and routine mercury injection measurement of pore size distributions is, in fact, obtained from the capillary pressure curve for the mercury-vacuum system.

A typical drainage capillary pressure curve is shown in Figure 1. The capillary pressure first rises abruptly as nonwetting fluid displaces wetting fluid from the largest pores. As more nonwetting phase is forced into the medium, the capillary pressure increases gradually, then again abruptly as a saturation corresponding to maximum displacement is reached. No further wetting phase can be displaced beyond this saturation, because the remaining wetting phase exists as discontinuous regions occupying only the smallest crevices.

Capillary pressure curves cannot yet be predicted accurately *a priori*, but methods for correlation and extrapolation of experimental data have been proposed. For geometrically similar solids and fluid pairs exhibiting the same wetting angle, it follows from dimensional analysis [Miller and Miller (*31*)] that the scaling criterion is

$$\frac{\delta p_c}{\gamma} = f_1\,(S) \tag{21}$$

where δ is some length characteristic of the microscopic dimensions of the porous solid, and $f_1(S)$ is a function of saturation only.

Leverett (*30*) proposed the semiempirical correlation

$$\left(\frac{k_0}{\epsilon}\right)^{1/2} \frac{p_c}{\gamma} = J(S) \tag{22a}$$

where $J(S)$ is a correlation function (Leverett function), ϵ is the porosity of the solid, and k_0 is the single-phase permeability of the medium. Leverett was able to correlate data for a variety of unconsolidated media, the data falling on one curve. Rose and Bruce (*47*) similarly correlated data for a variety of consolidated media, the data again falling on a single curve, but one different from that for unconsolidated media. Leverett's hypothesis in formulating Equation 22*a* was based on the observation that both permeability and capillarity could be characterized by means of the hydraulic radius.

For nondescript porous solids such as porous rock or crushed material, the single-phase permeability k_0 is most easily determined through direct measurement. If, however, the solid geometry is sufficiently well defined, as for beds of fibers or spheres having uniform diameters, then k_0 can be estimated theoretically or by means of an appropriate correlation. Davies (*8*) found that one-phase permeabilities of fibrous media were adequately correlated by the equation

$$k_0 = d_F^2/64\alpha^{3/2}\,(1 + 56\,\alpha^3)$$

over the range of solids volume fraction ($0.006 < \alpha < 0.3$) where d_F is the mean fiber diameter. This expression is in excellent agreement with theoretical calculations presented by Spielman and Goren (*54*). Substitution of Davies' expression into Equation 22*a* gives

$$\frac{p_c d_F}{\gamma}\,\frac{1}{8\alpha^{3/4}(1 - \alpha)^{1/2}(1 + 56\,\alpha^3)^{1/2}} = J(S) \tag{22b}$$

Application of Equations 22 to fibrous media that typically have porosities greater than 90% is certainly questionable, since it is well known that the hydraulic radius does not adequately characterize the permeability of fibrous media in that range, as indicated by the failure of the Kozeny equation [Sullivan and Hertel (*58*); Carman (*3*)]. For sparse fibrous media, as normally encountered, the distribution of two given phases in the pores is governed primarily by the axial configuration of the fibers and does not depend directly on the distribution of fiber diameters, since the fiber-fiber surface separations are roughly just the fiber-fiber axis separations. For similarly prepared fibrous media consisting of relatively unbroken fibers, only the length of fiber per unit volume pore space should characterize the microscopic fluid distribution (except at very low saturations of the wetting phase, when the wetting fluid concentrates at fiber junctions). Dimensional analysis then gives

$$\frac{p_c}{\gamma}\left(\frac{\epsilon}{l_F}\right)^{1/2} = f_2(S) \tag{23a}$$

where l_F is the length of fiber per unit total volume of medium, and $f_2(S)$ is a function of saturation alone for a specified wetting angle. For media prepared from fibers having uniform diameters Equation 23*a* reduces to

$$\frac{p_c d_F}{\gamma}\left[\frac{\pi(1 - \alpha)}{4\alpha}\right]^{1/2} = f_2\,(S) \tag{23b}$$

The dependence on fiber volume fractions as given by Equation 23*b* is not greatly different from that of Equation 22*b*.

Parker (*38*) published limited capillary pressure data for glass fibrous media. Drainage curves for the air-water system were measured using two different fiber diameters and porosities. Although wetting for the air-water system is likely to be somewhat different from that for an oil-water system, the capillary pressure curves for different wetting and nonwetting fluid pairs are often similar [Purcell (*41*)]. This suggests the possibility

Table II. Capillary Pressure and Relative Permeability Data of Parker (38)

S_w Water saturation, (wetting phase)	P_c Capillary pressure, (mg Hg)		J Equation 22b		f_2 Equation 23b		k_w/k_0 Relative permeability, (wetting phase)	
Nominal fiber diameter (μ)	1.0	3.0	1.0	3.0	1.0	3.0	1.0	3.0
1.00							1.0	1.0
0.95	9.14	5.85	0.262	0.414	0.861	1.44	0.10	0.30
0.90							0.090	0.20
0.80	11.9	7.18	0.342	0.508	1.12	1.77	0.055	0.11
0.70							0.040	0.075
0.60	14.5	8.25	0.416	0.583	1.37	2.03	0.028	0.047
0.50							0.020	0.030
0.40	17.0	8.78	0.488	0.621	1.60	2.16	0.013	0.017
0.30							0.007	0.008
0.20	20.6	10.1	0.591	0.714	1.94	2.48	0.0024	0.0027
0.10							0.0003	0.0003
0.05	30.5	14.6	0.875	1.03	2.87	3.59	0.00002	0.000028
0.00							0.0	0.0

Nominal fiber diameter (μ)	Mean diameter (μ)	α
1.0	1.41	0.057
3.0	4.54	0.083

Surface tension: 72 dyn/cm

of extrapolating the available air-water data to different porosities, fiber diameters, and fluid pairs by means of Equations 22 or 23. Values of p_c taken from Parker's curves are given in Table II, along with values of J and f_2 calculated from Equations 22*b* and 23*b*. The lack of agreement between values of the respective functions for the two different fiber diameters suggests that the fiber size distributions for the two nominal fiber sizes were different, or that the packing configurations were different. Parker observed the larger diameter fiber to contain "considerable fines and debris." Better agreement might be obtained by using Equation 23*a* rather than 23*b*, but the quantity l_F could not be evaluated from the data given.

Relative permeabilities. Typical relative permeability curves are also presented in Figure 1. Both wetting and nonwetting curves are shown to exhibit limiting saturations of the respective fluids below which the permeability vanishes, because the phase under consideration cannot form sufficiently continuous channels to sustain flow. As the nonwetting phase saturation is increased above its limiting value, its permeability rapidly rises and normally approaches unity before the wetting phase is completely displaced. This occurs because the wetting phase, present in relatively small amount, retreats by capillary action to very small pores or crevices that contribute negligibly to nonwetting phase flow, even when occupied by nonwetting fluid. In contrast, a relatively small amount of nonwetting phase can often greatly decrease the permeability to the wetting phase, since capillary action then forces the nonwetting fluid into large pores that contribute significantly to the flow. Scheidegger (*50*), Collins (*6*), Muskat (*33*), and other standard texts give detailed discussions interpreting the behavior of relative permeability.

Values of the limiting saturation for flow of the wetting or nonwetting phase are very specific to the particular porous solid under consideration. Typical limiting saturations reported for granular solids are 10 to 20%, although values as large as 40% have been encountered.

Jones (*26*) was able to correlate limited relative permeability data reasonably well using the expressions

$$\frac{k_n}{k_0} = 1 - 1.11\,S_w = 1.11\,S_n - 0.11 \qquad (24a)$$

$$\frac{k_w}{k_0} = S_w^{\,3} \qquad (24b)$$

where n and w refer to the nonwetting and wetting phases, respectively. Equation 24*a* predicts a constant slope of the relative permeability to the nonwetting phase *vs.* saturation in the neighborhood of the limiting nonwetting phase saturation, which occurs at $S_{nc} = 0.10$. Jones obtained Equations 24 from experiments with oil as the nonwetting phase, and water as the wetting phase in granular solids. Observe that Equation 24*b* shows no permeability cut-off for the wetting phase.

Theoretical treatments usually attempt to relate relative permeability to the capillary pressure function, since the latter is interpreted as an experimental measure of pore size distribution and provides information concerning pore occupation by the two phases.

Fatt and Dykstra (*14*) represent the porous solid as a bundle of capillaries having different radii and tortuosities depending on pore radius. The smallest pores are considered the most tortuous and occupied preferentially by the wetting phase. Their final expression for the wetting phase relative permeability is

$$\frac{k_w}{k_0} = \frac{\int_0^{S_w} \frac{dS_w}{p_c^{2(1+b)}}}{\int_0^1 \frac{dS_w}{p_c^{2(1+b)}}} \qquad (25)$$

where b is an empirical parameter incorporating tortuosity effects. Fatt and Dykstra correlated data with $b = {}^1/_2$. Gates and Lietz (*18*) earlier presented a result corresponding to Equation 25 with $b = 0$.

Wyllie and Spangler (*65*) developed an extended Kozeny-Carman equation and give for the wetting and nonwetting phase relative permeabilities:

$$\frac{k_w}{k_0} = \frac{1}{I_w^2 S_w^2} \frac{\int_0^{S_w} \frac{dS_w}{p_c^2}}{\int_0^1 \frac{dS_w}{p_c^2}} \qquad (26)$$

$$\frac{k_n}{k_0} = \frac{1}{I_n^2(1 - S_w)^2} \frac{\int_{S_w}^1 \frac{dS_w}{p_c^2}}{\int_0^1 \frac{dS_w}{p_c^2}}$$

Here I_w and I_n are the respective wetting and nonwetting phase resistivity indexes reflecting shape and tortuosity, and are obtained independently through electrical measurements.

Rapoport and Leas (*43*) developed expressions for upper and lower bounds on wetting phase permeability by using arguments based on interfacial thermodynamics. Their expressions also involve the capillary pressure function.

Good reviews of theoretical treatments of relative permeability are given by Scheidegger (*50*) and Wyllie and Gardner (*64*).

Parker (*38*) experimentally determined the relative permeability of glass fibrous media to water (wetting phase) using the air-water system. As data for fibrous media are rare and extremely relevant to the present study, points taken from Parker's curves have been presented in Table II, along with the corresponding capillary pressures discussed earlier. Two different nominal fiber diameters, in mats having unequal porosities, were used in the investigation. Introduction of very small amounts of nonwetting phase produced a more marked decrease in the permeability to the wetting phase than is usually observed for typical granular media. Parker correlated his data by means of a theory employing experimentally determined capillary pressures. No nonwetting phase permeabilities were measured.

Although Parker's results are limited to the air-water system, two fiber diameters, and porosities, it is anticipated that oil-water systems exhibiting small wetting angles should yield similar results. It follows from established scaling criteria [Miller and Miller (*37*)] that relative permeabilities for fluid pairs exhibiting the same relative wettability in geometrically similar solids are functions of saturation only.

Single-phase permeability. As mentioned earlier, channel models are usually unable to describe single-phase pressure drop through fibrous media because of the high porosities that typically occur in fibrous media. More realistic models represent fibrous solids as ensembles of individual cylinders, the flow around which is influenced by the neighboring fibers. The cell models of Happel (*21*) and Kuwabara (*29*) account for the influence of the fibers surrounding a central fiber by means of an artificial envelope, the size of which is established through the macroscopic voids. Another model, developed by Brinkman (*7*) to describe flow through a swarm of spheres, has been applied to cylinders by Spielman and Goren (*54*). This model assumes that the neighboring fibers damp the ensemble average macroscopic flow near the central fiber in precisely the same way that the fibers of the medium damp the average flow through the medium.

The inability of channel models to describe single-phase flow through sparse fibrous media recommends caution in extending such models to describe two-phase flow. One might speculate, however, that channel models would be better applied to two-phase flow than to single-phase flow, since the pore space occupied by one of two phases is effectively decreased by the presence of the other, and the "effective" porosity may fall within the range of applicability of channel theories.

Filter coefficient. Understanding of aerosol filtration has been advanced through theoretical calculations of the efficiency of particle removal by individual fibers when various capture mechanisms are operative. The individual fiber efficiency η is the predicted particle capture rate divided by the rate that would be obtained if particles moved in rectilinear flow with constant velocity until contacting a fiber. For clean fibrous mats, the filter coefficient is related to the individual fiber efficiency by

$$\lambda = \frac{4\alpha\eta}{\pi d_F} \qquad (27)$$

For a bed of spherical collectors of diameter d_s the relation is

$$\lambda = \frac{3\alpha\eta}{2\, d_s} \qquad (28)$$

Mechanisms of particle capture include interception, Brownian diffusion, inertial impaction, gravity settling, and long-range attractive forces, each giving rise to different expressions for η. Excellent reviews of the progress in aerosol filtration have been given by Chen (*4*), Fuchs (*17*), Pich (*40*), and Dorman (*10*).

Magnitude estimates indicate that interception, diffusion, and gravity settling are likely the dominant capture mechanisms in filtration of liquids. When operating individually, the predicted individual fiber efficiencies for these mechanisms are

Interception: $$\eta = 2\,A\,(d_p/d_F)^2 \tag{29}$$

Brownian diffusion: $$\eta = 2.3\,A^{-1/3}\,(d_F U/\mathfrak{D})^{-2/3} \tag{30}$$

Gravity settling: $$\eta = u_T/U \tag{31}$$

Here U is the superficial velocity, $\mathfrak{D}$ is the Brownian diffusion coefficient of the particles, u_T is the terminal velocity of the particles, and A is a parameter that characterizes the influence of neighboring fibers on the flow field near an individual fiber. Similar equations pertain for spherical collectors.

When more than one capture mechanism is important, other formulas must be used. Stechkina and Fuchs (*57*) and Friedlander (*15*) treated particle capture by fibers for combined interception and diffusion, the former basing their calculations on the velocity profiles of Happel (*21*) and Kuwabara (*29*). Spielman and Goren (*54*) adapted these treatments to their velocity profile. Natanson (*36*) performed calculations for capture of small particles under the influence of long-range electrostatic and adhesive forces combined with interception. Because of the presence of ions dissolved in the aqueous phase of oil-in-water suspensions, any electrostatic effects must be confined to narrow double layers in proximity to particle and obstacle surfaces, and therefore cannot be properly considered as long-range attractive or repulsive forces. Recently Yao and O'Melia (*66*) numerically calculated particle capture by a spherical collector under the combined effects of Brownian diffusion, interception, and gravity settling.

By contrast with aerosol filtration, theories predicting capture of particulates from water by sand beds have been considerably less successful. Much of the difficulty no doubt results from the apparent lack of high quality data to evaluate the effects of the numerous system variables on filter performance. A feature that seems to distinguish water from air filtration is the more significant role that surface chemical forces play in the former by preventing or promoting adherence of particles to the filter solid [*e.g.*, Ives and Gregory (*23*)]. Reviews of water filtration theory have been given by Sakthivadivel and Irmay (*48*) and O'Melia and Stumm (*37*). Friedlander (*16*) gives a technique for incorporating lyophobic colloid stability theory to treat capture by Brownian diffusion. In this interesting analysis, chemical effects enter through a boundary condition on the equation of convective diffusion. Particle capture by London-van der Waals attraction, taking particle-collector hydrodynamic interactions into account, is treated elsewhere [Spielman and Goren (*56*)]. Previous analyses have ignored such interactions, but these effects are thought to be important in determining particle capture from liquids.

Empirical correlations for water filtration are extremely specific to the particular operating conditions and systems. None of these correlations has been presented in useful dimensionless form for extrapolation to other conditions. Differences existing among various correlations can likely be traced to different mechanisms influencing deposition under the different operating conditions employed, as well as to experimental inaccuracy. Because of the uncertainty in estimating the filter coefficient from correlations, it is normally determined directly from pilot filtration experiments. Qualitatively, the various expressions indicate that deposition increases with decreased medium grain size, decreased flow rate, decreased viscosity, and increased suspended particle size.

Of the various empirical expressions for the filter coefficient, only those that give a dependence on the local deposit of accumulated solids (analogous to held-up oil) are presented. Iwasaki (*25*) summarized his data on sand filtration of *E. coli* (reported as *B. coli*) with the expression

$$\lambda = \lambda_0 + \lambda_1 D \text{ cm}^{-1} \tag{32}$$

with

$$\lambda_0 = \frac{1.297}{q^{0.299}}\left(\frac{341}{d_m} - 1\right) \text{cm}^{-1}$$

$$\lambda_1 = \frac{0.475 \times 10^{-9}}{q^{0.365}}\left(1 + \frac{923}{d_m}\right) \text{cm}^{-1}$$

D is the volume deposit per unit volume filter, d_m is in microns, and q is in m/day.

Ives and Sholji (*24*) summarized their findings on sand filtration of polyvinylchloride spheres (1.3-μ diameter) with the expression

$$\lambda = \lambda_0 + \lambda_1 D - \frac{\lambda_2 D^2}{(\epsilon - D)} \tag{33}$$

with

$$\lambda_0 = \frac{4 \times 10^{-8}}{q d_m \mu^2} \text{ cm}^{-1}$$

$$\lambda_1 = \frac{0.9 \times 10^{-4}}{q d_m \mu^{1.2}} \text{ cm}^{-1}$$

$$\lambda_2 = \frac{5.7 \times 10^{-6}}{q d_m \mu^2} \text{ cm}^{-1}$$

q is in cm/sec, d_m is in cm, and μ is in g/cm-sec. This last expression shows that in general filter performance passes through a maximum with increasing deposit.

Summary

The literature dealing with coalescence induced by flow through porous media is reviewed. A theoretical framework for coalescence of liquid-liquid dispersions by flow through porous media is given through a set of coupled equations in which concepts from air and water filtration describe the local capture of suspended droplets within the medium and two-phase flow theory, as developed in soil physics and petroleum reservoir engineering, describes the flow of coalesced liquid through the matrix. Within this framework the filter coefficient, capillary pressure, and fluid permeabilities appear explicitly. The literature concerning the determination of these coefficients is reviewed. Future communications in this series concern predictions drawn from the above theoretical framework of coalescer performance, including certain interesting capillary phenomena, theoretical prediction of the filter coefficient, and experimental investigation of the coalescence process for controlled and systematic variation of the parameters.

Nomenclature

a_p = particle radius (cm)
d_F = fiber diameter (cm)
d_p = particle diameter (cm)
f = volume fraction of finely dispersed oil (dimensionless)
$J(S)$ = correlation function (Leverett function) (dimensionless)
k_i = Darcy permeability to phase i (cm^2)
k_i/k_0 = relative permeability to phase i (dimensionless)
k_0 = single-phase permeability (cm^2)
$n(a_p)$ = number concentration distribution of particle radius a_p (particles/cm^4)
$p_c(S)$ = capillary pressure function of saturation (dyn/cm^2)
p_i = local pressure in phase i (dyn/cm^2)
q_i = superficial velocity of phase i (cm/sec)
R_G = coalesced globule radius (cm)
S = saturation (dimensionless)
S_i = saturation of phase i (fraction voids occupied by phase i)
x = distance into porous medium (cm)
α = fiber volume fraction (dimensionless)
γ = interfacial tension (dyn/cm)
ϵ = voids volume fraction (dimensionless)
η = collection efficiency for obstacle (dimensionless)
$\lambda(S_2, q_1)$ = local filter coefficient (cm^{-1})
μ_i = viscosity of phase i (g/cm-sec)

Subscripts

1 = aqueous phase
2 = coalesced oil phase
c = limiting saturation for flow of a given phase (except p_c)
w = wetting
n = nonwetting
0 = inlet

REFERENCES

(1) Brinkman, H. C., "On the Permeability of Media Consisting of Closely Packed Porous Particles," *Appl. Sci. Res., Sect. A*, 81 (1947).
(2) Burtis, T. A. and C. G. Kirkbride, "Desalting of Petroleum by Use of Fiberglass Packing," *Trans. Amer. Inst. Chem. Eng.*, **42**, 413 (1946).
(3) Carman, P. C., "Flow of Gases Through Porous Media," Academic Press, New York, 1956.
(4) Chen, C. Y., "Filtration of Aerosols by Fibrous Media," *Chem. Rev.*, **55**, 595 (1955).
(5) Clayton, W., "The Theory of Emulsions and Their Technical Treatment," Blakiston Co., New York, 1954.
(6) Collins, R. E., "Flow of Fluids Through Porous Materials," Reinhold, New York, 1961.
(7) Cordes, R., "Microscopic Study of the Mechanisms of Capture and Coalescence in Fibrous Media," M.S. Thesis, University of California, Berkeley, Calif., 1968.
(8) Davies, C. N., "The Separation of Airborne Dust and Particles," *Proc. Inst. Mech. Eng. (London)*, **1B**, 185 (1952).
(9) Davies, C. N., Ed., "Aerosol Science," Academic Press, New York, 1966.
(10) Dorman, R. G., "Filtration," *in* Aerosol Science, C. N. Davies, Ed., Academic Press, New York, 1966.
(11) Douglas, E. and Elliot, I. G., "Developments in Oily-Water Separator Design," *Trans.. Inst. Mar. Eng.*, **74**, 164 (1962).
(12) Farley, R. and Valentin, F. H. H., "Coagulation as a Means of Separating Oil from Effluents," *A.I.Ch.E.-I.E.C. Symp. Ser. 1*, 1 (1965).
(13) Fatt, I., private communication, 1968.
(14) Fatt, I. and Dykstra, H., "Relative Permeability Studies," *Trans. AIME*, **192**, 249 (1951).
(15) Friedlander, S. K., "Particle Diffusion in Low-Speed Flows," *J. Colloid Interface Sci.*, **23**, 157–64 (1967).
(16) Friedlander, S. K., private communication, and talk presented at Gordon Research Conference on Environmental Sciences and Engineering-Water, New Hampton, N.H., 1968.
(17) Fuchs, N. A., "The Mechanics of Aerosols," Macmillan Co., New York, 1964.
(18) Gates, J. I. and Tempelaar-Lietz, W., "Relative Permeabilities of California Cores by the Capillary Pressure Method," presented at the API Meeting, Los Angeles, Calif., May 11–12, 1950.
(19) Graham, R. J., "Separation of Immiscible Liquids by Gravity Settling and Induced Coalescence," M.S. Thesis, University of California, Lawrence Radiation Laboratory, Berkeley, Calif., 1962.
(20) Gudesen, R. C., "Coalescence of Petroleum Compounds in Mixed Fibrous Beds," M.S. Thesis, Illinois Institute of Technology, Chicago, Ill., 1964.
(21) Happel, J., "Viscous Flow Relative to Arrays of Cylinders," *A.I.Ch.E. J.*, **5**, 174 (1959).
(22) Hayes, J. G., Hays, L. A., and Wood, H. S., "Commercial Desalting Unit Employing Fiberglass as Contacting Agent," *Chem. Eng. Progr.*, **45**, 235 (1949).
(23) Ives, K. J. and Gregory, J., "Surface Forces in Filtration," *Proc. Soc. Water Treat. Exam.*, **15**, 93 (1966).
(24) Ives, K. J. and Sholji, I., "Research on Variables Affecting Filtration," *J. Sanit. Eng. Div., Amer. Soc. Civil Eng.*, **91**, 1 (August 1965).
(25) Iwasaki, T., "Some Notes on Sand Filtration," *J. Amer. Water Works Ass.*, **29**, 1591 (July–December 1937).
(26) Jones, P. J., "Petroleum Production," Reinhold, New York, 1946.
(27) Jordan, G. V., "Separation of Immiscible Liquids by Means of Porous Membranes," *Trans. ASME*, *77*, 393 (1955).
(28) Kintner, R. C., "Drop Phenomena Affecting Liquid Extraction," *Advan. Chem. Eng.*, **3**, 52 (1963).
(29) Kuwabara, S., "The Forces Experienced by Randomly Distributed Parallel Circular Cylinders or Spheres," *J. Phys. Soc. Jap.*, **14**, 527 (1954).
(30) Leverett, M. C., "Capillary Behavior in Porous Solids," *Trans. AIME*, **142**, 152 (1941).
(31) Miller, R. D. and Miller, E. E., "Physical Theory for Capillary Flow Phenomena," *J. Appl. Phys.*, **27**, 324 (1956).
(32) Morrow, N. R. and Harris, C. C., "Capillary Equilibrium in Porous Materials," *Soc. Petrol. Eng. J.*, **5**, 15 (1965).
(33) Muskat, M., "Physical Principles of Oil Production," McGraw-Hill, New York, 1949.
(34) Muskat, M. and Meres, M. W., "The Flow of Heterogeneous Fluids Through Porous Media," *J. Appl. Phys.*, **7**, 346 (1936).
(35) Naar, J., Wygal, R. J., and Henderson, J. H., "Imbibition Relative Permeability in Unconsolidated Porous Media," *Soc. Petrol. Eng. J.*, **2**, 13 (1962).
(36) Natanson, G. L., "The Condensation of Aerosol Particles by Electrostatic Attraction on a Cylinder Around Which They Are Streaming," *Dokl. Akad. Nauk SSSR*, **112**, 696 (1957).
(37) O'Melia, C. R. and Stumm, W., "Theory of Water Filtration," paper presented at Annual Conference of American Water Works Association, Atlantic City, N. J., 1967.
(38) Parker, J., "Partially Saturated Glass Fiber Beds," IND. ENG. CHEM., **52**, 247 (1960).
(39) Permutit Company, "Research and Development for a Shipboard Oil and Water Separation System," *PB 174639*, U. S. Department of Commerce Maritime Administration, Contract No. MA-2722 (December 1966).
(40) Pich, J., "Theory of Aerosol Filtration by Fibrous and Membrane Filters," *in* Aerosol Science, C. N. Davies, Ed., Academic Press Library, New York, 1966.
(41) Purcell, W. R., "Capillary Pressures—Their Measurement Using Mercury and the Calculation of Permeability Therefrom," *Trans. AIME*, **186**, 39 (1949).
(42) Raimondi, P. and Torcaso, M., "Distribution of the Oil Phase Obtained upon Imbibition of Water," *Soc. Petrol. Eng. J.*, **4**, 49 (1964).
(43) Rapoport, L. A. and Leas W. J., "Relative Permeability to Liquid in Liquid-Gas Systems," *Trans. AIME*, **192**, 83 (1951).
(44) Richards, L. A., "Capillary Conduction of Liquids Through Porous Mediums," *Physics (New York)*, **1**, 318 (1931).
(45) Richardson, J. G., Kerver, J. K., Hafford, J. A., and Osoba, J. S., "Laboratory Determination of Relative Permeability," *Petrol. Trans. AIME*, **195**, 187 (1952).
(46) Rose, P. R., "Mechanisms of Operation of a Fibrous Bed Coalescer," M.S. Thesis, Illinois Institute of Technology, Chicago, Ill., 1963.
(47) Rose, W. and Bruce, W. A., "Evaluation of Capillary Character in Petroleum Reservoir Rock," *Trans. AIME*, **186**, 127 (1949).
(48) Sakthivadivel, R. and Irmay, S., "A Review of Filtration Theories," Hydraul. Eng. Lab., *Rep. HEL 15-4*, University of California, Berkeley, Calif., 1966.
(49) Sareen, S. S., Rose, P. M., Gudesen, R. C., and Kintner, R. C., "Coalescence in Fibrous Beds," *A.I.Ch.E. J.*, **12**, 1045 (1966).
(50) Scheidegger, A. E., "The Physics of Flow through Porous Media," University of Toronto Press, Toronto, Can., 1960.
(51) Shackleton, L. R. B., Douglas, E., and Walsh, T., "Pollution of the Sea by Oil," *Trans. Inst. Mar. Eng.*, **72**, 409 (1960).
(52) Slattery, J. C., "Multiphase Viscoelastic Flow Through Porous Media," *A.I.Ch.E. J.*, **14**, 50 (1968).
(53) Spielman, L. A., "Separation of Finely Dispersed Liquid-Liquid Suspensions by Flow through Fibrous Media," Ph.D. Dissertation, University of California, Berkeley, Calif., 1968.
(54) Spielman, L. A. and Goren, S. L., "Model for Predicting Pressure Drop and Filtration Efficiency in Fibrous Media," *Environ. Sci. Technol.*, **2**, 279 (1968).
(55) Spielman, L. A. and Goren, S. L., "I. Theory of Coalescence by Flow Through Porous Media; II. Experiments on Coalescence by Flow Through Fibrous Media," in preparation.
(56) Spielman, L. A. and Goren, S. L., "Capture of Small Particles by London Forces from Low-Speed Liquid Flows," *Environ. Sci. Technol.*, **4**, 135 (1970).
(57) Stechkina I. B. and Fuchs N. A. "Studies of Fibrous Filters, I," *Ann. Occup. Hyg.*, **9**, 59 (1966).
(58) Sullivan, R. R. and Hertel, K. L., "The Permeability Method for Determining Specific Surface of Fibers and Powders," *Advan. Colloid Sci.*, **1**, 37 (1942).
(59) Sweeney, W. F., "Some Observations on Liquid-Liquid Settling," M.S. Thesis, University of California, Lawrence Radiation Laboratory and Department of Chemical Engineering, Berkeley Calif., 1964.
(60) Treybal, R. E., "Liquid Extraction," McGraw-Hill, New York, 1951, pp 447–450.
(61) Vinson, C. G. "The Coalescence of Micron-size Drops in Liquid-Liquid Dispersions in Flow Past Fine Mesh Screen," Doctoral Dissertation, University of Michigan, Ann Arbor, Mich., 1965.
(62) Voyutskii, S. S., Kal'yanova K. A. Panich, R., and Fodiman, N., "Mechanism of Separation of the Disperse Phase of Emulsions During Filtration," *Dokl. Akad. Nauk SSSR*, **91**, 1155 (1953); *CA*, **49**, 12053d (1955).
(63) Voyutskii, S. S., Fodiman, N. M., and Panich, R. "Filtration of Emulsions," *Izv. Vyssh. Ucheb. Zaved. Khim. Khim. Tekhnol.*, **2**, 1970 (1958); *CA* **52**, 19266a (1958).
(64) Wyllie, M. R. J. and Gardner, G. H. F., "The Generalized Kozeny-Carman Equation," *World Oil*, Parts I and II, 121–128, 210–226 (March–April 1958).
(65) Wyllie, M. R. J. and Spangler, M. B., *Amer. Ass. Petrol. Geol., Bull.* **36** (2), 359 (1952).
(66) Yao, K. and O'Melia, C. R., "Particle Transport in Aqueous Flow Through Porous Media," presented at the 16th Annual Conference of the Hydraulics Division of ASCE, M.I.T., Cambridge, Mass., August 20–22, 1968.

11

Reverse Osmosis

S. SOURIRAJAN
J. P. AGRAWAL

Reverse Osmosis

The reverse osmosis process is discussed with particular reference to systems involving aqueous solutions and Loeb-Sourirajan-type porous cellulose acetate membranes. Mechanisms of the process and porous cellulose acetate membrane technology are briefly reviewed. Based on a general capillary diffusion model for the transport of solvent and solute through the membrane, transport equations applicable for the entire range of solute separation are presented. The results of the analysis and correlations of experimental reverse osmosis data are illustrated. On the basis of the above equations and correlations, methods of membrane specification, expressing membrane selectivity, and predicting membrane performance are outlined. Reverse osmosis is then treated as a unit operation in chemical engineering. A set of general equations for reverse osmosis process design is then derived for reverse osmosis systems specified in terms of membrane specifications and operating conditions. The utility of these equations for studying system performance and process parameters is indicated. Reverse osmosis as a general concentration process and the separation of mixed solutes in aqueous solution are then briefly discussed from the points of view of process design and the predictability of membrane performance, respectively.

The reverse osmosis process is a general and widely applicable technique for the separation, concentration, or fractionation of inorganic or organic substances in aqueous or nonaqueous solutions in the liquid or the gaseous phase (*141–156*, *171*). The technique consists in letting the fluid mixture flow, under pressure, through an appropriate porous membrane, and withdrawing the membrane permeated product generally at atmospheric pressure and surrounding temperature; the product is enriched in one or more constituents of the mixture, leaving a concentrated solution on the upstream side of the membrane (*71–94*). No heating of the membrane and no phase change in product recovery are involved.

This process is a new development in the field of solute-solvent separation. Its most significant current application is in the field of water treatment, in general, and saline water conversion, in particular, for which porous cellulose acetate membranes have been found appropriate. In view of the recent advances in the above field, this paper is limited to a discussion of the reverse osmosis process involving aqueous solution systems and cellulose acetate membranes.

"Osmosis" and "Reverse Osmosis"

Under isothermal conditions, in both "osmosis" and "reverse osmosis" the preferential transport of material through the membrane is always in the direction of lower chemical potential. This is a thermodynamic requirement, and it does not, and cannot, specify which component, if any, of a solution is preferentially transported through a given membrane and the mechanism by which such transport takes place. In the literature, "osmosis" and "reverse osmosis" are associated with the existence of a "semipermeable" membrane; the origin of such semipermeability is still an open question. From a practical standpoint it is important to note that the reverse osmosis process is not restricted to the passage of water from aqueous solutions, that it is not restricted to 100% solute separation, that neither "osmosis" nor "reverse osmosis" is an explanation of the mechanism of the process involved, and that the distinction between the two terms is entirely one of arbitrary convention and popular usage.

Mechanisms

Several mechanisms of the reverse osmosis process have been proposed; some of these are discussed extensively by Merten (*109–111*). According to Reid and Breton (*16*, *17*, *124*), the "semipermeability" of cellulose acetate is caused by regions of bound water within the membrane, and the transfer of water and ions through the membrane is governed by two different mechanisms.

Reverse osmosis has practical significance in water cleanup and food processing

Those ions and molecules which can associate with the membrane through hydrogen bonding actually combine with the membrane and are transported through it by alignment-type diffusion; those which cannot enter into hydrogen bonding with the membrane are transported by hole-type diffusion. Consequently, according to this mechanism, the existence of pores in the membrane contributes only to the leakage of the solute through the membrane and hence must be avoided for practical desalination applications of the reverse osmosis process.

A solution-diffusion mechanism is favored by Lonsdale *et al.* (*95–98*, *128*), whose transport equations are apparently limited to their concept of "perfect" membranes, which are presumably those which have a completely nonporous surface structure (*128*). Banks and Sharples (*9–11*) also consider that the mechanism of reverse osmosis is one of diffusive flow through the pore-free layer on the membrane surface. According to Michaels *et al.* (*116*), water transport in reverse osmosis is by molecular diffusion through the polymer matrix, and solute transport is by parallel mechanisms involving sorption, activated diffusion, and hydrodynamic flow. According to Sherwood *et al.* (*138*), water and solute cross the membrane by parallel processes of diffusion and pore flow. The reverse osmosis process has also been interpreted in terms of nonequilibrium thermodynamics by several workers (*43*, *44*, *158*).

Preferential sorption-capillary flow mechanism. This has been discussed in detail by Sourirajan (*141*, *149*, *150*, *152*). According to this mechanism, reverse osmosis separation is the combined result of an interfacial phenomenon and fluid transport under pressure through capillary pores, and the appropriate membrane is a porous medium at all levels of solute separation. This approach is adopted here because it has proved useful in the development of the subject, and it fits in with the theme of this symposium.

Figure 1 is a conceptual model of the above mechanism for recovering fresh water from aqueous salt solutions. Here the solution is in contact with a solid material in the form of a porous membrane whose surface has a preferential sorption for water and/or preferential repulsion for the solute. A continuous removal of the preferentially sorbed interfacial water is effected by letting it flow under pressure through the membrane capillaries.

The above model also gives rise to the concept of a critical pore diameter for maximum separation and permeability. This is obviously twice the thickness of the preferentially sorbed interfacial water layer (Figure 2). On the basis of the above model, an appropriate chemical nature of the film surface in contact with the solution, as well as the existence of pores of appropriate size on the area of the porous film at the interface, constitutes the indispensable twin requirement for the success of this separation process.

The above mechanism gives rise to two basic characteristics of this separation process in its most general

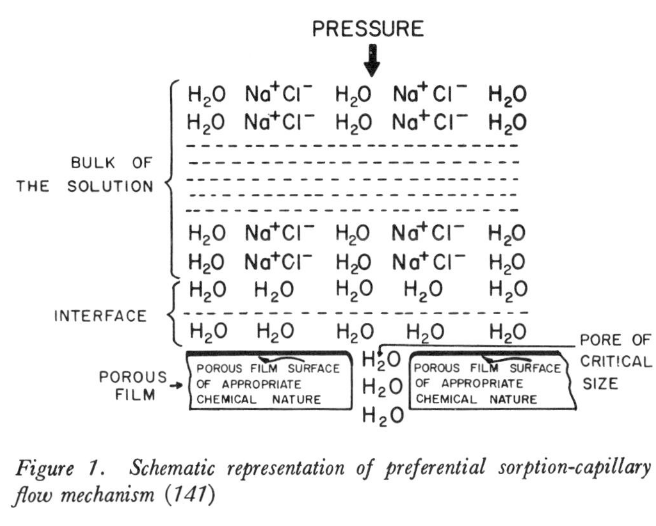

Figure 1. Schematic representation of preferential sorption-capillary flow mechanism (141)

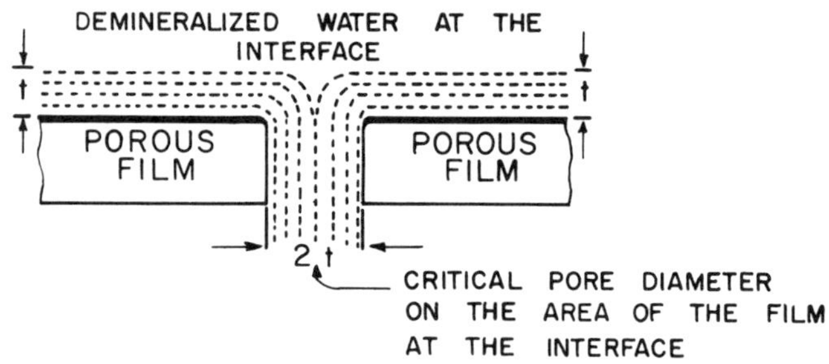

Figure 2. Critical pore diameter for maximum separation and permeability (141)

application (*153*): (1) with respect to a given solution containing components **A** and **B**, the membrane permeated product can be enriched either in **A** or **B**, depending upon the chemical nature of the membrane surface in contact with the solution; and (2) with respect to a given membrane solution system, wherein the membrane surface has a preferential sorption for the solvent, practically any degree of solute separation is possible, depending on the porous structure of the membrane and other operating conditions. These two characteristics are borne out experimentally (*146*).

From an industrial standpoint, the application of the reverse osmosis technique for a given separation problem involves essentially the problem of choosing the appropriate chemical nature of the film surface and developing methods for making films containing the largest number of pores of the required size on the area of the film at the interface. This approach formed the basis of the successful development of the Loeb-Sourirajan-type porous cellulose acetate membranes for saline water conversion and similar applications (*164*).

Porous Cellulose Acetate Membrane Technology

Finding that cellulose acetate material has a preferential sorption for water in aqueous salt solutions (*85–89, 155*), Loeb and Sourirajan developed a method of making porous cellulose acetate membranes capable of giving both high permeabilities and high degrees of solute separation from aqueous sodium chloride solutions (*90–94*). It is this particular development which has led to the current world-wide interest in the reverse osmosis process. Their technique, developed at the University of California, Los Angeles, during 1959–60, is essentially as follows (*164*).

The film-casting solution contains cellulose acetate dissolved in acetone, to which is added an additive soluble in water and not affecting the solubility of cellulose acetate in acetone. With the above solution, membranes are cast on flat or tubular surfaces (for example, glass plates or metal tubes) either at laboratory or lower temperatures. After it is cast, part of the solvent is allowed to evaporate for a definite period from the surface of the membrane at the casting temperature; this step in the film-making technique is called the "evaporation" step. This is followed by immersing the membrane in ice cold water for at least 1 hr; during this time, the film sets to a gel, from which the additive and the solvent (acetone) are leached out, leaving a tough solid porous film on the flat or the tubular surface, from which it can be removed easily. This step in the technique is called the "gelation" step. All details relating to the casting-solution composition, evaporation, and gelation steps are important in developing successful reverse osmosis membranes. Some typical experimental conditions for casting flat porous cellulose acetate membranes are given in Table I. These membranes are always stored at laboratory temperature either under water or dilute alcohol solution, or in a humid atmosphere to preserve their porous structure. On drying in air at the laboratory temperature, their porous structure generally changes irreversibly. Merten *et al.* (*113*) showed that the above membranes could be held "dry" after treatment with a surface active agent such as a mixture of Tergitol 15S7 and glycerol in water.

Films cast in the above manner have a relatively dense microporous structure (pore size probably <50Å) on an extremely thin layer (about 10^{-5} in.) of the film surface which was exposed to the atmosphere during casting; this side of the film is called the "surface layer side," and the other side, the "back side." The remainder of the film material underneath this thin surface layer is a spongy porous mass containing, comparatively, very big pores (∼4000Å) (*127, 129*). Consequently, the above films are said to have an asymmetric porous structure. It is the microporous structure of the dense surface layer which governs the level of solute

Table I. Typical Casting Conditions For Loeb-Sourirajan-Type Porous Cellulose Acetate Flat Membranes

Film Casting Details	Film type identification no.		
	CA-NRC-25 or Batch 25	CA-NRC-47 or Batch 47	CA-NRC-18 or Batch 18
Casting solution composition (wt %)			
Cellulose acetate (acetyl content 39.8%)	22.2	25	17
Acetone	66.7	45	68
Magnesium perchlorate	1.1	...	1.5
Water	10.0	...	13.5
Formamide	...	30	...
Casting temperature, °C	−10	Lab. temp. ∼24	−10
Evaporation time, min	4	<1	4
Gelation time in ice cold water, hour	∼1	∼1	∼1
Nominal film thickness	0.003 to 0.006 inch in all cases		

separation, and hence is held in contact with the feed solution during reverse osmosis operation.

In the as-cast condition, the surface pores are generally too big to give any significant solute separation. But the membrane can be shrunk and the pore size reduced by simply keeping the film immersed in hot water for about 10 min (*90*, *168*). Thus, by adjusting the shrinkage temperature, different porosities on the surface layer side of the membrane, capable of giving different levels of solute separation, can be obtained.

Batch-25-type films were the ones originally developed by Loeb and Sourirajan (*90*, *91*). Batch-47 and Batch-18-type films were later improvements reported by Manjikian *et al.* (*101–103*), and Sourirajan and Govindan (*153*), respectively. For water desalination application, the performances of the latter two types of films are about the same, and very much better than the performance of Batch-25-type films. Since Batch-47-type films are cast at laboratory temperature, they have an apparent simplicity in their fabrication, and, probably for this reason, they seem currently to be more popular. On the other hand, Batch-18-type films can be made with better quality control, and their performance in long-time experiments is more predictable; for this reason, Batch-18-type films have been extensively studied. In view of the common structural and fabrication characteristics of these membranes, they are generally called the Loeb-Sourirajan-type porous cellulose acetate membranes. A set of performance data on the above membranes is illustrated in Figures 3, 4, and 5, and Table II (*152*, *153*).

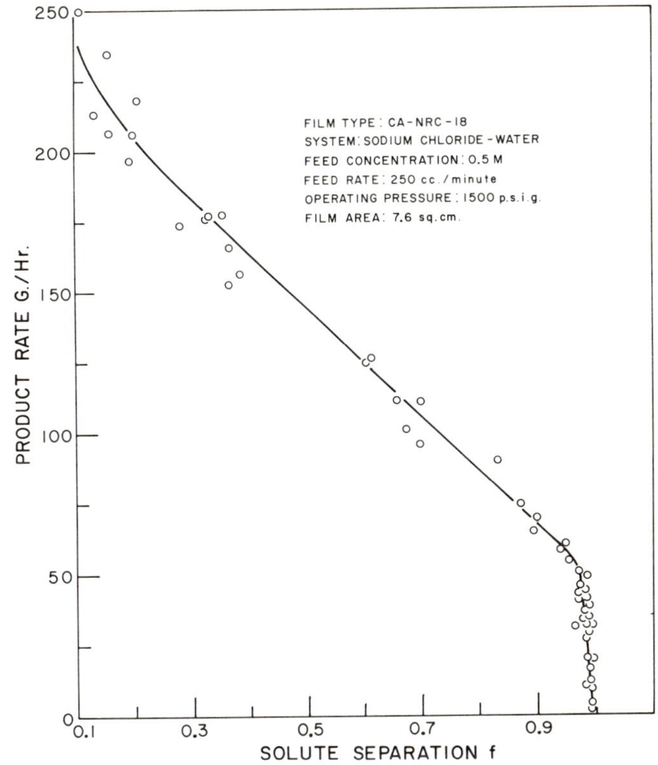

Figure 4. General performance of Batch-18-type porous cellulose acetate membranes (*152*)

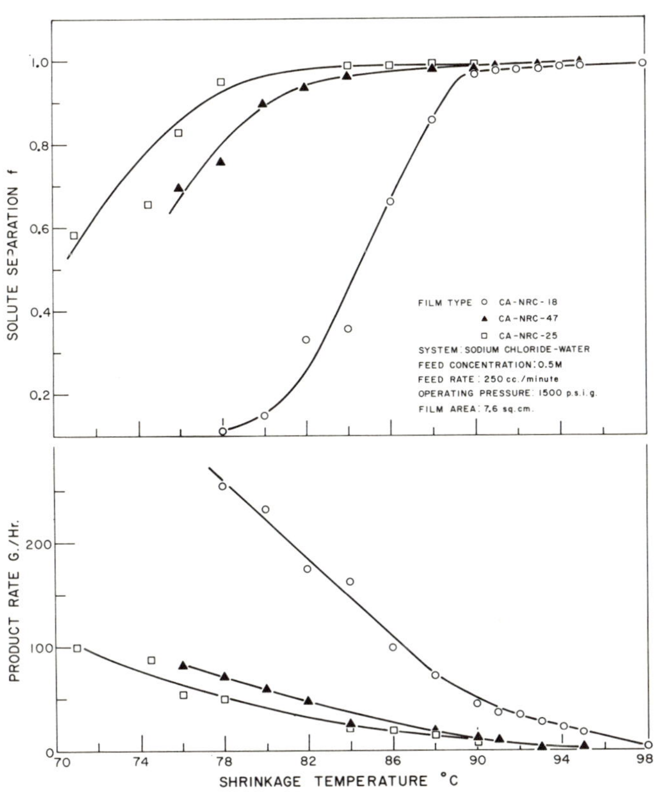

Figure 3. Effect of shrinkage temperature on the performance of different porous cellulose acetate membranes

Data of Kopeček and Sourirajan (*60*)

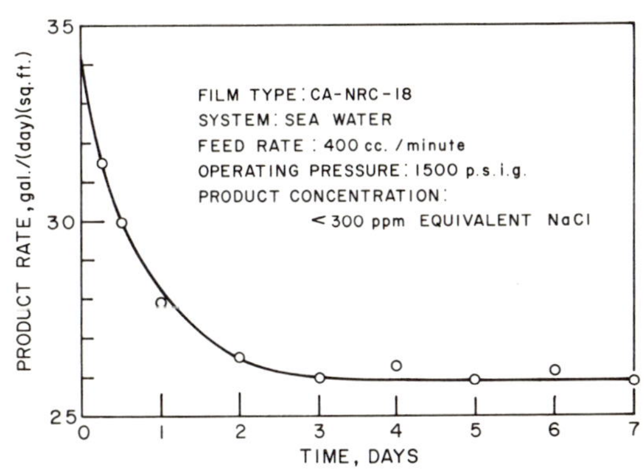

Figure 5. Performance of Batch-18-type porous cellulose acetate membranes for sea water conversion (*152*)

Cellulose acetate material is subject to hydrolysis whose rate at 25°C is least in solutions whose pH is about 5 (*166*, *167*); hence the cellulose acetate membrane may be expected to have the longest life with feed solutions whose pH is about 5.

Tubular membranes. One of the most significant developments in the field of porous membrane technology is the technique of casting the Loeb-Sourirajan-type cellulose acetate membranes in the tubular form. The use of such membranes integrally lined within a porous Fiberglas-reinforced tube was first disclosed by Havens Industries (*39*, *40*); however, no details of the Havens

Table II. Performance of Batch-18-Type Porous Cellulose Acetate Membranes for Saline Water Conversion

(Data of Sourirajan and Govindan (*153*))

Film pretreatment							
Shrinkage temp., °C	Pressure psig	Solute	Feed rate, cc/min	Operating pressure, psig	Solute concn. in feed, wt %	Solute concn. in product, ppm	Product rate, gal/(day)(ft²)
87	700	NaCl	300	600	0.25	400	52.4
88	700	NaCl	300	600	0.50	300	34.0
87	700	Na_2SO_4	300	600	0.25	150	53.3
87	700	Na_2SO_4	300	600	0.50	300	50.7
90	1700	NaCl	250	1500	1.44	<100	38.5
90	1700	NaCl	250	1500	2.84	280	31.3
90	1700	NaCl	250	1500	4.20	420	24.1
90	1700	NaCl	250	1500	5.53	550	18.2

technique were available in open literature. At UCLA, Loeb pioneered the gravity drop technique for making tubular porous cellulose acetate membranes and their subsequent assembly in reverse osmosis pilot plants (*74*). Loeb's technique is a model of simplicity and ingenuity, and it is fully described in the literature (*74–76, 84, 160*).

Other Porous Membranes for Reverse Osmosis Application

A wide variety of other porous membranes having separation characteristics in reverse osmosis has been produced (*8, 31, 113*). Among such membranes, the graphitic oxide membranes developed by Flowers *et al.* (*30*), polysalt complex membranes developed by Michaels *et al.* (*114, 115*), and dynamically formed membranes developed by Kraus *et al.* (*45, 64–66, 104, 123*) and Kuppers *et al.* (*67, 68*) are particularly significant.

Some Basic Questions

From the point of view of the preferential sorption-capillary flow mechanism, there are three basic questions in relation to the full scientific and industrial development of the reverse osmosis process: (1) What are the precise physicochemical criteria of preferential sorption or preferential repulsion at interfaces? (2) What is the mechanism of pore formation in the process of casting high permeability porous membranes? (3) What are the factors governing flow of fluids through porous membranes having separation characteristics?

No unequivocal answer to any of these questions can be given at this time. However, interfacial forces, such as London dispersion forces, dipole-dipole interactions, and hydrogen bonding are obviously of importance in connection with the first question. Glueckauf, for example, analyzed, at least partially successfully, Sourirajan's experimental data on the basis that reverse osmosis desalination was due to the repulsion which ions experienced in the close vicinity of membrane materials of low dielectric constant (*36, 37, 145*).

Even though the steps involved in the process of casting Loeb-Sourirajan-type porous cellulose acetate membranes have been critically examined by several workers (*9–11, 47, 50–52, 165, 168*), the origin of pores in the surface layer of the membrane is not yet well-defined (*59, 108*). Nevertheless, it seems reasonable to expect that further developments in porous membrane technology can lead to the production of films capable of giving at least a ten-fold increase in product rates at any given level of solute separation and set of operating conditions, compared to those obtained with the best present-day films.

The third question is of fundamental importance in engineering. It is concerned ultimately with the problem of predicting membrane performance from a minimum of experimental data. The Kimura-Sourirajan analysis of reverse osmosis experimental data offers a satisfactory solution to the above problem, at least with respect to the Loeb-Sourirajan-type porous cellulose acetate membranes and binary aqueous solution systems, such as sodium chloride–water in concentration ranges of practical interest. Hence this analysis is discussed below in detail.

Some Experimental Details

All the experimental data presented in this paper are for Batch-18-type membranes used in a small laboratory cell (Figure 6) (effective membrane area = 7.6 cm²) with the experimental arrangement shown in Figure 7. The apparatus and experimental procedure employed in obtaining the data are given in detail in the literature (*144, 145, 147*). Each preshrunk film was subjected to a pressure treatment before use in the reverse osmosis experiments; this treatment consisted of pumping distilled water over the surface of the film mounted in the reverse osmosis cell at a pressure about 15% higher than the maximum intended operating pressure, for at least an hour. This treatment helped to stabilize the overall porous structure of the film. In each experiment, the fraction solute separation, f, defined as

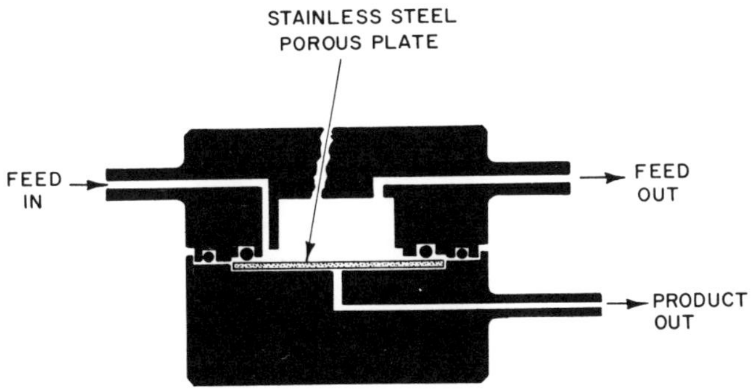

Figure 6. Reverse osmosis cell (152)

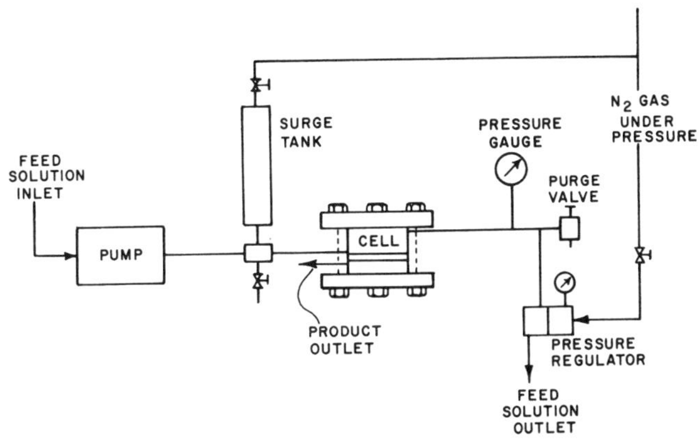

Figure 7. Schematic flow diagram of apparatus (152)

$$f = \frac{\text{molality of feed } (m_1) - \text{molality of product } (m_3)}{\text{molality of feed } (m_1)}$$

the product rate, *PR*, and the pure water permeability, *PWP*, in grams per hour per given area of film surface, were determined at the preset operating conditions of pressure, feed molality, and feed rate. In all cases, the terms "product" and "product rate" refer to the membrane permeated solutions. The concentrations of the solute in the feed and product solutions were determined by refractive index measurements using a Bausch and Lomb refractometer, or by specific resistance measurements using a conductivity cell.

Transport Through Reverse Osmosis Membranes

The following discussion is restricted to isothermal reverse osmosis separation process involving binary aqueous solution systems and Loeb-Sourirajan-type porous cellulose acetate membranes having a preferential sorption for water from the aqueous solution under consideration. Unless otherwise stated, all data presented in this paper are for 25°C.

Osmotic pressure. The osmotic pressure, π, of a solution is given by the relationships (*130*) of:

$$\pi = -\frac{RT}{\bar{V}_w} \ln a_w \tag{1}$$

$$= \frac{\Sigma_i \, RT \, M_B}{1000 \; \bar{V}_w} m\phi \tag{2}$$

The values of π for many solutions can be calculated from the activity (a_w) or the osmotic coefficient (ϕ) data available in the literature (*131*).

Concentration polarization. The existence and the continuous withdrawal of the preferentially sorbed interfacial layer along with the bulk feed solution through the porous membrane give rise to a product solution less concentrated than the feed solution and a more concentrated boundary solution between the interfacial region on the membrane surface and the bulk feed solution. Consequently, under steady-state operating conditions, there arises a concentration gradient between the boundary solution and the bulk feed solution (Figure 8). This concentration gradient, called "concentration polarization," has important effects on the performance of reverse osmosis systems. These effects have been the subject of extensive theoretical studies by several workers (*18–21, 24–26, 29, 32–35, 41, 43, 44, 54, 111–113, 136, 137, 159, 162*).

With respect to a given membrane-solution system, the magnitude of the interfacial region is a function of the concentration of the boundary solution, and the effective driving pressure (ΔP) for fluid flow through the membrane is the operating pressure, P, minus the difference between the osmotic pressure of the concentrated boundary solution, $\pi(X_{A2})$, and that of the membrane permeated product solution, $\pi(X_{A3})$.

Maximum separation. Under the reverse osmosis operating conditions,

$$\Delta P = P - \Delta\pi \tag{3}$$

and

$$\Delta\pi = \pi(X_{A2}) - \pi(X_{A3}) \tag{4}$$

For the limiting case, when the mass transfer coefficient on the high pressure side of the membrane tends to infinity, $\pi(X_{A2})$ approaches the osmotic pressure of the feed solution, $\pi(X_{A1})$.

When P is equal to or less than $\pi(X_{A2})$, Equation 3 also gives the maximum solute separation possible in this process whatever membrane is used, under which condition

$$\Delta\pi = P, \text{ and } \Delta P = 0 \tag{5}$$

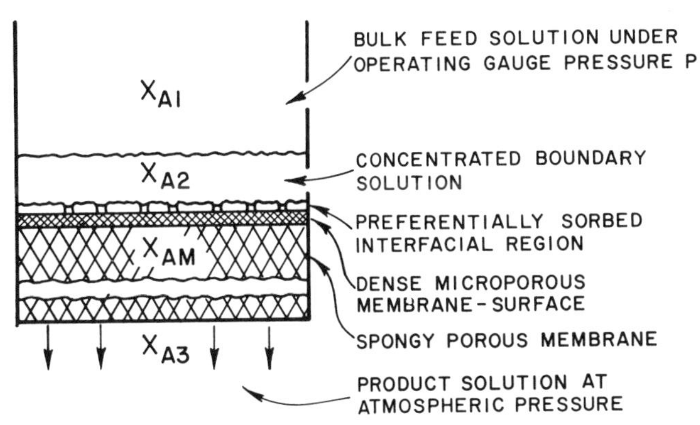

Figure 8. Reverse osmosis process under steady-state conditions (152)

If we use Equation 5 and let $\pi(X_{A2}) = \pi(X_{A1})$, Figure 9 gives the plot of the maximum solute separation possible (f_{max}) for the system sodium chloride–water at different operating pressures and feed concentrations. Figure 9 illustrates the thermodynamic significance of osmotic pressure as applied to this separation process.

When P is equal to or less than $\pi(X_{A1})$, the values of f_{max} given in Figure 9 correspond to the condition that the product rate is zero; when P is greater than $\pi(X_{A1})$, there is no thermodynamic limitation to the extent of solute separation and product rate obtainable in this process.

While Equation 3 is true whatever the value of the osmotic pressure of the feed solution, it says nothing about the separation characteristics of any membrane. In other words, the actual performance of any particular membrane with respect to a given solution system depends, not only on the osmotic pressure of the feed solution, but also on the physical and chemical nature of the membrane.

Kimura-Sourirajan analysis of reverse osmosis experimental data. This analysis (*53, 152*) is applicable for the entire possible range of solute separations in the reverse osmosis process. In this analysis, *PWP* is directly proportional to the operating pressure, and the proportionality constant obtained is called the pure water permeability constant, A; the transport of solvent water, N_B, through the membrane is proportional to the effective pressure where the proportionality constant is A; the transport of solute, N_A, through the membrane is due to pore diffusion through the membrane capillaries and hence proportional to the concentration difference across the membrane; and the mass transfer coefficient, k, applicable to the high pressure side of the membrane is obtained from a straightforward application of the film theory. The basic transport equations can be derived as follows.

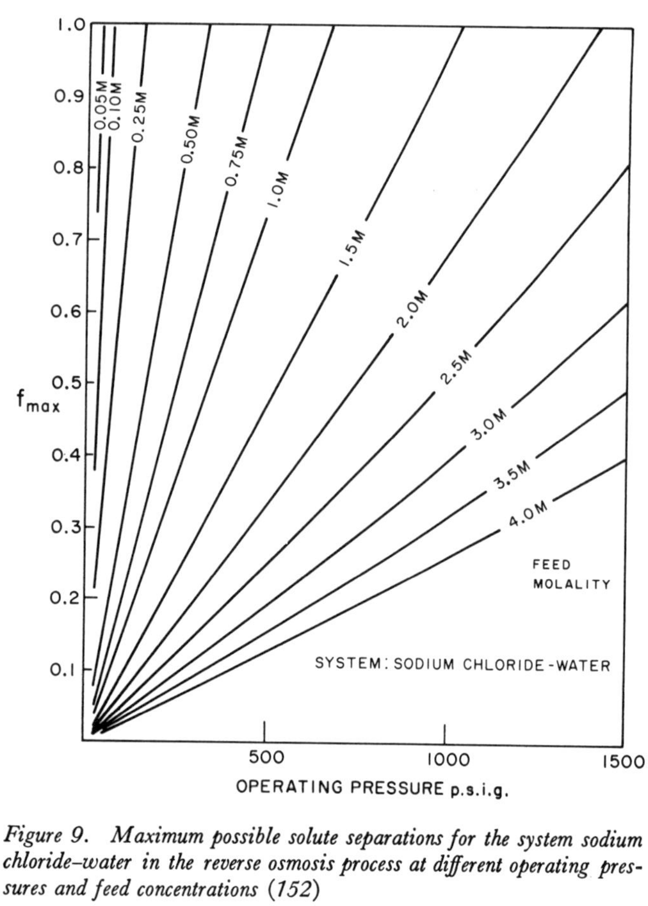

Figure 9. Maximum possible solute separations for the system sodium chloride–water in the reverse osmosis process at different operating pressures and feed concentrations (152)

Pure water permeability constant, A:

$$A = \frac{PWP}{M_B \times S \times 3600 \times P} \qquad (6)$$

In this analysis, A is a fundamental quantity; it is a measure of the overall porosity of the film; it corresponds to conditions of no concentration polarization; and it is independent of any solute under consideration.

Transport of solvent water, N_B, through the porous membrane:

$$N_B = A\Delta P = A\{P - [\pi(X_{A2}) - \pi(X_{A3})]\} \qquad (7)$$

$$= A[P - \pi(X_{A2}) + \pi(X_{A3})] \qquad (8)$$

At a given operating pressure, P, from any set of *PWP*, *PR*, and f data, the values of A, N_B, and X_{A3} are obtained, and X_{A2} can be calculated from Equation 8; further, while A is a constant at a given P, N_B and X_{A3} are functions of feed concentration and feed flow rate (or degree of turbulence on the membrane surface). Equation 8, by itself, does not suggest the existence of any unique relationship (independent of feed concentration and feed flow rate) between X_{A2} and X_{A3}.

Equation 8 is applicable for systems where the kinematic viscosity of the product solution is not too different from that of pure water. This condition is reasonably satisfied in most cases of practical interest.

There is no easy method to determine X_{A2} experimentally. Hence, one cannot be certain that X_{A2} calculated from Equation 8 is the true value of X_{A2}. This uncertainty, however, is no limitation in the Kimura-Sourirajan analysis so long as the correlation between the solute transport parameter (discussed below) and X_{A2} can be specified on the basis of Equation 8.

Transport of solute (N_A) through membrane phase:

$$N_A = \frac{D_{AM}}{\delta}(c_{M2}X_{AM2} - c_{M3}X_{AM3}) \qquad (9)$$

None of the quantities on the right side of Equation 9 are known, and the dividing line in the membrane phase between the regions corresponding to X_{AM2} and X_{AM3} is only conceptual. Equation 9 can be transformed into one containing measurable quantities and a group of unknown quantities by assuming a simple linear relationship between X_A (concentration in the solution phase) and X_{AM} (concentration in the membrane phase).

Thus let

$$cX_A = Kc_MX_{AM} \qquad (10)$$

so that

$$c_2X_{A2} = Kc_{M2}X_{AM2} \qquad (11)$$

and $$c_3X_{A3} = Kc_{M3}X_{AM3} \tag{12}$$

Equation 9 may now be written as

$$N_A = \left(\frac{D_{AM}}{K\delta}\right)(c_2X_{A2} - c_3X_{A3}) \tag{13}$$

Since

$$X_{A3} = \frac{N_A}{N_A + N_B} \tag{14}$$

then,

$$N_A = \left(\frac{X_{A3}}{1 - X_{A3}}\right) N_B \tag{15}$$

Equation 13 can be written as

$$N_B = \left(\frac{D_{AM}}{K\delta}\right)\left(\frac{1 - X_{A3}}{X_{A3}}\right)(c_2X_{A2} - c_3X_{A3}) \tag{16}$$

from which the quantity $(D_{AM}/K\delta)$, called the solute transport parameter, can be calculated from any given set of *PWP*, *PR*, and *f* data at any operating pressure, *P*.

The parameter $(D_{AM}/K\delta)$ plays the role of a mass transfer coefficient with respect to solute transport through the membrane; hence it is treated as a single quantity for purposes of analysis. It must, however, be understood that $(D_{AM}/K\delta)$ is not a single function, but it is a combination of several interrelated functions, none of which is, or need be, precisely known for chemical engineering calculations.

Since X_{A2} and X_{A3} are bound to the membrane phase concentrations by the same relation, Equations 10, 11, and 12 imply that, with respect to any membrane, the values of X_{A2} and X_{A3} must be uniquely related, and this relationship must be independent of feed concentration and feed flow rate past the surface of the membrane. That this is true has been shown experimentally (*57, 119, 154*), as illustrated in Figure 10, for the system sodium chloride–water (*152*).

Mass transfer on the high-pressure side of the membrane: The solute transfer from the concentrated boundary solution may be represented by the relation (*12*)

$$N_A = X_A(N_A + N_B) - D_{AB}c_1\frac{dX_A}{d\zeta} \tag{17}$$

or, by use of Equation 14,

$$\frac{dX_A}{d\zeta} - \frac{(N_A + N_B)}{c_1D_{AB}}X_A = -\frac{(N_A + N_B)}{c_1D_{AB}}X_{A3} \tag{18}$$

where ζ represents normal distance toward membrane measured from edge of the concentrated boundary solution. Equation 17, expressing N_A as the resultant of two vector quantities, is a form of Fick's first law of diffusion. Under steady-state conditions, $(N_A + N_B)$ is the total flux through the membrane, and Equation 17 is applicable at any point in the reverse osmosis system. The boundary conditions for Equation 18 are when

$$\zeta = 0, X_A = X_{A1} \tag{19}$$

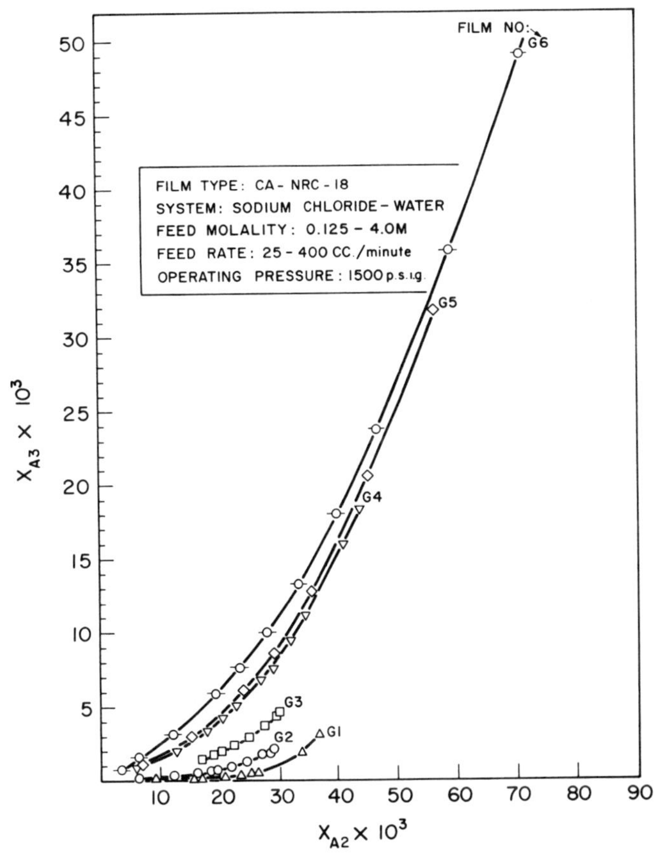

Figure 10. Concentration of solute in the boundary solution vs. that in the product solution for the system sodium chloride–water (152)

and when

$$\zeta = l, X_A = X_{A2} \tag{20}$$

Solving Equation 18:

$$X_{A2} = X_{A3} + (X_{A1} - X_{A3}) \exp\left[\frac{(N_A + N_B)}{c_1}\frac{l}{D_{AB}}\right] \tag{21}$$

or

$$\ln\left[\frac{X_{A2} - X_{A3}}{X_{A1} - X_{A3}}\right] = \frac{(N_A + N_B)}{c_1}\frac{l}{D_{AB}} \tag{22}$$

Defining the mass transfer coefficient, *k*, on the high-pressure side of the membrane in the conventional manner of the film theory (*135, 139, 163*):

$$k = \frac{D_{AB}}{l} \tag{23}$$

and by use of Equations 14 and 15

$$\ln\left[\frac{X_{A2} - X_{A3}}{X_{A1} - X_{A3}}\right] = \frac{N_B}{kc_1(1 - X_{A3})} \tag{24}$$

or

$$N_B = kc_1(1 - X_{A3}) \ln\left[\frac{X_{A2} - X_{A3}}{X_{A1} - X_{A3}}\right] \tag{25}$$

From the foregoing analysis, Equations 6, 8, 16, and 25 emerge as a set of basic equations describing the solute and solvent transport in a reverse osmosis process involving binary aqueous solutions and membranes having a preferential sorption for water from such aqueous solutions. For many solutions of practical interest, such as sodium chloride–water, the molar density (c) may be assumed constant—*i.e.*,

$$c = c_1 = c_2 = c_3$$

so that the basic transport equations in reverse osmosis may be summarized as follows:

$$A = \frac{PWP}{M_B \times S \times 3600 \times P} \tag{6}$$

$$N_B = A[P - \pi(X_{A2}) + \pi(X_{A3})] \tag{8}$$

$$= c\left[\frac{D_{AM}}{K\delta}\right]\frac{(1 - X_{A3})}{X_{A3}}(X_{A2} - X_{A3}) \tag{26}$$

$$= kc(1 - X_{A3}) \ln\left[\frac{X_{A2} - X_{A3}}{X_{A1} - X_{A3}}\right] \tag{27}$$

and

$$N_A = c\left(\frac{D_{AM}}{K\delta}\right)(X_{A2} - X_{A3}) \tag{28}$$

From any set of experimental PWP, PR, and f data at a given P, one can calculate A, $(D_{AM}/K\delta)$, and k, as well as N_A, N_B, and X_{A2}. The following useful relations can be easily derived:

$$f = \frac{m_1 - m_3}{m_1} = 1 - \left(\frac{X_{A3}}{1 - X_{A3}}\right)\left(\frac{1 - X_{A1}}{X_{A1}}\right) \tag{29}$$

and,

$$PR = \frac{N_B \times M_B \times S \times 3600}{\left\{1 - \dfrac{1}{\left[1 + \dfrac{1000}{m_1(1 - f)M_A}\right]}\right\}} \tag{30}$$

Correlations of experimental reverse osmosis data. The correlations of A, $(D_{AM}/K\delta)$, and k with operating pressure and temperature, feed concentration, feed flow rate, the nature of the solute, and membrane compaction during continuous operation are of practical interest from the point of view of specifying membranes and predicting membrane performance under different operating conditions. Extensive experimental studies (*3*, *53*, *55*, *119*, *154*) have established the following correlations for the Loeb-Sourirajan-type porous cellulose acetate membranes for a number of solution systems, such as sodium chloride–water and glycerol–water.

Correlations of A: At a given temperature

$$A = A_0 e^{-\alpha P} \tag{31}$$

and, at a given pressure,

$$A\mu_w = \text{constant} \tag{32}$$

Equations 31 and 32 give the effects of pressure and temperature, respectively, on A, which are illustrated in Figures 11 and 12.

Correlations of $(D_{AM}/K\delta)$: *At a given temperature,*

$$\left(\frac{D_{AM}}{K\delta}\right) \propto P^{-\beta} \tag{33}$$

and, at a given pressure,

$$\left(\frac{D_{AM}}{K\delta}\right) \propto e^{0.005\,T} \tag{34}$$

At a given temperature and pressure, $(D_{AM}/K\delta)$ is independent of feed concentration, feed flow rate, and membrane compaction; these correlations are illustrated in Figures 13 to 17.

Correlations of k: At a given temperature, k is a function of feed flow rate and feed concentration, and essentially independent of operating pressure; a set of such

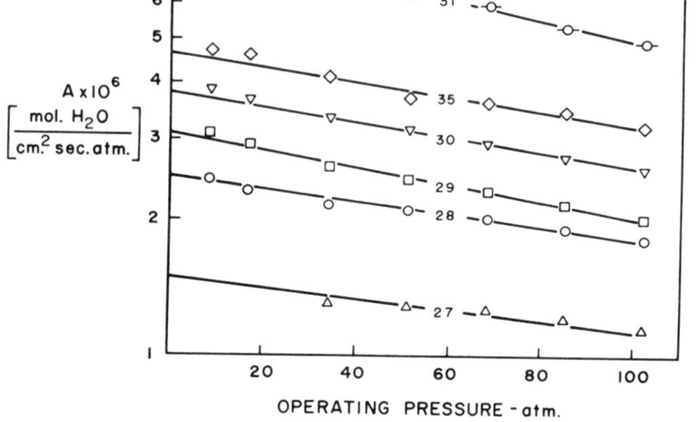

Figure 11. Effect of operating pressure on pure water permeability constant

Data of Sourirajan and Kimura (154)

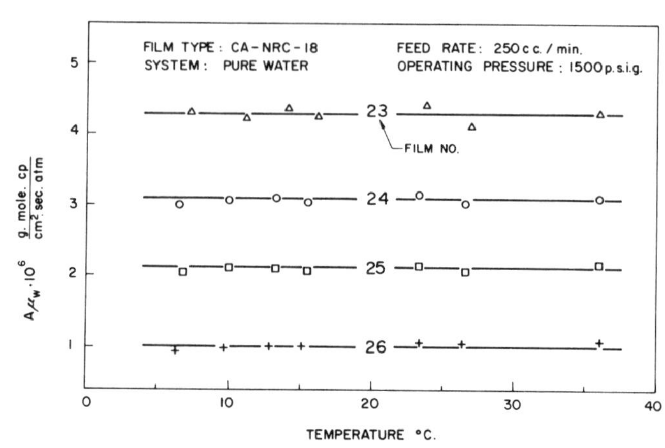

Figure 12. Effect of temperature on pure water permeability constant

Data of Kopeček and Sourirajan (60)

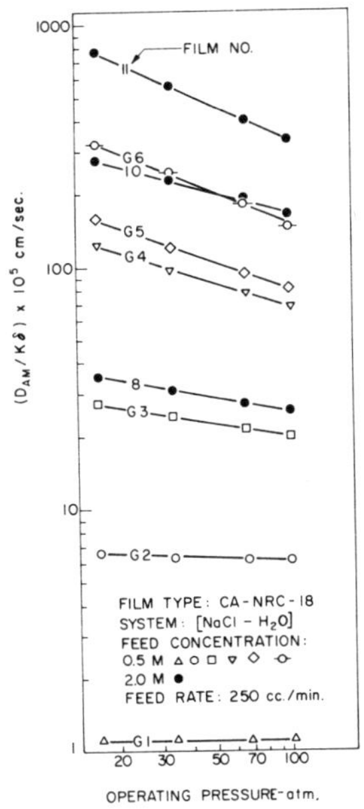

Figure 13. Effect of operating pressure on solute transport parameter for the system sodium chloride–water (152)

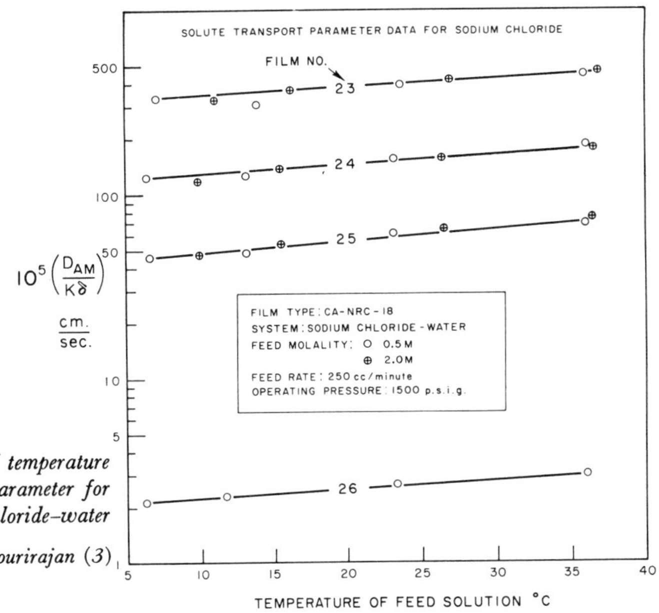

Figure 14. Effect of temperature on solute transport parameter for the system sodium chloride–water

Data of Agrawal and Sourirajan (3)

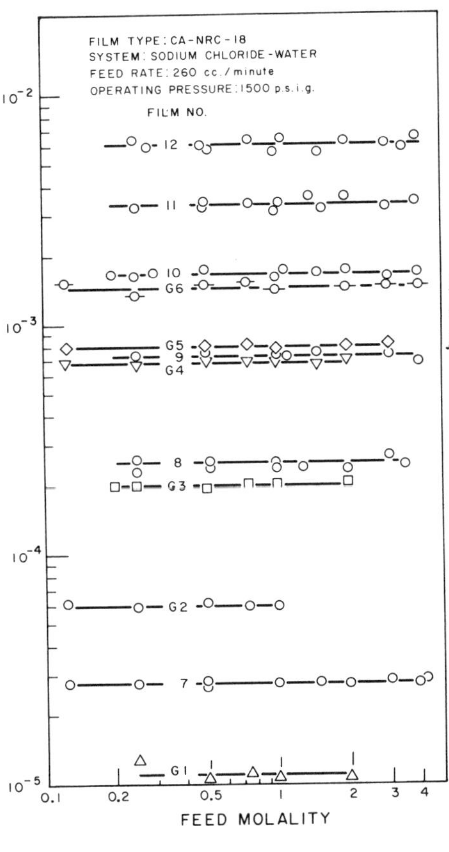

Figure 15. Effect of feed concentration on solute transport parameter for the system sodium chloride–water (152)

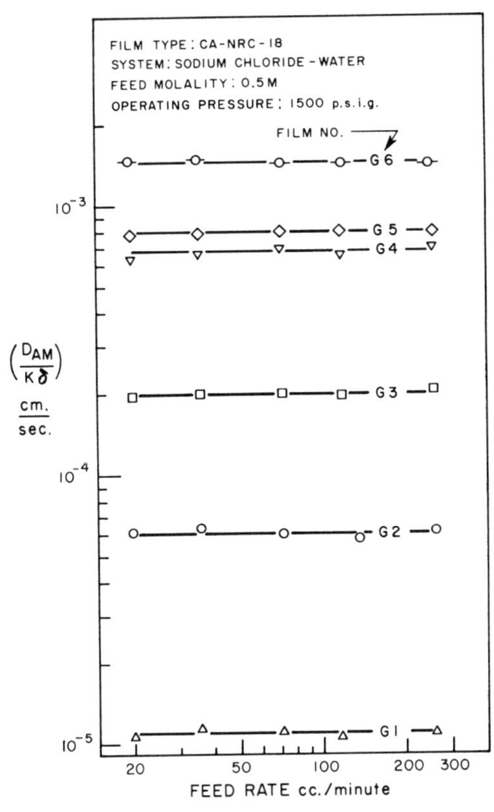

Figure 16. Effect of feed flow rate on solute transport parameter for the system sodium chloride–water (152)

correlations is illustrated in Figure 18 for the system glycerol–water. Further, k is well correlated by a generalized plot of N_{Re} *vs.* $N_{Sh}/N_{Sc}^{0.33}$, which coincides reasonably well with the same plot obtained in diffusion current experiments using the same apparatus (*53*, *154*); the effect of temperature on k is given by the relation

$$k \propto e^{0.005\,T} \tag{35}$$

These latter correlations are illustrated in Figures 19 and 20 for the system sodium chloride–water.

While the general form of the mass transfer coefficient correlations illustrated in Figures 18, 19, and 20 are probably applicable for any operating reverse osmosis system, the specific correlations given in the above figures are applicable only for the apparatus and experimental arrangement used in these studies. It is necessary that such specific correlations have to be experimentally obtained and reported for every other apparatus and experimental arrangement used in reverse osmosis work so that membrane performance data obtained with different operating systems can be compared. Further, the Reynolds numbers given in Figure 19 were obtained using an arbitrary relation [N_{Re} = feed flow rate/(depth of cell) (kinematic viscosity of feed solution)], and hence they do not have any reference to the nature of feed flow (laminar or turbulent) in the cell. The form of the correlations given in Figures 18–20 seem applicable for both laminar and turbulent flow.

Predictability of membrane performance. On the basis of the foregoing transport equations and correlations (which are particularly applicable for the range of feed concentration where the osmotic pressure is directly proportional to mole fraction of solute), a single set of experimental *PWP*, *PR*, and *f* data at any operating pressure and temperature enables the prediction of both solute separation and product rate at that pressure and temperature, for all feed concentrations and feed flow rates for which $(D_{AM}/K\delta)$ remains constant. The procedure for such prediction involves the following steps.

From the experimental data, the values of A and $(D_{AM}/K\delta)$ are calculated using Equations 6, 8, and 26; these values remain constant for all feed concentrations and feed flow rates.

The value of k is calculated for any preset conditions of the experiment from the generalized correlation shown in Figure 19 using the diffusivity data for the solute available from the literature; or k may also be obtained from predetermined experimental data of the type plotted in Figure 18.

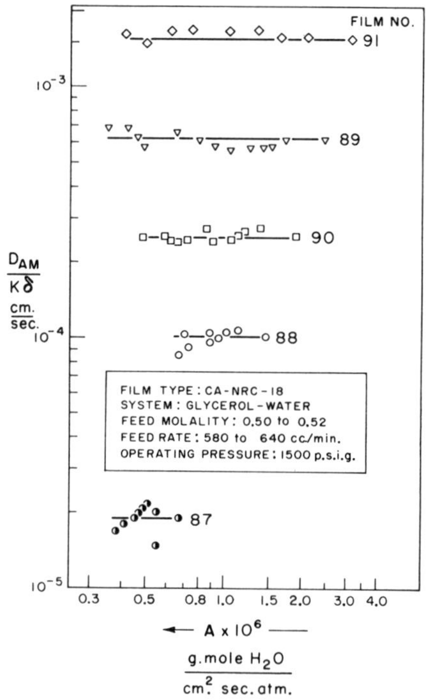

Figure 17. Variation of solute transport parameter with decrease in A during 30-day continuous test runs with the system glycerol-water—effect of membrane compaction on $(D_{AM}/K\delta)$

Data of Kimura and Sourirajan (55)

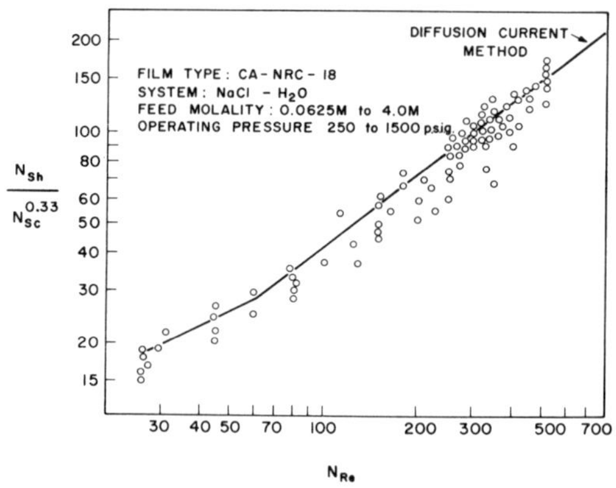

Figure 19. Comparison of experimental data for the system sodium chloride–water with the generalized mass transfer coefficient correlation

Data of Agrawal and Sourirajan (3)

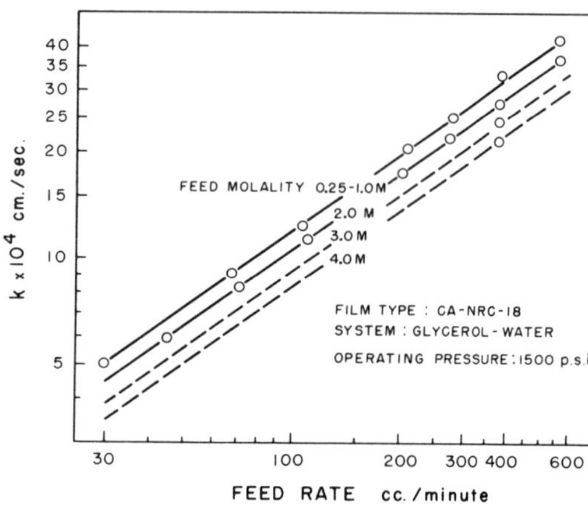

Figure 18. Experimental mass transfer coefficient data for the system glycerol–water

Data of Sourirajan and Kimura (154)

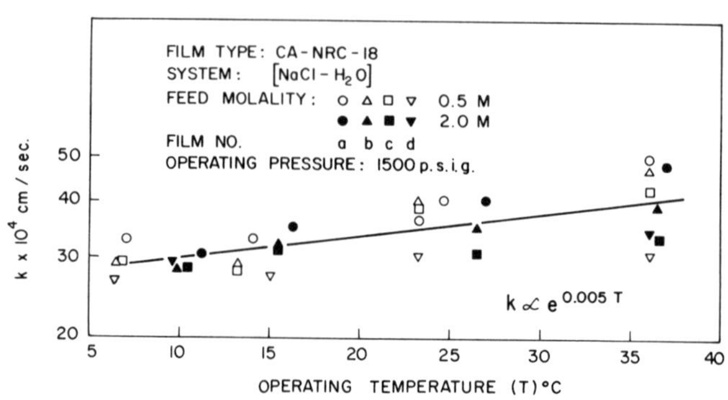

Figure 20. Effect of temperature on mass transfer coefficient for the system sodium chloride–water

Data of Agrawal and Sourirajan (3)

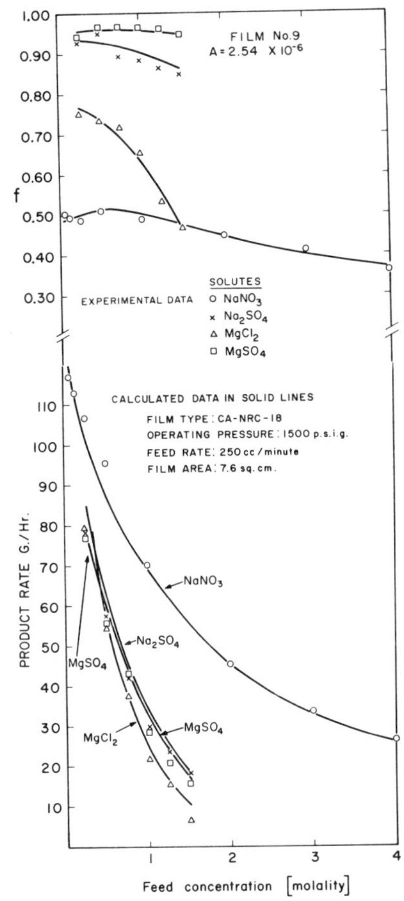

Figure 21. Comparison of experimental and calculated data on membrane performance for different solution systems

Data of Kimura and Sourirajan (53)

Table III. Membrane Specifications

Film Type: CA-NRC-18

Film no.	Solute	Operating pressure, psig	$A \times 10^6$ G-Mol H_2O / cm² sec atm	$(D_{AM}/K\delta) \times 10^5$ cm/sec
9	NaCl	1500	2.54	73.2
F1	NaCl	600	3.773	18.67
F2	NaCl	600	2.670	4.423
F3	NaCl	1500	1.295	0.891
F4	NaCl	600	2.670	5.825

AUTHORS *S. Sourirajan and J. P. Agrawal are with the Division of Chemistry, National Research Council of Canada, Ottawa 7, Canada.*

Combining Equations 26 and 27

$$\left(\frac{D_{AM}}{K\delta}\right)\left(\frac{X_{A2} - X_{A3}}{X_{A3}}\right) = k \ln \left(\frac{X_{A2} - X_{A3}}{X_{A1} - X_{A3}}\right) \quad (36)$$

By trial and error, the combination of values of X_{A2} and X_{A3} which satisfy simultaneously Equation 36 and the equality of Equations 8 and 26 is found, which fixes the value of N_B.

Using the values of X_{A3} and N_B determined above, and the known value of X_{A1} corresponding to the preset operating conditions, solute separation and product rate can be calculated from Equations 29 and 30.

Membrane specifications. The Kimura-Sourirajan analysis discussed above leads to a useful method of membrane specification for isothermal operation. With respect to a given solution system, such as sodium chloride–water, the separation and product rate characteristics of a particular membrane are completely specified for the entire applicable range of operating pressures, feed concentrations, and feed flow rates by specifying the four parameters, α, β, and A and $(D_{AM}/K\delta)$ at a particular operating pressure. The specification of the latter two parameters, only, is sufficient to specify a membrane at that pressure for the entire applicable range of feed concentrations and feed flow rates. The determination of all the four parameters requires the experimental *PWP*, *PR*, and f data at two different operating pressures; the determination of the latter two parameters alone requires only a single set of such experimental data at a specified operating pressure. Figure 21 illustrates a comparison between experimental f and *PR* data with those calculated using the membrane specifications given in Table III and the generalized mass transfer coefficient correlation given in Figure 19.

Effect of membrane compaction. During continuous reverse osmosis operation for extended periods of time, the spongy mass underneath the surface layer of the membrane compacts; under such conditions, A decreases and $(D_{AM}/K\delta)$ remains constant, provided the surface pore structure of the membrane remains the same (*55*). The above experimental observation, along with the generalized mass transfer coefficient correlation (Figure 19), offers a means of predicting membrane performance as a function of A-factor (= value of A at any time/initial value of A) which is a measure of membrane compaction. Some results of such prediction calculations are illustrated in Figures 22 and 23 for a set of films whose specifications are also given in Table III.

Relative scale of membrane selectivity. Between the quantities A and $(D_{AM}/K\delta)$ for a solute which specify a membrane at a given operating pressure, while A may change due to membrane compaction, $(D_{AM}/K\delta)$ remains constant for many solutes as long as the surface pore structure remains the same. Hence, for such solutes,

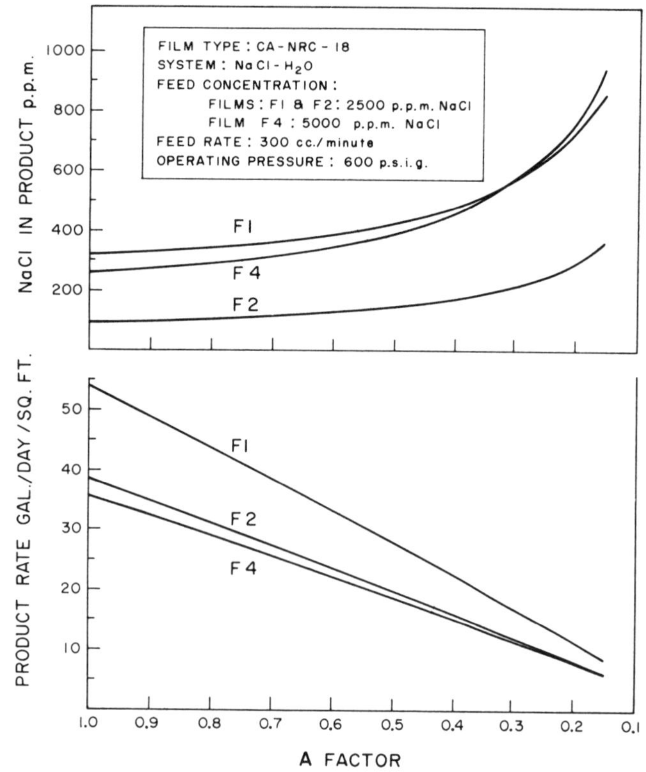

Figure 22. Effect of membrane compaction on performance of porous cellulose acetate membranes suitable for brackish water conversion

Data of Kimura and Sourirajan (55)

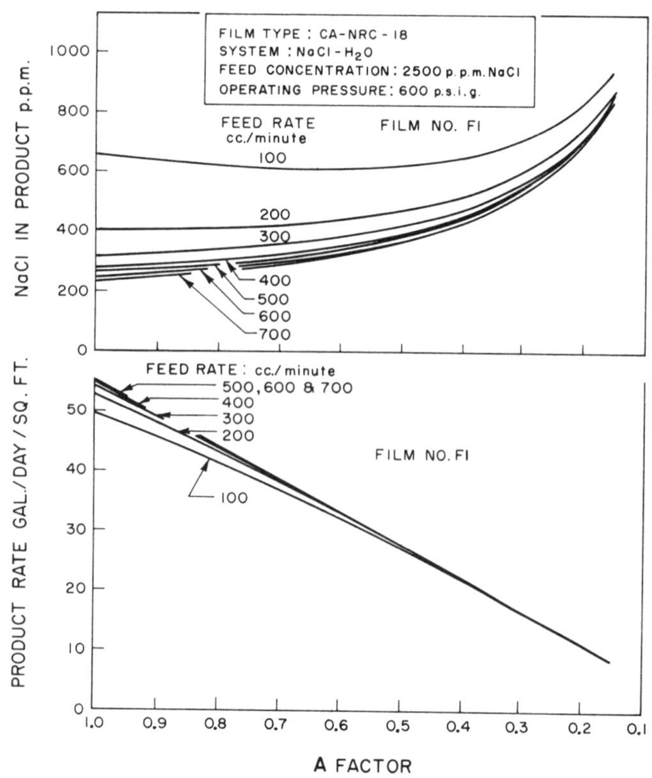

Figure 23. Effect of feed flow rate on the effect of membrane compaction

Data of Kimura and Sourirajan (55)

Table IV. Effect of Feed

Film type: CA-NRC-18
Film area: 7.6 cm²

Film	M1				
$(D_{AM}/K\delta)$ $\times 10^5$ for NaCl cm/sec	1				
Feed molality	0.05	0.1	0.5	1	2
Solute					
$MgSO_4$	99.9	99.9	99.9	99.9	99.8
Na_2SO_4	99.9	99.9	99.9	99.8	...
$BaCl_2$	99.8	99.8	99.7	99.3	...
$CaCl_2$	99.6	99.6	99.4	99.5	...
$MgCl_2$	99.5	99.5	99.3	97.8	...
LiCl	99.3	99.3	99.1	98.7	...
NaCl	99.2	99.2	99.0	98.6	93.4
KCl	99.0	98.9	98.7	98.3	94.6
NH_4Cl	98.6	98.6	98.4	97.8	93.7
$LiNO_3$	98.6	98.6	98.3	97.6	86.3
KNO_3	98.1	98.0	97.8	97.3	...
$NaNO_3$	97.6	97.5	97.1	96.3	92.4
Glycerol	97.2	97.2	97.0	96.6	95.5
Urea	74.25	74.1	73.2	72.1	69.5
Solute					
$MgSO_4$	45.78	45.04	39.01	31.15	11.24
Na_2SO_4	44.98	43.20	32.41	21.69	...
$BaCl_2$	44.80	42.93	28.51	11.98	...
$CaCl_2$	44.78	42.79	27.44	8.85	...
$MgCl_2$	44.72	42.73	26.83	7.12	...
LiCl	45.31	43.83	32.88	20.06	...
NaCl	45.25	43.87	33.55	22.16	3.84
KCl	45.32	43.92	34.01	23.16	6.00
NH_4Cl	45.42	43.89	34.07	23.30	6.49
$LiNO_3$	45.41	43.88	33.20	20.79	2.72
KNO_3	45.33	44.09	35.39	27.10	...
$NaNO_3$	45.36	43.95	34.59	24.75	10.00
Glycerol	45.91	45.19	39.36	32.58	20.72
Urea	46.25	45.60	41.33	36.70	30.00

the variation in the values of $(D_{AM}/K\delta)$ obtained with a given membrane offers a method of expressing membrane selectivity for different solutes. Figure 24 illustrates a relative scale of membrane selectivity for different solutes for the Batch-18-type porous cellulose acetate membranes.

General specification and performance of porous cellulose acetate membrane. Figure 25 gives the A and $(D_{AM}/K\delta)$ data obtained with the system NaCl–H_2O for a number of Batch-18-type porous cellulose acetate membranes. Considering the solid line in Figure 25 as the general specification for the type of films used at the

Concentration on Membrane Performance For Different Solution Systems

Data of Agrawal and Sourirajan (3)

Operating pressure: 1500 psig
Mass transfer coefficient: $k = 50 \times 10^{-4}$ cm/sec

M2							M4						
10							100						
0.05	0.1	0.5	1	2	3	4	0.05	0.1	0.5	1	2	3	4
% Solute Separation													
99.8	99.8	99.7	99.7	99.3	...	...	98.7	98.7	98.7	98.5	96.8	...	...
99.4	99.4	99.3	99.1	...	...	...	96.4	96.4	96.3	95.5	...	...	...
97.9	97.9	97.3	95.0	...	...	...	81.2	81.5	80.2	74.0	...	...	...
97.0	96.9	96.0	91.3	...	...	...	76.0	76.2	74.6	66.0	33.7	...	...
96.8	96.7	95.7	89.5	...	...	...	77.1	77.3	75.4	65.4	29.3	...	...
94.5	94.4	93.5	91.2	73.7	...	...	62.2	62.4	62.3	59.6	48.4	34.5	23.5
94.4	94.3	93.5	91.7	81.0	57.3	...	64.4	64.6	64.6	62.5	54.2	43.1	32.7
92.5	92.4	91.5	89.5	80.8	63.7	47.9	55.4	55.6	56.0	54.8	49.6	42.8	36.0
90.6	90.5	89.4	87.2	78.4	63.1	48.9	49.6	49.8	50.5	49.7	45.7	40.4	35.1
92.6	92.5	91.4	88.8	74.0	49.4	...	62.5	62.7	62.8	60.3	50.6	38.8	29.2
85.9	85.8	84.8	83.2	...	...	...	42.9	43.3	44.0	44.3	...	...	...
88.7	88.6	87.4	85.4	78.5	68.3	58.0	54.5	54.7	55.2	54.4	51.0	46.5	42.2
89.2	89.2	88.8	88.1	85.6	81.3	74.8	60.1	60.3	61.6	62.0	60.9	58.3	54.80
55.5	55.4	55.0	54.4	53.1	51.6	49.9	...	...	...	...	...	...	...
Product Rate Grams per Hour													
86.44	84.39	70.38	52.46	16.74	...	...	163.07	157.20	118.46	78.66	23.40	...	...
84.30	80.19	56.86	36.52	...	...	...	156.45	145.67	94.29	57.99	...	...	...
84.03	79.50	48.98	20.47	...	...	...	157.74	146.27	85.28	45.23	...	...	...
83.98	79.29	46.94	16.23	...	...	...	157.97	147.18	83.87	40.94	8.86	...	...
83.95	79.19	45.70	13.78	...	...	...	157.75	145.98	80.49	34.98	6.66	...	...
85.33	82.17	58.45	35.25	7.55	...	...	162.79	155.67	111.22	75.09	35.85	17.77	9.70
85.40	82.25	60.06	39.16	12.14	3.72	...	162.86	155.89	113.46	81.76	43.77	24.46	14.75
85.47	82.30	61.31	41.81	16.46	6.54	3.36	164.16	157.65	120.47	90.29	57.82	37.90	26.08
85.50	82.40	61.98	42.74	18.11	7.88	4.18	164.40	158.88	124.23	95.01	62.35	42.17	30.02
85.69	82.16	59.31	37.23	10.51	3.57	...	162.91	155.65	111.65	78.25	39.42	21.42	13.11
85.74	83.06	66.14	52.14	...	...	...	165.41	161.02	134.76	121.16	...	...	...
85.61	82.74	63.32	45.98	23.54	13.13	8.12	164.11	158.30	123.31	99.02	67.39	49.20	37.91
87.12	85.29	72.35	58.70	37.63	24.10	14.66	166.92	162.73	135.25	113.12	79.67	57.94	43.15
87.75	86.64	78.73	72.22	60.29	51.21	44.01	...	...	...	...	...	...	...

operating pressure of 1500 psig, one can predict the performance of the films for various solution systems, using the selectivity data given in Figure 24 and the generalized mass transfer coefficient correlation given in Figure 19. A set of such predicted data is given in Table IV, which shows the effects of pore size on the membrane surface, nature of solute, and feed solution molality on membrane performance—*i.e.*, solute separation and product rate. The data given in Table IV show that a higher value of $(D_{AM}/K\delta)$ means a bigger surface pore size, and hence a lower level of solute separation; but the order of solute separation with respect to any two solutes does not always correspond to the order of their $(D_{AM}/K\delta)$ values. At a given operating pressure, the values of $(D_{AM}/K\delta)$ for different solutes studied (Table IV) are independent of feed concentration and feed flow rate; hence, their relative values offer a firm basis for expressing membrane selectivity. The order of solute separation, however, is a function not only of the relative values of $(D_{AM}/K\delta)$ but also of the operating conditions of feed concentration and the mass transfer coefficient, k. Hence, in reverse osmosis, the order of $(D_{AM}/K\delta)$ and that of solute separation with respect to any two solutes must be considered separately.

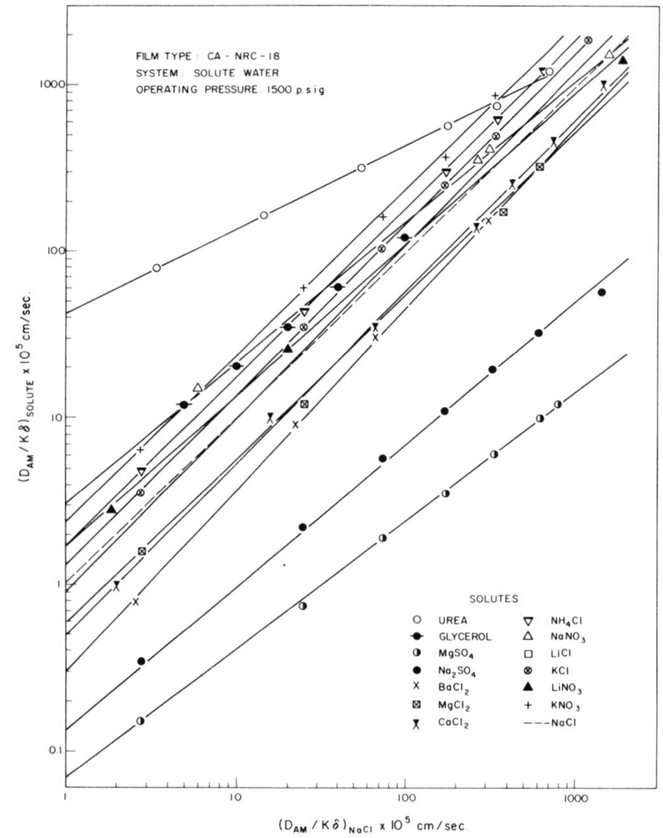

Figure 24. A relative scale of membrane selectivity for different solutes for the Batch-18-type porous cellulose acetate membranes

Data of Agrawal and Sourirajan (3)

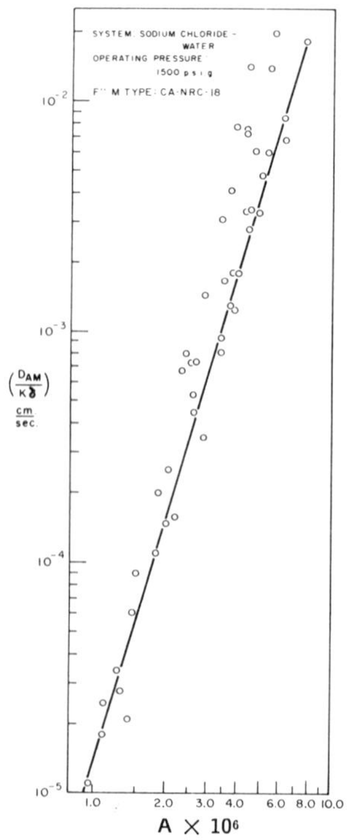

Figure 25. General specification for Batch-18-type porous cellulose acetate membranes

Data of Agrawal and Sourirajan (3)

Reverse Osmosis System Specification and Process Design—Ohya-Sourirajan Analysis

System specification. In any reverse osmosis process, the solution membrane-operating system may be specified by three fundamental nondimensional parameters—γ, θ, and λ—defined as follows:

$$\gamma = \frac{\pi(X_{A1}^0)}{P} = \frac{\text{osmotic pressure of feed solution}}{\text{operating pressure}} \quad (37)$$

$$\theta = \frac{(D_{AM}/K\delta)}{v_w^*} = \frac{\text{solute transport parameter}}{\text{pure water permeation velocity}} \quad (38)$$

and

$$\lambda = \frac{k}{(D_{AM}/K\delta)} = \frac{\text{mass transfer coefficient on the high pressure side of membrane}}{\text{solute transport parameter}} \quad (39)$$

where

$$v_w^* = \frac{AP}{c} \quad (40)$$

and $\pi(X_{A1}^0)$ refers to the osmotic pressure of the feed solution at membrane entrance or at the start of the operation. For reverse osmosis systems so specified, the basic transport equations discussed above can be rewritten in forms suitable for a generalized approach to reverse osmosis process design. This is illustrated by the following analysis (*54*, *58*, *120*).

System analysis. For this purpose, the dimensionless concentration C is defined as

$$C = \frac{X_A}{X_{A1}^0} = \frac{c_A}{c_{A1}^0} \quad (41)$$

so that $C_1 = X_{A1}/X_{A1}^0$ or c_A/c_{A1}^0, $C_2 = X_{A2}/X_{A1}^0$ or c_{A2}/c_{A1}^0, $C_3 = X_{A3}/X_{A1}^0$ or c_{A3}/c_{A1}^0, and so on, and the following approximations are assumed valid with respect to solution properties: (1) the molar density of the solution is essentially constant; (2) the osmotic pressure of solution is proportional to mole fraction of solute—*i.e.*,

$$\pi(X_A) = BX_A \quad (42)$$

and (3) the solute flux across the membrane is small compared to solvent flux, *i.e.*

$$N_A \ll N_B \text{ or } X_{A3} \ll 1 \quad (43)$$

so that

$$X_{A3} = \frac{N_A}{N_A + N_B} \approx \frac{N_A}{N_B} \quad (44)$$

Further, $(D_{AM}/K\delta)$ for the solute is assumed independent of feed concentration and feed flow rate. These restrictions do not limit the scope of the analysis; they only simplify the analytical expressions involved. All the above assumptions are reasonably valid for solutions such as sodium chloride–water in concentration ranges of practical interest.

By use of the above approximations, Equations 8, 26, and 27 may be written as

$$v_w = \frac{N_B}{c} = v_w^*[1 - \gamma(C_2 - C_3)] \tag{45}$$

$$= \left(\frac{D_{AM}}{K\delta}\right)\left[\frac{C_2 - C_3}{C_3}\right] \tag{46}$$

$$= k \ln\left[\frac{C_2 - C_3}{C_1 - C_3}\right] \tag{47}$$

Combining Equations 45 and 46,

$$C_2 = \frac{(1 + \theta)C_3 + \gamma C_3^2}{(\gamma C_3 + \theta)} = C_3\left[1 + \frac{1}{(\gamma C_3 + \theta)}\right] \tag{48}$$

Equation 48 gives the unique relationship between C_2 and C_3 at any point in the system, as illustrated in Figure 10; this relationship is independent of λ. Further, from Equations 46–48,

$$\frac{C_2 - C_3}{C_3} = \lambda \ln\left[\frac{C_2 - C_3}{C_1 - C_3}\right] \tag{49}$$

$$= \frac{1}{\gamma C_3 + \theta} \tag{50}$$

Combining Equations 49 and 50,

$$C_1 = C_3 q \tag{51}$$

where

$$q = 1 + \frac{1}{(\gamma C_3 + \theta)} \exp\left[-\frac{1}{\lambda(\gamma C_3 + \theta)}\right] \tag{52}$$

Equation 51 is an important one; it expresses directly the relationship between C_1 and C_3 at any point in the system as a function of γ, θ, and λ.

From Equations 28, 44, and 45,

$$C_3 = \frac{X_{A3}}{X_{A1}^0} = \frac{\theta(C_2 - C_3)}{[1 - \gamma(C_2 - C_3)]} \tag{53}$$

or

$$\gamma C_3^2 + (1 + \theta - \gamma C_2)C_3 - \theta C_2 = 0 \tag{54}$$

Solving Equation 54 for C_3, and omitting the inapplicable negative root,

$$C_3 = \frac{\sqrt{(1 + \theta - \gamma C_2)^2 + 4\gamma\theta C_2} - (1 + \theta - \gamma C_2)}{2\gamma} \tag{55}$$

Equations 45, 48, 51, and 55 express the relationships among the quantities v_w, C_1, C_2, and C_3 at any point in a reverse osmosis system specified by the parameters γ, θ, and λ. Using the above relationships, Ohya and Sourirajan (*120*) derived the following equations applicable for the batch or flow-type reverse osmosis process design.

Batch process. Let V_1 and C_1 be the volume and concentration, respectively, of the bulk solution on the high pressure side of the membrane, and let V_3 and C_3 be the corresponding values for the membrane permeated product solution on the atmospheric pressure side of the membrane at any time, t. At time $t = 0$,

$$V_1 = V_1^0 \tag{56}$$

and from Equations 41, 45, 48, and 55

$$C_1 = C_1^0 = 1 \tag{57}$$

$$(v_w)^0 = v_w^*[1 - \gamma(C_2^0 - C_3^0)] \tag{58}$$

$$C_2^0 = C_3^0\left[1 + \frac{1}{(\gamma C_3^0 + \theta)}\right] \tag{59}$$

and

$$C_3^0 = \frac{\sqrt{(1 + \theta - \gamma C_2^0)^2 + 4\gamma\theta C_2^0} - (1 + \theta - \gamma C_2^0)}{2\gamma} \tag{60}$$

At any time t,

$$-dV_1 = dV_3 \tag{61}$$

and

$$-d(V_1C_1) = C_3dV_3 = -C_3dV_1 \tag{62}$$

Therefore

$$\frac{dV_1}{V_1} = -\frac{dC_1}{C_1 - C_3} \tag{63}$$

or

$$\ln\frac{V_1}{V_1^0} = -\int_1^{C_1}\frac{dC_1}{C_1 - C_3} = -Z \tag{64}$$

where

$$Z = \int_1^{C_1}\frac{dC_1}{C_1 - C_3} \tag{65}$$

From Equation 51,

$$C_1 - C_3 = C_3(q - 1) \tag{66}$$

By differentiating Equations 51 and 52 and using Equation 66, Z may be expressed as

$$Z = \int_{C_3^0}^{C_3}\left[\frac{\gamma}{\lambda(\gamma C_3 + \theta)^2} - \frac{\gamma}{(\gamma C_3 + \theta)} + \left\{\frac{(\gamma C_3 + \theta)}{C_3}\exp\frac{1}{\lambda(\gamma C_3 + \theta)}\right\} + \frac{1}{C_3}\right]dC_3 \tag{67}$$

The above integration can be done either analytically or numerically.

The permeating velocity of solvent water (v_w) through the membrane can be expressed as

$$v_w = -\frac{1}{S}\frac{dV_1}{dt} \quad (68)$$

Combining Equations 45 and 68,

$$-\frac{1}{S}\frac{dV_1}{dt} = v_w^*[1 - \gamma(C_2 - C_3)] \quad (69)$$

From Equations 64 and 65,

$$V_1 = V_1^0 \exp(-Z) \quad (70)$$

or

$$dV_1 = -V_1^0 \exp(-Z)dZ \quad (71)$$

Combining Equations 69 and 71,

$$\frac{Sv_w^*}{V_1^0}dt = \frac{\exp(-Z)dZ}{1 - \gamma(C_2 - C_3)} \quad (72)$$

Let a dimensionless parameter τ, representing time, be defined as

$$\tau = \frac{Sv_w^* t}{V_1^0} \quad (73)$$

From Equation 72,

$$\tau = \int_{C_3^0}^{C_3} \frac{\exp(-Z)\dfrac{dZ}{dC_3}}{1 - \gamma(C_2 - C_3)} dC_3 \quad (74)$$

From Equation 48,

$$1 - \gamma(C_2 - C_3) = \frac{\theta}{(\gamma C_3 + \theta)} \quad (75)$$

Substituting for dZ/dC_3 from Equation 67,

$$\tau = \int_{C_3^0}^{C_3} \left\{\frac{\gamma}{\lambda(\gamma C_3 + \theta)} - \gamma + \frac{(\gamma C_3 + \theta)}{C_3} \times \left[(\gamma C_3 + \theta) \exp\frac{1}{\lambda(\gamma C_3 + \theta)} + 1\right]\right\} \frac{\exp(-Z)}{\theta} dC_3 \quad (76)$$

The value of τ may be evaluated from Equation 76 by numerical integration as functions of γ, θ, and λ for any assumed value of C_3; for this purpose C_3^0 is obtained from the simultaneous solutions of Equations 59 and 60.

Flow process. Consider the reverse osmosis process for feed flow through a channel. The geometry of the channel is immaterial for this discussion. Let $1/h$ be defined as membrane area per unit volume of channel space. Let $\bar{u}^0$ and $\bar{u}$ be the average velocity of feed at channel entrance and at a longitudinal distance x from channel entrance, respectively. Assume that there is no longitudinal diffusion. From the water and solute material balance over a differential length dx at distance x from channel entrance, the following equations are obtained

$$\bar{u} = \bar{u} + \frac{d\bar{u}}{dx}dx + \left(\frac{v_w}{h}\right)dx \quad (77)$$

Therefore

$$-\frac{d\bar{u}}{dx} = \frac{v_w}{h} \quad (78)$$

$$\bar{u}C_1 = \bar{u}C_1 + \frac{d}{dx}(\bar{u}C_1)\,dx + C_3\left(\frac{v_w}{h}\right)dx \quad (79)$$

Therefore

$$-\frac{d}{dx}(\bar{u}C_1) = \frac{C_3 v_w}{h} = -C_3\frac{d\bar{u}}{dx} \quad (80)$$

From Equation 80,

$$\frac{d\bar{u}}{\bar{u}} = -\frac{dC_1}{C_1 - C_3} \quad (81)$$

The right side of Equation 81 is identical to that of Equation 63.

Let a dimensionless length parameter X be defined as

$$X = \frac{v_w^*}{\bar{u}^0}\frac{x}{h} \quad (82)$$

Comparing the definitions of X (Equation 82) and τ (Equation 73), it is evident that at time $t = 0$, $1/h$ in Equation 82 is the same as S/V_1^0 (membrane area per unit initial volume of feed solution) in Equation 73, and $x/\bar{u}^0$ (distance from channel entrance divided by velocity at channel entrance) in Equation 82 is the same as t (time elapsed from the start of the operation) in Equation 73. Consequently, when (S/V_1^0) and t in the batch process are substituted by $1/h$ and $(x/\bar{u}^0)$, respectively, for the flow process, τ for the batch process corresponds identically to X for the flow process. With the above substitution, the correlations developed for the batch process, when expressed as functions of τ, become applicable for the flow process, when expressed as functions of X.

Basic correlations for reverse osmosis process design. The quantities C_1 (solute concentration in the concentrated effluent from the reverse osmosis unit), C_2 (concentration polarization), C_3 and $\bar{C}_3$ (local and average concentration of the product solution), and τ or X (defined by Equations 73 and 82, respectively), as functions of Δ (fraction product recovery), are the basic correlations of practical interest with reference to the performance of any reverse osmosis system defined in terms

Table V. Reverse Osmosis System Specifications and Performance Data—An Illustration

System specification: $\gamma = 0.3$, $\theta = 0.01$, and $\lambda = 100$

Performance Data (152)

Δ	C_1	C_2	C_3	$\bar{C}_3$	τ or X
0.10	1.1075	1.7682	0.0361	0.0339	0.2010
0.20	1.2408	1.9053	0.0422	0.0365	0.4177
0.30	1.4126	2.0701	0.0513	0.0398	0.6574
0.40	1.6369	2.2700	0.0652	0.0442	0.9292
0.50	1.9513	2.5257	0.0908	0.0507	1.2597
0.60	2.4070	2.8652	0.1474	0.0613	1.7001
0.70	3.1343	3.3894	0.3482	0.0844	2.4700
0.80	4.3640	4.4516	1.2098	0.1611	4.6747
0.90	6.6274	6.6602	3.3602	0.3796	11.1769

of γ, θ, and λ. The quantities $\bar{f}$ (average solute separation) and $\bar{v}_w$ (average velocity of product water) are also of practical interest. The equations establishing τ *vs.* V_1/V_1^0, C_1, C_2, and C_3 correlations are given by the combination of Equations 48, 51, 55, 59, 60, 64, 67, and 76. The quantities Δ, $\bar{C}_3$, $\bar{f}$, and $\bar{v}_w$ can be obtained from the following relationships

$$\Delta = 1 - \frac{V_1}{V_1^0} = 1 - \frac{\bar{u}}{\bar{u}^0} \tag{83}$$

From solute material balance

$$1 = C_1(1 - \Delta) + \bar{C}_3\Delta \tag{84}$$

$$\bar{f} \approx 1 - \bar{C}_3 \tag{85}$$

$$\bar{v}_w = -\frac{1}{S}\frac{\int_{V_1^0}^{V_1} \frac{dV_1}{dt}\,dt}{\int_0 dt} = \frac{V_1^0}{St}\Delta \tag{86}$$

or

$$\frac{\bar{v}_w}{v_w^*} = \frac{V_1^0}{Sv_w^* t}\Delta = \frac{\Delta}{\tau} \tag{87}$$

From the above correlations, it is clear that, for any reverse osmosis system specified by the parameters γ, θ, and λ, if any one of the quantities C_1, C_2, C_3, $\bar{C}_3$, τ or X, and Δ is given, the others are fixed and can be calculated using the equations derived above. A set of such correlations is illustrated in Table V.

Some special cases. The limiting cases of $\lambda = \infty$ and $\gamma = 0$ are of practical interest; the former corresponds to the limiting case of no concentration polarization, and the latter corresponds to the limiting case to which all systems approach when the osmotic pressure of the feed solution becomes small compared to the operating pressure. Some of the general equations of reverse osmosis process design discussed above assume the following forms for the special cases indicated below.

Case 1. All values of γ and θ, and $\lambda = \infty$

$$C_1 = \left[1 + \frac{1}{(\gamma C_3 + \theta)}\right] C_3 \tag{88}$$

$$C_2 = C_1 \tag{89}$$

$$C_2^0 = 1 \tag{90}$$

$$C_3 = \frac{\sqrt{(1 + \theta - \gamma C_1)^2 + 4\gamma\theta C_1} - (1 + \theta - \gamma C_1)}{2\gamma} \tag{91}$$

$$C_3^0 = \frac{\sqrt{(1 + \theta - \gamma)^2 + 4\gamma\theta} - (1 + \theta - \gamma)}{2\gamma} \tag{92}$$

$$\tau = \int_{C_3^0}^{C_3} \left[\frac{(\gamma C_3 + \theta)^2}{C_3\theta} + \frac{1}{C_3}\right] \exp(-Z)dC_3 \tag{93}$$

where

$$Z = \gamma(C_3 - C_3^0) + (1 + \theta)\ln\frac{C_3}{C_3^0} - \ln\left[\frac{C_3 + \frac{\theta}{\gamma}}{C_3^0 + \frac{\theta}{\gamma}}\right] \tag{94}$$

Case 2. $\gamma = 0$, any value of θ and $\lambda = \infty$

$$C_1 = C_3\left[\frac{1 + \theta}{\theta}\right] \tag{95}$$

$$C_2 = C_1 \tag{89}$$

$$C_2^0 = 1 \tag{90}$$

$$C_3^0 = \frac{\theta}{1 + \theta} \tag{96}$$

$$\frac{C_3}{C_3^0} = \left[\frac{1}{1 - \tau}\right]^{\frac{1}{1 + \theta}} \tag{97}$$

$$\Delta = 1 - C_1^{-(1+\theta)} \tag{98}$$

$$\tau = \Delta \tag{99}$$

$$\bar{C}_3 = \frac{1 - (1 - \Delta)^{\frac{\theta}{1+\theta}}}{\Delta} \tag{100}$$

Case 3. $\gamma = 0$, all values of θ and λ

$$C_1 = C_3 \left[1 + \frac{1}{\theta} \exp\left(-\frac{1}{\lambda\theta} \right) \right] \tag{101}$$

$$Z = \left[1 + \theta \exp\left(\frac{1}{\lambda\theta} \right) \right] \ln \frac{C_3}{C_3^0} \tag{102}$$

and

$$\frac{C_3}{C_3^0} = \left[\frac{1}{1 - \tau} \right]^{\frac{1}{1 + \theta \exp\left(\frac{1}{\lambda\theta}\right)}} \tag{103}$$

The foregoing analysis enables reverse osmosis to be treated as a unit operation in chemical engineering.

Some applications. It has been shown (*120*) that, for low values of θ, the system performance data are essentially the same for constant values of $\lambda\theta$; for such cases, $\lambda\theta$ is a useful design parameter. The utility of the general equations of reverse osmosis process design derived above has been illustrated by Ohya and Sourirajan for parametric studies on water desalination (*121*), and for water renovation from aqueous urea solutions in a stagewise process (*119*) using the Loeb-Sourirajan-type porous cellulose acetate membranes. The original papers must be consulted for details of data and analysis. As an example, in the water desalination problem, the power and membrane area requirements have been calculated as functions of operating pressure, $\lambda\theta$, and A-factor for a specified film for obtaining product water containing 500 ppm of salt from 3.5% aqueous sodium chloride feed solutions. The general method of calculation is as follows.

From the membrane specifications given in terms of A, $(D_{AM}/K\delta)$, α, and β, the γ and θ parameters for the reverse osmosis system are calculated for different operating pressures; for each assumed value of $\lambda\theta$, the reverse osmosis system is then completely specified in terms of γ, θ, and λ. From the feed and product concentrations, $\bar{C}_3$ is fixed; the corresponding Δ and τ values are then calculated from the general equations discussed above. The work and membrane area requirements are calculated using the relations:

$$\left.\begin{array}{l}\text{work required per unit vol-}\\ \text{ume of product recovered}\end{array}\right\} = \frac{P\Delta + (1 - \eta)P(1 - \Delta)}{\Delta} \tag{104}$$

where η represents fraction power recovery from the high pressure concentrated brine, and

$$S = \frac{\tau}{v_w^*} \frac{V_1^0}{t} \tag{105}$$

where (V_1^0/t) = (product rate/Δ). A set of results of Ohya and Sourirajan (*121*) is illustrated in Figures 26 and 27. Such results point out, among other things, the importance of reducing membrane compaction and maintaining a high value of λ, and hence a high value of k, to obtain the best membrane performance. Methods of calculating k, and hence λ, for the reverse osmosis process, have been discussed by Kimura and Sourirajan (*56*) for simple membrane geometries. Extensive experimental data on mass transfer coefficients, obtainable from different apparatus configurations, are very much needed for process design calculations.

Effect of Longitudinal Diffusion in Reverse Osmosis

Longitudinal diffusion has only a second order, and often negligible, effect on membrane performance in the reverse osmosis process.

Ohya and Sourirajan (*122*) have derived the analytical expressions describing the effect of longitudinal mixing in reverse osmosis. Their basic equations are given below.

Let L be the longitudinal length over which mixing is considered, and let a parameter M be defined as

$$M = \frac{h}{v_w^*} \frac{\bar{u}^0}{L} \tag{106}$$

so that

$$XM = \frac{x}{L} \tag{107}$$

At fluid inlet $XM = 0$, and at fluid exit $XM = 1$. Let E be the average longitudinal diffusion coefficient (assumed constant), and $\bar{u}$ be the fluid velocity at any longitudinal position x. From material balance considerations, the following equations can be derived:

$$-E \frac{d^2C_1}{dx^2} + \bar{u} \frac{dC_1}{dx} = -(C_1 - C_3) \frac{d\bar{u}}{dx} \tag{108}$$

$$-\frac{d\bar{u}}{dx} = \frac{v_w}{h} = \frac{v_w^*}{h} [1 - \gamma(C_2 - C_3)] \tag{109}$$

Let longitudinal diffusion and fluid velocity be expressed in dimensionless forms as follows:

$$PeB = \frac{\bar{u}^0 L}{E} \tag{110}$$

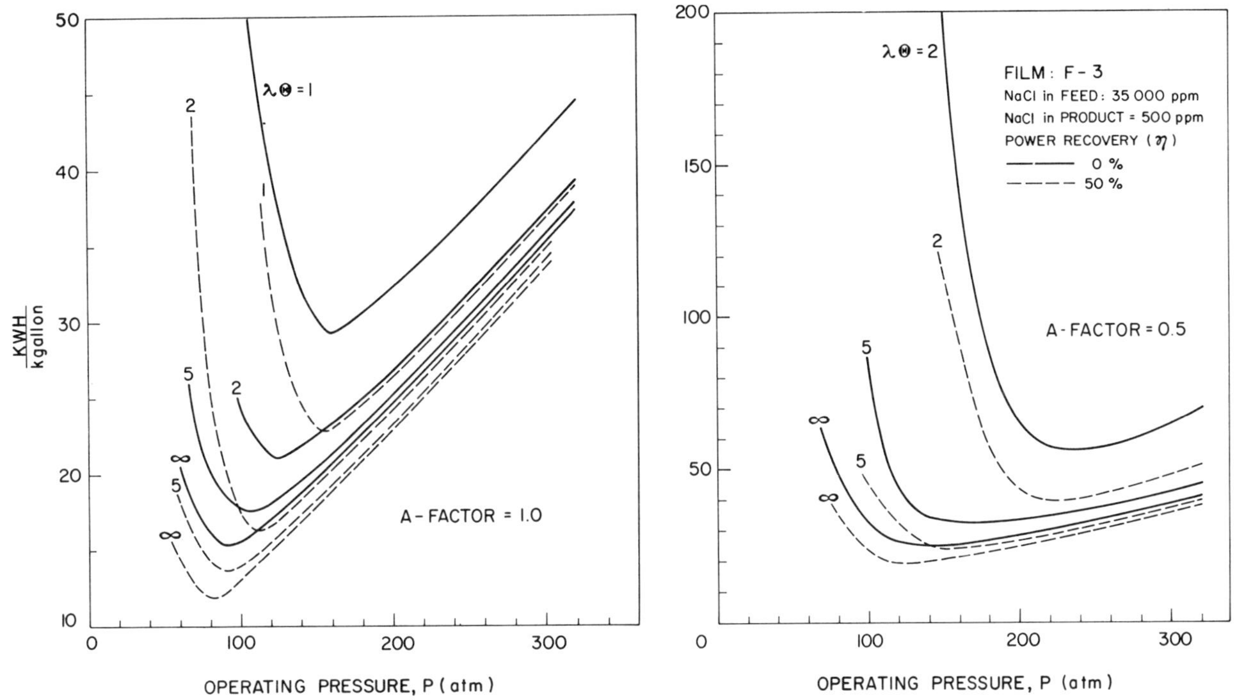

Figure 26. Power requirements at different operating pressures using Film F3 for obtaining product water containing 500 p.p.m. of NaCl from 3.5% aqueous sodium chloride solutions

Data of Ohya and Sourirajan (121)

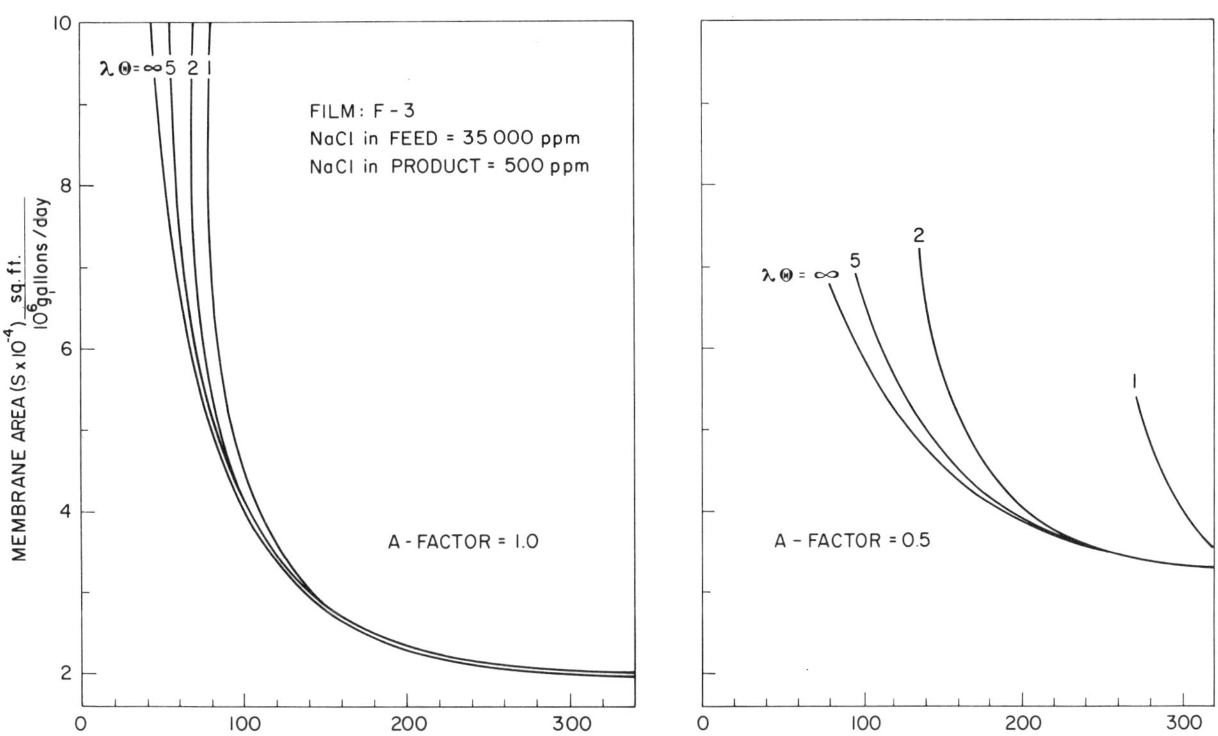

Figure 27. Membrane area requirements for the case represented in Figure 24

Data of Ohya and Sourirajan (121)

$$\bar{U} = \frac{\bar{u}}{\bar{u}^0} \tag{111}$$

The limiting values of $PeB = 0$ and ∞ correspond, respectively, to complete mixing and zero longitudinal diffusion.

Changing the variables, Equations 108 and 109 can be written as

$$-\frac{d^2C_1}{d(XM)^2} + PeB\bar{U}\frac{dC_1}{d(XM)} = -PeB(C_1 - C_3)\frac{d\bar{U}}{d(XM)} \tag{112}$$

$$-\frac{dU}{d(XM)} = \frac{\theta}{M(\gamma C_3 + \theta)} \tag{113}$$

On integration, Equation 113 becomes

$$\bar{U} = 1 - \int_0^{XM} \frac{\theta}{(\gamma C_3 + \theta)} \frac{d(XM)}{M} \tag{114}$$

The boundary conditions of Equation 112 are

at

$$XM = 0 \text{ (fluid inlet)}, \quad -\frac{dC_1}{d(XM)} = PeB(1 - C_1) \tag{115}$$

at

$$XM = 1 \text{ (fluid exit)}, \quad -\frac{dC_1}{d(XM)} = 0 \tag{116}$$

It should be noted that $C_1 \neq 1$ at fluid inlet due to longitudinal diffusion. Equations 115 and 116 follow from simple material balance considerations. Equation 112 is nonlinear and it can be solved numerically on a computer to obtain the values of C_1 and (XM) for specified values of PeB and M for any reverse osmosis system; the corresponding Δ, C_3, and $\bar{C}_3$ values can then be obtained from Equations 114, 51, and 84, respectively; some examples of such calculations are given by Ohya and Sourirajan (*122*).

With the present state of knowledge of the reverse osmosis process, no definite statement can be made on the value of E to be used in any practical case. If only the applicable value of E, and hence PeB, is known, its effect on membrane performance can be predicted exactly on the basis of the above analysis.

Reverse Osmosis Process as a General Concentration Technique

The development of the reverse osmosis process as a practical concentration technique has many potential applications in the fields of food processing, waste recovery, and water pollution control (*6, 7, 38, 42, 57, 99, 106, 123, 132, 148, 156, 169, 170*). Some useful parameters of process design for such applications have been developed by Sourirajan and Kimura (*154*). Their analysis is based on the availability of experimental solute separation and product rate data as a function of feed solution molality; no osmotic pressure data are needed to obtain the design parameters.

Volume change and solute recovery during concentration in batch process. Let the process be carried out at constant operating pressure, and let solute separation be some function of solution molality represented by the equation

$$f = f(m) \tag{117}$$

At any time, for the differential flow of water and solute across the membrane,

$$[1 - f(m)]\frac{dW_w}{W_w} = \frac{dW_s}{W_s} \tag{118}$$

where W_w and W_s are the weight in grams of water and solute, respectively, on the high pressure side of the membrane. Since

$$m = \frac{W_s}{M_s}\frac{1000}{W_w} \tag{119}$$

$$\frac{dW_s}{W_s} = \frac{dm}{m} + \frac{dW_w}{W_w} \tag{120}$$

Combining Equations 119 and 120

$$-\frac{dW_w}{W_w} = \frac{dm}{m\,f(m)} \tag{121}$$

which, on integration, gives

$$\ln\frac{(W_w)_i}{(W_w)_f} = \int_{(m)_i}^{(m)_f} \frac{dm}{m\,f(m)} \tag{122}$$

where suffixes i and f represent, respectively, the initial and final states of the given feed solution on the high pressure side of the membrane.

When we integrate Equation 120, the fraction solute recovery is given as

$$\ln\frac{(W_s)_f}{(W_s)_i} = \ln\frac{(m)_f}{(m)_i} - \int_{(m)_i}^{(m)_f} \frac{dm}{m\,f(m)} \tag{123}$$

Also, since

$$(V)_i = \frac{(W_w)_i + (W_s)_i}{(\rho_1)_i} \tag{124}$$

and

$$(V)_f = \frac{(W_w)_f + (W_s)_f}{(\rho_1)_f} \tag{125}$$

where $(\rho_1)_i$ and $(\rho_1)_f$ denote the initial and final density of the feed solution being concentrated,

$$\frac{(W_w)_i}{(W_w)_f} = \frac{(V)_i}{(V)_f}\frac{(\rho_1)_i}{(\rho_1)_f}\frac{\left(1 + \frac{(m)_f M_s}{1000}\right)}{\left(1 + \frac{(m)_i M_s}{1000}\right)} \tag{126}$$

Combining Equations 122 and 126

$$\ln\frac{(V)_i}{(V)_f} = \int_{(m)_i}^{(m)_f}\frac{dm}{m\,f(m)} - \ln\left\{\frac{(\rho_1)_i\left[1 + \frac{(m)_f M_s}{1000}\right]}{(\rho_1)_f\left[1 + \frac{(m)_i M_s}{1000}\right]}\right\} \tag{127}$$

Thus, the variations of volume change $(V)_i/(V)_f$, water removal $(W_w)_f/(W_w)_i$, and solute recovery $(W_s)_f/(W_s)_i$ with the concentration ratio $(m)_f/(m)_i$ are expressed by Equations 127, 122, and 123, respectively. The right side of Equation 122 can be evaluated either analytically or graphically from the experimental f *vs.* m data.

Processing capacity of membrane for solute concentration in batch process. The processing capacity of a membrane, $[=(V)_i/S\cdot t]$, is defined as the volume of charge (feed solution) that 1 ft^2 of the film surface can handle per day in a batch concentration process to increase the solute molality from $(m)_i$ to $(m)_f$ at a given operating pressure.

Let W and z represent, respectively, the total weight of solution and the weight fraction of solute on the high pressure side of the membrane; and, let Q and a represent, respectively, the product rate in weight units per unit area per unit time (t), and the weight fraction of solute in the membrane permeated product. Then

$$Q = -\frac{1}{S}\frac{dW}{dt} \tag{128}$$

and

$$Qa = -\frac{1}{S}\frac{d}{dt}(Wz) \tag{129}$$

Substituting Equation 128 in Equation 129, and differentiating the latter, we get

$$-\frac{dW}{W} = \frac{dz}{z-a} \tag{130}$$

which, on integration, gives

$$\ln\frac{(W)_i}{W} = \int_{(z)_i}^{z}\frac{dz}{z-a} \tag{131}$$

or

$$W = (W)_i \exp\left[-\int_{(z)_i}^{z}\frac{dz}{z-a}\right] \tag{132}$$

Equation 128 can then be written as

$$Q = -\frac{(W)_i}{S}\frac{d}{dt}\left\{\exp\left[-\int_{(z)_i}^{z}\frac{dz}{z-a}\right]\right\} \tag{133}$$

or

$$\frac{Sdt}{(W)_i} = -\frac{1}{Q}\,d\left\{\exp\left[-\int_{(z)_i}^{z}\frac{dz}{z-a}\right]\right\} \tag{134}$$

Integrating Equation 134

$$\frac{S\cdot t}{(W)_i} = \int_{(z)_i}^{(z)_f}\frac{\exp\left[-\int_{(z)_i}^{z}\frac{dz}{z-a}\right]}{Q(z-a)}dz \tag{135}$$

Since

$$(W)_i = (V)_i(\rho_1)_i \tag{136}$$

$$\frac{(V)_i}{S\cdot t} = \frac{1}{(\rho_1)_i\int_{(z)_i}^{(z)_f}\frac{\exp\left[-\int_{(z)_i}^{z}\frac{dz}{z-a}\right]}{Q(z-a)}dz} \tag{137}$$

For the special case $a = 0$ (complete solute separation), Equation 137 becomes

$$\frac{(V)_i}{S\cdot t} = \frac{1}{(\rho_1)_i(z)_i\int_{(z)_f}^{(z)_i}\frac{1}{Q}d\left(\frac{1}{z}\right)} \tag{138}$$

The application of Equations 122, 123, 127, 137, and 138 for the reverse osmosis concentration process has been illustrated in the literature (*148*, *154*) using the experimental data for the systems sucrose–water and glycerol–water.

Reverse Osmosis Separation of Mixed Solutes in Aqueous Solution

This subject is obviously of great practical importance, in view of the fact that most natural waters and industrial aqueous solutions contain more than one solute. The early investigations on the subject (*27*, *28*, *143*, *145*) have been mostly qualitative in nature. Agrawal and Sourirajan (*4*) have recently analyzed the reverse osmosis experimental data obtained with mixed solute systems. Their analysis leads to a simple method of predicting membrane performance for aqueous feed solutions containing two inorganic salts with a common ion, from the specifications of the membrane given in terms of A and the respective $(D_{AM}/K\delta)$ values along with the mass transfer coefficient correlation applicable for single solute systems.

Agrawal-Sourirajan analysis. Let $A1$–$A2$–H_2O represent the mixed solute system where $A1$ and $A2$ are the solutes under consideration, containing a common ion. The analysis is based on the experimental fact that the graph of product rate at total molality *vs.* fractional molality in the mixed system (*e.g.*, molality of $A2$/total molality) is a straight line at moderate total molalities; for the system $NaNO_3$–$NaCl$–H_2O for example, the above correlation is true for total molalities up to $2M$. The above correlation offers a means of calculating both product rate and solute separations for mixed solute systems using the following relationships.

From the solute separation and product rate data obtained for the single solute systems ($A1$–H_2O) and ($A2$–H_2O) at the given total molality, N_{A1} and N_{B1}, and N_{A2} and N_{B2} can be calculated. Defining

$$x_1 = \frac{(m_1)_{A1}}{(m_1)_{A1} + (m_1)_{A2}} \tag{139}$$

and

$$x_2 = \frac{(m_1)_{A2}}{(m_1)_{A1} + (m_1)_{A2}} \tag{140}$$

The product rate correlation may be expressed as

$$PR_{\text{mix}} = x_1 PR_{A1} + x_2 PR_{A2} \tag{141}$$

where PR_{mix}, PR_{A1}, and PR_{A2} represent, respectively, the product rate for the mixed solute system, and the product rates for the single solute systems ($A1$–H_2O) and ($A2$–H_2O) at the given operating pressure using the same total feed molality. Since product rate is proportional to the effective pressure, Equation 141 means that, in a mixed solute system involving no ionic interference, the difference between the osmotic pressure of the boundary solution on the high pressure side of the membrane and that of the product solution on the atmospheric pressure side of the membrane is a linear function of the corresponding values for the single solute systems at constant total feed concentration.

Since

$$PR_{A1} = k_{A1} N_{A1} + k_B N_{B1} \tag{142}$$

and

$$PR_{A2} = k_{A2} N_{A2} + k_B N_{B2} \tag{143}$$

Table VI. Reverse Osmosis Data for Some Mixed Solute Systems

Film type : Batch 18
System type : A1-A2-H_2O
Solute molality in feed : $(m_1)_A$
Feed rate : 360 cc per minute
Operating pressure : 1500 psig
Film area : 7.6 cm^2

Run no.	Film no.	Total feed molality	$\frac{(m_1)_A}{(m_1)_{A1} + (m_1)_{A2}}$		Solute Separation				Product rate, g/hr	
					A1		A2			
			A1	A2	Exptl.	Calcd.	Exptl.	Calcd.	Exptl.	Calcd.
				System: KCl–$NaCl$–H_2O						
570	22	2.0	0.8	0.2	64.7	63.6	72.3	71.0	37.2	37.6
570	26	2.0	0.8	0.2	43.8	43.2	52.4	54.0	44.0	44.4
571	22	2.0	0.6	0.4	62.0	61.8	68.9	69.6	35.5	35.8
571	26	2.0	0.6	0.4	42.8	40.8	50.8	52.1	41.8	42.5
572	22	2.0	0.4	0.6	59.8	59.9	65.2	68.1	32.8	34.0
572	26	2.0	0.4	0.6	37.0	38.1	45.2	49.9	39.8	40.6
573	22	2.0	0.2	0.8	59.1	57.7	63.7	66.3	32.2	32.2
				System: $NaNO_3$–$NaCl$–H_2O						
552	22	3.03	0.505	0.495	42.3	39.6	57.2	57.5	24.1	23.9
552	25	3.03	0.505	0.495	16.8	13.9	22.3	19.5	63.1	61.6
556	25	3.03	0.262	0.738	13.3	13.4	18.3	16.3	60.3	61.0
563	22	3.08	0.746	0.254	48.2	47.0	60.9	61.5	26.8	26.2
563	25	3.08	0.746	0.254	16.4	14.8	21.5	18.0	63.2	62.8
563	26	3.08	0.746	0.254	34.5	32.8	45.8	49.4	33.3	32.6

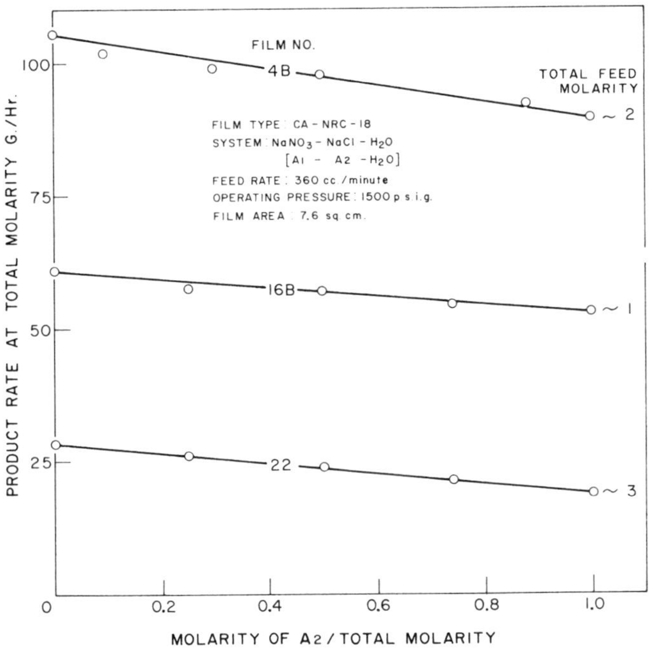

Figure 28. Product rate vs. fractional solute molarity correlation for the system sodium nitrate–sodium chloride–water

Therefore

$$PR_{\text{mix}} = x_1 k_{A1} N_{A1} + x_2 k_{A2} N_{A2} + k_B(x_1 N_{B1} + x_2 N_{B2}) \quad (144)$$

The molalities of the respective solutes in the product solution from the mixed solute system are then given by

$$(m_3)_{A1} = 1000 \times \frac{x_1 k_{A1} N_{A1}}{M_{A1} k_B (x_1 N_{B1} + x_2 N_{B2})} \quad (145)$$

$$= 55.5 \times \frac{x_1 N_{A1}}{x_1 N_{B1} + x_2 N_{B2}} \quad (146)$$

$$(m_3)_{A2} = 1000 \times \frac{x_2 k_{A2} N_{A2}}{M_{A2} k_B (x_1 N_{B1} + x_2 N_{B2})} \quad (147)$$

$$= 55.5 \times \frac{x_2 N_{A2}}{x_1 N_{B1} + x_2 N_{B2}} \quad (148)$$

using the relations

$$\frac{k_{A1}}{k_B} = \frac{M_{A1}}{M_B} \quad (149)$$

$$\frac{k_{A2}}{k_B} = \frac{M_{A2}}{M_B} \quad (150)$$

and

$$\frac{1000}{M_B} = 55.5 \quad (151)$$

The solute separations in the mixed solute systems can then be calculated from the relations:

$$\text{separation with respect to } A1 = \frac{(m_1)_{A1} - (m_3)_{A1}}{(m_1)_{A1}} \quad (152)$$

and

$$\text{separation with respect to } A2 = \frac{(m_1)_{A2} - (m_3)_{A2}}{(m_1)_{A2}} \quad (153)$$

Experimental product rate and solute separation data for the system $NaNO_3$–NaCl–H_2O at total molality 1*M* and 2*M* and the systems NaCl–$MgCl_2$–H_2O and NaCl–$BaCl_2$–H_2O at total molalities of 1*M*, have been reported (*4*), and they are in good agreement with those calculated using Equations 141, 146, 148, 152, and 153. Similar data for the system KCl–NaCl–H_2O (total molality ~2*M*) are given in Table VI.

Prediction technique. The foregoing analysis and the agreement between the experimental and calculated separation and product rate data offer a simple means of predicting membrane performance for moderate concentrations of aqueous feed solution systems involving two inorganic solutes with a common ion. For such systems, the prediction technique is as follows:

From the specifications of the membrane given in terms of A and $(D_{AM}/K\delta)$, and the applicable mass transfer coefficient correlation, calculate solute separation and product rate for the single solute systems at the given total molality following the Kimura-Sourirajan analysis.

Calculate N_{A1}, N_{B1}, N_{A2}, and N_{B2} from the above solute separation and product rate data for the single solute systems.

Calculate the product rate for the mixed solute system using Equation 141.

Finally, calculate solute separations with respect to each solute using Equations 146, 148, 152, and 153.

Some experimental results at high total concentrations. To test the validity of the above analysis for high feed concentrations, the system $NaNO_3$–NaCl–H_2O was studied recently at a total feed molality of ~3*M*. It was found that, under such conditions, Equation 141 was valid only when x_1 and x_2 (Equations 139 and 140) were defined in terms of molarity, instead of molality, as illustrated in Figure 28. Since molarity and molality are not too different up to about 2*M* for the systems $NaNO_3$–H_2O and NaCl–H_2O, it is probably more general to express the product rate correlation in terms of molarities, as illustrated in Figure 28. Under such conditions, the prediction procedure is slightly different and it is illustrated below with some numerical data.

First, calculate the effective total molality for the single solute systems corresponding to the total molarity of the mixed solute system. For example, referring to Table VI, Run no. 556, Film no. 25, for the system $NaNO_3$–NaCl–H_2O [A1–A2–H_2O].

$$(m_1)_{A1} = 0.794 \text{ molal} \equiv 0.78 \text{ molar}$$
$$(m_1)_{A2} = 2.236 \text{ molal} \equiv 2.14 \text{ molar}$$

total molality = 3.03; total molarity = 2.92
2.92 molar $NaNO_3 \equiv$ 3.28 molal $NaNO_3$
2.92 molar NaCl $\equiv$ 3.10 molal NaCl

Therefore, the effective total molalities for the single solute systems ($NaNO_3$–H_2O) and (NaCl–H_2O) are 3.28 and 3.10, respectively.

Next, calculate solute separations, and product rates, PR_{A1} and PR_{A2}, for the single solute systems at their corresponding total effective molalities calculated above from data on membrane specifications and the applicable mass transfer coefficient correlation; or such data can be obtained experimentally.

Then calculate N_{A1}, N_{B1}, N_{A2}, and N_{B2} from the above solute separation and product rate data obtained for the single solute systems.

Find x_1 and x_2 from Equations 139 and 140, using molarity instead of molality; for the above examples,

$$x_1 = \frac{0.78}{0.78 + 2.14} = 0.267$$

$$x_2 = \frac{2.14}{0.78 + 2.14} = 0.733$$

Finally, calculate as before the product rate for the mixed solute system, PR_{mix}, from Equation 141, and solute separations with respect to each solute using Equations 146, 148, 152 and 153.

A set of such results for the system $NaNO_3$–NaCl–H_2O at the total molality of $\sim 3M$ is illustrated in Table VI, covering a wide range of solute separations which show good agreement between experimental and calculated data.

General problem of predictability of membrane performance for mixed solute systems. The foregoing technique, though a very useful one, is only a partial answer to this general problem. Even for the type of solution systems tested, the possible applicability of the above technique for a wide range of solute systems and feed concentrations, and the extension of the technique for aqueous feed solutions containing more than two solutes with a common ion, need experimental verifications. Some preliminary experimental results with aqueous feed solutions containing mixed inorganic solutes with no common ion show that the above prediction technique is not applicable to such systems probably due to ionic interactions. Such and other systems involving inorganic and/or organic mixed solutes need further experimental and analytical studies.

Systems Such as Sucrose–Water

For the system sucrose–water, $(D_{AM}/K\delta)$ is a unique function of X_{A2}, irrespective of the particular combination of feed concentration and feed flow rate used in conjunction with the Loeb-Sourirajan-type porous cellulose acetate membranes (*57*). It is possible that several similar systems exist. For such systems, it is necessary to establish the $(D_{AM}/K\delta)$ *vs.* X_{A2} correlation experimentally, and include the result in the procedure for predicting membrane performance as illustrated for the system sucrose-water (*57*).

Conclusion

Reverse osmosis is a fast growing field of applied science. It is now being extensively studied for various water treatment applications, such as sea water and brackish water conversion, water softening, water pollution control, water renovation, and waste recovery (*6, 15, 38, 42, 69, 77–80, 84, 101, 105, 132–134, 140, 156, 169*). Several parametric studies of the reverse osmosis process for sea water and brackish water conversion have been made (*13, 14, 46, 48, 49, 58, 121, 160*) and several pilot plants have been designed, built, and operated successfully using the Loeb-Sourirajan-type porous cellulose acetate membranes (*1, 2, 22, 23, 70, 76, 82, 84, 100, 117, 125, 126, 157, 161*). The Coalinga pilot plant, built by the University of California, Los Angeles, for brackish water conversion, has maintained a record of outstanding performance since its commission in June 1965 (*76, 84, 107*). There is also now growing activity on reverse osmosis studies for food processing applications (*7, 57, 99, 106, 118, 148, 170*).

From the authors' point of view, there are several areas of reverse osmosis work which have potentially far-reaching beneficial consequences to chemical industry. Studies on the physicochemical criteria of preferential sorption and/or preferential repulsion at solid-solution interfaces and microcapillary hydrodynamics at such interfaces, the mechanism of pore formation in the process of making asymmetric porous membranes, the methods of creating and maintaining pores of appropriate size on the surface layer of the membrane, quality control of the Loeb-Sourirajan-type porous cellulose acetate membranes in their large-scale manufacture and the utilization of their fullest potentialities (*60, 63, 121*), integration of the science and engineering of reverse osmosis, development of different membranes for different applications, and the application of reverse osmosis for the separation of gaseous mixtures (*5, 142*), organic liquid mixtures (*61, 62, 146*), and highly acid or alkaline aqueous solutions are all some of the areas of enormous practical interest. Further work in this field can lead to the growth of reverse osmosis in all its applications and contribute significantly to the economic prosperity and physical well-being of all mankind. Such is the scope of reverse osmosis (*152*).

Acknowledgments

The authors are grateful to Logos Press Ltd., London, England, for permission to use in this paper several illustrations from the book on "Reverse Osmosis" by S. Sourirajan. One of the authors (J. P. A.) thanks the National Research Council of Canada for the award of a postdoctoral fellowship.

NOMENCLATURE

A = pure water permeability constant, $\frac{\text{g-mol H}_2\text{O}}{\text{cm}^2\text{ sec atm}}$

A_0 = extrapolated value of A at $P = 0$

a = weight fraction of solute in product

a_w = activity of water in aqueous solution

B = proportionality constant defined by Equation 42

C = (c_A/c_{A1}^0) or (X_A/X_{A1}^0)

C_1, C_2, C_3 = (c_{A1}/c_{A1}^0), (c_{A2}/c_{A1}^0), (c_{A3}/c_{A1}^0), respectively, or (X_{A1}/X_{A1}^0), (X_{A2}/X_{A1}^0), (X_{A3}/X_{A1}^0), respectively

C_1^0, C_2^0, C_3^0 = values of C_1, C_2, and C_3, respectively, at time $t = 0$ in batch process or at membrane entrance in flow process

$\bar{C}_3$ = $(\bar{c}_{A3}/c_{A1}^0)$ or $(\bar{X}_{A3}/X_{A1}^0)$

c = molar density of solution, g-mol per cc

c_1, c_2, c_3 = molar density of feed solution, concentrated boundary solution, and the product solution, respectively, g-mol per cc

c_A = solute concentration, g-mol per cc or mole fraction

c_{A1}, c_{A2}, c_{A3} = solute concentration in the bulk solution and the concentrated boundary solution on the high pressure side of the membrane, and in the product solution on the atmospheric pressure side of the membrane, respectively, at any point on the membrane surface at any time, g-mol per cc or mole fraction

$\bar{c}_{A3}$ = average solute concentration in product solution corresponding to a given Δ, g-mol per cc or mole fraction

c_M = molar density of solution in the membrane phase, g-mol per cc

c_{M2}, c_{M3} = molar density of solution in the membrane phase in equilibrium with c_2 and c_3, respectively, g-mol per cc

D_{AB} = diffusivity of solute in water, cm² per sec

D_{AM} = diffusivity of solute in membrane phase, cm² per sec

$(D_{AM}/K\delta)$ = solute transport parameter, cm per sec

d = diameter of cell, cm

E = longitudinal diffusion coefficient, cm² per sec

f = fraction solute separation

$\bar{f}$ = average fraction solute separation

f_{max} = maximum solute separation

H = depth of cell, cm

$1/h$ = membrane area per unit volume of fluid space, cm^{-1}

K = proportionality constant defined by Equation 10

k = mass transfer coefficient on the high pressure side of membrane, cm per sec

k_{A1} = $S \times M_{A1} \times 3600, \frac{(\text{cm}^2)(\text{sec})(\text{g})}{(\text{g-mol})}$

k_{A2} = $S \times M_{A2} \times 3600, \frac{(\text{cm}^2)(\text{sec})(\text{g})}{(\text{g-mol})}$

k_B = $S \times M_B \times 3600, \frac{(\text{cm}^2)(\text{sec})(\text{g})}{(\text{g-mol})}$

L = longitudinal length over which mixing is considered, cm

l = thickness of the concentrated boundary solution

M = parameter defined by Equation 106 or concentration of solution expressed in molality unit as indicated

M_{A1} = molecular weight of solute $A1$

M_{A2} = molecular weight of solute $A2$

M_B = molecular weight of water

M_s = molecular weight of solute

m = solute molality

m_1 = solute molality in feed solution or bulk solution on the high pressure side of membrane

$(m_1)_{A1}$ = molality of feed solution with respect to solute $A1$

$(m_1)_{A2}$ = molality of feed solution with respect to solute $A2$

m_3 = solute molality in product solution

$(m_3)_{A1}$ = molality of product solution with respect to solute $A1$

$(m_3)_{A2}$ = molality of product solution with respect to solute $A2$

N_A = solute flux through membrane, $\frac{\text{g-mol}}{\text{cm}^2\text{ sec}}$

N_{A1} = flux of solute $A1$ through membrane for feed system ($A1$-H_2O), $\frac{\text{g-mol}}{\text{cm}^2\text{ sec}}$

N_{A2} = flux of solute $A2$ through membrane for feed system (A_2-H_2O), $\frac{\text{g-mol}}{\text{cm}^2\text{ sec}}$

N_B = solvent water flux through membrane, $\frac{\text{g-mol}}{\text{cm}^2\text{ sec}}$

N_{B1} = flux of solvent water through membrane for the feed system ($A1$-H_2O), $\frac{\text{g-mol}}{\text{cm}^2\text{ sec}}$

N_{B2} = flux of solvent water through membrane for the feed system ($A2$-H_2O), $\frac{\text{g-mol}}{\text{cm}^2\text{ sec}}$

N_{Re} = Reynolds number = $\frac{\text{feed flow rate}}{H\nu}$

N_{Sc} = Schmidt number = $\frac{\nu}{D_{AB}}$

N_{Sh} = Sherwood number = $\frac{kd}{D_{AB}}$

P = operating pressure, atm

PeB = quantity defined by Equation 110

PR = product rate, g per hr per S cm² of film area

PR_{A1} = product rate for feed system (A1-H_2O), g per hr per S cm² of film area

PR_{A2} = product rate for feed system ($A2$-H_2O), g per hr per S cm² of film area

PR_{mix} = product rate for feed system ($A1$-$A2$-H_2O), g per hr per S cm² of film area

PWP = pure water permeability, g per hr per S cm² of film area

Q = product rate in batch concentration process, grams per unit area of film surface per unit time expressed by Equation 128

q = quantity defined by Equation 52

R = gas constant

S = effective area of film surface, cm²

T = absolute temperature, °K

t = time, sec or day

$\bar{U}$ = $\bar{u}/\bar{u}^0$

$\bar{u}$ = average fluid velocity at any given position at any time, cm per sec

$\bar{u}^0$ = value of $\bar{u}$ at time = 0

V_1, V_3 = volume of feed solution on the high pressure side of membrane and volume of product solution at the atmospheric pressure side of membrane, respectively, at any time in a batch process, cc

V_1^0 = value of V_1 at time = 0

$(V)_i, (V)_f$ = initial and final volume of solution on the high pressure side of membrane during the batchwise concentration process, cc

$\bar{V}_w$ = partial molal volume of water

v_w = permeating velocity of solvent water, cm per sec
$(v_w)^0$ = value of v_w at time = 0
v_w^* = quantity defined by Equation 40
W = weight of solution on the high pressure side of membrane during the batchwise concentration process, g
W_s = weight of solute in the above solution, g
W_w = weight of water in the above solution, g
X = parameter defined by Equation 82
X_A = mole fraction of solute
X_{A1}, X_{A2}, X_{A3} = mole fraction of solute in the bulk solution and the concentrated boundary solution on the high pressure side of the membrane, and in the product solution on the atmospheric pressure side of the membrane, respectively, at any point on the membrane surface at any time
X_{A1}^0 = value of X_{A1} at t = 0 in batch process or at membrane entrance in flow process
X_{AM} = mole fraction of solute in membrane phase
X_{AM2}, X_{AM3} = values of X_{AM} in equilibrium with X_{A2} and X_{A3}, respectively
x = longitudinal distance along the length of membrane from channel entrance, cm
x_1, x_2 = quantities defined by Equations 139 and 140, respectively
Z = quantity defined by Equation 65
z = weight fraction of solute in the solution on the high pressure side of the membrane during the batchwise concentration process

Greek Letters

α = constant
β = constant
γ = parameter defined by Equation 37
Δ = fraction product recovery defined by Equation 83
δ = effective thickness of membrane, cm
ζ = normal distance towards membrane measured from edge of the concentrated boundary solution, cm
η = fraction power recovery
θ = parameter defined by Equation 38
λ = parameter defined by Equation 39
μ_w = viscosity of water, CP
ν = kinematic viscosity of solution, cm² per sec
π = osmotic pressure of solution, atm
$\pi(X_A)$ = osmotic pressure of solution corresponding to mole fraction X_A of solute, atm
ρ_1 = density of solution, g per cc
Σ_i = total number of moles of ions given by one mole of electrolyte
τ = parameter defined by Equation 73
ϕ = osmotic coefficient

Suffixes

i = initial state of feed solution
f = final state of feed solution

REFERENCES

(1) Aerojet General Corp., "Design and Construction of a Desalination Pilot Plant (A Reverse Osmosis Process)," U.S. Dept. Interior, Office of Saline Water, Research and Development Progress Report No. 86 (1964).
(2) Aerojet General Corp., "Operation of a Reverse Osmosis Desalination Pilot Plant," U.S. Dept. Interior Office of Saline Water, Research and Development Progress Report No. 213 (1966).
(3) Agrawal, J. P., and Sourirajan, S., "Specification, Selectivity, and Performance of Porous Cellulose Acetate Membranes in Reverse Osmosis," Ind. Eng. Chem. Process Design Develop. (in press).
(4) Agrawal, J. P., and Sourirajan, S., "Reverse Osmosis Separation of Some Inorganic Salts in Aqueous Solution Containing Two Solutes With a Common Ion," paper submitted to Ind. Eng. Chem. Process Design Develop. (1969).
(5) Agrawal, J. P., and Sourirajan, S., *J. Appl. Polym. Sci.*, **13,** 1065 (1969).
(6) Ammerlaan, A. C. F., Lueck, B. F., and Wiley, A. J., *Tappi*, **52** (1), 118 (1969).
(7) Andelman, J. B., and Suess, M. J., *Anal. Chem.*, **38,** 351 (1966).
(8) Baldwin, W. H., Holcomb, D. L., and Johnson, J. S., *J. Polym. Sci.*, **A3,** 833 (1965).
(9) Banks, W., and Sharples, A., *J. Appl. Chem.*, **16,** 28 (1966).
(10) Banks, W., and Sharples, A., *ibid.*, **16,** 94 (1966).
(11) Banks, W., and Sharples, A., *ibid.*, **16,** 153 (1966).
(12) Bird, R. B., Stewart, W. E., and Lightfoot, E. N., "Transport Phenomena," p 520, John Wiley, New York, 1960.
(13) Bray, D. T., in "Desalination by Reverse Osmosis," U. Merten, Ed., Chap. 6, The M.I.T. Press, Cambridge, Mass., 1966.
(14) Bray, D. T., and Menzel, H. F., "Design Study of a Reverse Osmosis Plant for Sea Water Conversion," U.S. Dept. Interior, Office of Saline Water, Research and Development Progress Report No. 176 (1966).
(15) Bray, D. T., Merten, U., and Augustus, M., "Reverse Osmosis For Water Reclamation," Gulf General Atomic Report No. GA-6337 (1965).
(16) Breton, E. J., Jr., "Water and Ion Flow Through Imperfect Osmotic Membranes," U.S. Dept. Interior, Office of Saline Water, Research and Development Progress Report No. 16 (1957).
(17) Breton, E. J., Jr., and Reid, C. E., AIChE Chem. Eng. Symp. Ser., **24,** 171 (1959).
(18) Brian, P. L. T., Ind. Eng. Chem. Fundam., **4,** 439 (1965).
(19) Brian, P. L T., "Concentration Polarization in Reverse Osmosis Desalination with Variable Flux and Incomplete Salt Rejection," M.I.T. Desalination Research Laboratory, Report No. 295–7, DSR-4647 (1965).
(20) Brian, P. L. T., Proc. First International Symp. on Water Desalination, Washington, D.C., 1965. (Pub. U.S. Dept. Interior, Office of Saline Water, Washington, D.C., Vol. 1, p 349–70).
(21) Brian, P. L. T., in "Desalination by Reverse Osmosis," U. Merten, Ed., Chap. 5, The M.I.T. Press, Cambridge, Mass., 1966.
(22) Cheng, C., Cheng, S., and Fan, L., *AIChE J.*, **13,** 438 (1967).
(23) Cheng, C., and Fan, L., "A Flow Work Exchanger for Desalination Processes," U.S. Dept. Interior, Office of Saline Water, Research and Development Progress Report No. 357 (1968).
(24) Dresner, L., "Boundary Layer Build-up in the Demineralization of Salt Water by Reverse Osmosis," Report ORNL-3621, Oak Ridge National Laboratory, 1964.
(25) Dresner, L., *J. Phys. Chem.*, **69,** 2230 (1965).
(26) Dresner, L., and Kraus, K. A., ibid., **67,** 990 (1963).
(27) Erickson, D. L., "Cellulose Acetate Membranes as a Means of Removing Scale Forming Ions of Natural Saline Waters," Department of Engineering, University of California, Los Angeles, Report No. 66-7 (1966).
(28) Erickson, D. L., Glater, J., and McCutchan, J. W., Ind. Eng. Chem. Prod. Res. Develop., **5,** 205 (1966).
(29) Fisher, R. E., Sherwood, T. K., and Brian, P. L. T., "Salt Concentration at the Surface of Tubular Reverse Osmosis Membranes," U.S. Dept. Interior, Office of Saline Water, Research and Development Progress Report No. 141 (1965).
(30) Flowers, L. C., Sestrich, D. E., and Berg, D., "Reverse Osmosis Membranes Containing Graphitic Oxide," U.S. Dept. Interior, Office of Saline Water, Research and Development Progress Report No. 224 (1966).
(31) Francis, P. S., and Cadotte, J. E., "Second Report on Fabrication and Evaluation of New Ultrathin Reverse Osmosis Membranes," U.S. Dept. Interior, Office of Saline Water, Research and Development Progress Report No. 247 (1967).
(32) Gill, W. N., Tien, C., and Zeh, D. W., Ind. Eng. Chem. Fundam., **4,** 433 (1965).
(33) Gill, W. N., Tien, C., and Zeh, D. W., *Int. J. Heat Mass Transfer*, **9,** 907 (1966).
(34) Gill, W. N., Zeh, D. W., and Tien, C., Ind. Eng. Chem. Fundam., **5,** 367 (1966).
(35) Gill, W. N., Zeh, D. W., and Tien, C., *AIChE J.*, **12,** 1141 (1966).
(36) Glueckauf, E., Proc. First International Symp. on Water Desalination, Washington, D.C., 1965. (Pub. U.S. Dept. Interior, Office of Saline Water, Washington, D.C., Vol. **1,** p 143–56).
(37) Govindan, T. S., and Sourirajan, S., Ind. Eng. Chem. Process Design Develop., **5,** 422 (1966).
(38) Hauck, A. R., and Sourirajan, S., "Performance of Porous Cellulose Acetate Membranes for the Reverse Osmosis Treatment of Hard and Waste Waters," Environ. Sci. Technol. (in press).
(39) Havens, G. G., and Guy, D. B., "The Use of Porous Fiberglas Tubes for Reverse Osmosis," AIChE Meeting, Salt Lake City, Utah, 1967.
(40) Havens Industries, "Sea Water Conversion by Reverse Osmosis," Technical Brochure, San Diego, Calif., 1964.
(41) Hoffer, E., and Kedem, O., *Desalination*, **2,** 25 (1967).
(42) Ironside, R., and Sourirajan, S., *J. Water Res.*, **1,** 179 (1967).
(43) Jagur-Grodzinski, J., and Kedem, O., *Desalination*, **1,** 327 (1966).
(44) Johnson, J. S., Jr., Dresner, L., and Kraus, K. A., in "Principles of Desalination," K. S. Spiegler, Ed., Chap. 8, Academic Press, New York, 1966.
(45) Johnson, J. S., and Kraus, K. A., "Hyperfiltration by Dynamically Formed Membranes," Paper presented before Division of Water, Air, and Waste Chemistry, ACS, San Francisco Meeting, March 31–April 5, 1968.
(46) Johnson, K. D. B., Grover, J. R., and Pepper, D., *Desalination*, **2,** 40 (1967).
(47) Keilin, B., "The Mechanism of Desalination by Reverse Osmosis," U.S. Dept. Interior, Office of Saline Water, Research and Development Progress Report No. 84 (1963).
(48) Keilin, B., in "Membrane Processes for Industry," p 80–100, Southern Research Institute, Birmingham, Ala., 1966.
(49) Keilin, B., and DeHaven, C. G., Proc. First International Symp. on Water Desalination, Washington, D.C., 1965. (Pub. U.S. Dept. Interior, Office of Saline Water, Washington, D.C., Vol. **2,** p 367–380).
(50) Keilin, B., Saltonstall, C. W., Higley, W. S., Kesting, R. E., and Vincent, A. L., "Structure and Properties of Reverse Osmosis Membranes," U.S. Dept. Interior, Office of Saline Water, Research and Development Progress Report No. 154 (1965).
(51) Kesting, R. E., *J. Appl. Polym. Sci.*, **9,** 663 (1965).
(52) Kesting, R. E., Barsh, M. K., and Vincent, A. L., *ibid.*, **9,** 1873 (1965).
(53) Kimura, S., and Sourirajan, S., *AIChE J.*, **13,** 497 (1967).
(54) Kimura, S., and Sourirajan, S., Ind. Eng. Chem. Process Design Develop., **7,** 41 (1968).
(55) Kimura, S., and Sourirajan, S., *ibid.*, **7,** 197 (1968).
(56) Kimura, S., and Sourirajan, S., *ibid.*, **7,** 539 (1968).
(57) Kimura, S., and Sourirajan, S., *ibid.*, **7,** 548 (1968).
(58) Kimura, S., Sourirajan, S., and Ohya, H., Ind. Eng. Chem. Process Design Develop., **8,** 79 (1969).
(59) Klem, K., and Friedman, L., *J. Amer. Chem. Soc.*, **54,** 2637 (1932).
(60) Kopeček, J., and Sourirajan, S., *J. Appl. Polym. Sci.*, **13,** 637 (1969).
(61) Kopeček, J., and Sourirajan, S., "Performance of Porous Cellulose Acetate Membranes for the Reverse Osmosis Separation of Mixtures of Organic Liquids," Paper submitted to Ind. Eng. Chem. Process Design Develop. (1969).
(62) Kopeček, J., and Sourirajan, S., *Can. J. Chem.* **47,** 3467 (1969).
(63) Kopeček, J., and Sourirajan, S., Ind. Eng. Chem. Prod. Res. Develop. **8,** 274 (1969).
(64) Kraus, K. A., Marcinkowsky, A. E., Johnson, J. S., and Shor, A. J., *Science*, **151,** 194 (1966a).

(65) Kraus, K. A., Phillips, H. O., Marcinkowsky, A. E., Johnson, J. S., and Shor, A. J., *Desalination*, **1**, 225 (1966).

(66) Kraus, K. A., Shor, A. J., and Johnson, J. S., *Desalination*, **2**, 243 (1967).

(67) Kuppers, J. R., Harrison, N., and Johnson, J. S., Jr., *J. Appl. Polym. Sci.*, **10**, 969 (1966).

(68) Kuppers, J. R., Marcinkowsky, A. E., Kraus, K. A., and Johnson, J. S., *Separation Science*, **2** (5), 617 (1967).

(69) Larson, T. J., "Purification of Subsurface Waters by Reverse Osmosis (Pilot Plant Operations)," Gulf General Atomic Report No. GA-7819 (1967).

(70) Larson, T. J., Nusbaum, I., Reidinger, A. B., and Astl, J., "Reverse Osmosis Membrane Module (Spiral Wound Concept)," U.S. Dept. Interior, Office of Saline Water, Research and Development Progress Report No. 338 (1968).

(71) Loeb, S., "Sea Water Demineralization By Means of a Semipermeable Membrane," Department of Engineering, University of California, Los Angeles, Report No. 61-42 (1961).

(72) Loeb, S., *ibid.*, Report No. 62-26 (1962).

(73) Loeb, S., *ibid.*, Report No. 63-32 (1963).

(74) Loeb, S., *Desalination*, **1**, 35 (1966).

(75) Loeb, S., in "Desalination by Reverse Osmosis," U. Merten, Ed., Chap. 3, The M.I.T. Press, Cambridge, Mass., 1966.

(76) Loeb, S., and Johnson, J. S., *Chem. Eng. Progr.*, **63**, No. 1, 90 (1967).

(77) Loeb, S., and Manjikian, S., "Brackish Water Desalination by an Osmotic Membrane," Department of Engineering, University of California, Los Angeles, Report No. 63-22 (1963).

(78) Loeb, S., and Manjikian, S., "Brackish Water Desalination by an Osmotic Membrane," Department of Engineering, University of California, Los Angeles, Report No. 63-37 (1963).

(79) Loeb, S., and Manjikian, S., "Field Tests on Osmotic Desalination Membranes," Department of Engineering, University of California, Los Angeles, Report No. 64-34 (1964).

(80) Loeb, S., and Manjikian, S., Ind. Eng. Chem., Process Design Develop., **4**, 207 (1965).

(81) Loeb, S., and McCutchan, J. W., Ind. Eng. Chem. Prod. Res. Develop., **4**, 114 (1965).

(82) Loeb, S., and Milstein, F., *Dechema-Monographien*, **47**, 707 (1962).

(83) Loeb, S., and Nagaraj, G. R., U.S. Patent 3,170,867 (Feb. 23, 1965).

(84) Loeb, S., and Selover, E., *Desalination*, **2**, 75 (1967).

(85) Loeb, S., and Sourirajan, S., in "Sea Water Research," Department of Engineering, University of California, Los Angeles, Report No. 59-3 (1958).

(86) Loeb, S., and Sourirajan, S., in "Sea Water Research," Department of Engineering, University of California, Los Angeles, Report No. 59-28 (1959).

(87) Loeb, S., and Sourirajan, S., *ibid.*, Report No. 59-46 (1959).

(88) Loeb, S., and Sourirajan, S., *ibid.*, Report No. 60-5 (1960).

(89) Loeb, S., and Sourirajan, S., *ibid.*, Report No. 60-26 (1960).

(90) Loeb, S., and Sourirajan, S., "Sea Water Demineralization By Means of a Semipermeable Membrane," Department of Engineering, University of California, Los Angeles, Report No. 60-60 (1961).

(91) Loeb, S., and Sourirajan, S., *Advan. Chem. Ser.*, No. **38**, 117 (1963).

(92) Loeb, S., and Sourirajan, S., U.S. Patent 3,133,132 (May 12, 1964).

(93) Loeb, S., and Sourirajan, S., U.S. Patent 3,223,424 (Dec. 14, 1965).

(94) Loeb, S., Sourirajan, S., and Weaver, D. E., U.S. Patent 3,133,137 (May 12, 1964).

(95) Lonsdale, H. K., in "Desalination by Reverse Osmosis," U. Merten, Ed., Chap. 4, The M.I.T. Press, Cambridge, Mass., 1966.

(96) Lonsdale, H. K., Merten, U., and Riley, R. L., *J. Appl. Polym. Sci.*, **9**, 1341 (1965).

(97) Lonsdale, H. K., Merten, U., Riley, R. L., Vos, K. D., and Westmoreland, J. C., "Reverse Osmosis for Water Desalination," U.S. Dept. Interior, Office of Saline Water, Research and Development Progress Report No. 111 (1964).

(98) Lonsdale, H. K., Merten, U., and Tagami, M., *J. Appl. Polym. Sci.*, **11**, 1807 (1967).

(99) Lowe, E., Durkee, E. L., and Morgan, A. I., Jr., *Food Technol.*, **22**, 915 (1968).

(100) Mahon, H. I., Proc. Desalination Research Conference, Nat. Acad. Sci. Publication, **942**, 345 (1963).

(101) Manjikian, S., "Optimization of a Reverse Osmosis Sea Water Desalination System," Report No. CR67.027 for U.S. Naval Civil Engineering Laboratory, Port Hueneme, Calif., 1966.

(102) Manjikian, S., Ind. Eng. Chem. Prod. Res. Develop., **6**, 23 (1967).

(103) Manjikian, S., Loeb, S., and McCutchan, J. W., Proc. First International Symp. on Water Desalination, Washington, D.C., 1965. (Pub. U.S. Dept. Interior, Office of Saline Water, Washington, D.C., Vol. **2**, p 159–73).

(104) Marcinkowsky, A. E., Kraus, K. A., Phillips, H. O., Johnson, J. S., Jr., and Shor, A. J., *J. Amer. Chem. Soc.*, **88**, 5744 (1966).

(105) Markley, L. L., Cross, R. A., and Bixler, H. J., "Membranes for Desalination by Reverse Osmosis," U.S. Dept. Interior, Office of Saline Water, Research and Development Progress Report No. 281 (1967).

(106) Marshall, P. G., Dunkley, W. L., and Lowe, E., *Food Technol.*, **22**, 969 (1968).

(107) McCutchan, J. W., in "Saline Water Research Progress Summary," p 13, Department of Engineering, University of California, Los Angeles, Report No. 68-1 (1968).

(108) Meares, P., *Eur. Polym. J.*, **2**, 241 (1966).

(109) Merten, U., Ind. Eng. Chem. Fundam., **2**, 229 (1963).

(110) Merten, U., in "Desalination by Reverse Osmosis," U. Merten, Ed., Chap. 2, The M.I.T. Press, Cambridge, Mass., 1966.

(111) Merten, U., *Desalination*, **1**, 297 (1966).

(112) Merten, U., Lonsdale, H. K., and Riley, R. L., Ind. Eng. Chem. Fundam., **3**, 210 (1964).

(113) Merten, U., Lonsdale, H. K., Riley, R. L., and Vos, K. D., "Reverse Osmosis Membrane Research," U.S. Dept. Interior, Office of Saline Water, Research and Development Progress Report No. 265 (1967).

(114) Michaels, A. S., Ind. Eng. Chem., **57**, No. 10, 32 (1965).

(115) Michaels, A. S., Bixler, H. J., Hausslein, R. W., and Fleming, S. M., "Polyelectrolyte Complexes as Reverse Osmosis and Ion-Selective Membranes," U.S. Dept. Interior, Office of Saline Water, Research and Development Progress Report No. 149 (1965).

(116) Michaels, A. S., Bixler, H. J., and Hodges, R. M., Jr., *J. Colloid Sci.*, **20**, 1034 (1965).

(117) Miller, C. S., "Design, Construction, and Operation of a Pilot System for the Production of Continuous Reverse Osmosis Membranes," M. Eng. Thesis, The Thayer School of Engineering, Dartmouth College, Hanover (1968).

(118) Morgan, A. I., Jr., Lowe, E., Merson, R. L., and Durkee, E. L., *Food Technol.*, **19**, 1790 (1965).

(119) Ohya, H., and Sourirajan, S., Ind. Eng. Chem. Process Design Develop., **8**, 131 (1969).

(120) Ohya, H., and Sourirajan, S., "Some General Equations for Reverse Osmosis Process Design," *AIChE J.*, (in press).

(121) Ohya, H., and Sourirajan, S., *Desalination*, **6**, 153 (1969).

(122) Ohya, H., and Sourirajan, S., *AIChE J.*, **15**, 780 (1969).

(123) Perona, J. J., Butt, F. H., Fleming, S. M., Mayr, S. T., Spitz, R. A., Brown, M. K., Cochran, H. D., Kraus, K. A., and Johnson, J. S., Jr., *Environ. Sci. Technol.*, **1**, 991 (1967).

(124) Reid, C. E., and Breton, E. J., *J. Appl. Polym. Sci.*, **1**, 133 (1959).

(125) Riedinger, A. B., Laughlin, J. K., and Sudak, R. G., "Large Reverse Osmosis System Technology and Module Development," U.S. Dept. Interior, Office of Saline Water, Research and Development Progress Report No. 341 (1968).

(126) Riedinger, A., and Schultz, J., "Acid Mine Water Reverse Osmosis Test at Kittanning, Pennsylvania," U.S. Dept. Interior, Office of Saline Water, Research and Development Progress Report No. 217 (1966).

(127) Riley, R. L., Gardner, J. O., and Merten, U., *Science*, **143**, 801 (1964).

(128) Riley, R. L., Lonsdale, H. K., Lyons, C. R., and Merten, U., *J. Appl. Polym. Sci.*, **11**, 2143 (1967).

(129) Riley, R. L., Merten, U., and Gardner, J. O., *Desalination*, **1**, 30 (1966).

(130) Robinson, R. A., and Stokes, R. H., "Electrolyte Solutions," 2nd ed., p 30, Butterworths, London, 1959.

(131) Robinson, R. A., and Stokes, R. H., *ibid.*, p 476–490, Butterworths, London, 1959.

(132) Schultz, J., "Potamac River Reverse Osmosis Test," U.S. Dept. Interior, Office of Saline Water, Research and Development Progress Report No. 216 (1966).

(133) Schultz, J., and Newby, G. A., "Desalination by Reverse Osmosis," Gulf General Atomic Report GA-7153 (1966).

(134) Schultz, J., Riedinger, A., and McCracken, H., "Brackish Well Water Reverse Osmosis Tests at Midland, Fort Stockton, and Kermit, Texas," U.S. Dept. Interior, Office of Saline Water, Research and Development Progress Report No. 237 (1967).

(135) Sherwood, T. K., "Mass Transfer Between Phases," p 38, Thirty-Third Annual Priestley Lectures, The Pennsylvania State University, 1959.

(136) Sherwood, T. K., Brian, P. L. T., and Fisher, R. E., "Salt Concentration at Phase Boundaries in Desalination Processes," U.S. Dept. Interior, Office of Saline Water, Research and Development Progress Report No. 95 (1964).

(137) Sherwood, T. K., Brian, P. L. T., Fisher, R. E., and Dresner, L., Ind. Eng. Chem. Fundam., **4**, 113 (1965).

(138) Sherwood, T. K., Brian, P. L. T., and Fisher, R. E., *ibid.*, **6**, 2 (1967).

(139) Sherwood, T. K., and Pigford, R. L., "Absorption and Extraction," p 53, McGraw-Hill, New York, 1952.

(140) Sieveka, E. H., in "Desalination by Reverse Osmosis," U. Merten, Ed., Chap. 7, The M.I.T. Press, Cambridge, Mass., 1966.

(141) Sourirajan, S., Ind. Eng. Chem. Fundam., **2**, 51 (1963).

(142) Sourirajan, S., *Nature*, **199**, 590 (1963).

(143) Sourirajan, S., "Separation of Some Inorganic Salts in Aqueous Solution By Flow, Under Pressure, Through Porous Cellulose Acetate Membranes" National Research Council of Canada, NRC No. 7498 (1963).

(144) Sourirajan, S., Ind. Eng. Chem. Fundam., **3**, 206 (1964).

(145) Sourirajan, S., *J. Appl. Chem.*, **14**, 506 (1964).

(146) Sourirajan, S., *Nature*, **203**, 1348 (1964).

(147) Sourirajan, S., Ind. Eng. Chem. Prod. Res. Develop., **4**, 201 (1965).

(148) Sourirajan, S., Ind. Eng. Chem. Process Des. Develop., **6**, 154 (1967).

(149) Sourirajan, S., in "Water Resources of Canada," C. E. Dolman, Ed., p 154–82, University of Toronto Press, Toronto, 1967.

(150) Sourirajan, S., *The Canadian Scientist*, **1**, No. 3, 22 (1967).

(151) Sourirajan, S., Canadian Patent 797,604 (Oct. 29, 1968).

(152) Sourirajan, S., "Reverse Osmosis," Logos Press, London (in press).

(153) Sourirajan, S., and Govindan, T. S., Proc. First International Symp. on Water Desalination, Washington, D.C., 1965. (Pub. U.S. Dept. Interior, Office of Saline Water, Washington, D.C., Vol. **1**, p 251–74).

(154) Sourirajan, S., and Kimura, S., Ind. Eng. Chem. Process Design Develop., **6**, 504 (1967).

(155) Sourirajan, S., and Loeb, S., in "Sea Water Research," Department of Engineering, University of California, Los Angeles, Report No. 58-65 (1958).

(156) Sourirajan, S., and Sirianni, A. F., Ind. Eng. Chem. Prod. Res. Develop., **5** 30 (1966).

(157) Spatz, D. D., "The Design, Development, and Field Testing of Home Reverse Osmosis Desalination Units," M. Eng. Thesis, The Thayer School of Engineering, Dartmouth College, Hanover (1968).

(158) Spiegler, K. S., and Kedem, O., *Desalination*, **1**, 311 (1966).

(159) Srinivasan, S., Tien, C., and Gill, W. N., *Chem. Eng. Sci.*, **22**, 417 (1967).

(160) Stevens, D., and Loeb, S., *Desalination*, **2**, 56 (1967).

(161) Sudak, R. G., and Nusbaum, I., "Pilot Plant Operation of Spiral Wound Reverse Osmosis Systems," Gulf General Atomic Report No. GA-8515 (1968).

(162) Tien, C., and Gill, W. N., *AIChE J.*, **12**, 722 (1966).

(163) Treybal, R. E., "Mass Transfer Operations," p 38, McGraw-Hill, New York 1955.

(164) University of California, Office of Public Information Press Release, "New Water Desalting Process Developed at UCLA," Aug. 23, 1960.

(165) Vincent, A. L., Barsh, M. K., and Kesting, R. E., *J. Appl. Polym. Sci.*, **9**, 2363 (1965).

(166) Vos, K. D., Burris, F. O., Jr., and Riley, R. L., *J. Appl. Polym. Sci.*, **10**, 825 (1966).

(167) Vos, K. D., Hatcher, A. P., and Merten, U., Ind. Eng. Chem. Prod. Res. Develop., **5**, 211 (1966).

(168) Wasilewski, S., "Microwave Investigation of Temperature Induced Transient Changes in Cellulose Acetate Membranes," Department of Engineering, University of California, Los Angeles, Report No. 65-10 (1965).

(169) Wiley, A. J., Ammerlaan, A. C. F., and Dubey, G. A., *Tappi*, **50** (9), 455 (1967).

(170) Willits, C. O., Underwood, J. C., and Merten, U., *Food Technol.*, **21**, 24 (1967).

(171) Yuster, S. T., Sourirajan, S., and Bernstein, K., "Sea Water Demineralization by the Surface-Skimming Process," Department of Engineering, University of California, Los Angeles, Report No. 58-26 (1958).

INDEX

F

G

H

I

J

K

L

T

U

V

W

X

Y